Innovating in Practice

Tiziana Russo-Spena • Cristina Mele •
Maaria Nuutinen
Editors

Innovating in Practice

Perspectives and Experiences

Springer

Editors
Tiziana Russo-Spena
University of Naples Federico II
Naples, Italy

Cristina Mele
University of Naples Federico II
Naples, Italy

Maaria Nuutinen
VTT Technical Research Centre of Finland
VTT, Finland

ISBN 978-3-319-43378-3 ISBN 978-3-319-43380-6 (eBook)
DOI 10.1007/978-3-319-43380-6

Library of Congress Control Number: 2016958509

Printed on acid-free paper

This Springer imprint is published by Springer Nature
The registered company is Springer International Publishing AG Switzerland

Foreword

As the coeditors, Russo-Spena, Mele, and Nuutinen, emphasize, this book is not about innovation; it is about innovating. Innovating, as the editor and contributing authors convincingly demonstrate, is an inherently dynamic and ongoing social-cultural process. This idea unifies the 22 chapters, contributed by 46 internationally recognized experts, with a transcending theme.

Various perspectives presented by these experts depart from the traditional focus on sole inventors/entrepreneurs. Schumpeter, for instance, attributed innovation to individual entrepreneurs who through creative destruction dislodge existing market offerings, enterprises, and industries. Furthermore, the linear, stage-gate model that was dominant for most of the twentieth century, and continues to be on some of the academic literature and a number of organizations, focuses on how isolated firms manage R&D and invention, often in tightly closed departments. The fundamental contribution of this book is that this long-standing, dominant view of innovation represents a very restricted view. Arguably, in the development of service-dominant (S-D) logic, we, along with hundreds of scholars around the world, have been able to capture a new frame of reference, providing a more accurate lens for viewing the phenomenon.

"Invention" and "innovation" are nouns, versus "innovating" a verb. Innovating neither begins nor ends with invention. It is a part of the process of applying knowledge and skills to provide a benefit to a beneficiary, what S-D logic defines as "service." It thus provides the foundation for economic growth and human well-being, implying, as we have argued for a dozen years, that all economies are service economies, with innovating as a central driver of service provision. In unraveling this somewhat inverted logic, there have been divergent approaches to the study of innovating. Importantly, however, they typically focus on a community of actors that extends well beyond organizational boundaries and often spans long periods of time. As Johnson notes in his 2010 book, *Where Good Ideas Come From: The Natural History of Innovation*: "Good ideas are not conjured out of thin air; they are built out of a collection of existing parts, the composition of which expands (and, occasionally, contracts) over time." Brian Arthur, in his 2009 book on the nature of technology, more generally refers to "combinatorial evolution," and all innovation comes from recombinations of the outcomes of previous innovating. It is in this social-cultural environment of various actors and the cumulative, time-released interactive processes that ideas are derived and nurtured into fruition. Arthur (p. 120) describes the implications of the Wright Brothers' 1903 historic powered flight this way: "[The flight] was not so much a demonstration of 'an invention'; it was a marker along a lengthy path trodden by others before them."

As we initially developed what is now called S-D logic, we found ourselves synthesizing over a century of literature with special focus on the last quarter century, where we observed many marketing practices to be quite different from, and more effective than, academic literature and textbook knowledge. In brief, theory lagged practice. Similarly, this book convincingly demonstrates that relatively recent trends in innovating practices unequivocally demonstrate that theory is once again lagging practice, furthering validating its need. Some of the more contemporary practices that lag theory include (see evidence of these and other practices leading theory at the end of this foreword):

1. Open Innovation. Another widely advocated practice, open innovation suggests that companies should look outside organizational boundaries for ideas and intellectual property (IP) and license their underutilized IP to other organizations. Open innovation and open business models that support it have increasingly become a key innovation approach.
2. Use-Led Innovation. This practice challenges a strongly held economic assumption that producers, not users, are the ones who innovate the most, and thus user-dominated processes of innovating should be at the center of innovation strategies.
3. Cocreation. Cocreation is based on the idea of working collaboratively with many actors in the innovation. Business or service ecosystem to jointly create innovation.
4. Wisdom of Crowds. This approach advocates the value of insights and ideas from "groups," both inside and outside the organization.
5. Lean Start-Up. The lean start-up approach argues that innovation is a learning process between entrepreneurs and their customers. Thus, the innovation at successful start-ups should be geared at accelerating the feedback loop between ideas and customer responses.

These practices for innovating share a common view: innovation and innovating can be more productively carried out when treated as a social-cultural process that integrates a diversified set of market actors beyond traditional organizational boundaries. This implies the need for what we have called "zooming out" to see the nested, overlapping service ecosystems and institutions—norms, rules, meanings, and other heuristic tools—and institutional arrangements, interrelated institutions that influence perception and evaluation, that represent the fabric of S-D logic and locus of innovating.

Leaders from all industries and geopolitical areas will find that this book offers a needed service. We believe the serious reader of this book will discover that it can help with the discovery of new approaches to strategic benefit (organizational, customer, and societal) and also help to guide public policy decisions. Although the chapters can be read in any order, we suggest that the first four chapters be read in order. For those not familiar with S-D logic, Chap. 3 will provide exposure to the service-ecosystems framework and how it informs innovating as a social-cultural and economic process of value cocreation by resource integrating actors, coordinated in part by institutions and institutional arrangements. Additionally, it suggests that institutional innovation is a vital part of the innovating process. It also provides a perspective for the creation and maintenance of well-being for individuals, organizations, industries, and nations.

At the same time, this book is a valuable aid to scholars from the many disciplines conducting research on innovation and innovating. Collectively and individually, the chapters bring forth new conceptual frameworks and a language that can provide a blueprint for additional research, practice, and policy. It can also trigger new research ideas or programs that can help us better understand the social nature of innovation. Finally, since 75 % of the chapters include concepts from S-D logic, it furthers the advancement of S-D logic, which is of benefit not only to us but, more importantly, to the increasing number of scholars around the world who are using S-D logic foundationally, including advocates of an emerging service science, which is, itself, primarily focused on innovating.

Tucson, AZ — Robert F. Lusch
Honolulu, HI — Stephen L. Vargo
2016

References

Arthur, B. W. (2009). *The nature of technology: What it is and how it evolves*. New York: The Free Press.

Beinhocker, E. D. (2006). *The origin of wealth*. Cambridge, MA: Harvard Business School Press.

Chesbrough, H. W. (2006). *Open business models*. Boston: Harvard Business School Press.

Johnson, S. (2010). *Where good ideas come from: The natural history of innovation*. New York: Riverhead Books.

Martinelli, R., & Galluzzo, C. (2010). How the 'wisdom of crowds' concept inspired Intel to improve its innovation transfer process through the valley of death. *Vision, 34*(2), 29–33.

Prandelli, E., Sawhney, M., & Verona, G. (2008), *Collaborating with customers to innovate: Conceiving and marketing products in the networking age*. Cheltenham, UK: Edward Elgar Publishing.

Ramaswamy, V., & Gouillart, F. (2010). *The power of co-creation*. New York: The Free Press.

Ries, E. (2011). *The Lean startup: How today's entrepreneurs use continuous innovation to create radically successful businesses*. New York: Crown Business.

Surowiecki, J. (2004). *The wisdom of crowds: Why the many are smarter than the few and how collective wisdom shapes business*. New York: Doubleday.

Tapscott, D., & Williams, A. (2006). *Wikinomics: How mass collaboration changes everything*. New York: Portfolio.

Von Hippel, E. (1988). *The sources of innovation*. New York: Oxford University Press.

Contents

1 **Introduction** 1
Tiziana Russo-Spena, Cristina Mele, and Maaria Nuutinen

Part I Perspectives: Conceptual Insights

2 **Schools of Innovation Thought** 13
Cristina Mele, Tiziana Russo-Spena, Maaria Nuutinen, and Katri Kallio

3 **Extending the Context of Innovation: The Co-creation and Institutionalization of Technology and Markets** 43
Melissa Archpru Akaka, Stephen L. Vargo, and Heiko Wieland

4 **Market Innovation: Renewal of Traditional Industrial Networks** 59
Suvi Nenonen, Kristian Möller, and Kaj Storbacka

5 **Practice-Based Approaches to Learning and Innovating** 83
Katri Kallio, Tiziana Russo-Spena, Inka Lappalainen, and Cristina Mele

Part II Perspectives: A Practice-Based View

6 **Innovating as a Texture of Practices** 113
Cristina Mele and Tiziana Russo-Spena

7 **Innovating as an Interaction of Practices: The Case of Customers' Use of a New Website** 139
Stefano Pace, Antonella Carù, and Bernard Cova

8 Practices of Service Innovation Diffusion 159
Daniela Corsaro, Roberta Sebastiani, and Cristina Mele

9 Innovation with Effectuation: An Opportunity for the Public Sector 179
Kirsti Mäensivu, Marja Toivonen, and Karo Tammela

Part III Perspectives: Conceptual Merging

10 Framing the New Social–Service Innovation Mindset 205
Tiziana Russo-Spena, Maria Colurcio, and Monia Melia

11 Conceptualizing Resource Integration to Advance Service Innovation 237
Maria Colurcio, Angela Caridà, and Bo Edvardsson

12 Service Innovation and Smart Cities: Linking the Perspectives . . . 261
Francesco Bifulco and Marco Tregua

Part IV Experiences: Customer and Value

13 Value Proposal Co-Creation in Online Community-Based Idea Contests 291
Carol Kelleher, Aonghus Ó. Céilleachair, Anu Helkkula, and Joe Peppard

14 Customer-Centric Service Innovation in Public–Private Value Networks 317
Inka Lappalainen

15 Business Model Design and Value Co-creation: Looking for a New Pattern 339
Angela Caridà, Monia Melia, and Maria Colurcio

Part V Experiences: Multi Actors and Context

16 Seeking New Ways of Innovating in Industry-Research Collaboration Practice 365
Maaria Nuutinen, Marko Seppänen, Anssi Smedlund, and Eija Kaasinen

17 Practice-Based 'Inside-Out' Innovation in Public Service: A Regional Child Welfare Agency 393
Monica Pedrazza and Sabrina Berlanda

18 **Service Innovation in Translational Medicine** 417
Sergio Barile, Francesco Polese, Marialuisa Saviano, and Luca Carrubbo

19 **Stakeholder Engagement in Sustainability Innovation: Experiences in the Jewellery Business** 439
Alessandra De Chiara

Part VI Experiences: Tools

20 **Co-development and Retooling as New Roles for Applied Research** 463
Eveliina Saari, Mervi Hasu, Laura Honkaniemi, Katri Kallio, Tiina Tuominen, and Mikko Lehtonen

21 **Inter-professional Collaboration: An Evaluation Study** 487
Monica Pedrazza, Riccardo Sartori, and Sabrina Berlanda

22 **Investigating Elderly Care Contexts According to Stakeholders' Needs** 509
Elena Luppi

List of Contributors

Melissa Archpru Akaka University of Denver, Denver, CO, USA

Sergio Barile Department of Management, La Sapienza University of Rome, Rome, Italy

Sabrina Berlanda Department of Philosophy, Education and Psychology, University of Verona, Verona, Italy

Francesco Bifulco Department of Economics Management Institution, University of Naples Federico II, Naples, Italy

Angela Caridà Department of Legal, Historical, Economic and Social Sciences, University Magna Graecia of Catanzaro, Catanzaro, Italy

Luca Carrubbo University of Cassino, Cassino, Italy

Antonella Carù Department of Marketing, Bocconi University, Milan, Italy

Aonghus Ó. Céilleachair Department of Management & Marketing, University College Cork, Cork, Ireland

Maria Colurcio Department of Legal, Historical, Economic and Social Sciences, University Magna Graecia of Catanzaro, Catanzaro, Italy

Daniela Corsaro Department of Economics and Business Management, Catholic University of Scared Heart, Milan, Italy

Bernard Cova Department of Marketing, Kedge Business School, Marseille, France

Alessandra De Chiara Department of Human and Social Science, University of Naples "L'Orientale", Naples, Italy

Bo Edvardsoon CTF, Karlstad University, Karlstad, Sweden

Mervi Hasu Finnish Institute of Occupational Health, Helsinki, Finland

Anu Helkkula CERS Research Centre, Hanken School of Economics and Aalto School of Science, Helsinki, Finland

Laura Honkaniemi Finnish Institute of Occupational Health, Helsinki, Finland

Eija Kaasinen VTT Technical Research Centre of Finland, Helsinki, Finland

Katri Kallio VTT Technical Research Centre of Finland, Helsinki, Finland

Carol Kelleher University College Cork, Cork, Ireland

Inka Lappalainen VTT Technical Research Centre of Finland, Helsinki, Finland

Mikko Lehtonen Aalto University, Helsinki, Finland

Elena Luppi Department of Education Studies, University of Bologna, Bologna, Italy

Kirsti Mäensivu Hämeenlinna, Finland

Cristina Mele Department of Economics Management Institution, University of Naples Federico II, Naples, Italy

Monia Melia Department of Legal, Historical, Economic and Social Sciences, University Magna Graecia of Catanzaro, Catanzaro, Italy

Kristian Möller Aalto University, School of Business, Helsingfors, Finland

Suvi Nenonen University of Auckland, Auckland, New Zealand

Maaria Nuutinen VTT Technical Research Centre of Finland, Helsinki, Finland

Stefano Pace Kedge Business School, Marseille, France

Monica Pedrazza Department of Philosophy, Education and Psychology, University of Verona, Verona, Italy

Joe Peppard EMST European School of Management and Technology, University of Berlin, Berlin, Germany

Francesco Polese Department of Business Management & Innovation Systems, University of Salerno, Salerno, Italy

Tiziana Russo-Spena Department of Economics Management Institution, University of Naples Federico II, Naples, Italy

Eveliina Saari Finnish Institute of Occupational Health, Helsinki, Finland

Riccardo Sartori Department of Philosophy, Education and Psychology, University of Verona, Verona, Italy

Marialuisa Saviano Department of Business Management & Innovation Systems, University of Salerno, Salerno, Italy

Roberta Sebastiani Department of Economics and Business Management, Catholic University of Scared Heart, Milan, Italy

Marko Seppänen Tampere University of Technology, Tampere, Finland

Anssi Smedlund Aalto University, Helsinki, Finland

Kaj Storbacka University of Auckland Business School, Auckland, New Zealand

Karo Tammela Aalto University, School of Business, Helsingfors, Finland

Marja Toivonen VTT Technical Research Centre of Finland, Helsinki, Finland

Marco Tregua Department of Economics Management Institution, University of Naples Federico II, Naples, Italy

Tiina Tuominen Aalto University, Aalto, Finland

Stephen L. Vargo University of Hawaii at Manoa, Honolulu, HI, USA

Heiko Wieland California State University, Seaside, CA, USA

Chapter 1
Introduction

Tiziana Russo-Spena, Cristina Mele, and Maaria Nuutinen

1.1 A New Book on Innovation: Why?

Innovation is one of the most hotly debated topics in contemporary research disciplines.

Until the 1960s, the debate on innovation research was anecdotal and biographical (Toivonen 2010); it developed within the domains of technology and economics. Scholars recognized the great importance of innovation for growth and competitive performance of industries and nations and identified the roles of technological and industrial determinants as antecedents of innovation.

During the next two decades, studies concentrated on innovation as a process related to individual firms' decisions about developing strategy and achieving competitive advantage. These studies aimed to identify the characteristics that led to commercial and technical success. Over time, they covered both hard factors (technology, R&D) and soft factors (market, organization, human resources, knowledge, commitment) as factors of success.

In the 1990s, the increasing focus of researchers on innovation dynamics contributed to the rise of the service research debate. Three conceptual positions grew from the debate about similarities and differences between new product development and new service development: (1) assimilation, (2) distinction, and (3) synthesis between product and service innovation (Carlborg et al. 2014; Drejer 2004; Mele et al. 2014; Russo-Spena and Mele 2012).

T. Russo-Spena (✉) • C. Mele
Department of Economics Management Institutions, University of Naples Federico II, Naples, Italy
e-mail: russospe@unina.it; cristina.mele@unina.it

M. Nuutinen
VTT Technical Research Centre of Finland, Helsinki, Finland
e-mail: maaria.nuutinen@vtt.fi

T. Russo-Spena et al. (eds.), *Innovating in Practice*,
DOI 10.1007/978-3-319-43380-6_1

More recently, changes in the innovation landscape have produced a renewed discussion of the need for a broader conceptualization of the phenomenon. Scholars practicing within various research streams (Carlborg et al. 2014; Chesbrough 2011; Rubalcaba et al. 2012; Vargo et al. 2015) have called for the development of new ideas and frameworks that explain the complex and multifaceted nature and structure of innovation. Service scholars in particular have provided fresh insights on innovation (Edvardsson and Tronvoll 2013; Vargo et al. 2015). The divide between products and services is no longer seen as important (Michel et al. 2008); innovation is now viewed as an essential component of a firm's value proposition (Lusch and Nambisan 2015; Vargo et al. 2015). Everyday business activities are considered to be an important arena for innovation emergence (Toivonen 2010), with a multiplicity of actors becoming involved in the processes of cocreation of novelty (Michel et al. 2008; Sundbo and Toivonen 2011; Vargo et al. 2015). The linear innovation model has been strongly criticized by alternative factions; authors highlight the complex, networked, and recursive nature of innovation in a changing business landscape (Sundbo and Toivonen 2011; Russo-Spena and Mele, 2012; Vargo et al. 2015). As ICT evolves, boundaries between industries is disappearing, and the divide between manufacturers and service firms is blurring. Many established industries are now focusing on the involvement of users and their experiences. They are adopting technology to expand opportunities for virtual interaction and grasp the benefits of harnessing inventiveness from outside their corporate walls. External innovation inputs continue to prevail, even in high-tech and research-based industries.

Notwithstanding recent changes and challenges, a fresh conceptualization of innovation that overcomes the traditional boundaries of business, social, and economics disciplines is still in its infancy stage. Interesting insights are being presented in journal articles and conference proceedings, but books on the subject are rare. Published volumes focus mostly on conceptual frameworks that are well established within research streams that include economics studies, regional science, technological and engineering research approaches, management and marketing research traditions, and policy and educational studies. The number and diversity of such studies underscores the significance given to innovation but shows a limited recognition of the key transformations that have emerged in the innovation landscape. Most works provide a perspective on innovation that is focused strictly on industries and sectors and takes a privileged, dyadic relationship approach (firm-user or customer-centric view). The studies are still anchored to a more traditional, manageable/linear, process-oriented approach; although they pay increasing attention to knowledge as the main innovation resource, they regard it as a simple commodity that can be stored and transferred from one context to another.

More serious effort is needed to apply the understandings of economic, social, and business traditions to the complexities of innovation in an ever-changing context.

New conceptualizations can lead researchers, managers, decision makers, practitioners, and students to a deeper understanding of innovation and help them tackle the challenges of innovating in our economy and society.

1.2 Novelty of the Book: Innovating in Practice

The purpose of this book is to devise an alternative conceptual vocabulary for innovation that makes sense of the complex and dynamic reality of business and societal phenomena. This vocabulary is designed to escape the positivistic, paradigmatic view of science by drawing on the ontological assumption that innovation is above all a matter of value cocreation (Vargo and Lusch 2011) that takes place in a multiple, constructed, and socially embedded reality (Gherardi 2012).

Our attempt to capture the new complexity of innovation aligns with business and economic phenomena that engage researchers from fields such as organization and knowledge (Gherardi 2000), strategy (Whittington 2003), leadership (Carroll et al. 2008), markets (Kjellberg and Helgesson 2007), marketing (Storbacka and Nenonen 2011), and consumer research (Schau et al. 2009).

These studies are rooted in the sociology of science (Latour 1987; Callon 1998) and converge toward the epistemological view of seeing and understanding phenomena in society and business adopting a practice lens (Corradi et al. 2010). This approach exhorts business and economics studies to move beyond the myopia of the neoclassical view of a reality as objectively given and known; it calls for a social domain of science where social practices—the ordering of human activity across space and time (Giddens 1984; Schatzki et al. 2001)—make the world continuously constructed and reconstructed by individuals and groups in interaction with the nonhumans (physical environment, objects) (Callon 1998; Gherardi 2009).

Our perspective on innovation is focused on social as well economic and business factors. Our aim is to contribute to an integrative approach to innovating that is emerging from service-based and practice-based perspectives of market and business phenomena (Gherardi 2000, 2009; Mele et al. 2014; Russo-Spena and Mele 2012).

Our focus is on *innovating* as an ongoing and dynamic process; it goes far beyond the idea of innovation as a mere output of a process (e.g., a new product or service) and regards practices as natural loci of novelty. Our use of the word "innovating" in the title of this book, rather than "innovation," is deliberate. The verb form indicates that innovating is action that is happening, not only the result of action. Innovating is a system of ongoing activities, emerging from actors in networking and socializing. Actors participate in innovating by developing activities, using and integrating resources and tools (e.g., products, information, knowledge, competence and skills, language, meaning), and interacting within a physical and social context, in a continuous struggle to meet identified demands and embrace possibilities to create something new and better.

We propose to understand innovating as a social process of construction by a group of actors in which borders between companies and distinctions between the internal and the external give way to an actor-to-actor view. This landscape requires continuous exchanges between various actors, ongoing collaboration, and mutual learning. Innovating is not simply the making of novel units of output but rather the designing and creating of new markets, contexts, meanings, resources, activities, and values.

This new vocabulary for studying innovation takes seriously the role of social, contextual, and cultural topics in economics and business. This book does not describe the features of successful innovation, identify its processes, or outline how innovation can be efficiently controlled and managed; rather, it reveals the conditions in which it can emerge and the elements that support it.

The book offers fresh insights by taking a multidisciplinary perspective that includes service, learning, social, and cultural streams of research. Our aim is to tackle key issues in the realm of innovation concepts, including service and market innovation, knowledge and learning processes, innovation practices, social and sustainable innovation, smart service innovation, value (co)creation and resource integration, innovation activities, and collaborative innovation.

By addressing these issues, our goal is to provide a common and integrated way of examining innovating from different perspectives and explain how individuals, organizations, networks, and societies continuously foster creative ways of knowing and doing.

This book provides a resource for researchers, managers, decision makers, and highly skilled students who are interested in exploring innovation from a multidisciplinary perspective.

1.3 Structure

The book is a collection of 21 chapters written by invitation to internationally recognized experts from academic and research institutions. A total of 46 experts from more than ten countries make theoretical, empirical, and applied contributions with regard to conceptual insights, basic concepts, experiences, methods, and challenges related to innovating in practice.

The book consists of six parts grouped under two main subjects: perspectives and experiences.

In Part 1, four chapters are devoted to new theoretical and conceptual insights that outline innovating in practice-based theorizing. Chapter 2 aims to start a debate about the underlying assumptions of innovation schools (research traditions) in which various frameworks and models have been developed. By identifying the social science research dimensions in innovation studies, it highlights three schools of innovation thought: (1) linear and planned, (2) iterative and interactive, and (3) practice-based. A common perspective emerges within each school with regard to how scholars recognize themselves as a research community (sociological) and set meanings and purposes in focusing on social phenomena (teleological) according to specific beliefs about the nature and the structure of these phenomena (ontological) that in turn influence the knowledge scholars can obtain about them (epistemological). The overall process of grasping social reality is affected by researchers' values and axiological skills (ethical). The analysis provides a path toward new frameworks that reflect the complexity of innovation phenomena and create better conditions for innovation theorizing.

Chapter 3 extends the systemic views of innovation by advancing an ecosystems perspective. The authors argue for technology and market innovations that are shaped by value cocreation and the ongoing negotiation and recombination of overlapping and intersecting institutions. A cocreative and institutional approach extends the context of innovation beyond firms' activities and provides a robust, parsimonious, and dynamic context for studying how collaborative practices that drive the cocreation of value can guide innovation.

Chapter 4 introduces the theme of new market creation by adopting a constructionist approach. The authors posit that markets are socially constructed human artifacts resulting from the actions and cognitive framing of the involved actors. The chapter provides a conceptual framework for categorizing market definitions by investigating empirically how practitioners frame novel markets and by identifying challenges associated with creating innovative market definitions in traditional industrial networks.

Chapter 5 brings new conceptual insights to the topical phenomena of innovating in practice through an in-depth investigation of practice-based learning studies. By deeply analyzing literature, the authors distinguish between three practice-based learning approaches: knowing-in practices, knowing-in-between practices, and expansive learning. They focus on increasing conceptual understanding of learning processes in each perspective by examining how the processes take place, how they enhance renewal in work activity, and how they contribute to the creation of innovations. By conceptualizing the intrinsic reproductive and creative dynamics of the practice-based learning approaches, the chapter sheds light on the processes of innovation in the practice-based approach.

In Part 2, in four chapters, the authors present and discuss the challenges and chances of innovating in practice.

By applying the concept of a practical accomplishment, Chap. 6 discusses innovating as "a texture of practices," that is, a set of practices resting on other practices. Just as fabric texture is made of a weft and a warp, innovating as texture has a weft—the practices of cocreating—and a warp, the practices of weaving. Emphasis is on the social-contextual nature of innovating in which knowing practices are sustained through networking practices. The focus is on activities, actors, and resources involved in sociomaterial and cultural practices through which innovating occurs.

Chapter 7 addresses "innovating" as innovation in action, taking place within the normal consumer practices surrounding a new product or service. The chapter focuses on the actual use of new web services by customers. It analyzes the practices of Internet users that unfold when they face a website that is new to them and addresses the question of which practices customers adopt to allow an innovation to enter their daily practices. The authors argue that the use of a new service is an interaction between the consumer practices required by the design of the new service and the regular daily practices of the customers. The outcome of this interaction determines the success of the innovation; its failure requires an adjustment to the innovation design.

Chapter 8 proposes the service innovation diffusion (SID) as an emergent process of co-construction and sense-making by companies, customers, and other partners. Such actors enact practices by interacting and integrating resources in a social, economic, and cultural context. The study identifies specific practices of SID in which actors make sense of and share meanings about innovation, both individually and collectively. These practices enable confidence in service innovation, creation of shared meanings among actors, and leveraging of actors' competencies. The research also highlights the dual role of users, who can simultaneously act as adopters and rejecters.

Chapter 9 addresses the entrepreneurial innovation processes combined with the perspectives of learning and networking. Within the realm of public organization, the authors examine the effectual approach (closely linked to service-dominant logic) as a theoretical construct for innovation. A case study describes an "effectual-type" process in which a city administration collaborates with citizens' groups to foster mutual expansion of resources and identifies characteristics that demonstrate analytical generalizability. The study highlights two findings that are particularly relevant to effectual innovation process in the public context. First, public innovations are seen as collective activities of partially autonomous actors, and second, effectuation reached in the individual or group level innovations becomes more general via institutional work.

Part 3, consisting of three chapters, features contributions from authors who attempt to build conceptual and theoretical bridges between varying streams of innovation literature.

Chapter 10 presents a novel conceptual model that depicts the connections between service and social innovation. Building on insights and categories from existing literature on social innovation and service innovation, the authors identify five issues that are integral to a proposed social-service innovation framework. Their proposed framework aims to assist scholars in developing new ideas, concepts, and tools that capture more deeply the challenges of new social innovation issues within the contemporary landscape of the service economy.

Chapter 11 provides a literature review of resource integration as a key topic of debate that links management, marketing, and service literature. A bibliometric analysis defines resource integration according to how it has been portrayed in literature. Further, it offers a deeper understanding of the six conceptual dimensions that characterize the resource integration debate. The review distinguishes between two main research areas and approaches related to resource integration: the "old school" of goods-based logic and the more recent school of service-based logic. From the service-based logic perspective, it emphasizes the connections between innovation and practices that integrate resources into attractive value propositions and presents a view of innovation as a novel integration of resources and capabilities that supports value cocreation.

Chapter 12 links smart city and service innovation from a theoretical point of view. The authors use content analysis to highlight how interventions by smart service providers are regarded by service innovation literature and how smart cities operate in connection with the key elements of service innovation developed by Lusch and Nambisan (2015). The results of the analysis are useful to describe service innovations in cities and how roles played by actors involved in smart services are relevant. Such actors possess knowledge of needs to be satisfied; they contribute resources and skills, evaluate services, and provide data. At the core of smart services provision, data plays a crucial role, helping technology support service innovation in the city context.

Part 4 introduces the "experiences" section. A series of empirical studies focuses on cases and results and presents in-depth examination of specific applications of co-innovation experiences according to different actors' perspectives.

Chapter 13 examines why and how participants cocreate value proposals in online community-based idea contests (OIBICs). More specifically, the authors explore participant motivations, roles, and behaviors as they cocreate ideas that emerge as value proposals. They identify three propositions: (1) participants are intrinsically and extrinsically motivated to compete and collaborate to cocreate value proposals in online community-based idea contests, (2) participants iteratively adopt a number of diverse and overlapping roles when co-creating value proposals in online community-based idea contests, and (3) participant behavior in community-based idea contests involves complex negotiations between the contest rules enforced by the host organization and the community norms and values. The context of investigation is the global automotive sector.

Chapter 14 focuses on customer-centric service innovation in the value networks of public and private actors. Innovation is regarded as complex networked value cocreation produced by resource rebundling, embedded in an everyday service exchange among different actors. The findings extend service concepts and platforms and stress the need for transparent value-creation processes and strategic management of novel resource integration at different levels of the service ecosystem. The chapter also includes an examination of the contributions made by external facilitators, who act as intermediaries of innovation activity in value chains and networks.

Chapter 15 provides a framework for designing a business model based on networked value creation. The study frames the dimensions of a co-creative business model by depicting the dimensions of the model and describing how they interact to integrate resources for enhancing value creation and capture. The study provides a detailed and granular approach to cocreation design to sustain the competitiveness of small and medium-sized enterprises in a complex and increasingly digital-based ecosystem. More specifically, it offers new knowledge about mechanisms and methods for aligning the contributions of different actors to better understand how value is cocreated through actors' resource-integrating efforts and captured by the lead company.

Part 5 focuses on case studies and results; it reports experiences within different business contexts.

Chapter 16 sheds light on the dynamics of innovation activity and its prerequisites for industry–research collaboration practice. Three theoretical lenses are used to examine collaboration practice and learning in the industry–research innovation program. The findings show how the characteristics of the program and its management practices construct favorable conditions for co-innovating. The chapter provides a framework that aims to support practical efforts to create better prerequisites and conditions for industry–research collaboration in innovation activity.

Chapter 17 focuses on inside-out, open innovation in daily evaluation practices, on the part of a third-sector organization providing residential care to foster youth.

The empirical setting is a regional child welfare agency. The study demonstrates that different professional groups are involved in the interinstitutional decision-making process; local and role-specific knowledge is needed to resolve problems, such as the joint evaluation of children's behaviors, that arise at a more complex interinstitutional level. The authors suggest that practitioners need to reduce rates of intra-team conflict to activate innovation in this context.

Chapter 18 addresses innovation in healthcare. It proposes "translational medicine" as a new approach to innovation that aims to lower barriers between clinical research and medical treatments. Translational medicine allows integration and optimization of inputs in basic research by improving patient care and advancing the process of offering new protocols, therapies, and practices. The chapter increases comprehension of relationships and complexity reduction processes in translational medicine and in service innovation in healthcare. By seeking strategic solutions with a holistic view of healthcare networks, it improves our understanding of the constraints to innovation and supports the translational medicine approach.

Chapter 19 studies engagement in the context of innovative projects related to social and environmental sustainability of the jewelry business. The paper explores the experience of the Ethical Committee on Coloured Gemstones, by Assogemme, in Italy. Focusing on a multi-stakeholder committee for the creation of ethical guidelines, it investigates both opportunities and associated ethical problems (e.g., child labor, environmental havoc, social inequality). The study of the committee provides meaningful insights into the practice of multi-stakeholder engagement as well as the creation of a managerial system for the traceability of jewelry products.

Finally, Part 6 presents tools and methods that support or enable the practices of innovating in various contexts.

In Chap. 20, the authors discuss the role of researchers in a new way. The researcher is regarded not only as a constructor of scientific fact but also as a codeveloper of practical tools. The chapter analyzes how codevelopment occurs between researchers and practitioners as a learning process in the context of a multi-actor, application-orientated project. It provides in-depth analysis of the researchers' various methods for enhancing learning between actors and practice contexts. Such codevelopment is characterized as an interplay between local practice-specific and generalized knowledge that enables practitioners to retool for future innovation.

Chapter 21 (linked to Chap. 17) provides formalized tools and shared protocols to investigate innovation in a regional child welfare agency. Here, innovation consists of the introduction of a new tool for the direct, systematic observation of children's behavior and attitudes. Residential youth workers, an important part of the child welfare agency's workforce, introduce the new tool to improve the quality of their contribution to the information flow on young people in residential care. Follow-up research addresses the introduction of a new tool that improves team performance and reduces interpersonal and team conflicts at the intraservice educative service level.

Chapter 22 presents a method for innovating eldercare services through evaluation, self-evaluation, and participatory planning. The research is guided by an approach of educational evaluation and action research. The professionals are actively involved during both the definition and evaluation of quality standards. Specific evaluation tools are developed, and the data gathered is constantly fed back to the professionals to stimulate their reflexive practice and foster innovation by improving the quality of care.

Acknowledgment This book is the result of a collaborative effort undertaken over time and distance.

On February 2013, seminal ideas on what it means to examine innovation from a fresh and renewed perspective came from a collaborative workshop organized by the University of Naples Federico II and VTT research on the theme of "co-innovation." During 2 days of meetings, we shared interests to provide a comprehensive collection of new contributions in this developing research area of innovation and conceived the project of a joint book.

Our initial thoughts have been developed in close collaboration with our colleagues Katri Kallio and Inka Lappalainen from the VTT Technical Research Centre of Finland Ltd., Evelina Saari from the Finnish Institute of Occupational Health, Karo Tammela from Aalto University, Monica Pedrazza from the University of Verona, and Maria Colurcio and Angela Caridà from the University of Catanzaro. Other colleagues added to our initial idea of a new book on innovation (see List of Contributors). Without their expertise and work, this book's format could not have been developed. We express our deepest gratitude and appreciation to all authors for their outstanding contributions.

Our intellectual debts are too numerous to detail at length, but we would like to express our particular gratitude to Bob Lusch and Steve Vargo. First of all, they agreed to be contributors when the idea for this book was only in its seminal stages and gave us a special gift by accepting our request that they write the foreword to the book. But above all, their amazing service-dominant logic significantly inspired us to take on the project of the new book, which arose from questions on how traditional approaches have understood innovation to date. Our special thanks also go to Silvia Gherardi. We started to make our first steps into the jungle of studies on "practice" some years ago, and we were intrigued about the different view of society and economics that the practice approach conveys. Our reading of Silvia's articles and books, as well as meeting with her and further debating the topic, greatly influenced the way we formulated our concepts and helped us to see innovation in a different way. Finally, we would like to thank Marja Toivonen for her encouragement and broad scientific insight during the process.

We also extend our sincere thanks to all reviewers who provided insightful input and suggestions. Our special thanks go to, Annika Ravald, Arho Suominen, Chiara Cantù, Elina Jaakkola, Hanna Lehtimäki, Harri Laihonen, Helge Löbler, Jaqueline Pels, Jennifer Chandler, Kirsi Hyytinen, Krista Keränen, Laura Seppänen, Lerzan Aksoy, Linda Peters, Luise Li Langergaard, Marie Taillard, Montserrat Diaz Mendez, Päivi Ristimäki, Patrik Gottfridsson, Samuel Sebhatu, Stefan Holmlid, and Stefano Consiglio for their support and donation of their precious time to providing constructive and critical reviews.

The preparation of this book has been made possible through the support of Valeria Improta; her contribution to our editing work has been gratefully acknowledged.

We also thank the publisher, Springer, and especially Hemalatha Gunasekaran and Prashanth Mahagaonkar who guided us through the publishing process.

References

Callon, M. (1998). *The laws of the markets. Sociological review monographs.* Oxford: Basil.

Carlborg, P., Kindstrom, D., & Kowalkowski, C. (2014). The evolution of service innovation research: A critical review and synthesis. *Service Industries Journal, 34*(5), 373–398.

Carroll, B., Levy, L., & Richmond, D. (2008). Leadership as practice: Challenging the competency paradigm. *Leadership, 4*(4), 363–379.
Chesbrough, H. W. (2011). Bringing open innovation to services. *MIT Sloan Management Review, 52*(2), 85.
Corradi, G., Gherardi, S., & Verzelloni, L. (2010). Through the practice lens: Where is the bandwagon of practice-based studies heading? *Management Learning, 41*(3), 265–283.
Drejer, I. (2004). Identifying innovation in surveys of services: A Schumpeterian perspective. *Research Policy, 33*(3), 551–562.
Edvardsson, B., & Tronvoll, B. (2013). A new conceptualization of service innovation grounded in SD logic and service systems. *International Journal of Quality and Service Sciences, 5*(1), 19–31.
Gherardi, S. (2000). Practice-based theorizing on learning and knowing in organizations. *Organization, 7*(2), 211–223.
Gherardi, S. (2009). Introduction: The critical power of the 'practice lens'. *Management Learning, 40*, 115–128.
Gherardi, S. (2012). *How to conduct a practice-based study: Problems and methods*. Cheltenham: Edward Elgar.
Giddens, A. (1984). *The constitution of society*. Cambridge: Polity.
Kjellberg, H., & Helgesson, C. F. (2007). On the nature of markets and their practices. *Marketing Theory, 7*(2), 137–162.
Latour, B. (1987). *Science in action. How to follow scientists and engineers through society*. Cambridge, MA: Harvard University Press.
Lusch, R. F., & Nambisan, S. (2015). Service innovation: A service-dominant logic perspective. *MIS Quarterly, 39*(1), 155–175.
Mele, C., Colurcio, M., & Russo-Spena, T. (2014). Research traditions of innovation: Goods-dominant logic, the resource-based approach, and service-dominant logic. *Managing Service Quality, 26*(4), 612–642.
Michel, S., Brown, S. W., & Gallan, A. S. (2008). An expanded and strategic view of discontinuous innovations: Deploying a service-dominant logic. *Journal of the Academy Marketing Science, 36*, 54–66.
Rubalcaba, L., Michel, S., Sundbo, J., Brown, S. W., & Reynoso, J. (2012). Shaping, organizing, and rethinking service innovation: A multidimensional frame work. *Journal of Service Management, 23*(5), 696–715.
Russo-Spena, T., & Mele, C. (2012). Five Co-s' in innovating: A practice-based vie. *Journal of Service Management, 23*(4), 527–553.
Schatzki, T. R., Knorr-Cetina, K., & von Savigny, E. (2001). *The practice turn in contemporary theory*. London: Routledge.
Schau, H. J., Muñiz, A. M., & Arnould, E. J. (2009). How brand community practices create value. *Journal of Marketing, 73*(5), 30–51.
Storbacka, K., & Nenonen, S. (2011). Markets as configurations. *European Journal of Marketing, 45*(1/2), 241–258.
Sundbo, J., & Toivonen, M. (2011). *User-based innovation in services*. Cheltenham: Edward Elgar.
Toivonen, M. (2010). Different types of innovation processes in services and their organizational implications. In F. Gallouj & F. Djellal (Eds.), *The handbook of innovation and services* (pp. 221–249). Cheltenham: Edward Elgar.
Vargo, S. L., & Lusch, R. F. (2011). It's all B2B and beyond...: Toward a systems perspective of the market. *Industrial Marketing Management, 40*(2), 181–187.
Vargo, S. L., Wieland, H., & Akaka, M. A. (2015). Innovation through institutionalization: A service ecosystems perspective. *Industrial Marketing Management, 44*, 63–72.
Whittington, R. (2003). The work of strategizing and organizing: For a practice perspective. *Strategic organization, 1*(1), 117–126.

Part I
Perspectives: Conceptual Insights

Chapter 2
Schools of Innovation Thought

Cristina Mele, Tiziana Russo-Spena, Maaria Nuutinen, and Katri Kallio

Abstract This chapter aims to start a debate about the underlying assumptions of innovation schools (research traditions) in which various frameworks and models have been developed. By identifying the social science research dimensions in innovation studies, we highlight three schools of innovation thought: (1) linear and planned, (2) iterative and interactive, and (3) practice based. A common perspective emerges within each school with regard to how scholars recognize themselves as a research community (sociological) and set meanings and purposes in focusing on social phenomena (teleological) according to specific beliefs about the nature and the structure of these phenomena (ontological) that in turn influence the knowledge scholars can obtain about them (epistemological). The overall process of grasping social reality is affected by researchers' values and axiological skills (ethical). Our main claims are that we should be more aware of the epistemological and ontological assumptions of our research, and we should make those assumptions more explicit. In these ways, we cannot only recognize limitations but also be confident when various stances are working "well enough" for practical purposes. Such knowledge creates paths toward new innovation frameworks that reflect the complexity of innovation phenomena and create better conditions for innovation theorizing.

2.1 Introduction

> We must understand our roots and foundations of knowledge as we go forward toward generating new and innovative theories.
>
> —Sheth et al. (1988: 2)

C. Mele (✉) • T. Russo-Spena
Department of Economics Management Institutions, University of Naples Federico II, Naples, Italy
e-mail: cristina.mele@unina.it; russospe@unina.it

M. Nuutinen • K. Kallio
VTT Technical Research Centre of Finland, Helsinki, Finland
e-mail: maaria.nuutinen@vtt.fi; katri.kallio@vtt.fi

T. Russo-Spena et al. (eds.), *Innovating in Practice*,
DOI 10.1007/978-3-319-43380-6_2

Innovation is a hot topic in business studies. Scholars are calling for the development of new concepts and methods; they maintain that previous guidelines are no longer valid (Carlborg et al. 2014; Vargo et al. 2015). However, although it is crucial for researchers to create new frameworks for comprehending increasingly complex phenomena, our first step should be to obtain a deeper understanding of the philosophical foundations of innovation studies. Even though innovation is a multidisciplinary topic (Grønhaug and Kaufmann 1988; Randhawa and Scerri 2015) that crosses the domains of social and natural sciences, there is scant debate on the epistemological and ontological aspects of the subject. Literature well recognizes the importance of such debate in the marketing field (Helkkula et al. 2012; Löbler 2011; Möller 2013; Peters et al. 2013; Tadajewski 2004). Scholars address the advantages of a paradigmatic discussion to provide direction for ongoing research (Möller et al. 2009; Tronvoll et al. 2011) and stress the need to pay attention to the ontological assumptions upon which the theorizing process is based (Peters et al. 2014).

In this chapter, we focus on innovation schools of thought (research traditions) in which frameworks and models have been developed. A conceptual framework is "a network of interlinked concepts that together provide a comprehensive understanding of a phenomenon" (Jabareen 2009: 51) that has ontological, epistemological, and methodological assumptions. Ontological assumptions relate to "the nature of reality," "real" existence, and "real" action. Epistemological assumptions relate to "how things really are" and "how things really work" in an assumed reality (Guba and Lincoln 1994). Methodological assumptions relate to the process of developing the conceptual framework in order to grasp the "real" world (Jabareen 2009).

Conceptual frameworks are often illustrated by graphic models—depictions of scholarly understanding and representations of phenomena. A model is "a schematic description of a system, theory, phenomenon that accounts for its known or referred properties and may be used for further study of its characteristics" (The Free Dictionary 2014). In innovation literature, models have been developed to depict the innovation process. As Mahajan and Muller (1979: 55) have pointed out, "a number of models have been developed to represent the spread of a new product in the marketplace." The most common types of models include the Stage-Gate (Cooper 1990, 2001), the development funnel (Urban and Hauser 1980; Wheelwright and Clark 1992), the service blueprint (Bitner et al. 2008), open innovation models (Chesbrough 2003), models that emphasize networked innovation (Möller and Rajala 2007), and, most recently, practices (Russo-Spena and Mele 2012b).

Although frameworks and models rest on certain basic assumptions, these assumptions are not often explicitly stated in innovation studies. This inexplicitness limits clear understanding of studies and restricts development of scientific debate. Furthermore, it leads scholars to build their works on risky foundations in which results are characterized by high inconsistency and low trustworthiness and the context is outside the researchers' assumptions and research traditions (Mele et al. 2014).

In this chapter, we begin the debate on the assumptions underlying research in innovation. We refer to these assumptions as metatheoretical; they rest on the sociological, teleological, ontological, epistemological, and ethical dimensions of social science research (Mouton and Marais 1990). Until now, in innovation literature, these five dimensions have been unexamined and unchallenged. Scholars have not advanced theories of knowledge or views of reality to support theoretical perspective and methodology, and in research studies, philosophical stances have not been linked to practical aims.

Therefore, in this chapter, we examine social science research dimensions in innovation studies by highlighting three schools of innovation thought: (1) linear and planned, (2) iterative and interactive, and (3) practice based.

We are aware that "any categorization of meta-theories deals with ideal types and operates at a high level of abstraction" (Löbler 2013: 51) and that our classification is partial and does not take into account rich literature on innovation in business studies. We do not wish to—and cannot—be exhaustive nor do we wish to offer a detailed analysis of every conceptual framework and model. Our aim is to make a first attempt to capture the essential features of three schools. Although the orientations of the schools are highly varied, the variation is much greater among the three we have identified. We examine the cohesion of each research tradition "which binds the work of a group of theorists together in such a way that they can be usefully regarded as approaching social theory within the bounds of the same problematic" (Burrell and Morgan 1979: 23).

Our attempt to summarize these schools of innovation thought is a crucial step in generating new and innovative theories in innovation. To have a fresh perspective on innovation, we must know our roots and understand our foundations of knowledge. In this way, we can frame a philosophical debate in innovation.

In the following sections, we first clarify the dimensions of social science research. Second, we propose a schema to be used in reviewing major research traditions. Third, we present the three schools of innovation thought, and finally, we offer conclusions.

2.2 Dimensions of Social Science Research

When approaching a research topic, scholars begin with methodology—the ways in which valid knowledge is acquired—to make decisions about research strategy (quantitative vs. qualitative or mixed method) and method (specific procedures and techniques used in the study). However, these choices are only part of the researcher's paradigm. We can use the "iceberg" metaphor (Lor 2011) to describe the relationship between metatheory, methodology, and methods. Although the latter two aspects form the part of the iceberg that rises above the surface of the sea, they are a mere tenth of the volume of the floating iceberg (see Fig. 2.1).

The much larger, immersed portion of the iceberg is metatheory; it is comprised of key assumptions about the sociological, teleological, ontological, epistemological, and ethical dimensions of social research:

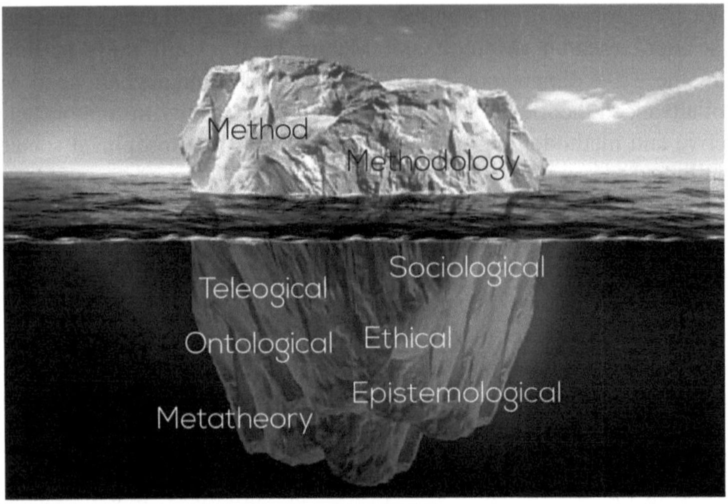

Fig. 2.1 Iceberg model of the dimensions of research

- The sociological dimension refers to the researchers' social factors: Where do the researchers come from? Who does the research and what is the context? Scholars investigate "within a research community which has shared goals and values, exercises control, and recognizes contributions" (Lor 2011: 6). By sharing identifiable disciplinary paradigms, researchers are linked through research networks (Mouton and Marais 1990) that are influenced by the environment, the academic reward system, and the availability of funding. Values, ideological (and non-epistemic) interests, power, and conflict are all important in understanding sociological mechanisms in scientific research.
- The teleological dimension refers to what we want to achieve through our research. According to Mouton and Marais (1990), the teleological dimension is concerned with purpose or aim; research is "intentional and goal directed." Researchers' metatheoretical orientations influence their purposes and the goals they set for their research. Research output cannot be understood without reference to the meanings and purposes attached by researchers to their activities (Guba and Lincoln 1994).
- The ontological dimension is about the nature of reality that is being investigated and the way the world operates (Guba and Lincoln 1994; Sanders et al. 1992, 1993). It is usually debated according to objectivism and subjectivism: What are the form and the nature of reality? What can we know? What is "out there" to know? How does it exist? What constitutes social reality? It is about the domain assumption and the "beliefs about the nature, structure and status of social phenomena" (Mouton and Marais 1990: 11–13).
- The epistemological dimension poses the question, "What is the nature of knowledge and the relationship between the knower and the would-be known?" (Mertens 1998: 6). How can we come to know reality? How can we

understand its existence? Epistemology is thus concerned with "the nature of knowledge, its possibility, scope, and general basis" (Honderich 2005: 260).
- The ethical dimension refers to the role of values played in the research process (Sanders et al. 1992, 1993). Axiological skills play a key role in the researcher's process of making judgments.

Various metatheories emerge from researchers' answers to questions about these five dimensions (Lor 2014). The concept of metatheory is associated with paradigms and research traditions. According to Kuhn (2012: 8), paradigms are "recognized scientific achievements that for a time provide models, problems and solutions to a community of practitioners." Paradigms arise from consensus among researchers who are committed to the same rules and standards of scientific practice. They are seen as being the most successful ways to solve problems. The idea of science with problem-solving effectiveness has been strongly advocated by Laudan (1978: 81), who defines research tradition as a set of "guidelines for the development of specific theories." These guidelines constitute an ontology that specifies the types of entities that exist in the domain; they also identify legitimate methods of inquiry. Laudan affirms (1978: 80) that "a research tradition is a set of general assumptions about entities and process in a domain of study and about the appropriate methods to be used for investigating the problems and constructing theories in that domain" and goes on to state that research tradition "is a set of ontological and methodological do's and don'ts."

If we regard paradigms, schools, research traditions, and metatheories as synonymous, we must also recognize that they are broader and less specific than theories. We acknowledge the existence of "a basic belief system or worldview that guides the investigator, not only in choices of method but in ontologically and epistemologically fundamental ways" (Guba and Lincoln 1994: 105). In other words, there are assumptions (both conscious and unconscious) that guide theoretical, empirical, and practical work.

The various expressions of paradigms, schools, research traditions, and metatheories include approaches such as positivism, post-positivism, interpretivism, critical theory, and pragmatism (Sanders et al. 1992, 1993). These approaches differ according to key assumptions about the dimensions of social research. In line with Guba and Lincoln (1994), we regard all of them as human constructions used to explain phenomena and leave it to the proponents of each approach to address their values in terms of usefulness and persuasiveness.

2.3 The Process of Schooling

We use the five dimensions of social science research to identify major innovation schools of thought (research traditions). Rather than comparing the schools and addressing the supremacy of one over the others, we choose to start a philosophical conversation by mapping the field—arguing for multiplicities and proliferations

Table 2.1 A schema for investigation

Sociological	Who does the research and with the context in which the research is conducted? What is the main research community? Does this community share goals and values, exercise control, and recognize contributions? Does the community operate within economic, political, and ideological frameworks?
Teleological	What is the purpose or aim? Is research "intentional and goal directed"? What about meanings and purposes involved in the research?
Ontological	What is the reality that is being investigated? In which research domain? What are the domain's assumptions and beliefs about the nature, structure, and status of social phenomena? What assumptions do the researchers make about the way in which the world works?
Epistemological	What about the nature of knowledge, its possibility, scope, and general basis? What is the relationship between the knower and the would-be known?
Methodological	Which are the research methods? Qualitative or quantitative?
Ethical	What is appropriate ethically? What about researchers' value judgments? What is the role of values?

(Lather 2006) instead of paradigm shifts (normal vs. revolutionary). By moving away from noncommensurable stances, we attempt to grasp the dominant and emerging knowledge that competes for acceptability and try to shed light on practices of knowing among innovation scholars (Lather 2006).

To address the variety that describes current approaches, we set up a schema (Table 2.1) with specific questions for investigation. This schema outlines scholars' assumptions about social science research dimensions and allows us to see the undersurface iceberg of innovation research.

In line with previous attempts to identify schools of thought in marketing (Möller et al. 2009; Sheth et al. 1988; Tadajewski 2004) and organization (Mckinley et al. 1999), we assert that an evaluation of current innovation schools enables scholars to be aware of actual knowledge and incorporate and combine previous and new knowledge to generate fresh theories in innovation. There is a particular need for a general theory of innovation that is able to zoom out from micro-issues (as single new artifacts) to take into account practices, complexity, and the service ecosystem context (Lusch and Vargo 2014). There is increasing interest in theorizing about the discipline (Mele et al. 2014) and conducting a metatheoretical debate (Möller et al. 2009; Tadajewski 2004).

2.4 Schools of Innovation Thought

In our quest to identify research traditions in innovation, we review literature in the area of business studies, searching for methodologies for developing schools of thought (Bartels 1962; Mckinley et al. 1999; Möller et al. 2009; Sheth et al. 1988).

A school of thought can be defined as an integrated set of theoretical framework "that provides a distinct viewpoint on innovation and that is associated with an active stream of empirical research" (Mckinley et al. 1999: 635). Schools of thought are one of the most salient features of contemporary innovation disciplines, "providing the basic intellectual structure within which ongoing theoretical and empirical work take place" (Mckinley et al. 1999: 635).

We identify three major schools of innovation thought: (1) linear and planned, (2) iterative and interactive, and (3) practice based.

All innovation schools share certain criteria (Möller et al. 2009; Sheth et al. 1988). First, there is a distinct focus on innovation goals. Second, there is a perspective on innovation activities (i.e., which activities are carried out, in which manner, and by whom). Third, there are a significant number of contributing scholars, and finally, there is novelty, continuity, and scope (McKinley et al. 1999):

- Novelty is the property of being new, unique, or different, particularly in relation to theoretical frameworks that have been central to a discipline in the past.
- Continuity is a property of affinity: it refers to a linkage with intellectual frameworks that are already familiar to scholars.
- Scope is the range of phenomena encompassed by the school.

The challenge for emerging schools is "to balance the tension between novelty and continuity: adequate levels of each are necessary for school development" (McKinley et al. 1999: 637). It is crucial for them to maintain enough novelty to warrant interest to fuel scholarship. At the same time, however, evolving schools cannot stray too far from existing conceptions. In other words, "continuity enfolds novelty with meaning, establishing a context within which the novel claims of an evolving school can be understood" (McKinley et al. 1999: 638).

We have chosen the three names for the schools' indicators on the basis of scholars' definitions and because they are expressions widely used within each school. Rather than seeking objectivity, we aim to grasp native categories that are part of the innovation lexicon. As Buckley and Chapman (1997: 289) note, "native categories are not just words; they are part of a lexicon, certainly, but they are also a living part of action and understanding."

2.4.1 Linear and Planned School

We refer to the first school or research tradition in innovation as linear and planned (Cooper 2001).

This school includes the research conceptualizations of the American community of engineering and business management; it represents a normative view of innovation, in which the firm operates as stand-alone actor (Table 2.2). It is the first school of innovation thought, established in the 1980s through the leadership of US researchers and practitioners who published in highly ranked business and technological journals (e.g., *Research Technology Management*, *Journal of Product Innovation Management*, *Business Horizons*, *Industrial Marketing Management*, *International Journal of Service Industry Management*).

Table 2.2 Linear and planned school

Dimensions	Main content
Sociological	Research community is represented by American scholars of engineering and business management
Teleological	Intentional and goal-directed nature of research—prediction and control Insights for decision-makers
Main models	Sequential models – NPD: Six-stage model (Booz et al. (1982), Stage-Gate model (Cooper 1990, 2001), development funnel (Wheelwright and Clark 1992)) – NSD: Eight-stage model (Bowers 1987), linear model (Scheuing and Johnson 1989), four-stage model (Johnson et al. 2000) Parallel models – NPD: Concurrent engineering (Wheelwright and Clark 1988; Brown and Eisenhardt 1995) – NSD: Service design (Ramaswamy 1996), service design (Edvardsson and Olsson 1996), ten-stage model (Alam and Perry 2002)
Ontological	Reality stable over time and predictable Research domain: New product development and new service development established as different research domains with NSD derived from assimilation with NPD conceptualizations
Epistemological	Knowledge based on information gained from observable and objective experience: – Market and customer needs known – Technology known and in the proprietary domain – Market or technical uncertainty, risks few General principles: – Innovation as normative process – Innovation as defined programmatic effort with clearly resource constraints
Methodological	Rigor, validity, reliability, objectivity Descriptive research based on: – Quantitative studies (survey: large sample, measurement) – Qualitative studies (multiple case studies)
Ethical	Value-free way Researcher independent of the data, maintains an objective stance

Much scientific debate has arisen from a variety of proposed models that capture the stages of new product development (NPD) process within firms. The models proposed by this school's group of researchers are designed to answer the question of how companies drive new products to market. Reflecting the teleological dimension, these studies are led primarily by intentional and goal-directed research that aims to determine the one best way to develop innovation.

One of the school's first models was developed in 1982 by the management consulting firm of Booz Allen Hamilton; it is known as the BAH model. The model divides the product innovation process into seven stages and proposes a simple unitary progression from the strategy development phase to the market phase. This framework has inspired most other new product development models, including the popular Stage-Gate model of Cooper (1990), the development funnel of Wheelwright and Clark (1992), and the product innovation process of Crawford (1994).

The Stage-Gate model was the first to provide a comprehensive conceptual and operational map of the new product development process, based on the progression of phases as separated by gates that act as decision points between the stages. On the map, research and development (R&D) activities, marketing insights, and a mixture of both components represent simple linear sequential events (Cooper 1988, 2001; Garcia and Calantone 2002).

Since then, through the assimilation of the product development process into new service development (NSD), researchers have provided new models that account for the intangible nature of service attributes. They include the eight-stage model (Bowers 1987), the linear model (Scheuing and Johnson 1989), and the four-stage model (Johnson et al. 2000). NSDs have been regarded as more likely to succeed when they are supported by systematic development and strong engineering know-how. In many cases that reflect the more complex and dynamic nature of innovation in service, NSD models outline a greater number of stages (e.g., Scheuing and Johnson (1989) counted 15 stages) with an active presence that supports internal inputs.

Linear models have generated considerable debate, much of which has been focused on questioning whether development stages actually proceed according to a sequential order. In the product domain, the concurrent engineering model (Wheelwright and Clark 1988) is a more advanced model in which different stages can run simultaneously, rather than consecutively, to reduce cost and improve market effectiveness. In the service domain, Edvardsson and Olsson (1996) describe simpler models that depict the NSD process as having three activities, starting with the development of the service concept and followed by the design of the service system and the service process. Similarly, Ramaswamy (1996) regards the process of service design as an integrated process, in which the management phase includes measurement and testing strategies that provide feedback for further improvements. Alam and Perry's (2002) model provides insight into activities that run in both linear and parallel ways to speed up NSD.

These models operate in different research domains and according to different approaches, but they generally agree on the substance of typical operations and share an engineering mind-set that identifies stages and provides labels for various contexts of investigation.

The ontological position is rarely declared in the frameworks and models, but a prevailing objectivism can be detected in the assumption of social entities existing in reality external to social actors concerned with their existence (Sanders et al. 1992, 1993). A typical feature of the linear and planned approach is the ontological view that the investigated reality is stable over time and that predictable phenomenon are "out there" (Löbler 2013: 56), that is, they can be observed, analyzed, and managed through prescribed tasks and gates. As Cooper (2001) notes, the reasoning behind innovation model research is not simply the understanding of a phenomenon but also the setting of an approach to finding problems and solutions. Resembling the objective-oriented view (Löbler 2013), the Stage-Gate approach is almost deterministic in its attempt to summarize the knowledge of the way things are in form of time and context, free of generalizations such as the law of cause and effect (Guba and Lincoln 1994). The scholarly view in this school stresses the structural aspects of the innovation process and assumes that the innovation model (e.g., Stage-Gate) is similar in all organizations.

The epistemological dimension of this school provides evidence of the idea that knowledge is based on information gained from observable and unbiased experience. The researcher is seen as an entity independent of the researched companies and separate from the process of innovation. The rules of the game in terms of factors such as technological possibilities, market demands, competitor behavior, and political context are fairly clear or knowable, and the process of acquiring information is closely related to the activities carried out in an innovation project. The normative nature of innovation clearly emerges; the final development goal is well articulated in advance as an output of the entire process and locked in prior to initiating development. The sequence of activities in the processes is treated as an ongoing, information-gathering effort that makes resource constraints and expected results progressively clearer.

The models of the linear and planned schools are suitable for describing the innovation experiences of large, close organizations (Cooper 2001) in which sequential choices are applied according to a cause-and-effect logic. The rigorous process for managing innovation—including a disciplined stage-by-stage approval process—has been well documented (Booz et al. 1982; Cooper 2001; Urban et al. 1993) as has the regular measurement of critical factors affecting innovation progress and end results (Ramaswamy 1996; Ulrich and Eppinger 2004).

Notwithstanding that many scholars in this school have warned of limits to these models (Cooper 2001; Trott 2008), the linear and planned innovation framework has for many years played a key role in business planning; it has been treated as a defined programmatic or planned task based on knowledge of clear resource constraints and well-defined outputs. Linear and planned innovation scholars remain well positioned to help companies manage innovation to reach profitability goals (Garcia and Calantone 2002; Godin 2006).

In the search for relationships between key innovation variables, the methodology at the bases of these models is represented mainly by the descriptive research design approach. Researchers work in empirical settings that include data from extensive surveys, in-depth interviews, and case studies (Yin 2002); they provide a fairly good representation of prevailing innovation best practices in the business (Griffin 1997). Reliable, replicable findings (Guba and Lincoln 1994) satisfy scholars' needs for validity, reliability, objectivity, and generalizability. Researchers are able to gain knowledge to discover uniformities that yield empirical regularities and lawlike generalizations (Hunt 1976; Löbler 2013). They adopt a reductionist approach that assumes that the object of research, no matter what the domain of study, can be broken into constituent parts.

By taking a logical–empiricist position, researchers in this school aim to produce practically useful knowledge. According to linear and planned scholars, the most desirable features of knowledge production include "a commitment to objective research undertaken by scientists who are able to distance themselves from the object of study through the careful use of the scientific method that sets out the rules and guidelines that delimit explanatory adequacy, prescribe inductive accuracy and insists on empirical testing" (Hunt 2003: 85).

2.4.2 Iterative and Interactive Innovation School

We refer to the second school or research tradition as iterative and interactive (Cooper 2008, 2014). It includes studies that promote a user-centered and relational conceptualization of innovation (Table 2.3). Scholars in this school have responded to the need to better integrate requirements of users, possibilities of technology, and opportunities for business success (Cooper and Kleinschmidt 2007; Tidd et al. 2005). A major part of this school's scientific community comes from the engineering and managerial tradition of linear and planned scholars but goes further in an attempt to frame innovation according to a more open, relational, and systemic approach. Its contributions come from publications in the mainstream of management, marketing, and organizational journals (e.g., *Journal of Marketing, Strategic Management Journal, Journal of Business Research, Journal of Service Management, Information Systems Research*).

In this tradition, scholars within management (Cooper and Kleinschmidt 2007; Tidd et al. 2005), marketing (Von Hippel 1988, 2005), and organizational learning (Cohen and Levinthal 1990; Teece et al. 1997) highlight innovation as a complex and multidimensional undertaking. The group includes the flourishing European service school, which has been widely recognized to be the origin of increased interest in a process-grounded view of innovation (den Hertog 2000; Sundbo 1997; Toivonen and Tuominen 2009; Toivonen 2010).

A substantial part of literature grouped under the iterative and interactive school supports a common effort to identify the antecedents and enablers of innovation. The idea of innovation as a deliberate affair in which organizations follow a formal, methodological procedure with well-defined and pre-ordered steps is challenged by these scholars, who view innovation as a phenomenon with deeper-lying processes. Here, the teleological dimension focuses on intentional and goal-setting-oriented aims, and research efforts are aimed at providing understanding of the hidden driving forces and mechanisms that support and develop innovation in continuous and systematic ways (Tidd et al. 2005).

Within the new product research tradition, the pioneering work of Von Hippel with regard to lead users (1988) has inspired a wide variety of customer-oriented models that view innovation as integral to business strategies that depend on a thorough understanding of customers' and users' voices. In this school, the scope of innovation is broader: it includes not only new products, services, and processes but also more strategic and business-oriented goals such as new market positions, segments, knowledge, and business (Chesbrough 2003; Möller and Rajala 2007; Tidd et al. 2005).

This body of research not only establishes the importance of feeding customer insights into the new product development process but also shows the advantages of

Table 2.3 Iterative and interactive school

Dimensions	Main content
Sociological	Research community is broader and includes technological, management, marketing, and organizational learning scholars both from American and European countries
Teleological	Research is goal-setting oriented, with an aim to provide understanding of hidden driving forces and mechanisms that govern and develop innovation; deep understanding of the users, their needs, and the context in which these needs will be fulfilled is at the heart of the innovation problem Scope of innovation includes strategic dimensions (e.g., create new position in the market, new market segments, new knowledge, new businesses)
Main models	*User-based interactions* – NPD: User-based innovation (Von Hippel 1986), iterative or next-generation Stage-Gate model (Cooper 2008, 2014) – NSD: New service development models (Johnson and Gustafsson 2003), iterative new service development (Miettinen 2011; Tuulaniemi 2011), agile or spiral development model (Schneider and Stickdorn 2011) – User-generation or community-generation models (Franke and Piller 2004; Franke and Shah 2003; Nambisan and Nambisan 2008) *Multi-actor interactions* – Open innovation (Chesbrough 2003; Chesbrough et al. 2006) – Connect & Develop model (Sakkab 2002) – Platform innovation (Gawer and Cusumano 2002) – Innovation value chain (Hansen and Birkinshaw 2007)
Ontological	Reality is complex, often unstable, and only partially known and predictable
Epistemological	Knowledge is often based on conjecture: – Customer needs not well known or articulated – Customer needs not only satisfied but also discovered – Multiple actors must be satisfied – Technology not necessarily under the main actor's control and domain – Many market and technological uncertainties and risks General principles: – Innovation process responsive in nature by being open to change – Innovation process conceived as a system of overlapping interactive spaces rather than a sequence of orderly steps – Innovation encompasses relationships and collaborative work – Not only process or market factors but also soft managerial and organizational factors come into focus (e.g., human resources, relationship management)
Methodological	Exploratory research, abduction Qualitative research: Case studies of a small number of cases conducted through methods such as in-depth interviews, participant observation, and use of informants Qualitative phase typically proceeds quantitative phase
Ethical	Research is value laden; researcher biased by factors such as worldview, cultural experience, and upbringing, which impact research

doing this throughout the entire innovation process, from the earliest opportunity-identification phase to the final launch phase. Researchers highlight the importance of more dynamic relationships between firms and customers in which customer inputs are incorporated into the innovation process, activities are developed through customer interaction, and the perspectives of "users" are taken into account (Franke and Piller 2004; von Hippel 2005). In this school, interest broadens to users and their communities, treating them as innovators and sources of valid information (Franke and Shah 2003).

As a result, major stages of new product development (NPD) are being integrated into creative marketing research techniques such as brainstorming, focus groups, workshops, channeled ideation to develop relationships between providers, and customers acting as two-way, overlapped information exchanges. In this sense, the next-generation Stage-Gate model is a recent advancement proposed by Cooper (2008, 2014). It is designed to emphasize the iterative nature of innovations that involves multiple circles of design and marketing and earlier program testing. As Cooper (2014: 25) reports, "the iterative development is about building something light, then testing it, getting feedback from it and finally revising it accordingly." The model maintains a dynamic approach to innovation by depicting arrows that loop back into certain stages of the process. Use of a complex network of relationships is encouraged, to reduce risk of failure, reveal underlying factors, and convey unexpected results. User-generated and community-generated models (Franke and Piller 2004; Franke and Shah 2003; Nambisan and Nambisan 2008) challenge the idea that innovation process knowledge should be protected; they encourage companies to build processes and spaces in which users and customers—with differing roles and motives—can participate. The models provide guidance for new ways to encourage innovation by customers and other actors, organize internal practices, and develop customer relationships.

New service development models (NSDs) stress the importance of iterative and strategic approaches to innovation. Johnson and Gustafsson (2003) have developed an NSD model based on strategic fit and cultural change fit as key criteria for selecting and screening new ideas. This model shows a closer relationship between the service provider and customers and emphasizes the role of managerial and organization factors to support and drive NSD processes. The iterative new service development model (Miettinen 2011; Tuulaniemi 2011) and Agile or spiral development model (Schneider and Stickdorn 2011) also capture the critical role of interaction among the costumer and service staff and the importance of skilled frontline employees and service interface design in defining how the customer or user experiences the service (Toivonen and Tuominen 2009; Toivonen 2010).

The iterative view of innovation is also at the core of many other conceptual elaborations of innovation models that emphasize the increasing complexity of innovation and the importance of the wider relational context of the firm.

The innovation network is a promising alternative coordination mechanism that allows firms to have access to capabilities and assets for innovation (Ritter and Gemünden 2003). Business researchers in the tradition of industrial marketing purchasing (IMP) suggest that innovation occurs mostly within relationships (Möller and Rajala 2007; Möller and Svahn 2006); thus, innovation results from interactions and ongoing improvements in structures and interaction capabilities (Johnson and Sohi 2003). The innovation network approach examines carefully selected relationships among different actors engaged in cooperation to develop technological design or boost commercial innovation (Möller and Svahn 2006). In innovation networks, the goal is to create radical innovations and generate new businesses and technological fields (Möller and Rajala 2007; Möller and Svahn 2006). These networks are often characterized by professional and social relationships and are not primarily business networks.

Networking and multi-actor interactions have inspired one of the most prominent frameworks in recent innovation and technology management literature: the open innovation model. Since the publication of Henry Chesbrough's book on the subject in 2003, open innovation has been affirmed as a concept that goes beyond the traditional idea of using external sources of innovation. According to Chesbrough (2003), open innovation moves away from outsourcing or insourcing of R&D activities to use both inflows and outflows of knowledge to improve internal innovation processes and expand markets. The model assumes the integration of external and the internal knowledge throughout the innovation process; it is a way of innovating with partners by sharing risks and rewards (Chesbrough 2003; Chesbrough et al. 2006). This idea of multi-actor innovation that requires high levels of interaction and is facilitated by IT-based networking is at the foundation of the platform innovation models (see the Connect & Develop model of Sakkab 2002; see also Gawer and Cusumano 2002).

Notwithstanding their different origins and conceptualizations, the studies grouped under this school share the *ontological* dimension, in which the understanding of reality is complex, often unstable, and only partially known and predictable. Distinctions between models can be traced to the natures of the investigated reality they describe. For example, various scholars within the service innovation tradition emphasize that services are entities distinct from products; they stress that any service innovation design involves a combination of aspects, including new concept, new client interface, new service delivery systems, and technological options (den Hertog and Bilderbeek 1999).

To further shift the boundaries of previous models, scholars in this school focus on ontological aspects of innovative phenomenon that are often regarded as dichotomies (e.g., internal/external, open/closed, market/technology, and users/stakeholders). Reality is investigated through differences. In addition to making ontological assumptions, they also make *epistemological* assumptions about the uncertainty of knowledge processes and the unpredictable nature of the market and technology development process; in these ways, a systemic view of innovation arises, encompassing the needs of multiple sources and actors taking part in the innovation process.

Current initiatives in designing and evaluating innovation continue to follow mainstream linear innovation thinking about the management process, even though such thinking oversimplifies the phases of innovation and the complex dynamics between actors. Although innovation models within the iterative and interactive approach build on a wider range of activities, tasks, and information flows, there is still an expectation that there is a performance advantage for firms that have a formalized innovation process in place. Projects are locked down no earlier than absolutely necessary so as not to miss newness arising from the market, technology, or business. The innovation process develops according to less articulated, lighter steps to allow teams to move rapidly to a finalized design through a series of "create-test-feedback-and-revise iterations" (Cooper 2008, 2014). The interactive approach to managing key innovation activities aims to create a climate for continuous innovation that captures the knowledge and learning achieved during the development process. The innovation process is described as an organizational learning process (Nonaka and Takeuchi 1995; von Krogh et al. 2000) in which the input is some kind of informational or knowledge stimuli and the output is new knowledge. According to this approach, reality—even if complex and uncertain—can be learned, and knowledge can be obtained and assimilated into the company's processes. The knowledge spiral is one of the key models used to explain the knowledge-creation mechanism that underlies and stimulates continuous innovation processes.

The position of ontological and epistemological studies within this school falls between post-positivism and critical realism. Scholars develop an approach "that falls between the object orientation and the subjective orientation, as it assumes a reality but recognizes the limitations of observers to fully perceive it" (Peters et al. 2014: 258). The critical realist's position that the social world exists amidst constant interrelation and change seems more aligned with the purpose of the iterative and interactive school, in which the aim of business and management research is to understand phenomena as a precursor to recommending change.

This school includes the industrial marketing and purchasing (IMP) tradition, in which we find clear ontological and epistemological positions that are close to the critical realist view (Bhaskar 1998); they inspire studies that recognize the importance of the multilevel approach (individual, group, organization). In line with critical realist scholars, they state that reality is a result of social conditioning (Peters et al. 2013). Their argument is that researchers are able to understand what is going on in the social world only if they understand the social structures that give rise to the phenomena to be understood. In other words, what scholars see is only part of a bigger picture. The critical realist recognizes the importance of multilevel study (Sanders et al. 1992) in which the firm, the dyad, and the network are recognized at each level by researchers who have the ability to change their understanding of what is being studied, according to "the existence of a greater variety of structures, procedures and processes and the capacity that these structures, procedures and processes have to interact with one another" (Saunders et al. 2011: 115).

The methodological approach of exploratory research is used most extensively to gain insights and comprehension rather than collect replicable data. Analytical studies frequently involve the use of in-depth interviews, analysis of a small number of case studies, and the use of informants and are likely to include studies in which hypotheses are developed as a result of research rather than research being guided by hypotheses. Often, the general research approach is not restricted to a single technique or method but is a combination of methods (Dubois and Gadde

2002). Researchers advocate use of the triangulation method to improve accuracy and reduce incidences of misinterpretation (Yin 2002).

By viewing innovation as a relational and network affair, this research tradition opens up the possibility of exploring completely new technology paths. For managers, the implication is that they can move beyond highly structured, risk-averse processes that are limited by fixed goals, budgets, and schedules. Researchers in this tradition are devoted to acquainting managers with the need to connect and mobilize relational contexts, both inside and outside their firms, to assure flexibility, continuous stimulation, and opportunities for innovation.

2.4.3 Practice-Based School

We refer to the third school in our review as the practice-based school (Gherardi 2006, 2009). This school embraces a large and heterogeneous community of researchers who define innovation as the understanding of dimensions of social and collective doings (Gherardi 2012a, b) (Table 2.4). The practice-based school focuses on construction of human activity, knowledge, learning, and practices. Its position is reflected in studies published in a wide variety of journals, with topics ranging from service research journals and organizational learning to innovation policy and design (e.g., *Journal of Workplace Learning*, *Research Policy*, *Organization Science*, *International Journal of Quality and Service Science*, *Service Science*, *Design Management Journal*, *Creativity Research Journal*).

Although the term "practice" conveys many concepts and spans many research domains (Corradi et al. 2010; Gherardi 2009, 2012a, b), we narrow our use of the term to three different tracks of research that contribute to the emerging practice-based school of thought.

First, we consider studies that seek inspiration from classical works of human knowledge and learning, such as "communities-of-practice" literature (Brown and Duguid 1991; Lave and Wenger 1991), theory of practice and reflective practice (Schön 1983), pragmatic philosophers (Dewey 1944), situated actions (Suchman 1987), cultural historical theories of activity (Leontyev 1977; Engeström 1987; Vygotsky 1997), and more recent work on organizational change and organizational culture (Tsoukas and Chia 2002). The core of this learning community comes from the tradition of social science and encompasses scholarly work on concepts of learning, practices, and activities as a starting point for building innovation approaches and criticizing mainstream innovation research (Russo-Spena and Mele 2012a; Russo-Spena et al. 2014). Researchers argue strongly for the value of acknowledging the social, contextual, and situated natures of human knowing and acting. They enrich and challenge the discussion by utilizing learning concepts and the classical works of social construction of knowledge and action.

Second, in this tradition, we identify scholars who are part of the service innovation research and managerial tradition of iterative and interactive school and who are supplementing these ideas by searching for deeper perspectives to

Table 2.4 The practice-based school

Dimensions	Main content
Sociological	Research community includes sociology of knowledge, design, and organizational learning researchers and service scholars
Teleological	Understanding and constructing; research aim is interpretive; aim of researchers is to open up the nature of innovation problems and ask hard questions, with researchers acting as facilitators of multi-voice reconstruction
Main models/Conceptual framework	Design-driven innovation (Brown 2008; Verganti 2009), creative thinking models (Carson 2010), critically reflective learning/expansive learning approach (Engeström 1987, 2001), developmental learning (Ellström 2001, 2010) Service ecosystems (Lusch and Nambisan 2015; Vargo et al. 2015) "Co-s" (Russo-Spena and Mele 2012b) and innovation practices (Mele and Russo-Spena 2015a, b)
Ontological	Reality is the result of social construction; no universal solutions to innovation problems and no ready answers but thoughtful guidelines and insights
Epistemological	Innovation can be interpreted in different ways and acquired only by participation in practice Constructivism view of conceiving innovation as emergent collective social phenomena that follows nonlinear dynamics Path-dependent phenomenon Role of social factors (norms, rules, shared meanings) in activities, contradictions, and negotiations between social groups
Methodological	Exploratory, phenomenological Action research, case studies, ethnography, netnography, narrative workshop Trustworthiness, authenticity
Ethical	Research is value bound; the researcher is part of what is being researched

understand and describe the logics of innovation and practices. Service-dominant logic (Vargo and Lusch 2004, 2008) proposes interpretive views in which innovation is seen as a process of new benefit provision rather than a tangible or intangible outcome (Edvardsson and Tronvoll 2013; Lusch and Nambisan 2015; Mele et al. 2014; Vargo et al. 2015). It emphasizes the role of relationships and knowledge and the dynamic nature of assets and regards innovation as a cocreational and ongoing process that takes the form of complex and interdependent systems (i.e., service ecosystems) (Lusch and Vargo 2014; Vargo et al. 2015).

Third, this school includes a community of design-practitioner researchers and researchers focused on design-driven innovation. They work at the core of innovation; it is both their job and their identity. In contrast to 1960s-era academic discussions related to design and design management, the positions of these practitioners include teleological, ontological, and epistemological standpoints, with clear contradictions occurring in more recent discussions of design management as an academic field. Such explorations of the design line of research provide insight into innovation activity in terms of its ability to construct subjective knowledge (Kimbell 2011) as well as its approach to culturally and historically formed structures, interpretations, and meanings that characterize and constrain the activity of designing.

Notwithstanding the great heterogeneity among the scholars and research domains of this tradition, their common approach to problems seems to be motivated by their joint teleological view. Thus, a constructing view of innovation emerges, allowing us to understand innovation as a developmental path from some initial state rather than a teleological progress toward some final state (Burman 2007). This view further challenges the definitions of innovating and innovation and alters the focus of innovation research.

In this tradition, models and concepts are used to analyze and support the understanding of innovation phenomenon in a particular context rather than to provide normative descriptions on how things are or should be. This teleological position produces new concepts that help scholars, market actors, and nonmarket actors grasp social reality not as an objective reality "out of there." By focusing on practices, scholars can too detect and analyze social connections among individuals, collectives, organizations, and institutions and describe the social, cultural, and economic contexts in which these connections are formed and institutionalized (Mele and Russo-Spena 2015a; Lusch and Vargo 2014).

One motive that drives research in this tradition is the quest to compare actual human behavior with the behavioral assumptions used in managerial guidance. Scholars share a genuine concern for the consequences if this distinction is not acknowledged. A broader view of business emerges from this line of argument: when people aim to construct better conditions for work, life, and societal value, they also provide business value. Even if not explicit, this approach resembles the aims of teleological interpretivists, who critique and transform structures that constrain and exploit humankind by engaging in confrontation and even conflict (Guba and Lincoln 1994).

Within this school, models are less codified than the Stage-Gate model or open innovation models, and there is a greater variety of emergent models and new conceptual frameworks. Researchers aim to understand innovation according to ontological and epistemological assumptions that differ from those made in the other two schools. There is also an intrinsic tension between conceptualizations of innovation and the unavoidable consequences of providing further "constraints" to our thinking.

The motivation to enhance innovation research through in-depth discussions of the role of learning in the context of innovation is one of the strongest drivers of the practice learning community (Ellström 2010; Kallio and Lappalainen 2014, 2015; Russo-Spena and Mele 2012a; Russo-Spena et al. 2014). The focus is on examining innovation in practice and grasping the changes in how people interact, work, and discover new ways of doing things and creating value that go beyond business benefits to improve human life and society as a whole. The expansive learning framework (Engeström et al. 1999; Engeström 2001) and the development learning framework (Ellström 2001) approach the process of learning as collective development that is focused on the "expansion of the object" (Engeström 1987). The expression refers to the process in which the learner focuses not only on the problems at hand but also on the wider context that generates the problems. By creating new skills, activities, and practices, a broader horizon of opportunities (Engeström 2001) becomes possible.

Studies within service-dominant logic regard innovation as arising from the interplay of multiple actors who find new ways to integrate resources, resulting in mutual value creation (Edvardsson and Tronvoll 2013; Lusch and Nambisan 2015; Michel et al. 2008). From a contextual, process-oriented point of view, innovation ceases to be simply the product of a company's processes and becomes a set of cocreation practices viewed as collective actions and accomplishments (Russo-Spena and Mele 2012b; Mele and Russo-Spena 2015a, b; Vargo et al. 2015). Collaboration and mutual gain are facilitated within integrating, representational, and normalizing practices through actor-to-actor (A2A) interaction in a larger service ecosystem (Vargo et al. 2015). Innovation encompasses not only business and economics but also the intricacy of institutions; it is related to sociocultural practices and processes that drive value cocreation (Edvardsson and Tronvoll 2013; Vargo and Lusch 2015, 2016).

By building on the practice-based approach and service-dominant logic, some studies (Hasu et al. 2015; Korhonen and Kaarela 2015; Russo-Spena and Mele 2012b; Mele and Russo-Spena 2015a, b) frame innovation as co-innovation in which actors collaborate and integrate resources. By addressing the role of cocreation, they propose the "Co-s" model of co-ideation, co-valuation, co-design, co-test, and co-launch. Each "Co-" represents a phase of the innovation process resulting from ongoing interactions among resource actions and a group of actors interrelated via a dense network (Russo-Spena and Mele 2012b). These phases do not need to be performed sequentially because they are independent from the development of the other "Co-s"; each "Co-" configures a share of the network through which the potential of innovation and cocreation can be exploited. The study by Hasu et al. (2015) goes further in proposing an integrated approach based on S-D logic, expansive learning, and the practical view that enables exploring resource integration as a practical, dynamic activity which is sensitive to practice-based interaction and learning between users, employees, and other potential stakeholders.

Design-driven innovation models (Brown 2008; Verganti 2009) stress empathy with user needs. When these models are interpreted outside the context of consultancy jargon (resembling the iterative and interactive school), they offer a fresh view on how to go deeper into the meaning of innovation and how it benefits human life (Kimbell 2011; Verganti 2009). The emphasis is on empathy—the ability to feel with people rather than for people. Emphatic interaction is key to the interpretivist assumption as its methodological standpoint. It is the designer (as investigator) who interacts with the object of investigation and investigates the construction of reality. Interest in the activities and practices of "creative people" is established by the creative thinking model of Carson (2010: ix) who aims to unravel the mystery of how "to activate our brains in similar patterns to those of highly creative individuals."

The ontological and epistemological premises of this school are that being and knowing are grounded in social practices. The ontological dimension has an intersubjective orientation (Löbler 2013: 57) that is interpretative (understanding) and constructionist (cocreated). The social world has a precarious ontological status (Burrell and Morgan 1979); reality is viewed as an outcome of social relations (Gergen 1999; Tronvoll et al. 2011).

In this tradition, the social world is regarded not as an external fact (from the researcher's perspective) "but as a subjectively experienced construct" (Löbler 2013: 56) and furthermore as the "intersubjective experience of individuals" (Morgan 1980: 608). Society is understood from the standpoint of the "participating action" (Morgan 1980: 608). It follows that in place of assuming an external, concrete reality, interpretive researchers seek to investigate the social world at the level of subjective experience, with social reality regarded as being intersubjectively composed.

The implication of this ontological view is that epistemologically, knowledge is apprehended not from the standpoint of an external, objective position but from the individual participant in action or the lived experience of the co-participant. Knowledge is not an artifact or an outcome but a process linked to learning and doing: it is knowing in practice and collectively constructed (Gherardi 2012a, b).

The practice-based approach invokes a performative and relational epistemology in which "objects, artefacts, and technologies acquire meaning and agency only in a context of action, and therefore in relation to the human actors that interact with them" (Corradi et al. 2010: 25). Cook and Brown (1999) and Raelin (1997, 2007) address the epistemology of practice by arguing that "this approach recognizes that practitioners, in order to be proficient, need to bridge the gap between theory and practice [...] Practitioners use theories to frame their understanding of the context but simultaneously incorporate an awareness of the social processes in which organizational activity is embedded" (Raelin 1997: 572).

In this view, a practice is regarded not simply as empirical phenomena but also as epistemology (Feldman and Orlikowski 2011; Orlikowski 2010). According to Gherardi (2015: 5), "it is through epistemic practices that researchers construct both the objects of knowledge—ontology—and the methods for producing knowledge, that is, epistemology. Researchers are inside the practices that they study" (Table 2.4).

Theories of practices have multiple views on epistemological choices because in some studies, humans are carriers of practices and actions that start with individuals building the blocks of the social structure (Schatzki et al. 2001) and in others, equal weight is given to humans and nonhumans; agency is attributed to both, along with an investigation of sociomaterial relationships. Actions take place or happen, "as being performed through a network of connections-in-action" (Gherardi 2015: 6). The resulting epistemological implication is important for business studies; "the practice turn is part of the movement toward a relational, sociomaterial, or posthumanist epistemology because practice makes it possible to see and to represent a mode of ordering the social in which doing and knowing are not separated and the knowing subject and the known object emerge in their ongoing intra-action" (Gherardi 2015: 6). This view further divides ontology and epistemology; it expresses classic dualism (action/structure, human/nonhuman, mind/body) by proposing the cumbersome term of "epistemontology" (the study of practices of knowing in being), in which "things and people constitute one another in the here-and-now through intra-action; subjects and objects are dynamically and iteratively co-articulated in intra-action" (Gherardi 2015: 5).

Within these ontological and epistemological perspectives, researcher's actions are intertwined with objects of research. The course of investigation is derived from the constructivist and interpretivist school, in which innovation phenomena are understood by the opening up of innovation activity, and related challenges are met by asking hard questions and participating in the reconstruction of understanding and knowledge. Innovation emerges from the starting point of the social–cultural context, and innovation development depends on social factors; the focus is on activities and negotiations between social groups formed by sociocultural settings.

In this tradition, the research approach to innovation is open-ended. It is exploratory in nature and aims to interpret phenomena. Questions are not answered according to a problem-solving process; answers are uncovered. Researchers do not search for the truth; they become part of what is being sought. Their roles shift away from facilitation, and the objects of research become co-researchers. Thus, the study of innovation is approached according to a constructing view (Berger and Luckmann 1966; Gergen 1985) in which the main concern is explicating the process by which people come to describe, explain, or otherwise account for the world (including themselves) in which they live. Methods are derived from various methodological reflections: practice-based studies (Gherardi 2012a, b), in-depth interviews, ethnographic research (including netnography) (Kozinets 2009; Mele and Russo-Spena 2015a, b), deep case studies, action research, narrative workshops, peer debriefing, and check members (Creswell 1994). Trustworthiness and authenticity are used to grasp complexity of phenomena. Research criteria include credibility in lieu of internal validity, transferability in lieu of generalizability, dependability in lieu of reliability, and confirmability in lieu of objectivity (Guba and Lincoln 1994; Korhonen and Kaarela 2015).

2.5 Conclusion

In opening this chapter, we stated the need to know our roots. We asserted that to view innovation from a fresh perspective, we must first understand its theoretical foundations. To establish this foundation, we use a social science perspective to highlight three schools of innovation thought according to specific metatheoretical assumptions: (1) linear and planned, (2) iterative and interactive, and (3) practice based.

Although we acknowledge that each school contains a variety of frameworks and models, we capture the essential characteristics of each school by describing it according to sociological, teleological, ontological, epistemological, methodological, and ethical research dimensions. We show there is a common perspective within each school with regard to how scholars recognize themselves as a research community (sociological) and set meanings and purposes in focusing on social phenomena (teleological) according to specific beliefs about the nature and the structure of the phenomena (ontological) that in its turn influence the knowledge scholars can gain about them (epistemological). The entire process of grasping social reality is affected by researchers' values and axiological skills (ethical).

Although our categorization of schools includes historical analysis, in that the earliest studies occurred in the first school and the most recent studies come from the third, it is not our intention to infer a graduation among them. We do not evaluate the appropriateness of the schools but simply point out that each school uses assumptions specific to its character to explore the phenomena of interest. In absolute terms, no research tradition is better than the others. The determination of "better" depends on the researchers' weltanschauung and the research questions to be answered.

Our categorization of the three schools is our attempt to start a debate on innovation that takes philosophical stances into account. Our main claims are that we should be more aware of the epistemological and ontological assumptions of our research, and we should make those assumptions more explicit. In this way, we cannot only recognize limitations but also be confident that various stances are working "well enough" for practical purposes. This knowledge creates paths toward new innovation frameworks that reflect the complexity of innovation phenomena and create better conditions for theorizing about innovation.

We encourage researchers to explore innovation from a wide range of positions within the five social science research dimensions. By claiming a research philosophy, scholars can declare their assumptions about their worldviews and be aware of how those views influence research strategy, research practice, and individual understanding, thereby "reflecting upon, and explaining the philosophical assumption underpinning a research study provides stronger support for the process of theorizing" (Peters et al. 2014: 264).

We suggest that researchers seek opportunities to adopt both "theoretical and paradigmatic pluralism" and move away from being mental "prisoner[s] of a tradition" (Möller et al. 2009: 152). A researcher with a monoparadigm position "is likely to or has previously been found to foster a too narrow view of the complexity, and the multifaceted nature of business reality" (Tadajewski 2004: 325). In other words, the need to understand and handle complexity requires researchers to have comprehensiveness (Gioia and Pitre 1990) and recognize "a diverse range of paradigmatic views to foster an increased awareness of the insights and limitations associated with each approach to overcome paradigm myopia" (Tadajewski 2004: 323).

Although pluralism is absent in innovation studies, it can be advocated to achieve "a more comprehensive explanation and understanding of a social phenomenon" (Möller et al. 2009: 163). Getting fresh insights "from different ontological and epistemological assumptions can produce markedly different and uniquely informative theoretical views of events under study" (Gioia and Pitre 1990: 591). Peters et al. (2013) recognize that "one of the challenges of living with the real is that ontologies are neither mutually exclusive, nor wholly encompassing" (Peters et al. 2013: 337). Therefore, they suggest that rather than bridging specific paradigms, researchers should "explore phenomenon by means of comparing and contrasting differing ontological perspectives in order to develop an informed awareness of the respective contributions to knowledge generated by each of these different perspectives" (Peters et al. 2013: 253). By moving away from non-commensurability of research traditions, we can develop pluralistic and holistic approaches to handling complexity in innovation. Such philosophical and methodological pluralism, however, will require the development of new language with a consistent vocabulary. This is a challenge that offers the following new avenues for research in innovation:

- Adoption of philosophical pluralism to distinguish the potential contributions of different perspectives and foster greater self-reflection in the research process (Lewis and Grimes 1999; Lewis and Kelemen 2002; Scherer 1998)
- Embracing of methodological pluralism to achieve comprehensive understanding by using multiple methods (Jackson 1999; Mingers 2001; Morse and Chung 2003)
- Development of a consistent language that moves away from positivistic vocabulary to address complexity, pluralism, and holism (Eisenhardt 2000; Mele et al. 2015; Scherer 1998)

References

Alam, I., & Perry, C. (2002). A customer-oriented new service development process. *Journal of Services Marketing, 16*(6), 515–534.

Bartels, R. (1962). *The development of marketing thought*. Homewood, IL: Richard D. Irwin, Inc.

Berger, P. L., & Luckmann, T. (1966). *The social construction of reality: A treatise in the sociology of knowledge*. Garden City, NY: Anchor Books.

Bhaskar, R. (1998). Critical realism. Essential readings. In M. Archer, R. Bhaskar, A. Collier, T. Lawson, & A. Norrie (Eds.), *Centre for critical realism*. London: Routledge.

Bitner, M. J., Ostrom, A. L., & Morgan, F. N. (2008). Service blueprinting: A practical technique for service innovation. *California Management Review, 50*(3), 66–94.

Booz, E. E., Allen, J. L., & Hamilton, C. L. (1982). *New product management for the 1980s*. New York, NY: Booz-Allen & Hamilton.

Bowers, M. R. (1987). Developing new services for hospitals: A suggested model. *Journal of Health Care Marketing, 7*(2), 35–44.

Brown, T. (2008). Design thinking. *Harvard Business Review, 86*(6), 84.

Brown, J. S., & Duguid, P. (1991). Organizational learning and communities-of-practice: Toward a unified view of working, learning, and innovation. *Organization Science, 2*(1), 40–57.

Brown, S. L., & Eisenhardt, K. M. (1995). Product development: Past research, present findings, and future directions. *Academy of Management Review, 20*(2), 343–378.

Buckley, P. J., & Chapman, M. (1997). The use of native categories in management research. *British Journal of Management, 8*(4), 283–299.
Burman, J. T. (2007). Piaget no "remedy" for Kuhn, but the two should be read together: Comment on Tsou's 'Piaget vs. Kuhn on Scientific Progress. *Theory and Psychology, 17*, 721–732.
Burrell, G., & Morgan, G. (1979). *Sociological paradigms and organisational analysis*. London: Heinemann.
Carlborg, P., Kindstrom, D., & Kowalkowski, C. (2014). The evolution of service innovation research: A critical review and synthesis. *The Service Industries Journal, 34*(5), 373–398.
Carson, S. (2010). *Your creative brain: Seven steps to maximize imagination, productivity, and innovation in your life*. New York: Wiley.
Chesbrough, H. (2003). The logic of open innovation: Managing intellectual property. *California Management Review, 45*(3), 33–58.
Chesbrough, H., Vanhaverbeke, W., & West, J. (2006). *Open innovation: Researching a new paradigm*. Oxford: OUP.
Cohen, W. M., & Levinthal, D. A. (1990). Absorptive capacity: A new perspective on learning and innovation. *Administrative Science Quarterly, 35*(1), 128–152.
Cook, S. D., & Brown, J. S. (1999). Bridging epistemologies: The generative dance between organizational knowledge and organizational knowing. *Organization Science, 10*(4), 381–400.
Cooper, R. G. (1988). The new product process: A decision guide for management. *Journal of Marketing Management, 3*(3), 238–255.
Cooper, R. G. (1990). Stage-gate systems: A new tool for managing new products. *Business Horizons, 33*(3), 44–54.
Cooper, R. G. (2001). *Winning at new products: Accelerating the process from idea to launch* (3rd ed.). Reading, MA: Perseus Books.
Cooper, R. G. (2008). The stage-gate idea-to-launch process- update: What's new and next-gen systems. *Journal of Product Innovation Management, 25*(3), 213–232.
Cooper, R. G. (2014). What's next? After stage-gate. *Research Technology Management, 57*(1), 20–31.
Cooper, R. G., & Kleinschmidt, E. J. (2007). Winning businesses in product development: The critical success factors. *Research-Technology Management, 50*(3), 52–66.
Corradi, G., Gherardi, S., & Verzelloni, L. (2010). Through the practice lens: Where is the bandwagon of practice-based studies heading? *Management Learning, 41*(3), 265–283.
Crawford, C. M. (1994). *New products management* (4th ed.). Boston: Irwin, Burr Ridge.
Creswell, J. W. (1994). *Research design: Qualitative, quantitative, and mixed methods approaches*. Thousand Oaks, CA: Sage.
Den Hertog, P. (2000). Knowledge-intensive business services as co-producers of innovation. *International Journal of Innovation Management, 4*(4), 491–528.
Den Hertog, P., & Bilderbeek R. (1999). *Conceptualising service innovation and service innova tion patterns*. Research program on innovation in services (SIID) for the Ministry of Economic Affairs, Dialogia, Utrecht.
Dewey, J. (1944). *Democracy and education*. New York: The Free Press.
Dubois, A., & Gadde, L. E. (2002). Systematic combining: An abductive approach to case research. *Journal of Business Research, 55*(7), 553–560.
Edvardsson, B., & Olsson, J. (1996). Key concepts for new service development. *Service Industries Journal, 16*(2), 140–164.
Edvardsson, B., & Tronvoll, B. (2013). A new conceptualization of service innovation grounded in SD logic and service systems. *International Journal of Quality and Service Sciences, 5*(1), 19–31.
Eisenhardt, K. M. (2000). Paradox, spirals, ambivalence: The new language of change and pluralism. *Academy of Management Review, 25*(4), 703–705.
Ellström, P. E. (2001). Integrating learning and work: Problems and prospects. *Human Resource Development Quarterly, 12*(4), 421–435.

Ellström, P. E. (2010). Practice-based innovation: A learning perspective. *Journal of Workplace Learning, 22*(1/2), 27–40.
Engeström, Y. (1987). *Learning by expanding*. Helsinki: Orienta-Konsultit Oy.
Engeström, Y. (2001). Expansive learning at work: Toward an activity theoretical reconceptualization. *Journal of Education and Work, 14*(1), 133–156.
Engeström, Y., Miettinen, R., & Punamäki, R. L. (1999). *Perspectives on activity theory*. Cambridge: Cambridge University Press.
Feldman, M. S., & Orlikowski, W. J. (2011). Theorizing practice and practicing theory. *Organization Science, 22*(5), 1240–1253.
Franke, N., & Piller, F. (2004). Value creation by toolkits for user innovation and design: The case of the watch market. *Journal of Product Innovation Management, 21*(6), 401–415.
Franke, N., & Shah, S. (2003). How communities support innovative activities: An exploration of assistance and sharing among end-users. *Research Policy, 32*(1), 157–178.
Garcia, R., & Calantone, R. (2002). A critical look at technological innovation typology and innovativeness terminology: A literature review. *Journal of Product Innovation Management, 19*(2), 110–132.
Gawer, A., & Cusumano, M. A. (2002). *Platform leadership: How Intel, Microsoft, and Cisco drive industry innovation*. Boston: Harvard Business School Press.
Gergen, K. J. (1985). The social constructionist movement in modern psychology. *American Psychologist, 40*(3), 266–275.
Gergen, K. J. (1999). Agency social construction and relational action. *Theory and Psychology, 9* (1), 113–115.
Gherardi, S. (2006). *Organizational knowledge: The texture of workplace learning*. Oxford: Blackwell.
Gherardi, S. (2009). Organizational knowledge: *The texture of workplace learning*. Wiley.
Gherardi, S. (2012a). *Why do practices change and why do they persist? Models of explanations. In practice, learning and change*. Netherlands: Springer.
Gherardi, S. (2012b). *How to conduct a practice-based study: Problems and methods*. Cheltenham: Edward Elgar.
Gherardi, S. (2015). To start practice theorizing anew: The contribution of the concepts of agencement and formativeness. *Organization, Published online before print September 21*.
Gioia, D. A., & Pitre, E. (1990). Multiparadigm perspectives on theory building. *Academy of Management Review, 15*(4), 584–602.
Godin, B. (2006). The linear model of innovation the historical construction of an analytical framework. *Science, Technology and Human Values, 31*(6), 639–667.
Griffin, A. (1997). PDMA research on new product development practices: Updating trends and benchmarking best practices. *Journal of Product Innovation Management, 14*(6), 429–458.
Grønhaug, K., & Kaufmann, G. (1988). *Innovation: A cross-disciplinary perspective*. Oxford University Press.
Guba, E. G., & Lincoln, Y. S. (1994). Competing paradigms in qualitative research. *Handbook of Qualitative Research, 2*, 163–194.
Hansen, M. T., & Birkinshaw, J. (2007). The innovation value chain. *Harvard Business Review, 85* (6), 121.
Hasu, M., Toivonen, M., Tuominen, T., & Saari, E. (2015). Employees and users as resource integrators in service innovation: A learning framework. In R. Agarwal, W. Selen, G. Roos, & R. Green (Eds.), *The handbook of service innovation* (pp. 169–192). London: Springer.
Helkkula, A., Kelleher, C., & Pihlström, M. (2012). Practices and experiences: Challenges and opportunities for value research. *Journal of Service Management, 23*(4), 554–570.
Honderich, T. (Ed.). (2005). *The Oxford companion to philosophy*. Oxford: Oxford University Press.
Hunt, S. D. (1976). The nature and scope of marketing. *Journal of Marketing, 40*(3), 17–28.
Hunt, S. D. (2003). *Controversy in marketing theory: For reason realism, truth and objectivity*. Armonk, NY: M.E. Sharpe.

Jabareen, Y. (2009). Building a conceptual framework: Philosophy, definitions, and procedure. *International Journal of Qualitative Methods, 8*(4), 49–62.

Jackson, M. C. (1999). Towards coherent pluralism in management science. *Journal of the Operational Research Society, 50*(1), 12–22.

Johnson, M. D., & Gustafsson, A. (2003). *Competing in a service economy: How to create a competitive advantage through service development and innovation*. Wiley.

Johnson, S. P., Menor, L. J., Roth, A. V., & Chase, R. B. (2000). A critical evaluation of the new service development process. In J. Fitzsimmons & M. Fitzsimmons (Eds.), *New service development*. Thousand Oaks: Sage.

Johnson, J. L., & Sohi, R. S. (2003). The development of interfirm partnering competence: Platforms for learning, learning activities, and consequences of learning. *Journal of Business Research, 56*(9), 757–766.

Kallio, K., & Lappalainen, I. (2014). Learning about service-orientation in KIBS: Understanding the customer as an active actor. *Service Science, 6*(2), 78–91.

Kallio, K., & Lappalainen, I. (2015). Organizational learning in an innovation network: Enhancing the agency of public service organizations. *Journal of Service Theory and Practice, 25*(2), 140–161.

Kimbell, L. (2011). Rethinking design thinking: Part I. *Design and Culture, 3*(3), 285–306.

Korhonen, H. M., & Kaarela, I. (2015). Practices for involving organizational customers in service innovation. In R. Agarwal, W. Selen, G. Roos, & R. Green (Eds.), *The handbook of service innovation* (pp. 591–615). London: Springer.

Kozinets, R. V. (2009). *Netnography: Doing ethnographic research online*. Thousand Oaks, CA: Sage.

Kuhn, T. S. (2012). *The structure of scientific revolutions*. Chicago: University of Chicago Press.

Lather, P. (2006). Paradigm proliferation as a good thing to think with: Teaching research in education as a wild profusion. *International Journal of Qualitative Studies in Education, 19*(1), 35–57.

Laudan, L. (1978). *Progress and its problems: Towards a theory of scientific growth*. Berkeley: University of California Press.

Lave, J., & Wenger, E. (1991). *Situated learning: Legitimate peripheral participation*. Cambridge: Cambridge University Press.

Leontyev, A. N. (1977). *Activity and consciousness. Philosophy in the USSR, problems of dialectical materialism*. Moscow: Progress Publishers.

Lewis, M. W., & Grimes, A. I. (1999). Metatriangulation: Building theory from multiple paradigms. *Academy of Management Review, 24*(4), 672–690.

Lewis, M. W., & Kelemen, M. L. (2002). Multiparadigm inquiry: Exploring organizational pluralism and paradox. *Human Relations, 55*(2), 251–275.

Löbler, H. (2011). Position and potential of service-dominant logic – Evaluated in an 'ism' frame for further development. *Marketing Theory, 11*(1), 51–73.

Löbler, H. (2013). Service-dominant networks: An evolution from the service-dominant logic perspective. *Journal of Service Management, 24*(4), 420–434.

Lor, P. J. (2011). International and comparative librarianship: A thematic approach. Chapter 4, *Methodology in comparative studies*. Draft available http://peterlor.com/the-book

Lor, P. J. (2014). Revitalizing comparative library and information science: Theory and metatheory. *Journal of Documentation, 70*(1), 25–51.

Lusch, R. F., & Nambisan, S. (2015). Service innovation: A service-dominant logic perspective. *MIS Quarterly, 39*(1), 155–175.

Lusch, R. F., & Vargo, S. L. (2014). *Service-dominant logic: Premises, perspectives, possibilities*. Cambridge: Cambridge University Press.

Mahajan, V., & Muller, E. (1979). Innovation diffusion and new product growth models in marketing. *The Journal of Marketing, 43*, 55–68.

McKinley, W., Mone, M. A., & Moon, G. (1999). Determinants and development of schools in organization theory. *Academy of Management Review, 24*(4), 634–648.

Mele, C., Colurcio, M., & Russo-Spena, T. (2014). Research traditions of innovation: Goods-dominant logic, the resource-based approach, and service-dominant logic. *Managing Service Quality, 24*(6), 612–642.
Mele, C., Pels, J., & Storbacka, K. (2015). A holistic market conceptualization. *Journal of the Academy of Marketing Science, 43*(1), 100–114.
Mele, C., & Russo-Spena, T. (2015a). Innomediary agency and practices in shaping market innovation. *Industrial Marketing Management, 44*, 42–53.
Mele, C., & Russo-Spena, T. (2015b). Eco-innovation practices. *Journal of Organizational Change Management, 28*(1), 4–25.
Mertens, D. M. (1998). *Research methods in education and psychology: Integration diversity with quantitative and qualitative approaches*. Thousand Oaks, CA: Sage.
Michel, S., Vargo, S. L., & Lusch, R. F. (2008). Reconfiguration of the conceptual landscape: A tribute to the service logic of Richard Normann. *Journal of the Academy of Marketing Science, 36*(1), 152–155.
Miettinen, S. (2011). Palvelumuotoilu – yhteissuunnittelua, empatiaa ja osallistumista. In S. Miettinen (Ed.), *Palvelumuotoilu – uusia menetelmiä käyttäjätiedon hankintaan ja hyödyntämiseen* (2nd ed.). Helsinki, Finland: Teknologiainfo Teknova Oy.
Mingers, J. (2001). Combining IS research methods: Towards a pluralist methodology. *Information Systems Research, 12*(3), 240–259.
Möller, K. (2013). Theory map of business marketing: Relationships and networks perspectives. *Industrial Marketing Management, 42*(3), 324–335.
Möller, K., Pels, J., & Saren, M. (2009). The marketing theory or theories into marketing? Plurality of research traditions and paradigms. In P. Maclaran, M. Saren, B. Stern, & M. Tadajewski (Eds.), *Handbook of marketing theory* (pp. 151–173). London: Sage.
Möller, K., & Rajala, A. (2007). Rise of strategic nets – New modes of value creation. *Industrial Marketing Management, 36*(7), 895–908.
Möller, K., & Svahn, S. (2006). Role of knowledge in value creation in business nets. *Journal of Management Studies, 43*(5), 985–1007.
Morgan, G. (1980). Paradigms, metaphors, and puzzle solving in organization theory. *Administrative Science Quarterly, 25*(4), 605–622.
Morse, J. M., & Chung, S. E. (2003). Toward holism: The significance of methodological pluralism. *International Journal of Qualitative Methods, 2*(3), 13–20.
Mouton, J., & Marais, H. C. (1990). *Basic concepts in the methodology of the social sciences* (Revised ed.). Pretoria, South Africa: Human Sciences Research Council.
Nambisan, S., & Nambisan, P. (2008). How to profit from a better virtual customer environment. *MIT Sloan Management Review, 49*(3), 53–61.
Nonaka, I., & Takeuchi, H. (1995). *The knowledge-creating company: How Japanese companies create the dynamics of innovation*. New York: Oxford University Press.
Orlikowski, W. J. (2010). The sociomateriality of organisational life: Considering technology in management research. *Cambridge Journal of Economics, 34*(1), 125–141.
Peters, L. D., Löbler, H., Brodie, R. J., Breidbach, C. F., Hollebeek, L. D., Smith, S. D., Sörhammar, D., & Varey, R. J. (2014). Theorizing about resource integration through service-dominant logic. *Marketing Theory, 14*(3), 249–268.
Peters, L. D., Pressey, A. D., Vanharanta, M., & Johnston, W. J. (2013). Constructivism and critical realism as alternative approaches to the study of business networks: Convergences and divergences in theory and in research practice. *Industrial Marketing Management, 42*(3), 336–346.
Raelin, J. A. (1997). A model of work-based learning. *Organization Science, 8*(6), 563–578.
Raelin, J. A. (2007). Toward an epistemology of practice. *Academy of Management Learning and Education, 6*(4), 495–519.
Ramaswamy, R. (1996). *Design and management of service processes: Keeping customers for life*. New York: Addison-Wesley.

Randhawa, K., & Scerri, M. (2015). Service innovation: A review of the literature. In R. Agarwal, W. Selen, G. Roos, & R. Green (Eds.), *The handbook of service innovation* (pp. 27–51). London: Springer.
Ritter, T., & Gemünden, H. G. (2003). Network competence: Its impact on innovation success and its antecedents. *Journal of Business Research, 56*(9), 745–755.
Russo-Spena, T., Kallio, K., Lappalainen, I., & Mele, C. (2014). Knowledge and learning in practice-based studies. In W. Cellary & L. Freund (Eds.), *Advances in human factors and service engineering*. Cracovia: CRC Press.
Russo-Spena, T., & Mele, C. (2012a). Creativity and learning in the practices of service innovation. In J. C. Spohrer & L. E. Freund (Eds.), *Advances in human factors and service engineering*. San Francisco: CRC Press; Taylor & Francis Group.
Russo-Spena, T., & Mele, C. (2012b). Five Co-s' in innovating: A practice-based view. *Journal of Service Management, 23*(4), 527–553.
Sakkab, N. Y. (2002). Connect and develop complements research and develop at P&G. *Research Technology Management, 45*(2), 38–45.
Sanders, T., Spooren, W., & Noordman, L. (1992). Toward a taxonomy of coherence relations. *Discourse Processes, 15*, 1–35.
Sanders, T., Spooren, W., & Noordman, L. (1993). Coherence relations in a cognitive theory of discourse representation. *Cognitive Linguistics, 4*, 93–133.
Saunders, M. N., Lewis, P., & Thornhill, A. (2011). Research methods for business students, 5/e. Pearson Education India.
Schatzki, T. R., Knorr Cetina, K., & von Savigny, E. (2001). *The practice turn in contemporary theory*. London: Routledge.
Scherer, A. G. (1998). Pluralism and incommensurability in strategic management and organization theory: A problem in search of a solution. *Organization, 5*(2), 147–168.
Scheuing, E. Z., & Johnson, E. M. (1989). A proposed model for new service development. *The Journal of Services Marketing, 3*(2), 25–34.
Schneider, J., & Stickdorn, M. (2011). *This is service design thinking: Basics, tools, cases*. Chichester: Wiley.
Schön, D. A. (1983). *The reflective practitioner: How professionals think in action*. New York: Basic Books.
Sheth, J. N., Gardner, D. M., & Garrett, D. E. (1988). *Marketing theory: Evolution and evaluation*. New York, NY: Wiley.
Suchman, L. A. (1987). *Plans and situated actions: The problem of human-machine communication*. Cambridge: Cambridge University Press.
Sundbo, J. (1997). Management of innovation in services. *Service Industries Journal, 17*(3), 432–455.
Tadajewski, M. (2004). The philosophy of marketing theory: Historical and future directions. *The Marketing Review, 4*(3), 307–340.
Teece, D. J., Pisano, G., & Shuen, A. (1997). Dynamic capabilities and strategic management. *Strategic Management Journal, 18*(7), 509–533.
The Free Dictionary. *Model*. Retrieved October (2014).
Tidd, J., Bessant, J., & Pavitt, K. (2005). *Managing innovation: Integrating technological, market and organizational change*. Chichester: Wiley.
Toivonen, M. (2010). Different types of innovation processes in services and their organizational implications. In F. Gallouj & F. Djellal (Eds.), *The handbook of innovation and services* (pp. 221–249). Cheltenham: Edward Elgar.
Toivonen, M., & Tuominen, T. (2009). Emergence of innovations in services. *The Service Industries Journal, 29*(7), 887–902.
Tronvoll, B., Brown, S. W., Gremler, D. D., & Edvardsson, B. (2011). Paradigms in service research. *Journal of Service Management, 22*(5), 560–585.
Trott, P. (2008). *Innovation management and new product development*. London: Pearson Education.

Tsoukas, H., & Chia, R. (2002). On organizational becoming: Rethinking organizational change. *Organization Science, 13*(5), 567–582.
Tuulaniemi, J. (2011). *Palvelumuotoilu*. Hämeenlinna, Finland: Talentum Media Oy.
Ulrich, K. T., & Eppinger, S. D. (2004). *Product design and development*. New York, NY: McGraw-Hill.
Urban, G. L., & Hauser, J. R. (1980). *Design and marketing of new products*. New York: Prentice-Hall.
Urban, G. L., Hauser, J. R., & Urban, G. L. (1993). *Design and marketing of new products*. Englewood Cliffs, NJ: Prentice Hall.
Vargo, S. L., & Lusch, R. F. (2004). Evolving to a new dominant logic for marketing. *Journal of Marketing, 68*(1), 1–17.
Vargo, S. L., & Lusch, R. F. (2008). Service-dominant logic: Continuing the evolution. *Journal of the Academy of Marketing Science, 36*(1), 1–10.
Vargo, S. L., & Lusch, R. F. (2016). Institutions and axioms: An extension and update of service-dominant logic. *Journal of the Academy of Marketing Science, 44*(1), 5–23.
Vargo, S. L., Wieland, H., & Akaka, M. A. (2015). Innovation through institutionalization: A service ecosystems perspective. *Industrial Marketing Management, 44*, 63–72.
Verganti, R. (2009). *Design driven innovation*. Boston: Harvard Business School Press.
von Hippel, E. (1986). Lead users: A source of novel product concepts. *Management Science, 32* (7), 791–805.
von Hippel, E. (1988). *The sources of innovation*. Oxford: Oxford University Press.
von Hippel, E. (2005). *Democratizing innovation*. Cambridge, MA: MIT Press.
Von Krogh, G., Ichijo, K., & Nonaka, I. (2000). *Enabling knowledge creation: How to unlock the mystery of tacit knowledge and release the power of innovation*. London: Oxford University Press.
Vygotsky, L. S. (1997). *The collected works of LS Vygotsky: Problems of the theory and history of psychology*. New York: Springer Science & Business Media.
Wheelwright, S. C., & Clark, K. B. (1988). *Dynamic manufacturing: Creating the learning organization*. New York: Simon and Schuster.
Wheelwright, S. C., & Clark, K. B. (1992). *Revolutionizing product development: Quantum leaps in speed, efficiency, and quality*. New York: Simon and Schuster.
Yin, R. K. (2002). *Case study research: Design and methods*. Thousand Oaks, CA: Sage.

Chapter 3
Extending the Context of Innovation: The Co-creation and Institutionalization of Technology and Markets

Melissa Archpru Akaka, Stephen L. Vargo, and Heiko Wieland

Abstract This chapter extends traditional, as well as systemic, views of innovation, by advancing an ecosystems perspective on innovation that is based on value co-creation. We draw on a service-ecosystems view, which is grounded in service-dominant (S-D) logic, to integrate and extend prior views on innovation. In particular, we argue that both technologies and markets are shaped by value co-creation and the ongoing negotiation and recombination of overlapping and intersecting institutions. This points toward institutionalization—the development, maintenance, and change of institutions—as a central process by which multiple actors, in dynamic social systems, contribute to the development of new technologies and markets. This co-creative and institutional approach extends the context of innovation beyond firms' activities and potentially provides a robust, parsimonious, and dynamic context for studying the underlying drivers of innovation. This systemic approach reveals important insights into how collaborative practices that drive the co-creation of value can guide innovation as well.

This chapter is based heavily on Vargo et al. (2015), published in *Industrial Marketing Management*.

M.A. Akaka (✉)
University of Denver, Denver, CO, USA
e-mail: Melissa.Akaka@du.edu

S.L. Vargo
University of Hawaii at Manoa, Honolulu, HI, USA
e-mail: svargo@hawaii.edu

H. Wieland
California State University, Monterey Bay, Seaside, CA, USA
e-mail: hwieland@csumb.edu

T. Russo-Spena et al. (eds.), *Innovating in Practice*,
DOI 10.1007/978-3-319-43380-6_3

3.1 Introduction

Traditionally, innovation literature has focused on firm-centric, product-development processes, as well as dyadic interactions through which value flows sequentially from innovation-creating firms to innovation-adopting customers. In this view, innovation is largely driven by firm activities and results in the development of new products and/or processes. Recent research on networked and systemic approaches to innovation has begun to broaden the context of innovation, beyond individual firm activities to the interactions and efforts of multiple actors, including "end users" (Nelson and Nelson 2002; Möller and Rajala 2007; von Hippel 2007). However, even this broader context of innovation generally remains focused on the development of new products, which limits the understanding of how new markets form.

This chapter extends traditional, as well as systemic, views of innovation, by advancing a perspective on innovation that is based on *value co-creation*. More specifically, we draw on a service-ecosystems approach (Vargo and Lusch 2011) that is rooted in service-dominant (S-D) logic (Vargo and Lusch 2004, 2008), to integrate and extend recent views on innovation (Geels 2004; Sundbo and Gallouj 2000). This approach draws attention to how value is co-created in systems of service-for-service exchange and is based on the premise that service—the application of one actor's resources for the benefit of another—is the foundation of all social and economic exchange. Thus, a service-ecosystems approach extends the context of innovation to include the ongoing interactions among multiple actors and the institutions—i.e., humanly devised rules, norms, and meanings that enable and constrain human action (Scott 2001)—that guide them.

In a service-ecosystems view, aggregated levels of interactions and institutions contribute to the co-creation of value as well as multiple "types" of innovative outcomes (i.e., technologies and markets) in dynamic systems of service exchange. In other words, we argue that both technologies and markets are shaped by value co-creation and the ongoing negotiation and recombination of overlapping and intersecting institutions. This points to *institutionalization*—the development, maintenance, and change of institutions—as a central process by which multiple actors, in dynamic social systems, contribute to the development of new technologies *and* markets. This co-creative and institutional approach potentially provides a robust, parsimonious, and dynamic framework for studying the underlying drivers of innovation as well as important insights into how collaborative processes that drive the creation of value can guide innovation.

The purpose of this chapter is to extend the context of innovation, beyond firms' product-development activities, to consider how the interactions and institutions of multiple actors guide, and are guided by, their co-creative, innovative efforts. To this end, we first discuss the context of innovation from a collaborative and dynamic service-ecosystems perspective. We then highlight the role of institutions in value co-creation in general, and innovation in particular, to emphasize the social and contextual aspects of developing new technologies and markets. The role of institutions in innovation is explicated through a discussion of institutionalization as an underlying process of innovation, through which new forms of value emerge. We elaborate how this process leads to the development of new technologies, or value propositions, as well as new markets, or institutionalized solutions. The chapter concludes with an outline of the implications of this extended context for innovation and directions for future research.

3.2 The Context of Innovation: A Service-Ecosystems View

A service-ecosystems view broadens the process of value creation beyond a firm's operational activities to include the active participation of multiple actors, including customers. This approach advances the research regarding networks of innovation (e.g., Freeman 1991), which has led the way for considering the contributions of multiple actors in the development and diffusion of new products. For example, Dodgson et al. (2008: 431) recognize that "innovation networks encompass a number of cooperative relationships between firms, with constituent members engaged in innovation-supporting activities, ranging from R&D to commercialization and diffusion." In addition, Teubal et al. (1991: 386) extend a network view in general, and the diffusion stage in particular, by arguing that innovation requires an "innovator" or "entrepreneur" to see beyond its individual viewpoint and consider future uses of a particular technology. These networked and systemic views emphasize the roles of multiple actors and processes that contribute to innovation.

It is important to note that, in addition to considering interactions among many actors, existing network and systems approaches have begun to explore the role of institutions in innovation. For example, Nelson and Nelson (2002: 267) discuss institutions as "social technologies" that contribute to "molding the technologies used by a society, and technological change itself." Similarly, Geels (2004: 915) argues that only an institutional perspective of innovation can provide "a dynamic sociological conceptualization which understands human action as structure, but leaves much room for intelligent perception and strategic action." Thus, systemic views on innovation have begun to recognize three common building blocks—networks of individuals, social practices, and institutions—that are essential for understanding collaboration in social systems (Peters et al. 2013). However, although this work points toward interaction and institutions among multiple actors as important components of innovation, these studies remain largely focused on the development of new products and often reflect a linear process model that separates processes of production and consumption, as well as innovation and adoption.

To underscore the contextual and systemic nature of value co-creation, Vargo and Lusch (2016: 11) define the concept of service ecosystem as a "relatively self-contained, self-adjusting system of resource-integrating actors connected by shared institutional arrangements and mutual value creation through service exchange." This dynamic and interactive perspective points "away from the fallacy of the conceptualization of the linear, sequential creation, flow and destruction of value and toward the existence of a much more complex and dynamic system of actors that relationally co-create value and, at the same time, jointly provide the context through which 'value' gains its collective and individual assessment" (Giddens 1984: 25; Slater 2002: 60) (Vargo and Lusch 2011: 182).

A service-ecosystems perspective contributes to developing a systemic view of innovation, which is based on the idea that value creation is driven by the integration, exchange, and application of resources among a variety of actors (Vargo and Lusch 2008), that is, *value is always co-created, often through markets*. This provides a more holistic and contextual lens for understanding value creation and, thus, innovation (Edvardsson et al. 2011; Lusch and Vargo 2014; Vargo and Lusch 2008, 2011). Specifically, it differs from dominant views on innovation in that a service-ecosystems approach removes not only the distinction between "producers" and "consumers" of value but also the distinction between "innovators" and "adopters." This is based on an actor-to-actor perspective (Vargo and Lusch 2011) and an emphasis on the joint creation (i.e., co-creation) of value, as well as markets (Vargo and Akaka 2012). Thus, a service-ecosystems perspective advances prior work on innovation networks that place the context of innovation in "loosely coupled systems of autonomous firms" (Dhanarag and Parkhe 2006: 659) or "a web of interactions among different actors and within a diverse economic, social, political, cultural and geographical context" (Corsaro et al. 2012: 782), by focusing on a wider array of actors as active participants in value creation, as well as innovation.

Advancing a systems approach, a service-ecosystems view extends the context of innovation beyond network configurations and interactions and further emphasizes the importance of institutions in value creation and innovation (Vargo and Lusch 2011). Recently, Vargo and Lusch (2016) describe how institutions—i.e., "routinized, coordinating mechanisms of various types"—are underlying drivers of value co-creation. Thus, a service-ecosystems view recognizes that institutions influence the interactions that contribute to the creation and evaluation of value among multiple actors, even those that result in the rejection of value propositions. Furthermore, service ecosystems are composed of multiple networks of actors as well as the institutional arrangements—i.e., sets of institutions that are nested in multiple levels of social systems (Kiser and Ostrom 1982)—that guide their actions and interactions. Institutional arrangements intersect and overlap and often create conflicting views on value and how value is derived. Because of this, the contextual and phenomenological nature of value, value-in-context (Vargo and Lusch 2008)—i.e., the evaluation of an experience based on the integration of a particular set of resources, at a specific place and time—becomes a central aspect to value creation and a critical factor in innovation.

The complexity of context highlighted in a service-ecosystems perspective (Akaka et al. 2013) broadens the dyadic views of interactions between firms and customers by emphasizing the application of resources (i.e., service) as the basis of technology and extending the context of innovation beyond firm activities to the collaborative value creation efforts among multiple actors. Furthermore, in this view, value is created, and innovation occurs, through multiple levels of interaction: micro (e.g., service encounters), meso (e.g., organizations, "industries," and brand communities), and macro (e.g., societal). However, these levels of context are not fixed; rather, they are relative levels of aggregation (i.e., one level is nested within and constitutes another) that evolve and change over time (Chandler and Vargo 2011). This approach extends networked and systemic views on innovation by emphasizing the influence of multiple (micro, meso, and macro) levels of interaction and institutions (i.e., social structures) in value creation.

Importantly, S-D logic recognizes two broad categories of resources that are continually integrated to create value: operand and operant resources. A service-ecosystems approach emphasizes the primacy of operant resources—those that are capable of acting on other resources to contribute to value creation, rather than operand resources, and those that require action taken upon them to be valuable (Vargo and Lusch 2004, 2008). In this view, the integration of resources is central to the generation of new resources, and, thus, innovation is ultimately driven by the "combinatorial evolution" (Arthur 2009) of new, useful knowledge (i.e., operant resources). Ultimately, within this view, innovation is driven by collaborative efforts to find or develop new ways to create value. However, in order for value to be created, it must be derived and determined by a beneficiary. Thus, value is proposed by one or more service providers, including firms, but always determined by one or more service beneficiaries (e.g., customers) through the integration of resources, in particular social contexts (Chandler and Vargo 2011; Vargo and Lusch 2008). Stated differently, value perceptions are always guided by socially constructed systems of shared norms, values, and beliefs (i.e., institutions) that influence a service beneficiary.

The dynamic nature of service ecosystems suggests that each instance of resource integration and value co-creation has the potential to change the system and, thus, the context for the next iteration and determination of value creation. In this way, value co-creation continually occurs as service is exchanged for service and actors integrate value propositions and enact various practices to adapt to contextual requirements. In addition, as actors engage in value-creating activities and interact with other actors, they simultaneously change contexts of value co-creation as well (Chandler and Vargo 2011; Edvardsson et al. 2011).

This is because value co-creation is driven through resource integration and service exchange among multiple stakeholders, with varying views on value (e.g., institutional arrangements), which determine what works and what does not work (i.e., what is valuable and what is not) (Lusch et al. 2007). Although the service-ecosystems approach outlined above emphasizes the critical role institutions play in innovation, the way in which institutions are maintained and change requires elaboration. In addition, further investigation regarding the nested relationship between technological and market innovation is also needed.

The following section elaborates a service-ecosystems approach to innovation by drawing on scholarly work on institutions, technology, and markets to provide insight into the underlying processes through which institutionalization takes place and innovation occurs.

3.3 Institutionalization as an Underlying Process of Innovation

The movement toward a deeper understanding of the dynamic, systemic, and service-driven nature of innovation is "liable to lead to nothing less than a widening or even complete reinterpretation of the concept [of innovation] itself" (Coombs and Miles 2000: 100). Along these lines, we have outlined a service-ecosystems approach to innovation that extends the context of innovation to dynamic ecosystems of interaction and institutions, which are continually reconstituted as multiple actors integrate resources, exchange service, and co-create value. However, in order to better understand innovation and the contexts in which it occurs, a discussion of how institutions (re)form—i.e., how *institutionalization* occurs—is needed.

Barley and Tolbert (1997) draw on structuration theory (Giddens 1984) and develop a framework for institutionalization to better understand "the processes by which existing institutions are maintained and modified." In this view, institutions are continually reproduced through the ongoing enactment of practices and redevelopment of relationships, and understanding the process of institutionalization requires "a conceptual framework that specifies the relations between interactional episodes and institutional principles" (Barley and Tolbert 1997: 100). This structuration approach to institutionalization provides critical insights into how practices (i.e., routine actions) contribute to the ongoing formation and reformation of social structures (i.e., institutions) and systems. However, some argue that structuration theory lacks the consideration of the dynamics of institutions, or social structures, which is needed to fully understand institutional change (Sewell 1992).

Institutional change is a central issue for innovation, because it addresses the issue of "how institutions influence actors' behaviour and also how these actors might, in turn, influence, and possibly change institutions" (Battilana et al. 2009: 66). In this view, both institutional maintenance and change are components of "institutional work," which is driven by the activities of diverse, spatially dispersed actors and their involvement in the political struggles and the interactions among them (Hardy and Maguire 2008; Lawrence and Suddaby 2006). This approach extends the analysis beyond the creation of new institutions by highlighting the influence of actors on purposefully maintaining and disrupting existing institutions (Lawrence et al. 2009). Lawrence and Suddaby (2006) draw on the seminal research from Giddens (1984), as well as Bourdieu (1977), DiMaggio (1988), and Oliver (1991), and emphasize the idea that institutional work involves transformative action as well as the repairing and concealing of tensions and conflicts within and across institutions.

This approach is imperative for the study of innovation because it highlights the importance of institutional *maintenance*, as well as *change*. It is important to recognize institutional maintenance as a critical aspect of innovation. This function of institutional work is often overlooked, especially with increasing attention being paid to radical or "disruptive" innovations (Bower and Christensen 1995) and "blue ocean" strategies (Kim and Mauborgne 2005). While these types of innovation represent major institutional shifts (e.g., blue ocean vs. red ocean market opportunities), even drastically different offerings require and build on existing institutions to enable potential customers to evaluate offerings in positive (or negative) ways.

For example, the introduction of "ridesharing," by companies such as *Uber*, created a drastically different form of transportation than had been previously available through traditional transportation options (e.g., taxi). Rather than relying on traditional taxi companies and their employed drivers for transportation, ridesharing enables customers to purchase transportation services from independent contractors using a third-party organization. Although the emergence of this market offering has led to many changes in transportation, it has also maintained many institutional norms that other, more traditional, transportation companies provide. Expectations of getting from one place to another in a safe and efficient manner remain, as do requirements of monetary exchange and pricing that is dependent on distance.

However, ridesharing has introduced many institutional changes as well. For example, because ridesharing enables customers to see and rate specific drivers, expectations of service differ. In addition, customers receive ratings from drivers as well, and, thus, customers who misbehave or do not provide sufficient tips may not be able to find a driver who wants to provide a ride as easily as others.

In this example, it is clear that the innovation of ridesharing as a market offering requires the cooperation of multiple actors, including the drivers, third-party organizations, and customers. Furthermore, this example emphasizes how the institutions within service-ecosystems influence, and are influenced by, the actions of multiple actors and the integration and use of new technologies and, ultimately, shape the formation of new markets. The way in which institutionalization drives the innovation of technologies and markets is elaborated below.

3.4 Technological Innovation as the Institutionalization of Value Propositions

Prior research regarding innovation traditionally focuses on new product development, or the design and production of goods, and more recently, "services" (i.e., intangible goods). Recently, however, the conceptualization of technology has been reconsidered in light of a more dynamic and social perspective, which has important implications for product-centered approaches to innovation. In other words, social approaches to technology shift attention away from tangible goods (i.e., operand resources) and on to the integration of knowledge and ideas required for innovative outcomes (i.e., operant resources). For example, Arthur's (2009: 28) definition of "technology as an assemblage of practices and components that are means to fulfil human purposes" suggests that products and processes do not need to be classified as disparate categories. Rather, this approach suggests that the term "technology" refers to a wide class of phenomena, both "software" (i.e., process or methods) and "hardware" (i.e., physical devices). Similarly, Layton (1974) equates technology, without referring to physical attributes, to knowledge that provides a rational basis for design. Importantly, in this view, design is considered as the reconfiguration of resources to solve a variety of problems (Simon 1996) rather than the physical components of a particular product. Along the same lines, Nelson and Nelson (2002) distinguish between "physical" and "social" technologies for which the latter are defined as "institutions."

Thus, based on these views, technology is conceptualized as both physical and social, and considered as potentially useful knowledge that may provide solutions for new or existing problems. This notion of technology as knowledge aligns with a service-ecosystems emphasis on the centrality of operant resources (i.e., resources that are capable of acting on other resources to create value) and highlights the idea that competences, and not physical things, lie at the heart of technology. Viewing knowledge as foundational to technology, however, does not diminish the importance of physical artifacts in technological innovation, since these artifacts can be viewed as vehicles that convey embedded knowledge and skills (Orlikowski 1992), as well as mechanisms of institutionalization.

The ridesharing example discussed above provides clear evidence of technology as knowledge, because the material components of this innovation are not new. Both traditional forms of taxi service and new forms of ridesharing rely on similar operand resources, such as cars, money, and communication devices. However, what makes ridesharing distinct from traditional taxi services is an understanding of how to coordinate independent drivers with customers who have unique needs. In this way, the innovation of this technology is not based on the development of new "hardware" or tangible products. Rather, innovation emerges as a new form of service provision that involves a deeper understanding of the needs and wants of customers, including convenience, cleanliness, and choice.

In the case of ridesharing, differences between the needs and wants among customers as well as contexts are reflected in differences in price. In general, ridesharing companies are able to charge higher prices because customers feel that they offer increased levels of cleanliness and quality of service. In addition, depending on the time, place, and other situational factors, ridesharing prices can increase dramatically based on demand (e.g., getting a ride home from downtown on New Year's Eve). These differences in demand provide anecdotal evidence for research on the socially constructed nature of technology (Orlikowski 1992; Pinch and Bijker 1984), which points out that different social groups can construct radically different meanings of technologies (e.g., people are willing to pay more for ridesharing services than taxi services although, from a transportation standpoint, they may be quite similar).

Thus, innovation not only is centered on technological advances (i.e., new knowledge for design processes). It is also influenced by the institutions (e.g., social rules, norms, values, meanings, and beliefs) that guide both the "design" and "use" phases associated with new and emerging technologies (i.e., potentially useful knowledge) (Orlikowski 1992).

In this social approach to technology, innovation is always a co-creational and ongoing process, which suggests that new technologies (i.e., potentially useful knowledge) can be conceptualized as new value propositions that always possess institutional components. Thus, value propositions or technologies are always the co-created outcomes of systemic human action and interactions among multiple actors. More specifically, from a service-ecosystems view, value-proposing actors rely on their institutional arrangements and their competences (i.e., operant resources) and engage in institutional work (maintenance and change) by recombining or proposing not only new integrative practices but also new normative and representational practices (Kjellberg and Helgesson 2006, 2007). In this way, it becomes clear that novel value propositions are not only evidence of institutional change, but they also reflect the institutional work of overlapping maintenance and disruption components (Creed et al. 2010).

This ecosystems approach to technology provides insight into how the integration of resources and co-creation of value drive the institutionalization (i.e., maintenance disruption and change) of new value propositions in dynamic socio-technical systems. In line with this shift toward social aspects of innovation, Coombs and Miles (2000: 100) argue that "we are moving away from a model of innovation that puts all the emphasis on artifacts and technological innovation; and toward a model which sees innovation in terms of changes in market relationships but with major artifact and technological dimensions."

This view of innovation suggests that technological advancement is always embedded within dynamic social systems and technological innovation can be considered as the "combinatorial evolution" of value propositions (Arthur 2009), which are endogenous to the formation and reformation of markets (Akaka and Vargo 2014). However, this discussion of technological innovation does not fully explain how the innovation of technologies can (potentially) lead to the innovation of new markets. The following section explores the relationship between technological and market innovation and sheds light on how institutionalization not only drives the innovation of technologies, or value propositions, but also the innovation of markets, or the institutionalization of new solutions.

3.5 Market Innovation as the Institutionalization of New Solutions

A service-ecosystems approach to innovation supports Orlikowski's (1992) idea of "duality of technology" and argument that technology is both an outcome and a medium of human action. Based on this, Akaka and Vargo (2014) extend a socially constructed model of technology (Orlikowski 1992) to include the consideration of technology as an operant resource because it is constituted by dynamic resources, such as knowledge and skills, and is central to influencing value creation and, thus, innovation.

However, in the same way, markets can also be considered as operant resources that dynamically and continually contribute to the co-creation of value (Vargo and Lusch 2013). Vargo and Akaka (2012) discuss value co-creation by extending a practice approach to market formation (Kjellberg and Helgesson 2006, 2007) that suggests that markets are continually "performed" by the enactment of interlinked sets of exchange, normative, and representational practices. In their markets-as-practice framework, Kjellberg and Helgesson (2006) explain that exchange practices are activities involved in the exchange of market offerings, normative practices are activities involved in forming normative expectations, and representational practices are activities that shape images of markets. Vargo and Akaka (2012) draw on Kjellberg and Helgesson's (2006) framework and broaden the classification of exchange practices to include all integrative practices (including exchange practices) to discuss the practices and processes by which value is co-created and markets are (re)formed. In this view, both technology and markets can change and be changed through the integration and application of resources and co-creation of value.

It is important to note that technological innovations do not necessarily lead to market formation—e.g., many new value propositions do not emerge as viable markets. In other words, market innovation does not automatically occur when actors (e.g., firms) or groups of actors (e.g., innovation networks) introduce new value propositions. The formation of a market only occurs when new practices (i.e., solutions) become institutionalized. For example, the only way that ridesharing was able to make the transition from a technological innovation (new value proposition) to a viable market was because customers adapted new practices—e.g., using an app to look up nearby drivers and setting up an account to pay electronically. In this way, a new solution (e.g., type of transportation) became institutionalized through the changing of norms and practices. Zietsma and McKnight (2009) describe this institutionalization process as a nonlinear process in which all actors engage in institutional work and co-create institutions through multiple iterations of institutional developments until common templates emerge that reflect shared conceptions of problems and solutions.

This view of innovation is compatible with Vargo and Lusch's (2011) normalized actor-to-actor (A2A) view and sheds light on how markets form and reform, through technological advancements and changes in institutions that are embedded within service ecosystems.

Market innovation therefore involves the ongoing and systemic maintenance, change, and disruptions of the institutional arrangements that enable and constrain integrative, normative, and representational practices. In this dynamic view, service ecosystems, with their overlapping institutions, or institutional arrangements, need to be viewed as the venues for enabling and constraining market innovation (Vargo and Lusch 2012). Loasby (2000), for example, describes institutions in the context of innovation as "mixed blessings." Service ecosystems (including markets) always possess path dependencies, which provide stability through continuity and repetitiveness. However, at the same time, these institutional arrangements also lead to problems and conflicts, through their intra-institutional inconsistencies and contradictions. These conflicts propel the ongoing emergence of new value propositions (i.e., technologies) and drive the institutionalization of new solutions (i.e., markets).

Consistent with the need to view markets as highly relational, this view highlights a systemic process in which all actors engage in "ongoing negotiations, experimentation, competition, and learning" (Zietsma and McKNight 2009: 145). It is important to reiterate the need to view service ecosystems, at least partly, as loosely coupled, interconnected, and nested. "Any particular social structure is viewed not as an isolated, abstract phenomenon but, rather, as part of a larger whole composed of multiple, interpenetrating social structures operating at multiple levels and in multiple sectors" (Seo and Creed 2002: 225). Benson (1977) highlights that social structures, due to their loosely coupled nature, are susceptible to incompatibilities both within and among institutional arrangements. The institutionalization of new solutions always involves the integration of multiple institutions and, therefore, is always an ongoing process, since the social order produced in the process of social construction always creates new institutional contradictions, ruptures, inconsistencies, and incompatibilities (Benson 1977).

In the case of ridesharing, existing institutions that surround traditional taxi services continue to shape customer expectations and norms for transportation problems and solutions. However, new knowledge and ideas on how to solve existing problems have contributed to the development of new value propositions. The continual integration of new practices surrounding this value proposition have led to the institutionalization of ridesharing as a viable solution and thereby formed a new market. Although conflict between taxi companies and new ridesharing services reveal how tensions can surface when one type of practice rubs up against other practices, it is clear that customers are embracing this new value proposition as an effective alternative solution. However, these market tensions (e.g., diverging practices among service providers and service beneficiaries) indicate that markets are never completely stabilized, but, instead, the institutionalization of new solutions is an ongoing process and markets are continually formed and reformed.

3.6 Discussion and Conclusion

In this chapter, we extend the context of innovation, beyond firm activities and new product development, to emphasize how value is co-created and innovation occurs, through dynamic systems of service exchange. This service-ecosystems approach points toward institutions as a central factor in the development of new forms of value (i.e., innovation). Furthermore, this extended context and emphasis on institutions in innovation suggests that institutionalization (change, disruption, and maintenance of institutions) is a central process of the innovation of both technology and markets.

This consideration of institutionalization as an underlying process of innovation provides insight into the dynamic nature of technologies and markets and provides a more comprehensive framework that focuses on the integration and application of operant resources for co-innovating both. In this view, innovation can be broadly conceptualized as the co-creation or collaborative recombination of practices that provide novel solutions for new or existing problems. Along these lines, technological innovation is the co-creation of new value propositions, or collective, combinatorial evolution that leads to the generation of new, potentially useful knowledge (i.e., operant resources). Market innovation, then, is the co-creation of institutionalized solutions (Lusch and Vargo 2014). This particular process is driven by and drives the development of new technologies but also requires the acceptance of a value proposition, as well as the continued exchange, interpretation, integration, and application of a particular technology among multiple actors, over time (i.e., institutionalization). In both cases, innovation is driven by value co-creation and the integration and application of operant resources (Vargo and Akaka 2012). However, value co-creation is enabled and constrained by a multitude of higher-level social structures, or institutions. Importantly, as new solutions become institutionalized (new markets are created), they recursively contribute to the exchange of service and the co-creation of value.

Foundational to this service-centered approach is the notion that the integration and reintegration (combinatorial evolution) of operant resources (useful knowledge) lie at the heart of technology (Arthur 2009; Mokyr 2004; Vargo and Lusch 2004). Drawing on Orlikowski's (1992) social approach to innovation, we argue that technology (i.e., institutionalized value propositions) is always endogenous to markets (i.e., institutionalized solutions). In other words, a service-ecosystems perspective allows researchers to view market and technological innovation not as separate sequential processes but rather as occurring through a single, ongoing process of institutionalization. Thus, value co-creation and innovation occur through combinatorial evolution in a massively collaborative, ongoing creation of new institutional arrangements. Through this lens, innovation is a collaborative process, rather than an output, which always involves the participation of all value co-creating parties and social, as well as technical, developments. To varying degrees, market innovation is driven by the institutionalization of a technology, which becomes integrated into the fabric of a particular socio-technical system, composed of rules, norms, values, meanings, and practices.

This consideration of the importance of institutions in ecosystems of service exchange can provide vital insights into both technological and the market components of innovation and can capture the systemic and relational nature of market (re) formation. Importantly, a service-ecosystems perspective directs our attention to the institutions that guide interaction among actors. Although we recognize a multitude of practices and institutions within any given market, the discussion of practices in markets and marketing (Kjellberg and Helgesson 2007; Schau et al. 2009; Warde 2005), as well as value co-creation and innovation, is in its infancy. The existing work on the role of institutions in markets has only begun to explore institutionalization processes in socio-technical and economic systems.

Furthermore, the consideration of a service-ecosystems approach for innovation emphasizes that the maintenance, disruption, and change of institutions (i.e., institutionalization) are always co-creational processes in which actors try to resolve the nested contradictions and inconsistencies that are foundational to all institutional arrangements. However, there are many opportunities for a deeper exploration of how actors resolve these contradictions and inconsistencies. We believe that the linkages between, or relationships among, various levels (micro, meso, and macro) of institutions are important components of this exploration and therefore encourage both conceptual and empirical investigations of these linkages.

Finally, this chapter suggests that traditional innovation approaches that center on the development of new products and corporate processes limit the understanding of the co-created, systemic, and dynamic nature of market innovation. As it relates to the study of innovation, a service-ecosystems framework extends the context of innovation beyond technologies to the study of new organizations and new markets.

Thus, this approach can be used to further develop theories of entrepreneurship, such as effectuation theory (Read et al. 2009), and provide a revised outlook for marketing strategy in general. In particular, this view of innovation redirects the focus of managers, as well as policy makers, beyond the individual development of new technologies to the joint development of lasting institutional change. Ultimately, the service-ecosystems approach to innovation provides an important avenue for reconsidering the context of value creation and innovation, from the management of firm activities in existing markets (e.g., segmenting, targeting, and positioning) to the maintenance, disruption, and change of institutions and the continual co-creation of new solutions.

References

Akaka, M. A., & Vargo, S. L. (2014). Technology as an operant resource in service (eco) systems. *Information Systems and e-Business Management, 12*(3), 367–384.

Akaka, M. A., Vargo, S. L., & Lusch, R. F. (2013). The complexity of context: A service ecosystems approach for international marketing. *Journal of Marketing Research, 21*(4), 1–20.

Arthur, W. B. (2009). *The nature of technology: What it is and how it evolves*. New York: Free Press.

Barley, S. R., & Tolbert, P. S. (1997). Institutionalization and structuration: STUDYING the links between action and institution. *Organization Studies, 18*(1), 93–117.

Battilana, J., Leca, B., & Boxenbaum, E. (2009). How actors change institutions: Towards a theory of institutional entrepreneurship. *The Academy of Management Annals, 3*(1), 65–107.

Benson, J. K. (1977). Organizations: A dialectical view. *Administrative Science Quarterly, 22*(1), 1–21.

Bourdieu, P. (1977). *Outline of a theory of practice*. Cambridge: The Press Syndicate of the University of Cambridge.

Bower, J. L., & Christensen, C. M. (1995). Disruptive technologies: Catching the wave. *Harvard Business Review, 73*, 43–53.

Chandler, J. D., & Vargo, S. L. (2011). Contextualization: Network intersections, value-in-context, and the co-creation of markets. *Marketing Theory, 11*(1), 35–49.

Coombs, R., & Miles, I. (2000). Innovation, measurement and services: The new problematique. *Economics of Science Technology and Innovation, 18*, 85–104.
Corsaro, D., Cantu, C., & Tunisini, A. (2012). Actors' heterogeneity in innovation networks. *Industrial Marketing Management, 32*(8), 703–716.
Creed, W. E., DeJordy, R., & Lok, J. (2010). Being the change: Resolving institutional contradiction through identity work. *Academy of Management Journal, 53*(6), 1336–1364.
Dhanarag, C., & Parkhe, A. (2006). Orchestrating innovation networks. *Academy of Management Review, 31*(3), 659–669.
DiMaggio, P. J. (Ed.). (1988). *Interest and agency in institutional theory*. Cambridge, MA: Ballinger.
Dodgson, M., Mathews, J., Kastelle, T., & Hu, M. C. (2008). The evolving nature of Taiwan's national innovation system: The case of biotechnology innovation networks. *Research Policy, 37*(3), 430–445.
Edvardsson, B., Tronvoll, B., & Gruber, T. (2011). Expanding understanding of service exchange and value co creation. *Journal of the Academy of Marketing Science, 39*(2), 327–339.
Freeman, C. (1991). Networks of innovators: A synthesis of research issues. *Research Policy, 20* (5), 499–514.
Geels, F. W. (2004). From sectoral systems of innovation to socio-technical systems: Insights about dynamics and change from sociology and institutional theory. *Research Policy, 33*(6), 897–920.
Giddens, A. (1984). *The constitution of society*. Berkeley and Los Angeles, CA: University of California Press.
Hardy, C., & Maguire, S. (Eds.). (2008). *Institutional entrepreneurship*. Thousand Oaks, CA: Sage.
Kim, W. C., & Mauborgne, R. (2005). *Blue ocean strategy*. Boston, MA: Harvard Business School Press.
Kiser, L. L., & Ostrom, E. (1982). The three worlds of action: A metatheoretical synthesis of institutional approaches. In E. Ostrom (Ed.), *Strategies of political inquiry*. Beverly Hills: Sage.
Kjellberg, H., & Helgesson, C.-F. (2006). Multiple versions of markets: Multiplicity and performativity in market practice. *Industrial Marketing Management, 35*(7), 839–855.
Kjellberg, H., & Helgesson, C.-F. (2007). On the nature of markets and their practices. *Marketing Theory, 7*(2), 137–162.
Lawrence, T. B., & Suddaby, R. (Eds.). (2006). *Institutions and institutional work*. London: Sage.
Lawrence, T. B., Suddaby, R., & Leca, B. (Eds.). (2009). *Introduction: Theorizing and studying institutional work*. New York: Cambridge University Press.
Layton, E. T., Jr. (1974). Technology as knowledge. *Technology and Culture, 15*, 31–41.
Loasby, B. J. (2000). Market institutions and economic evolution. *Journal of Evolutionary Economics, 10*(3), 297–309.
Lusch, R. F., & Vargo, S. L. (2014). *Service-dominant logic: Premises, perspectives, possibilities*. Cambridge: Cambridge University Press.
Lusch, R. F., Vargo, S. L., & O'Brien, M. (2007). Competing through service: Insights from service-dominant logic. *Journal of Retailing, 83*(1), 5–18.
Mokyr, J. (2004). *The gifts of Athena: Historical origins of the knowledge economy*. Princeton: Princeton University Press.
Möller, K., & Rajala, A. (2007). Rise of strategic nets – New modes of value creation. *Industrial Marketing Management, 36*(7), 895–908.
Nelson, R. R., & Nelson, K. (2002). Technology, institutions, and innovation systems. *Research Policy, 31*(2), 265–272.
Oliver, C. (1991). Strategic responses to institutional processes. *The Academy of Management Review, 16*(1), 145–179.
Orlikowski, W. J. (1992). The duality of technology: Rethinking of the concept of technology in organizations. *Organization Science, 3*(3), 398–427.

Peters, L. D., Pressey, A. D., Vanharanta, M., & Johnston, W. J. (2013). Constructivism and critical realism as alternative approaches to the study of business networks: Convergences and divergences in theory and in research practice. *Industrial Marketing Management, 42*, 336–346.

Pinch, T. J., & Bijker, W. E. (1984). The social construction of facts and artefacts: Or how the sociology of science and the sociology of technology might benefit each other. *Social Studies of Science, 14*, 399–441.

Read, S., Dew, N., Sarasvathy, S. D., Song, M., & Wiltbank, R. (2009). Marketing under uncertainty: The logic of an effectual approach. *Journal of Marketing, 73*(3), 1–18.

Schau, H. J., Muniz, A. M., Jr., & Arnould, E. J. (2009). How brand community practices create value. *Journal of Marketing, 73*(5), 30–51.

Scott, W. R. (2001). *Institutions and organizations* (2nd ed.). Newbury Park, CA: Sage.

Seo, M. G., & Creed, W. E. D. (2002). Institutional contradictions, praxis, and institutional change: A dialectical perspective. *The Academy of Management Review, 27*(2), 222–247.

Sewell Jr, W. H. (1992). A theory of structure: Duality, agency, and transformation. *American Journal of Sociology, July*, 1–29.

Simon, H. A. (1996). *The science of the artificial.* Cambridge, MA: The MIT Press.

Slater, D. (2002). Capturing markets from the economists. In P. du Gay & M. Pryke (Eds.), *Cultural economy* (pp. 59–77). London: Sage.

Sundbo, J., & Gallouj, F. (2000). Innovation as a loosely coupled system in services. *International Journal of Services Technology and Management, 1*(1), 15–36.

Teubal, M., Yinnon, T., & Zuscovitch, E. (1991). Networks and market creation. *Research Policy, 20*(5), 381–392.

Vargo, S. L., & Akaka, M. A. (2012). Value cocreation and service systems (re) formation: A service ecosystems view. *Service Science, 4*(3), 207–217.

Vargo, S. L., & Lusch, R. F. (2004). Evolving to a new dominant logic for marketing. *Journal of Marketing, 68*, 1–17.

Vargo, S. L., & Lusch, R. F. (2008). Service-dominant logic; further evolution. *Journal of the Academy of Marketing, 36*(1), 1–10.

Vargo, S. L., & Lusch, R. F. (2011). It's all B2B...and beyond: Toward a systems perspective of the market. *Industrial Marketing Management, 40*(2), 181–187.

Vargo, S. L., & Lusch, R. F. (2012). *Special issue: Toward a Better understanding of the role of value in markets and marketing.* Book Series: Review of Marketing Research (Vol. 9). Emerald Group Publishing.

Vargo, S. L., & Lusch, R. F. (2013). *Service-dominant logic: Premises, perspectives, possibilities.* Paper presented at the Naples Forum on Service, Ischia, 19–21 June, Italy.

Vargo, S. L., & Lusch, R. F. (2016). Institutions and axioms: An extension and update of service-dominant Logic. *Journal of the Academy of Marketing Science, 44*(1), 5–23.

Vargo, S. L., Wieland, H., & Akaka, M. A. (2015). Innovation through institutionalization: A service ecosystems perspective. *Industrial Marketing Management, 44*, 63–72.

von Hippel, E. (2007). Horizontal innovation network – By and for users. *Industrial and Corporate Change, 16*(2), 293–315.

Warde, A. (2005). Consumption and theories of practice. *Journal of Consumer Culture, 5*(2), 131–153.

Zietsma, C., & McKnight, B. (Eds.). (2009). *Building the iron cage: Institutional creation work in the context of competing proto-institutions.* New York: Cambridge University Press.

Chapter 4
Market Innovation: Renewal of Traditional Industrial Networks

Suvi Nenonen, Kristian Möller, and Kaj Storbacka

Abstract The study adopts a constructionist approach to markets, positing that markets are socially constructed human artifacts resulting from the actions and cognitive framing of the involved actors. The present research had three objectives: to develop a conceptual framework for categorizing managerial market definitions, to investigate empirically how practitioners frame novel markets, and to identify challenges associated with creating innovative market definitions in traditional industrial networks. The chapter draws on data collected within an extensive research program, carried out between March 2009 and December 2010. The program consisted of six academic research projects covering a wide range of theoretical backgrounds as well as interactions with 112 senior managers from ten major firms in a series of 52 workshops. The proposed conceptual framework supplements the current dominant product-based view of market definitions by positing that managers in industrial networks can also choose to define their markets based on (1) the firm's resources and capabilities, (2) the network's resources and capabilities, (3) the customers' processes, and (4) the end-consumers' practices. The results of the study additionally showed that current product-based theories of markets also influence managerial judgment: practitioners framed the vast majority of new market opportunities (73 %) around the product or service being sold.

S. Nenonen (✉) • K. Storbacka
University of Auckland Business School, Auckland, New Zealand
e-mail: s.nenonen@auckland.ac.nz; k.storbacka@auckland.ac.nz

K. Möller
Aalto University School of Business, Helsingfors, Finland
e-mail: kristian.moller@aalto.fi

T. Russo-Spena et al. (eds.), *Innovating in Practice*,
DOI 10.1007/978-3-319-43380-6_4

4.1 Introduction

This research posits that there are no objective markets "out there" (Jenkins and MacDonald 1997) in which firms compete for positions and market share. Markets are increasingly seen as "socially constructed human artifacts," created by the actors who populate a specific context and link resources within it (Granovetter 1992; Krippner et al. 2004; Peñaloza and Venkatesh 2006).

The constructionist perspective emphasizes the role of "market innovation" as a key opportunity for renewal in traditional industrial networks. Geroski (1998: 693) proposes that "thinking about market boundaries is important for a firm because redefining market boundaries is a fundamental part of the process of innovation." According to the constructionist view, markets, and consequently their boundaries, are the results of the actions and cognitive framing of the involved actors (Araujo 2007; Callon 1998). By introducing new subjective market definitions through reframing processes, firms influence the collective and continuous construction of markets. This performative action (Araujo 2007) presumes the identification of relevant network(s) in which to participate, both in terms of exploiting existing opportunities and looking out for new ones (March 1991; Möller and Svahn 2009).

However, the extant marketing research provides little support in understanding socially constructed markets. The majority of the marketing literature either neglects to define the market construct or implicitly adopts definitions from neoclassical economics where the market is defined in relation to the product and its exchange value (Lusch and Vargo 2006; Mele et al. 2014; Vargo and Lusch 2008; Venkatesh et al. 2006). Thus, research using the constructionist approach to understanding markets and market innovation is fragmented (Kjellberg et al. 2012; Vargo and Lusch 2008), and the majority of it focuses on the emergence of new markets or business fields. The present research therefore had three objectives:

1. To identify the basic dimensions or perspectives that managers currently use when pursuing new potential market definitions and, based on these, to develop a conceptual framework for categorizing managerial market definitions
2. To investigate empirically how practitioners frame potential novel markets
3. To identify challenges associated with creating innovative market definitions in traditional industrial networks

Two reasons can be given for focusing on a traditional or established industrial network context. First, as pointed out, the relatively recent constructive market-making research focuses primarily on emerging markets. Consequently, there is a substantial lack of knowledge on market innovation in mature industries. Second, because relatively established and ingrained network relationships characterize the mature industries, we can assume that market innovations will encounter severe barriers in such a context. We examine these barriers in this chapter and suggest ways of overcoming them.

The chapter draws on data collected within an extensive research program addressing innovation practices and opportunities within Finnish forest cluster companies. Titled the RAdical Market Innovations (RAMI) project, the program was carried out between March 2009 and December 2010. Within the program, market innovation opportunities were investigated via six research projects involving over 100 managers from ten companies as well as researchers representing a wide range of theoretical backgrounds, including, for example, industrial networks, practice theory, strategic management and cluster evolution, and management accounting. This chapter is the first attempt to create a synthesis of these relatively eclectic findings.

We begin by describing the research process and the six research projects that provide the material for this study. Second, we synthesize the findings of the six research projects into a conceptual framework and discuss the proposed framework in the light of existing literature. Third, we examine how practitioners frame novel markets and relate these findings to the proposed conceptual framework. Fourth, we describe the identified challenges in creating innovative market definitions in traditional industrial networks. Finally, we discuss the theoretical and managerial contributions and identify avenues for further research.

4.2 Research Process

The overview of the research process described in this section has two parts. In the first, we detail the RAMI research program forming the foundation of the present study. In the second, we describe the process of synthesizing the findings of the RAMI project and the principles governing it.

4.2.1 The RAdical Market Innovations (RAMI) Research Program

The RAMI project was initiated and financed by Forestcluster Limited, one of Finland's six strategic centers for science, technology, and innovation. Forestcluster is owned by key companies in the forest cluster as well as by main research institutes and universities in Finland (for an overview, see http://fibic.fi/about-fibic and http://fibic.fi/wp-content/uploads/2012/07/Forestcluster_prev1.pdf). RAMI's objective was to support Finnish forest cluster companies to produce radical innovations outside their current core businesses.

The RAMI program consisted of six academic research projects that involved 28 researchers from Finland and Sweden. Forestcluster Limited selected the overall research project themes through an iterative process involving the company's board and the research committee as well as RAMI's steering group. Once these people had defined the overall themes of the RAMI program, each research team defined the more precise research questions that it would address. The teams were

Table 4.1 Overview of the research projects in the RAMI program

Theme	Research questions
Base of the pyramid	Examine consumer practices in Base of the Pyramid (BOP) markets (disposable income under 5 €/day) related to packaging, mass communication, and primary education Identify and describe best practice BOP business models Describe BOP markets and their trends Identify opportunity spaces for the Finnish forest cluster
Capabilities and forest cluster evolution	Map the capabilities that the forest cluster companies possess and potential capability gaps Analyze strengths and weaknesses of Finland-based forest companies compared to international competitors Describe how and why capabilities interact with cluster evolution Investigate how capabilities enable (and hinder) innovation and renewal
Customer processes and cost structures	Analyze customers' revenue and cost structures Identify key processes that drive firm performance Map activities and identify critical interdependencies Explore potential innovation opportunities for the Finnish forest cluster
Multimedia evolution	Analyze multimedia behavior of young people using electronic media and print media Identify different reading profiles and their behavior regarding different media choices Identify trends related to media consumption
Paper in practice	Study, ethnographically, office work, and paper use Investigate the historical evolution of everyday practices related to paper Develop scenarios pertaining to future consumers and consumption
Value networks	Analyze and map the value networks and business models of the packaging industry, as well as the mental models of industry's leaders and experts Analyze benchmark cases that represent successful value system innovations outside the forest industry Explore potential for new value system innovations in the Finnish packaging industry

supported in this process by Forestcluster Limited. The leaders of the academic research projects, however, had substantial freedom in selecting their research methods and appointing the research team members. Table 4.1 provides a description of the research projects and their respective research questions.

In addition to the academic researchers, the RAMI program involved a total of 112 senior managers from ten major firms within the Finnish forest cluster. The intermediate and final results of the six academic research projects were discussed with the practitioner participants in a total of 52 workshops ranging from project-specific sparring sessions with selected industry representatives to large-scale seminars and innovation workshops. Each of us (the authors of this chapter) participated in the RAMI program in different and distinct roles. One of us was

RAMI's program manager, a role that involved overseeing the development of all six academic research projects and participating in most of the workshops. One of us was a member of the program steering group and a facilitator of the seven largest seminars and workshops. And one of us was the project leader of one of the academic research projects.

4.2.2 The Process and Principles of Synthesizing the Findings

The research process synthesizing the findings of the RAMI program focused on interpreting and reflecting instead of on data collection and processing (Alvesson and Sköldberg 2009). The research process was abductive in nature, combining induction and deduction in a nonlinear, nonsequential iterative process (Dubois and Gadde 2002; Locke 2010).

Abductive reasoning is particularly suitable when pursuing theory development, that is, when refining existing theories instead of inventing entirely new ones (Dubois and Gadde 2002; van Echtelt et al. 2008). This approach, involving interaction with a study's theoretical and empirical elements, was enabled by the extensive collaboration between managers and researchers. The entire research process, including both the RAMI program and the following synthesis phase, took 4 years. This time span allowed sufficient time for the reflective process, aimed at considering and combining elements in order to detect emergent patterns, and further develop the constructs used to portray reality (Eisenhardt 1989).

The process of synthesizing the findings of the RAMI program was divided into three phases. During the first phase, we individually familiarized ourselves with the reports and final seminar presentation materials from all six research projects. These materials consisted of six Word documents, totaling 613 pages, and ten PowerPoint documents, totaling 645 pages.

During the second phase, we shared our findings with one another, discussed them in the light of the extant literature, and created a first version of the conceptual framework. The third phase of the synthesizing process involved writing the research report, during which we continued to reflect on the synergy between the empirical findings and existing theories. All three of us were active participants throughout the synthesizing process and collaborated on creating the conceptual developments.

Explicitly defined principles, four in total, governed the process of synthesizing the relatively large set of empirical data. We drew these principles from criteria commonly used in interpretive research and grounded theory (see, in this regard, Flint et al. 2002). More specifically, the four principles, namely, pre-understanding, dependability, conformability, and integrity, referenced work by Lincoln and Guba (1985), Miles and Huberman (1994), Spiggle (1994), Strauss and Corbin (1990).

Pre-understanding, defined as the extent to which researchers are familiar with the empirical phenomena, was gained through our long-term academic experience in the area of B2B marketing. Also, because all of us were involved in the RAMI program, we gained firsthand exposure to the empirical material. Furthermore, all three of us have been involved in business development projects closely related to the topics of the present research, either through applied research projects or consulting assignments.

Dependability, defined as the extent to which there is consistency of explanations, and conformability, defined as the extent to which interpretations are subject to researcher biases, were improved through triangulation (Creswell and Miller 2005; Denzin 1978; Stake 1995). During the research process, we tapped into three forms of triangulation:

- Data triangulation, involving the use of several sampling strategies to collect (at different times, in different social situations and contexts and from various firms) the empirical data in the six underlying research projects
- Methodological triangulation, involving the use of various methods ranging from bibliometric studies and quantitative content analyses of secondary data to ethnographic observations and case studies
- Investigator triangulation, during which all three of us interpreted the data

Integrity, defined as the extent to which misinformation from study participants influences interpretations, was supported by the fact that all the companies and all individual managers participating in the RAMI program considered the program would benefit their businesses. Additionally, the fact that the in-depth academic research projects were conducted in a noncompetitive environment ensured that no competing companies were sponsoring the same projects. Finally, all the interviews were anonymous and the workshops participative and creative in nature, an approach that encouraged the industry representatives to express their true views.

4.3 Conceptual Framework for Creating Innovative Market Definitions

The majority of the extant literature defines markets in terms of the product to be sold and its exchange value (Lusch and Vargo 2006; Vargo and Lusch 2008; Venkatesh et al. 2006). Since the 1970s, practitioners have most commonly defined their markets in terms of the product–geography matrix (Boardman and Vining 1976) despite commentators maintaining that the framing processes based on products and geographies can lead firms to miss opportunities in the adjacencies (Rothschild 1984; Zook 2007). The dominance of product-based market definitions are further supported by the fact that most industry statistics are created for product markets and often for purposes related to national economics and/or the enforcement of antitrust legislation (Geroski 1998). The independent market research agencies also tend to investigate markets defined in product-based terms.

There are, however, also more innovative approaches to defining markets. Based on the synthesizing process described in the previous section and building on Zook (2007), we concluded that these more innovative market definitions in firms operating in traditional industrial networks can be classified using two dichotomies. These are the supply versus demand perspective and the dyad versus network perspective.

Most marketing literature has adopted a demand-oriented view, which means taking customers as the focal unit of analysis and defining markets in terms of segments of customers (Clarke and Freytag 2008; Sausen et al. 2005; Shapiro and Bonoma 1984). Jenkins and MacDonald (1997) critique this demand-oriented approach and propose that market definitions should be closely linked to specific supply characteristics, such as firm capabilities and the nature of the organization. In order to reconcile the supply- and demand-oriented views to markets, Datta (1996) and Geroski (1998) argue for integrating demand and supply-side factors into market definitions that simultaneously reflect both these viewpoints.

Continuing the tradition set forth by the neoclassical economics, the majority of current marketing literature implicitly adopts a dyadic view of the markets: there is "supply" provided by the providers, and there is "demand" created by the customers. However, some researchers argue that this dyadic view of markets is limited and that wider network perspective is needed. Henneberg et al. (2009), for example, argue for a segmentation of business networks instead of segmentation of only customers. Combining these supply versus demand and dyad versus network dichotomies creates a conceptual framework consisting of four alternative routes to surpassing the traditional product-based view of markets and creating market definitions based on:

- Firm's resources and capabilities
- Network's resources and capabilities
- Customers' processes
- End-consumers' practices

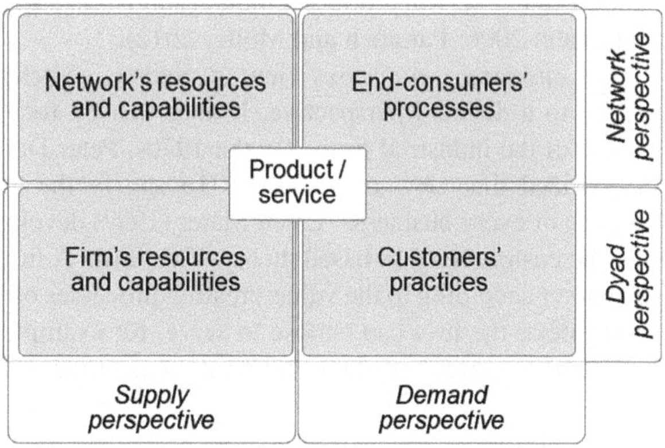

Fig. 4.1 Conceptual framework for creating innovative market definitions

Figure 4.1 illustrates the proposed conceptual framework. The *firm's resources and capabilities* view approaches market definition from supply-side and dyadic perspectives. Managers adopting this thinking endeavor to define markets through the resources and capabilities that a firm possesses. Depending on the firm's resource and capability base, the firm can operate within, for example, "paper production" or "fiber know-how" markets. Theoretically, this view is intimately linked to the resource-based view of the firm and its capability extension (Barney 1991; Teece 2010; Wernerfelt 1994), which defines the firm and explains the performance differences among firms based on their resource and capability endowments. Some marketing scholars have also adopted this view to market definition. For example, Jenkins and MacDonald (1997) advocate for market definitions and segmentation models that build on the capabilities and the nature of the firm. More recently, Daneels (2010) has shown that firms' resources and capabilities are based on managers' behavioral and cognitive practices, a finding that further supports the relevance of the constructionist perspective in understanding markets and market innovation.

The *network's resources and capabilities* view of markets also approaches market definition in terms of supply-side characteristics, but it expands the analysis level from the individual firm's resource and capability base to the resources and capabilities within the firm's network. The markets-as-networks approach (Brennan 2006; Johanson and Vahlne 2011; Mattsson 1997) developed within the IMP Group focuses on the actors, resources, and activities in industrial networks and proposes that this network system can be perceived as a market. Gadde et al. (2003) continued this line of thinking through their call for strategizing in industrial networks. They proposed that firms can enhance their performance by creatively investigating their network and by continuously seeking to combine and recombine existing resources to create new resource dimensions. In similar vein, Normann and Ramirez (1993) argue that the key task of a strategy is to create more value by reconfiguring the value-creating system or the value constellation in which the firm is embedded. Research also indicates that among intentionally designed coalitions of firms, so-called strategic nets (Möller et al. 2005), actors can jointly create shared cognitive views about the network and network strategy. This research lends support to the networked-based market innovation approach (Knight 2002; Möller and Svahn 2006; Partanen and Möller 2012).

The *customers' processes* view of markets switches market definitions from a supply to a demand perspective. It does this by focusing on the immediate customers of the industrial firms. By the 1970s, Peter Drucker (1973: 79) had already crystallized this view of markets: "To satisfy the customer is the mission and purpose of every business." Later, Slater (1997) developed the same line of thought into the customer value-based theory of the firm. A firm taking this view can define its market according to the value-creating processes of its customers. Depending on the context, the firm can choose to serve, for example, "product development" or "efficient logistics" markets. In theoretical terms, relationship marketing (Berry

1983; Grönroos 1994; Payne et al. 2008) means extensive investigation of the value-creating processes of the customers and how providers can support customers' value creation. This view of markets is also closely linked to service-dominant logic, which proposes that service is the fundamental basis for exchange and that the value of this service is always uniquely and phenomenologically determined by the customer (Vargo and Lusch 2004, 2008).

The *consumers' practices* view of markets also approaches market definition from the demand perspective, but it looks beyond the dyadic B2B customer relationships to end-consumers and their value-creating practices. Although defining the market for an industrial B2B firm through the value-creating processes of end-consumers may seem far-fetched, companies such as Intel have successfully adopted this approach. Among Intel's business units are those focused on "digital home" and "ultramobility" markets (http://www.intc.com/corpInfo.cfm). Theoretically, practice theory (Reckwitz 2002; Schatzki 2001) provides a valuable means of aiding understanding of end-consumers' value-creating practices. The concept of practice is not linear (as are many process descriptions), and it extends the unit of analysis to the system, including all of its relevant subjective and objective elements, thereby fostering action. Thus, practices are especially beneficial with regard to analyzing the consumers' contextual, mundane, subjective, and sometimes even partially subconscious value creation (Korkman 2006; Shove and Pantzar 2005). In the industrial network context, Helgesson and Kjellberg (2009) adopted a practice-oriented view when they investigated "use" as the core construct in understanding both the user and the objects being used. In the business-to-consumers context, consumer culture theory has allowed extensive investigation of consumer practices as social and cultural phenomena (Arnould and Thompson 2005; Featherstone 2007).

4.4 Managerial Framing of Market Innovations

After completion of the RAMI program academic research projects in May 2010, the people involved in them attended four two-day innovation workshops held in June and August 2010, respectively. During these workshops, attendees discussed the research findings and used these research results as a platform from which to identify new growth opportunities for the Finnish forest cluster outside its current core business. Table 4.2 sets out the research projects covered and the number of participants in each innovation workshop.

Each of the innovation workshops unfolded in a similar manner. During the first day, the workshop participants (excluding the facilitators) were grouped into teams of three and asked to familiarize themselves with the research results and to brainstorm possible new growth opportunities for the Finnish forest cluster outside its current core business. The total number of these "first day" ideas from all four innovation workshops amounted to 2323.

Table 4.2 Innovation workshops in the RAMI program

Research projects discussed	Date	Participants
Base of the pyramid	June 8–9, 2010	Sixteen manager-practitioners Four members of the research team One student Two employees of Forestcluster Ltd Five facilitators (not involved in identifying or describing new growth opportunities)
Customer processes and cost structures	June 15–16, 2010	Nineteen manager-practitioners One student Two employees of Forestcluster Ltd Five facilitators (not involved in identifying or describing new growth opportunities)
Multimedia evolution and paper in practice	August 19–20, 2010	Seventeen manager-practitioners Three members of the research teams and three other researchers One student Two employees of Forestcluster Ltd Five facilitators (not involved in identifying or describing new growth opportunities)
Capabilities and forest cluster evolution and value networks	August 24–25, 2010	Twenty-four manager-practitioners Four members of the research team and one other researcher Two students Two employees of Forestcluster Ltd Five facilitators (not involved in identifying or describing new growth opportunities)

After the first day, the facilitators categorized the identified ideas (with the number of ideas across the workshops ranging in total from 477 to 800) into four to six groups. The categorization process followed the tradition of grounded theory (Glaser and Strauss 1967), which meant that the facilitators coded the ideas and then grouped them into similar concepts and categories. During the second day, participants were divided into four to six groups, depending on the number of emerged idea categories, and asked to further synthesize and develop the individual ideas into more detailed idea descriptions. The total number of the detailed idea descriptions derived from the four innovation workshops was 146.

Because some of the ideas generated during the innovation workshops overlapped, two separate teams, consisting of four and seven practitioners, respectively, recategorized the 146 detailed idea descriptions and developed them further from October to November 2010. After this synthesis phase, the total number of truly individual ideas for new growth opportunities for the Finnish forest cluster came to 126. Each of these 126 ideas was then described from four different viewpoints reflected in four questions:

- What is the main idea behind the identified growth opportunity?
- Which actors could turn this idea into a viable business or new market?

- Which actors could be potential customers in this new market?
- What are the main benefits associated with the growth opportunity?

The 126 ideas were also assessed by the 11 practitioners involved in the October–November synthesis phase but this time from five different viewpoints: direct or indirect business potential for the Finnish forest cluster, level of "radicalism," level of challenge or risk, speed of realization, and which actors should be responsible for turning the idea into reality.

After this, we scrutinized the resulting 126 ideas for new growth opportunities and then used the proposed conceptual framework for defining markets in an industrial network context (illustrated in Fig. 4.2) to analyze those opportunities. We first determined whether the proposed growth ideas could be classified as markets. The question we asked in this regard was whether the ideas held potential for monetary exchange resulting in cash flows to the forest cluster firms. We concluded that 24 of the growth opportunity ideas did not meet this criterion of a market and removed them from the data set to be analyzed. We then categorized the remaining 102 ideas for new growth opportunities into prospective markets defined in the following terms: (1) a product or service, (2) the firm's resources and capabilities, (3) the network's resources and capabilities, (4) the customers' processes, and (5) the end-consumers' practices.

We classified the vast majority of the ideas (74 out of 102 or 73 % of the total) as pertaining to product or service markets. Examples of these ideas included "see-through cardboard," "modular school building," and "print-it solar cells." The second most common category contained ten ideas (10 % of the total). These related to markets defined in terms of customers' processes. Examples of these ideas included "water recycling for efficient water purification," "reforestation," and

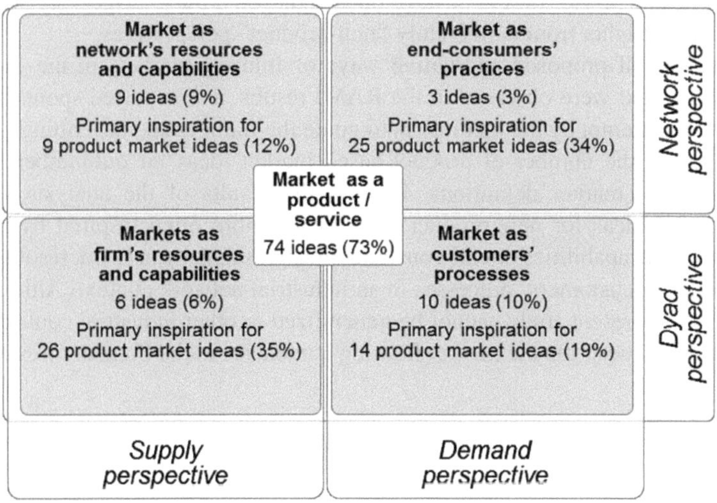

Fig. 4.2 Managers' framing of RAMI growth ideas as markets

"utilizing toilet waste for energy and fertilization." The third category (nine ideas or 9 % of the total) concerned markets defined in terms of a network's resources and capabilities, with examples including "EasyPrint global printshop network" and "closed loop recycling integration." The fourth and the second least common category contained six ideas (6 % of the total) and referenced markets defined according to firm's resources and capabilities. Examples of this category included "pharmaceutical products from forestry raw materials" and "water technology." The category with the fewest ideas (three, or 3 % of the total) related to markets defined according to end-consumers' practices. Ideas in this category included "water solutions for households to decrease water usage" and "local content (created for local media)."

Due to the dominance of product market ideas, we chose to analyze these ideas in more detail and to investigate which of the four proposed routes surpassing the product-based market definitions (i.e., firm's resources and capabilities, network's resources and capabilities, customers' processes, and end-consumers' practices) was the main influence behind each product market idea. This analysis was based on the idea-description documents written for each idea during the synthesis phase. This analysis led to the majority of the product market ideas (26 out of 74 or 35 % of the total) being classified as inspired primarily by a firm's resources and capabilities. Interestingly, the second most commonly detected influence behind the product market ideas (25 ideas or 34 % of the total) was end-consumers' practices, followed by customers' processes (14 ideas or 19 %), and network's resources and capabilities (nine ideas or 12 %).

Figure 4.2 summarizes the results of the above-described analysis of managers' framing of innovative growth ideas outside the current core business. The analysis indicates that the most common and intuitive way managers frame markets is to define them through the product or service being sold, even if the discussions relate to products and services yet to be developed and even if the managers are sensitized to data and insights from deliberately "non-product" perspectives.

However, all proposed alternative ways of framing markets in the industrial network context were observed in the RAMI results, and occurred spontaneously, without any attempt by the facilitators to guide the framing and describing of ideas. Nevertheless, the number of product-based market ideas far outnumbered these "non-product" market definitions. Finally, the results of the analysis seem to indicate that ideas for new product markets are more often inspired by a firm's resources and capabilities or end-consumer practices than by network resources and capabilities or customers' processes in an industrial network context. Although the results of the present study cannot be generalized to other industrial contexts, they most probably describe the market framing tendencies in the Finnish forest cluster.

4.5 Challenges in Creating Innovative Market Definitions in Traditional Industrial Networks

In this section we discuss the main identified challenges in creating innovative market definitions in traditional industrial networks that recurred in the findings of the six research projects and during the following innovation workshops. The identified challenges included the market size myopia, the embeddedness of relevant know-how, the negative impacts of market relationships to systemic innovations, and the need to develop new business models to match the new markets.

4.5.1 *Market Size Myopia*

The focal industry in our study, the forest cluster, and especially its paper and pulp production is characterized by large firm sizes and highly capital-intensive production technology. As such, it emphasizes the need to consider the economics of scale in manufacturing and distributing activities. Scale economy conditions are typical not only in many mature industries utilizing process technology but also in the manufacturing of durable goods such as cars and electronics.

This "scale effect" seemed to be deeply ingrained in the dominant logic of the RAMI program managers (Prahalad 2004; Prahalad and Bettis 1986) because they framed and evaluated the potential value of any new market from the perspective of their current business. It seems that, for the managers, ideas that did not match the current scale of their firm's volume and revenue potential failed to overcome "an invisible threshold of attraction and credibility" (von Krogh and Grand 2000). Because of this, practitioners within the incumbent Finnish forest cluster firms were dismissing many potentially profitable and fast-growing new markets as unfeasible. Now mindful, as an outcome of the RAMI program, of this cognitive bias, the firms in the Finnish forest cluster are looking at ways to circumvent it. One approach in this regard is creation of a platform that will bring together firms' new market ideas that are "suffering" from low (initial) business volume. Actors from outside the firms' traditional industry network are being brought in to help develop business models that enable profitable management of low-volume businesses. Similar solutions at the corporate level have been realized in large US corporations in order to overcome the barriers that development of new radical business ideas and innovations typically meet in the incumbent firms (O'Connor and Peters 2008).

4.5.2 *Embeddedness of Relevant Know-How*

Defining a market in terms of a network's resources and capabilities or customers' value-creating processes requires an in-depth understanding of those resources, capabilities, and processes. However, the people traditionally involved in making firms' market definitions (i.e., top management) rarely possess this understanding. Instead, this crucial know-how tends to be dispersed around the organization and even more extensively around the focal network, and it is generally middle-management individuals who hold it. Various knowledge management researchers have acknowledged the challenges associated with identifying and transferring tacit knowledge as well as the need to develop management practices that facilitate them (Bolisani and Scarso 2000; Dayasindhu 2002; Foos et al. 2006; Herschel et al. 2001).

This situation presents a true dilemma for the designers of strategy processes in terms of channeling this partially tacit understanding so that top management can use it. This process of distilling the know-how of middle management for use by top management can also be a time-consuming process. Consider, for example, KONE, the Finnish elevator and escalator company that changed its market definition from "elevators and escalators" into "People Flow" in 2008. Interestingly, the term "people flow" had been in relatively active use in KONE's R&D department for over a decade before top management discovered it and decided to use it as the new foundation for the company's market definition.

4.5.3 Negative Impacts of Market Relationships on Systemic Innovations

According to researchers involved in certain projects, those involving "customer processes and cost structures" and "value networks" especially, the presence of market-based relationships, that is, relationships characterized by competitive monetary exchange over a market mechanism, hinder the emergence of value-creating systemic innovations. Within the context of market innovations, it is interesting to contemplate whether this negative impact on systemic innovations are due to market relationships per se or to the fact that these market relationships are configured around a relatively stable product and may involve the use of strong price competition by the dominant actors. While several industrial marketing researchers maintain that the negative effects of strong and established relationships can diminish firms' capacity to change (Beverland 2005; Brennan and Turnbull 1999; Håkansson and Ford 2000), economic sociologists propose that products stabilize, at least temporarily, otherwise fluid exchanges between actors (Callon et al. 2002).

Lacking the empirical material to answer this question conclusively, we need to keep in mind, when seeking new, innovative market definitions, the inertia involved in market configurations. The current (most probably product-based) market definition is likely to cause resistance to change both internally within the organization and externally among the other actors in the network. Also, as the new market definition becomes more accepted as the norm, it too will inhibit new systemic innovations.

4.5.4 Need to Develop New Business Models to Match New Markets

All research projects within the RAMI program investigating end-consumers (i.e., "base of the pyramid," "multimedia evolution," and "paper in practice") emphasized that if the forest cluster firms were to be more relevant than currently with regard to their different end-consumer practices, they would probably need to embrace radically different business models. For example, consumers' multimedia consumption practices are not exhaustively supported by a business model that is built around production and distribution of newsprint and magazine paper.

In recent years, the business model construct has attracted an increasing number of studies (Demil and Lecocq 2010; Nenonen and Storbacka 2010; Teece 2010; Zott and Amit 2010). The identified challenges associated with changing the business model of an established actor range from unresolvable conflicts between the current and the new business models (Amit and Zott 2001; Johnson et al. 2008) through to inability to envision a viable new business model (Chesbrough 2010). Firms seeking to radically redefine their markets therefore need to acknowledge that their current business model might not be suitable in the new market configuration. Various researchers have characterized changing an existing business model as a challenging task (Amit and Zott 2001; Chesbrough 2010; Johnson et al. 2008).

4.6 Conclusions

In this section, we discuss the theoretical contributions of the study and its managerial implications and limitations. We also identify avenues for further research.

4.6.1 Theoretical Contribution

This chapter responded to calls for research in strategic marketing issues that are relevant to marketing practitioners (Jaworski 2011; Reibstein et al. 2009; Varadarajan 2010) by endeavoring to accomplish three objectives:

- Develop a conceptual framework for categorizing managerial market definitions.
- Investigate empirically how practitioners frame novel markets.

- Identify challenges associated with creating innovative market definitions in traditional industrial networks.

We began the chapter by presenting a conceptual framework that supplements the currently dominant product-based market definitions (Boardman and Vining 1976; Geroski 1998; Lusch and Vargo 2006; Vargo and Lusch 2008; Venkatesh et al. 2006) with two dichotomies relevant for defining markets in an industrial network context. These are the supply versus demand perspective and the dyad versus network perspective. According to the proposed framework, managers can choose to define their markets based on the firm's resources and capabilities, network's resources and capabilities, customers' processes, and end-consumers' practices (see Fig. 4.1 for an illustration). The framework also contributes to emerging theorizing on socially constructed markets and market innovations (Araujo 2007; Kjellberg et al. 2012; Mele et al. 2014; Storbacka and Nenonen 2011; Vargo and Lusch 2008) by providing a conceptualization on how managers can frame their current and prospective markets.

We then described how we used the developed conceptual framework in order to categorize the 126 ideas for new growth opportunities for the Finnish forest cluster outside its current core business developed during the RAMI: RAdical Market Innovations research program. Our aim here was to gain an understanding of how practicing managers from the Finnish forest cluster were framing new market opportunities. The results of this part of our study showed that managers described the vast majority of ideas for new markets (73 %) in terms of the product or service being sold. Even though researchers were able to identify instances when new market opportunities had been defined according to the firm's resources and capabilities, the network's resources and capabilities, customers' processes, or end-consumers' practices, their overall adherence to product-based market definitions indicated a clear dominant logic (Prahalad 2004; Prahalad and Bettis 1986). Markets were thus being defined in terms of products and services in an industrial network context. This research finding is especially interesting because the entire RAMI research program and all interactions within it were designed not only to sensitize the managers to data and insights from deliberately "non-product" perspectives but also to encourage them to engage in "out-of-the-box" thinking. These findings contribute to the embryonic existing understanding of how practitioners actually define markets in practice (see Azimont and Araujo 2007 for an example).

Our third step in the research process identified four challenges associated with creating innovative market definitions in traditional industrial networks. The first of the challenges, market size myopia, refers to the deeply ingrained dominant logic (Prahalad 2004; Prahalad and Bettis 1986) of managers in industries characterized by scale effects. This mindset saw the managers almost intuitively disregarding potentially profitable and fast-growing markets because they deemed their current scale insignificant compared to their current business volumes. Second, embeddedness of relevant know-how made it challenging for actors trying to create innovative market definitions to make the relevant, often tacit, information (on, for example, the network's resources and customers' value-creating processes) available to the people involved in the market redefinition process. Third, market-based relationships seemed to be hindering the emergence of value-creating systemic innovations such as market innovations, thereby creating inertia that actors needed to overcome when endeavoring to institutionalize a new innovative market definition. Fourth, the need to develop new business models to fit the new market was also obstructing realization of innovative market definitions.

Identification of these four barriers to creating innovative market definitions in traditional industrial networks contribute to the literature on market innovations (Johne 1999; Storbacka and Nenonen 2012), market-driving strategies (Jaworski et al. 2000; Kumar et al. 2000; Varadarajan 2010), and other nontechnical systemic innovations (Edquist 2011). This identification also contributes to literature on innovation in industrial networks (Capaldo 2007; Dhanaraj and Parkhe 2006; Lundgren 1995; Möller et al. 2005).

4.6.2 Managerial Implications

The main managerial implications of the present research are twofold. First, it provides managers with a simple conceptual framework for identifying innovative market definitions that are not necessarily framed according to the product or service being sold. Although business practitioners involved in the RAMI program acknowledged "finding growing markets" and "clarifying the business definition" as crucial strategic tasks, we identified a clear void in managerial tools supporting these themes. Because existing conceptual tools focus on visualizing the market definition in terms of product–geography matrices (Boardman and Vining 1976), practitioners are intuitively inclined to also define emerging market opportunities according to the product, as the results of the present study appear to confirm. Providing practitioners with alternative tools to define markets is therefore likely to facilitate strategic discussions on alternative market definitions. Also, the fact that markets are, unlike their frequent portrayal in strategy and marketing handbooks, malleable and socially constructed, they are receptive to actors' market-shaping efforts.

The second set of implications concerns the four identified challenges to creating innovative market definitions in traditional industrial networks. Managerially, the possible ways of overcoming these barriers are quite understandably more relevant than the barriers themselves. We therefore, at this point, offer some recommendations for practitioners.

Market size myopia can be surmounted by creating structures that enable the development of new, initially low-volume, business initiatives with actors other than the network incumbents affected by scale effect (O'Connor and Peters 2008). Another way of diminishing internal resistance caused by low initial business volumes involves quantifying future market size in terms of potential use value creation for customers (Peñaloza and Venkatesh 2006; Storbacka and Nenonen

2011). This approach can be useful in terms of illustrating long-term market potential when the initial exchange value derived from a particular market is low.

Embeddedness of the relevant know-how can be circumvented by designing strategizing processes that involve people also from outside the management team as well as from the firm's own middle management who may be from outside the firm's organizational boundaries. Furthermore, the extended and networked view to knowledge does not have to be limited to knowledge acquisition. Ideally, strategizing in a network will result in the creation of new relevant knowledge through network learning (Knight 2002; Knight and Pye 2005).

The *negative impacts of market relationships on systemic innovations* can be managed by fostering market relationships that are characterized by trust, openness, and reciprocity. In addition, firms can actively participate in emerging value systems (Möller et al. 2005) that are more future-oriented and often facilitate radical changes. Moreover, because, as Callon and colleagues (2002) suggest, products stabilize markets and exchanges between actors, we can hypothesize that changing the definition of the product being exchanged will be a powerful method of overcoming inertia in an established market configuration.

Finally, the challenges of developing new business models to fit the new market definition can be overcome by, for example, engaging in experimentation to define the new business model (Chesbrough 2010), unveiling the mechanisms and institutional norms protecting the current business model (Johnson et al. 2008), and organizing the new business model into a new business unit able to manage the tensions (Johnson et al. 2008).

4.7 Limitations and Further Research Avenues

The empirical material for the present study was gathered from only a singular context—the Finnish forest cluster. Despite this cluster covering several industries, consisting of several actors of varying sizes and business models, being truly global in nature, and (arguably) a good representation of a traditional industrial network, more empirical studies from varying industrial contexts are needed to advance our understanding of the managerial framing of markets and the barriers to creating innovative market definitions in practice.

Also, the present research focused solely on market definitions and managerial framing of new market opportunities. Thus, other aspects of market innovation were left outside the scope of this study. These aspects include influencing market practices (Andersson et al. 2008; Kjellberg and Helgesson 2006), the outcomes of subjective framing processes, such as market multiplicity (Kjellberg and Helgesson 2006), continuous identity construction and boundary work (Ellis et al. 2010), and competing market definitions (Rosa et al. 1999).

The identified challenges associated with creating innovative market definitions in traditional industrial networks open up interesting avenues for further research. For example, there is very little research on how to quantify and measure the use value creation for customers in different market configurations. Similarly, while experimentation as a route to finding a suitable new business model that matches the novel market definition is appealing, much of the extant experimentation and effectuation literature (Sarasvathy 2008; Wiltbank et al. 2006) is conducted in the entrepreneurial or start-up context. Thus, more insights are needed on how established actors in industrial networks can effectively facilitate experimentation or "nonpredictive" strategies.

Finally, the present study does not discuss the geographical dimension of market definitions even though this dimension was very prominent in the market definitions that the RAMI practitioners used. We excluded the geographical dimension of the study because we regard it as a primarily managerial limiting condition than a foundational premise of a market definition. As such, we consider it pertinent to regard geographical market entries more as extensions of existing markets than as market innovations.

References

Alvesson, M., & Sköldberg, K. (2009). *Reflexive methodology: New vistas for qualitative research*. London: Sage.

Amit, R., & Zott, C. (2001). Value creation in e-business. *Strategic Manage Journal, 22*(6/7), 493–520.

Andersson, P., Aspenberg, K., & Kjellberg, H. (2008). The configuration of actors in market practice. *Marketing Theory, 8*(1), 67–80.

Araujo, L. (2007). Markets, market-making and marketing. *Marketing Theory, 3*, 211–226.

Arnould, E. J., & Thompson, C. J. (2005). Consumer culture theory (CCT): Twenty years of research. *Journal of Consumer Research, 31*(4), 868–882.

Azimont, F., & Araujo, L. (2007). Category reviews as market-shaping events. *Industrial Market Management, 36*(7), 849–860.

Barney, J. (1991). Firm resources and sustained competitive advantage. *Journal of Management, 17*(1), 99–120.

Berry, L. L. (1983). Relationship marketing. In L. L. Berry, G. L. Shostack, & G. D. Upah (Eds.), *Emerging perspectives on services marketing* (pp. 25–28). Chicago: American Marketing Association.

Beverland, M. (2005). Adapting within relationships to adapt to market-led change: Does relationship success lead to marketplace inertia? *Industrial Market Management, 34*(6), 577–589.

Boardman, A. E., & Vining, A. R. (1976). Defining your business using product-customer matrices. *Long Range Planning, 2*(1), 38–49.

Bolisani, E., & Scarso, E. (2000). Electronic communication and knowledge transfer. *International Journal of Technology Management, 20*(1/2), 116–133.

Brennan, R. (2006). Evolutionary economics and the markets-as-networks approach. *Industrial Marketing Management, 35*(7), 829–838.

Brennan, R., & Turnbull, P. W. (1999). Adaptive behavior in buyer–supplier relationships. *Industrial Marketing Management, 28*(5), 481–495.

Callon, M. (1998). Introduction: The embeddedness of economic markets in economics. In M. Callon (Ed.), *The laws of the markets* (pp. 1–57). Oxford: Blackwell.

Callon, M., Méadel, C., & Rabeharisoa, V. (2002). The economy of qualities. *Economy and Society, 3*(2), 194–217.

Capaldo, A. (2007). Network structure and innovation: The leveraging of a dual network as a distinctive relational capability. *Strategic Management Journal, 28*(6), 585–608.
Chesbrough, H. (2010). Business model innovation: Opportunities and barriers. *Long Range Planning, 43*(2/3), 354–363.
Clarke, A. H., & Freytag, P. V. (2008). An intra- and inter-organisational perspective on industrial segmentation: A segmentation classification framework. *European Journal of Marketing, 42* (9/10), 1023–1038.
Creswell, J. W., & Miller, D. (2005). Determining validity in qualitative inquiry. *Theory into Practice, 39*(3), 124–130.
Daneels, E. (2010). Trying to become a different type of company: Dynamic capability at Smith Corona. *Strategic Management Journal, 32*(1), 1–31.
Datta, Y. (1996). Market segmentation: An integrated framework. *Long Range Planning, 29*(6), 797–811.
Dayasindhu, N. (2002). Embeddedness, knowledge transfer, industry clusters and global competitiveness: A case study of the Indian software industry. *Technovation, 22*(9), 551–560.
Demil, B., & Lecocq, X. (2010). Business model evolution: In search of dynamic consistency. *Long Range Planning, 43*(2/3), 227–246.
Denzin, N. K. (1978). The logic of naturalistic inquiry. In N. K. Denzin (Ed.), *Sociological methods: A sourcebook* (pp. 54–73). New York: McGraw-Hill.
Dhanaraj, C., & Parkhe, A. (2006). Orchestrating innovation networks. *Academy of Management Review, 31*(3), 659–669.
Drucker, P. F. (1973). *Management*. New York: Harper & Row.
Dubois, A., & Gadde, L. E. (2002). Systematic combining: An abductive approach to case research. *Journal of Business Research, 55*(7), 553–560.
Edquist, C. (2011). Design of innovation policy through diagnostic analysis: Identification of systemic problems (or failures). *Industrial and Corporate Change, 20*(6), 1725–1753.
Eisenhardt, K. (1989). Building theories from case study research. *Academy of Management Review, 14*(4), 532–551.
Ellis, N., Jack, G., Hopkinson, G., & O'Reilly, D. (2010). Boundary work and identity construction in market exchanges. *Marketing Theory, 10*(3), 227–236.
Featherstone, M. (2007). *Consumer culture and postmodernism*. London: Sage.
Flint, D., Woodruff, R., & Gardial, S. (2002). Exploring the phenomenon of customers' desired value exchange in business-to-business context. *Journal of Marketing, 66*(4), 102–117.
Foos, T., Schum, G., & Rothenberg, S. (2006). Tacit knowledge transfer and the knowledge disconnect. *Journal of Knowledge Management, 10*(1), 6–18.
Gadde, L. G., Huemer, L., & Håkansson, H. (2003). Strategizing in industrial networks. *Industrial Market Management, 32*(5), 357–364.
Geroski, P. A. (1998). Thinking creatively about markets. *International Journal of Industrial Organisation, 16*(6), 677–695.
Glaser, B. G., & Strauss, A. L. (1967). *The discovery of grounded theory: Strategies for qualitative research*. Chicago: Aldine.
Granovetter, M. (1992). Economic institutions as social constructions. *Acta Sociologica, 35*(1), 3–11.
Grönroos, C. (1994). Quo vadis, marketing? Toward a relationship marketing paradigm. *Journal of Marketing Management, 10*(5), 347–360.
Håkansson, H., & Ford, D. (2000). How should companies interact in business networks? In D. Ford (Ed.), *Understanding business marketing and purchasing* (pp. 248–258). London: Thomson.
Helgesson, C. F., & Kjellberg, H. (2009). Practical use: Enacting the user and that which is being used. In H. Håkansson, A. Waluszewski, F. Prenkert, & E. Baraldi (Eds.), *Use of science and technology in business: exploring the impact of using activity for systems, organizations, and people* (pp. 247–270). Bingley: Emerald.

Henneberg, S. C., Mouzas, S., & Naudé, P. (2009). Going beyond customers: A business segmentation approach using network pictures to identify network segments. *Journal of Business Marketing Management, 3*(2), 91–113.

Herschel, R. T., Nemati, H., & Steiger, D. (2001). Tacit to explicit knowledge conversation: Knowledge exchange protocols. *Journal of Knowledge Management, 5*(1), 107–116.

Jaworski, B. J. (2011). On managerial relevance. *Journal of Marketing, 75*(4), 211–224.

Jaworski, B. J., Kohli, A. K., & Sahay, A. (2000). Market-driven versus driving markets. *Journal of the Academy of Marketing Science, 28*(1), 45–54.

Jenkins, M., & MacDonald, M. (1997). Market segmentation: Organizational archetypes and research agendas. *European Journal of Marketing, 31*(1), 17–32.

Johanson, J., & Vahlne, J. E. (2011). Markets as networks: Implications for strategy-making. *Journal of the Academy of Marketing Science, 39*(4), 484–491.

Johne, A. (1999). Successful market innovation. *European Journal of Innovation Management, 2* (1), 6–11.

Johnson, M. W., Christensen, C. M., & Kagermann, H. (2008). Reinventing your business model. *Harvard Business Review, 86*(12), 50–59.

Kjellberg, H., & Helgesson, C. F. (2006). Multiple versions of markets: Multiplicity and performativity in market practice. *Industrial Marketing Management, 35*(7), 839–855.

Kjellberg, H., Storbacka, K., Akaka, M., et al. (2012). Market futures/future markets: commentary on future research directions in the study of markets. *Marketing Theory, 12*(2), 219–223.

Knight, L. (2002). Network learning: Exploring learning by inter-organizational networks. *Human Relations, 55*(4), 427–454.

Knight, L., & Pye, A. (2005). Network learning: An empirically derived model of learning by groups of organizations. *Human Relations, 58*(3), 369–392.

Korkman, O. (2006). *Customer value formation in practice: A practice-theoretical approach.* Doctoral dissertation, publications of the Swedish School of Economics and Business Administration No. 155, Helsinki.

Krippner, G., Granovetter, M., Block, F., Biggart, N., Beamish, T., Hsing, Y., et al. (2004). Polanyi symposium: A conversation on embeddedness. *Socio-Economic Review, 2*(1), 109–135.

Kumar, N., Scheer, L., & Kotler, P. (2000). From market driven to market driving. *European Management Journal, 18*(2), 129–142.

Lincoln, Y., & Guba, E. (1985). *Naturalistic inquiry*. Newbury Park: Sage.

Locke, T. (2010). Teachers as action researchers: Towards a model of induction. *L1 Educational Studies in Language and Literature, 10*(2), 41–66.

Lundgren, A. (1995). *Technological innovation and network evolution.* London: Routledge.

Lusch, R. F., & Vargo, S. L. (2006). Service-dominant logic as a foundation for a general theory. In R. F. Lusch & S. L. Vargo (Eds.), *The service-dominant logic of marketing: Dialog, debate, and directions* (pp. 251–265). Armonk: ME Sharpe.

March, J. G. (1991). Exploration and exploitation in organizational learning. *Organization Science, 2*(1), 71–87.

Mattsson, L. G. (1997). Relationship marketing and the markets-as-networks approach: A comparative analysis of two evolving streams of research. *Journal of Marketing Management, 13* (5), 447–461.

Mele, C., Pels, J., & Storbacka, K. (2014). A holistic market conceptualization. *Journal of the Academy of Marketing Science, 43*(1), 100–114.

Miles, M., & Huberman, M. (1994). *Qualitative data analysis: An expanded sourcebook.* Thousand Oaks: Sage.

Möller, K., Rajala, A., & Svahn, S. (2005). Strategic business nets: Their type and management. *Journal of Business Research, 58*(9), 1274–1284.

Möller, K., & Svahn, S. (2006). Role of knowledge in value creation in business nets. *Journal of Management Studies, 43*(5), 985–1007.

Möller, K., & Svahn, S. (2009). How to influence the birth of new business fields: Network perspective. *Industrial Marketing Management, 8*(4), 450–458.

Nenonen, S., & Storbacka, K. (2010). Business model design: Conceptualizing networked value co-creation. *International Journal of Quality Service Science, 2*(1), 43–59.
Normann, R., & Ramirez, R. (1993). From value chain to value constellation. *Harvard Business Review, 71*(4), 65–77.
O'Connor, G., & Peters, L. (2008). *Grabbling lightning: Building a capability for breakthrough innovation*. San Francisco: Jossey-Bass.
Partanen, J., & Möller, K. (2012). How to build a strategic network: A practitioner-oriented process model for the ICT sector. *Industrial Marketing Management, 41*(3), 481–494.
Payne, A., Storbacka, K., & Frow, P. (2008). Managing the co-creation of value. *Journal of the Academy of Marketing Science, 36*(1), 83–96.
Peñaloza, L., & Venkatesh, A. (2006). Further evolving the new dominant logic of marketing: From services to the social construction of markets. *Marketing Theory, 6*(3), 299–316.
Prahalad, C. K. (2004). The blinders of dominant logic. *Long Range Planning, 37*(2), 171–179.
Prahalad, C. K., & Bettis, R. A. (1986). The dominant logic: A new linkage between diversity and performance. *Strategic Management Journal, 7*(6), 485–501.
Reckwitz, A. (2002). Toward a theory of social practices: A development of culturalist theorizing. *European Journal of Social Theory, 5*(2), 243–263.
Reibstein, D. J., Day, G., & Wind, J. (2009). Guest editorial: Is marketing academia losing its way? *Journal of Marketing, 73*(4), 1–4.
Rosa, J. A., Porac, J. F., Runser-Spanjol, J., & Saxon, M. S. (1999). Sociocognitive dynamics in a product market. *Journal of Marketing, 63*(Special issue), 64–77.
Rothschild, W. E. (1984). Surprise and the competitive advantage. *Journal of Business Strategy, 4*(3), 10–18.
Sarasvathy, S. D. (2008). *Effectuation: Elements of entrepreneurial expertise*. Cheltenham: Edward Elgar Publishing.
Sausen, K., Tomczak, T., & Herrmann, A. (2005). Development of a taxonomy of strategic market segmentation: A framework for bridging the implementation gap between normative segmentation and business practice. *Journal of Strategic Marketing, 13*(3), 151–173.
Schatzki, T. R. (2001). Practice-minded orders. In T. R. Schatzki, K. Knorr-Cetina, & E. von Savigny (Eds.), *The practice turn in contemporary theory* (pp. 42–55). New York: Routledge.
Shapiro, B. P., & Bonoma, T. V. (1984). How to segment industrial markets. *Harvard Business Review, 62*(3), 104–110.
Shove, E., & Pantzar, M. (2005). Consumers, producers and practices: Understanding the invention and reinvention of Nordic Walking. *Journal of Consumer Culture, 5*(1), 43–64.
Slater, S. F. (1997). Developing a customer value-based theory of the firm. *Journal of the Academy of the Marketing Science, 25*(2), 162–167.
Spiggle, S. (1994). Analysis and interpretation of qualitative data in consumer research. *Journal of Consumer Research, 21*(3), 491–503.
Stake, R. E. (1995). *The art of case study research*. Thousand Oaks: Sage.
Storbacka, K., & Nenonen, S. (2011). Markets as configurations. *European Journal of Marketing, 45*(1/2), 241–258.
Storbacka, K., & Nenonen, S. (2012). Competitive arena mapping: Market innovation using morphological analysis in business markets. *Journal of Business-to-Business Marketing, 19*(3), 183–215.
Strauss, A., & Corbin, J. (1990). *Basics of qualitative research, grounded theory procedure and techniques*. Newbury Park: Sage.
Teece, D. J. (2010). Business models, business strategy and innovation. *Long Range Planning, 43*(2/3), 172–194.
van Echtelt, F., Wynstra, F., van Weele, A., et al. (2008). Managing supplier involvement in new product development: A multiple-case study. *Journal of Product Innovation Management, 25*(2), 180–201.

Varadarajan, R. (2010). Strategic marketing and marketing strategy: Domain, definition, fundamental issues and foundational premises. *Journal of the Academy of the Marketing Science, 38* (2), 119–140.

Vargo, S. L., & Lusch, R. F. (2004). Evolving to a new dominant logic for marketing. *Journal of Marketing, 68*(1), 1–17.

Vargo, S. L., & Lusch, R. F. (2008). Service-dominant logic: Continuing the evolution. *Journal of the Academy of the Marketing Science, 36*(1), 1–10.

Venkatesh, A., Peñaloza, L., & Firat, F. (2006). The market as a sign system and the logic of the market. In R. F. Lusch & S. L. Vargo (Eds.), *The service-dominant logic of marketing: Dialog, debate, and directions* (pp. 251–265). ME Sharpe: Armonk.

Von Krogh, G., & Grand, S. (2000). Justification in knowledge creation: Dominant logic in management discourses. In G. von Krogh, I. Nonaka, & T. Nishiguchi (Eds.), *Knowledge creation: A source of value* (pp. 13–29). London: McMillan.

Wernerfelt, B. (1994). The resource-based view of the firm: Ten years after. *Strategic Management Journal, 16*(3), 171–174.

Wiltbank, R., Dew, N., Read, S., et al. (2006). What to do next? The case for non-predictive strategy. *Strategic Management Journal, 27*(10), 981–998.

Zook, C. (2007). Finding your next core business. *Harvard Business Review, 85*(4), 66–75.

Zott, C., & Amit, R. (2010). Business model design: An activity systems perspective. *Long Range Planning, 43*(2/3), 216–226.

Chapter 5
Practice-Based Approaches to Learning and Innovating

Katri Kallio, Tiziana Russo-Spena, Inka Lappalainen, and Cristina Mele

Abstract This chapter examines how practice-based learning perspectives allow to approach the broad, service-based view of innovation and its processes as a collaborative learning process of creating something new in a better way. We focus on studying three practice-based learning approaches through their key concepts. More specifically, we are interested in how these theories interpret the creation of new social structures, learning processes, practices and tools in the cultural-historic context for the creation of new knowledge and activity. The study brings new conceptual insights to the topical phenomena about various learning mechanisms in order to contribute especially to the processes of innovation and enhance as well as facilitate innovation activities within and between organisations in a purposeful manner. Thus, as our conclusion, we state that, by increasing the in-depth understanding of different mechanisms of learning in action and between actors, we can create better conditions for innovation to take place.

5.1 Introduction

The innovation paradigm in general is undergoing changes from economic, technological and positivistic views towards collectivistic, process-oriented and dynamic paradigms (Edvardsson and Tronvoll 2013; Greer and Lei 2012; Lusch and Nambisan 2015). It has been realised that a large number of different actors outside a producing organisation are needed, such as users, and they can even be responsible for the processes of creating innovations (Chesbrough 2011; Michel et al. 2008; Sundbo and Toivonen 2011).

K. Kallio (✉) • I. Lappalainen
VTT Technical Research Centre of Finland, Espoo, Finland
e-mail: katri.kallio@vtt.fi; inka.lappalainen@vtt.fi

T. Russo-Spena • C. Mele
Department of Economics Management Institution, University of Naples Federico II, Napoli, Italy
e-mail: russospe@unina.it; crimele@unina.it

T. Russo-Spena et al. (eds.), *Innovating in Practice*,
DOI 10.1007/978-3-319-43380-6_5

When previously innovations were considered to be driven mainly by technological opportunities and marked push, currently novel societal and service needs are often seen to function as the driver (Lundvall 1992; Toivonen and Tuominen 2009). Novel perspectives on innovation processes are elaborated, for example, in studies based on new service development (NSD) (Alam and Perry 2002; Edvardson and Olsson 1996; Syson and Perks 2004) and value creation, especially that of service-dominant logic (Vargo and Lusch 2004, 2008; also Edvardsson and Tronvoll 2013; Michel et al. 2008; Ordanini and Parasuraman 2011), and organisational studies on practices (Dougherty 2004; Gherardi 2012b; Schatzki et al. 2001; Swan and Scarborough 2005). Even if they differ in their research domains, taken together, these studies open up the notion that the development of the new kind of ideas and frameworks is needed that more deeply explicates the complex and multifaceted nature, structure and processes of innovation.

In particular, NSD-based scholars draw attention to the practical, collaborative processes in service creation and delivery, in the management of such activities. In the latest NSD studies, these processes are found to imply complex, uncertain and iterative learning activity instead of predictable stages to be implemented (Gottfridsson 2010, 2012; Payne et al. 2008; Sundbo and Toivonen 2011). Also, the perspective on value creation has stressed the importance of intangible assets (such as relationships and knowledge) and the dynamic nature of these assets in action, which produces a service (Vargo and Lusch 2008). The perspective of service-dominant logic emphasises the importance of interlinked activities and relational interactions in service innovation (Edvardsson and Tronvoll 2013; Michel et al. 2008; Ordanini and Parasuraman 2011) for mutual value creation (Lusch and Nambisan 2015). Finally, drawn from practice-based theories (Dougherty 2004; Gherardi 2012a; Russo-Spena and Mele 2012a; Swan and Scarborough 2005), some recent studies on innovation introduce practices as fundamental units for understanding innovation and its processes. The idea of practice-based innovation (PBI) has found its way into a dynamic view of innovation that occurs as an emergent and social process (Cook and Brown 1999; Lave and Wenger 1991; Nooteboom 2012).

Importantly, all these perspectives challenge many of the dichotomies of the traditional view of innovation by suggesting some changes into the innovation conceptualisations. Firstly, broadening the research domain, the unit of analysis is not anymore the distinction between service and product but innovation integrating the overarching service concept (viz. new benefit for another party) (Michel et al. 2008; Vargo and Lusch 2008; Vargo et al. 2015). Also, the attention is being paid to network-centric perspective (Lusch and Nambisan 2015; Vargo et al. 2015) making innovation dependent on open processes and to value co-creation that is new way of co-creating value through more effective participation in resource integration from mutual value (Edvardsson and Tronvoll 2013; Mele et al. 2014; Vargo et al. 2015). Moreover the innovation processes are being found to imply more complex, uncertain, emergent activity (Toivonen 2010; Sundbo and Toivonen 2011) instead of predictable stages to be implemented (Gottfridsson 2010, 2012; Gherardi 2012b; Mele and Russo-Spena 2015).

However, while the issue of what innovation means has continued to receive great attention, less is understood of the mechanisms underlying this process. Particularly, in service research, dialogical studies are called for, explaining the processes and dynamics behind the broad service-based view on innovation (Edvardsson and Tronvoll 2013).

The contribution of learning to innovation is suggested but has remained rather obscure. Organisational scholars, such as Miettinen (2002), have raised concerns that currently learning and innovation concepts are even used as synonyms, without problematising or fully understanding what these are truly about. Thus, in both general innovation studies and service innovation studies, learning is referred to as important but is usually left as a "black-box".

General innovation research has focused a great deal on formal learning mechanisms (education, science and technology-based learning and training programmes) as the perspective has arisen from macroeconomic origins (Ellström 2010; Lundvall and Johnson 1994; Lundvall et al. 2002). Then again, some management researchers criticise the fact that too often learning and its results in organisations are seen as subjective and vague, whereas more objective, conceptual models are needed (Zollo and Winter 2002). We find that service scholars have so far not described in detail the mechanisms of how this complex social interaction in everyday work activity takes place. The contribution of learning in this sense has still only been discussed at the general level. On the other hand, we found that, even if practice-based innovation studies consider learning in its sociocultural context to be a topical phenomenon, the explicated connection points and in-depth contribution of practice-based learning mechanisms to the emergence of innovations have not been investigated to the same extent.

Consequently, studies of innovation would benefit from more in-depth discussion of the mechanisms of learning and interfaces of the learning and innovation phenomena (Ellström 2010; Gherardi 2000, 2009a; Kallio and Lappalainen 2014; Kallio 2015; Lundvall et al. 2002; Russo-Spena and Mele 2012b; Russo-Spena et al. 2014). Supporting this dialogical perspective, our aim in this chapter is to seek explanations regarding the dynamics of the innovation processes from the studies grouped under the wide umbrella of practice-based learning studies (Gherardi 2000). We address the potential contributions of the theories grounded within the tradition of social and cultural science (Nicolini et al. 2003; Vygotsky 1978). These studies see that neither knowledge, learning, nor innovation should any longer be seen solely as something we possess (e.g. scientific facts), but as something constructed in social interaction by utilising artefacts and tools in and between groups, communities and organisations in action and in practices (Engeström 2004, 2007; Gherardi 2000, 2009a; Orlikowski 2002; Svabo 2009).

Looking in-depth at practice-based learning studies, our chapter aims to first increase understanding conceptually of learning processes in sociocultural perspectives and how they take place, enhance renewal in work activity and contribute to the creation of service innovations. Second, it aims at creating an interpretative framework that can help both scholars and practitioners working with innovation to draw together relevant concepts and views from different innovation and practice-based learning fields and provide an interdisciplinary perspective on the topic. To research the aim, three research questions lead our study: What are the key concepts that characterise practice-based theories of learning? How do the practice-based learning key concepts contribute to the process of creating service innovation? What kind of (service) innovation conceptualisation emerges from a practice-based learning approach?

We structure the chapter as follows. First, we describe the research process, and second, we briefly present the three practice-based approaches on learning that we chose to study. Third, for understanding the learning mechanism of the three approaches in greater depth, we utilise as the main frame of this study a categorisation of the key learning features developed by Nicolini et al. (2003) (see also Svabo 2009). Importantly, around and through this frame, we discuss and ponder how the chosen learning approaches can contribute to the broadened understanding of innovation. This chapter ends in a discussion of its implications and a reflection on further research avenues.

5.2 Research Process

Below we examine three broad practice-based learning approaches in order to better clarify the different conceptual learning constructs underlying them. We take an in-depth look at the conceptual analysis of knowledge and learning as sociocultural phenomena and ponder what implications for organisations and their innovation activities can be made. We have found that a more holistic understanding of practice-based learning studies was needed, since over the last 30 years, the debate within these studies has given rise to a variety of perspectives, which sometimes comprise conflicting discourses, and no unified understanding of knowledge and learning process seems to exist (Corradi et al. 2010).

The debate started in organisation and learning literature in the 1990s to challenge the established knowledge theory within knowledge management (Nonaka and Takeuchi 1995). Altogether, many practice-based learning authors use the foundations of Lave and Wenger's (1991) community of practice to suggest that practices or actions should be the critical point of analysis, and this understanding also requires examining the context (Brown and Duguid 2001).

A contribution of practice scholars is that they question the separation of "knowledge" from the knower and suggest that we should discuss "knowing" instead. In other words, whereas in knowledge management literature, knowledge is defined as something that can be objectified and possessed, practice-based learning scholars argue that knowing implies a process and an activity situated in social, historical and cultural contexts (Engeström 1987; Nicolini et al. 2003).

The categorisation of knowledge and knowing makes a difference in terms of epistemology; Cook and Brown (1999) have distinguished between epistemology of possession and epistemology of practice. They have placed knowledge in the former case and seen it as a tool in the service of knowing. Knowing is about activity and relationships as opposed to a substance, commodity or piece of information (Brown and Duguid 2001; Svabo 2009). Practice-based studies have arisen within the so-called post-epistemologies or radical constructivist epistemology (Gherardi 2014). The practice thinking goes beyond problematic dualism (action/structure; human/non-human; mind/body) to define practice as an interweaving of elements that are shaped by being interconnected in action (Gherardi 2014). This introduces the helpful topic of knowledge—in practice-based studies as a situated knowing which connect one activity to the next. This study follows the epistemology of practice (Gherardi 2014), which sees knowing as situated in the system of ongoing practices, and it is different in the sense that is relational, mediated by artefacts and always rooted in a context of action, interaction, culture and history (Brown and Duguid 1991, 2001; Cook and Brown 1999; Engeström 1987). In order to better understand how the discourse on practice-based learning emerged and to provide better understanding regarding the key mechanisms, we conducted a three-step literature-review process. We followed the suggestion of Finfgeld (2003) for how to conduct a literature review.

In the first step, we made clear the specific literature interest of our qualitative study by deciding what were the relevant interests and inclusion criteria for the studies. We conducted a search of articles in which the title, keywords or abstract mentioned the keywords "learning", "knowledge", "innovation", "action" and "activity" in conjunction with "practice" (or "practices") (searched as net*). We found the combined keywords to be most useful in sourcing the relevant articles. The research databases we used were Google Scholar, ISIWeb of Science and SciVerse Scopus. Our search produced a result of more than 70 documents including articles and books covering a period of about 35 years. This period also represents the golden age within practice-based debate (Gherardi 2000).

In the second step, we aimed to provide the first distinguishing elements between the studies. In so doing, a critical reading of documents was conducted separately by researchers. It was not a one-off and stand-alone process, but it emerges more as an abductive, cyclical process of search and questioning on our insights and the key topics. Magnani (2001) acknowledges that the term abduction is sometimes used to denote different parts of the theorising process. He explains that there are two main epistemological meanings of the word abduction: generation of plausible hypotheses (the definition that he himself uses) or inference to the best explanation (the definition that we utilise to describe the reasoning process of this chapter).

In this study, having established the key concepts, researchers reread the relevant concepts and interpretations on the basis of a common schema. This proved a useful way to appreciate how the academic debate is different around the concepts and how they evolve. As a result of the second phase, a first synthesis of the different learning approaches was provided. It was inspired by the classification of practice-based learning approaches by Nicolini et al. (2003), who distinguished between four main traditions, namely, (1) cultural interpretive tradition presented by Yanow (2000), (2) social learning described by Wenger (2000), (3) cultural-historical activity theory presented by Engeström (2004) and Blackler et al. (2000) and

(4) the sociology of translation presented by Gherardi and Nicolini (2002) and Suchman (2000).

However, we soon found that this categorisation was not satisfying enough for us. In this first categorisation, the concepts still seemed to be too scattered for our purpose. By deeply analysing and integrating different authors' positions, we provided the second classification of the learning approaches utilised in this study. In this classification, we distinguish between three practice-based learning approaches that we have called *knowing-in* practice (Brown and Duguid 1991; Orlikowski 2002, 2006; Wenger 1998, 2000), *knowing-in-between* practices (Gherardi 2000, 2006, 2008; Nicolini et al. 2003, 2012) and *expansive learning* (Blackler et al. 2000; Engeström 2004, 2007). This tripartite categorisation became possible due to our notion that the approaches differ in terms of "where" learning takes place (Russo-Spena et al. 2014). Even if we know that other approaches and classifications could also exist, our choice was to favour a classification that enables us to put the evidence into the dynamic aspect of change in practices and learning.

In the third and final step of the process, we moved further within the approaches in clarifying and comparing the key learning concepts and elements in order for us also to determine how the studies on knowing and learning can contribute to the process and new concepts of innovation. As a frame to compare the key concepts of the three selected approaches, we utilise the characterisation by Nicolini et al. (2003) of the five common features of practice-based approaches to learning (also utilised by Svabo 2009). This framework facilitates our navigation among the research approaches and helps us (1) to make clearer what the "key learning features" are, (2) to identify their importance for innovation debate and (3) to explain how they provide direction for the development of the broader innovation perspective.

By following this line of questioning, a list of key ideas, concepts, phrases and their relationships is used to compare the three approaches. The following two paragraphs provide evidence of the main issues dealt with by each tradition and end with insights on progress in the innovation debate that we build on.

5.3 Three Practice-Based Learning Approaches

The spread of the practice concept has led to a wide variety of definitions and approaches, with some authors (Gherardi 2009a; Gherardi and Strati 2012; Nicolini 2011) leading the attempts towards a more complete depiction regarding the role of learning and practices in organisations. In particular, practice-based learning scholars question individual action as a building block of understanding knowledge and learning and focus rather on the continuity between learning, knowing and doing as social phenomena (Gherardi 2012b).

However, much remains unknown about how the dynamics of knowledge and learning has been addressed, and how development and changes in practice occur, incrementally or radically, and what happens in practice through the application of situated knowledge in the social context. In this chapter, we are interested in creating more in-depth understanding from the key concepts of practice-based learning approaches. We find that the focus we address on the dynamic perspective of change in "practices" is one of the central topics for the development of new ideas also to the broad service innovation perspective.

The first approach—*knowing-in* practice—is characterised by scholars that have attempted to explore the link between knowledge and the practices of organisational members, i.e. to examine what is termed a process of knowing in practices. This approach arises in a strand of the social learning tradition (Lave and Wenger 1991). It includes the studies by learning scholars (Brown and Duguid 1991, 2001; Cook and Brown 1999; Orlikowski 2002; Wenger 2000) that widely question the difference between knowledge and knowing. The situatedness and the sociality of knowledge come into focus by the assumption that, in a community, practitioners have a common understanding of what the community does, how to do it and how it relates to other communities and their practices (Lave and Wenger 1991; Wenger 1998). Align with the social learning approach (Wenger 2000), the cultural interpretive tradition developed especially by Yanow (see Nicolini et al. 2003) has also focused on those practice mechanisms that see individual acquiring knowledge in action in the form of participation in context. Importantly, the knowing-in practice approach stresses the practice as the locus of knowing and learning that takes place in social relationships, normally occurring at workplaces.

The second approach—*knowing-in-between* practices—groups scholars which have developed the debate on the role of knowing and learning in interconnected sociocultural contexts. This approach mainly embraces the views of authors within the sociology of translation studies (Gherardi 2006, 2008; Gherardi and Nicolini 2000, 2002; Nicolini et al. 2003, 2012). They look more at the range of mechanisms through which knowing and practices are constituted and connected within and across communities, rather than considering practices as a pure community accomplishment. The discussion of Gherardi (2008, 2009a) about practices as a "between in" concept is of particular interest so as to grasp the difference. The author argues that the concept of practices constitutes a bridge between antithetical concepts because "practice lies in the habitual features, but has also the characters of purposeful action" (Gherardi 2009b: 121), and the attention is more on activities which hold together a configuration of people, artefact and social relations (Gherardi 2009b: 121). Insights from the actor network theory (Latour 1987) have inspired the work of social translation authors (Gherardi 2008; Nicolini et al. 2012) in emphasising the constitutive power of relations in the nexus-shaped context of practices. What distinguishes knowing-in-between scholars is that they regard the "nexus of practices" as the loci of knowing and learning (Gherardi and Nicolini 2002).

The third approach—the *expansive learning*—was developed originally in Finland by Yrjö Engeström and his research group in the late 1980s (Engeström 1987; Engeström et al. 1999; Engeström and Sannino 2010). The approach was developed within the cultural-historical activity theory (sometimes called Scandinavian activity theory) by building the conceptual framework on Vygotsky's (1978) cultural-historical psychology and Leont'ev's (1978) activity theory. Engeström (1987, 2004, 2007) also combined ideas from pragmatism and constructivism and introduced the theory to the organisation studies in late 1980s. Expansive learning (or learning by expanding) refers more specifically to the process and mechanism where human beings emancipate their thought and activity by creating new skills, activity and practices. Its main starting point is that human activity is object oriented, and capacities develop and abilities emerge in action with the specific cultural, historical and social setting in the construct of social reality. Expansive learning interprets practice as an activity in its specific context and sees learning taking place in collective interaction, in or between activity systems (Blackler et al. 2000).

5.4 Key Mechanisms and Contribution to Innovation

Nicolini et al. (2003) (see also Svabo 2009) have made a preliminary characterisation of the five common features under which the key concepts of practice-based learning approaches can be grouped. They include first the assumption that the world is always in the making, and thus the focus should be on processes, i.e. what people actually do by talking and in their actions. Second, an interest in these studies is in the social aspects of knowing and learning, placing processes of knowing not in the mind of the individual but in the social subjects. Third, the interest is in the "unorderly", introduced by terms such as uncertainty, conflict and incoherence. Fourth, the view of knowledge is as situated in a spatio-temporal context, and finally, the theories use a range of object terms, referring both to the material artefacts and to specific historical conditions and a broader "object of human and work activity". In Table 5.1, we summarise as our framework the key concepts recognised in the Nicolini et al. (2003) work in order to clarify the key learning concepts and also ponder their potential contribution in terms of innovation.

Below, we utilise the frame shown in Table 5.1 in order to describe, element by element, the key concepts of the three learning approaches of knowing-in, knowing-in-between and expansive learning. We ponder the main similarities and differences, and at the end of each key element, we elaborate its potential contribution to the broadened view on innovation.

5.4.1 *Processes Orientation*

The knowing-in approach responds to the invitation of practice scholars to bring daily and working activities into the focus of learning (Cook and Brown 1999). The focal point of learning and its mechanism becomes the specific activities that people engage in while solving problems during their daily interactions. Knowing and

Table 5.1 Nicolini et al. (2003) framework: key elements and concepts to be clarified

Key elements	Main focus	The concepts to be clarified
Processes orientation	What people do in action	Practices, action, activity
Social aspects of knowing and learning	Understanding what learning means in a social context and in relation to the collective subject	Individual and collective subject of learning
Attention to the unorderly	The triggers and nature of the learning process	Uncertainty, conflict and incoherence
Knowledge situated in a spatio-temporal context	The importance of situated work activities inseparable from the "knowers"/"learners"	Context, space, history and patterns
Material and artefactual	Understanding how knowing and learning emerge in interaction with materiality	Tools, artefacts and object

learning are a mode of action (Cook and Brown 1999) that develop from and are manifested in practice, expertise and skills at work. They are iteratively sustained by a mechanism of social interactions with both the social and material world (Brown and Duguid 1991, 2001). Meanwhile, knowing deals with a matter of displaying competences defined in a social context (Lave and Wenger 1991), while learning combines personal transformation with the evolution of social structure (Brown and Duguid 1991). Competence and capacity represent only some aspects of the learning dynamic as Wenger (2000) stated; learning always takes place through the tension between competence and experience. Learning is not a static embedded capability (of acquiring and applying already existing knowledge), but it emerges as an ongoing social accomplishment, constituted and reconstituted in the context as actors engage the world of practice (Orlikowski 2002).

The knowing-in-between approach provides more evidence of learning as a process through which different actors anchor a fragmented set of knowledge. The process dimension of knowing and learning comes into focus alongside working activities that are distributed among the nexus of practices (Nicolini et al. 2003; Gherardi 2009b). Learning is conceptualised to require the activation of a fragmented and spatially distributed system of knowledge. In this perspective, plural terms are utilised, and it is stated that actors do not just act within a stated community or at the boundaries of connected communities; they act within a range of sociocultural contexts and through a nexus of interconnected practices (Nicolini 2009). These are not pre-existent to groups, but are formed and distributed in and between groups. The interest moves towards the analysis of social and collective learning that takes place in the interplay of communities (Nicolini 2011; Nicolini et al. 2012) not as single, local practices, but rather as a shift in the overall nexus of interconnected practices (Gherardi and Nicolini 2002).

The expansive learning approach in its turn sees learning taking place in collective interaction, in or between activity systems of certain cultural-historical context. In this way, the approach does not solely follow the actors, but takes into account the structures and interlinked system dynamics of activity systems. It provides concrete tools and frames of references for the examination of the daily life, development and learning practices in and between organisations from constructive and process-oriented perspectives (Engeström 1987, 2004, 2007). Expansion in the object of activity means tensions and a change of needs in some or all of the elements in the activity system that are the subjects (who), tools (by which), community (in which), rules and division of labour (what kind of social order). More specifically, it refers to a process in which the learner(s) focuses not just on the problems at hand but also on the wider context that generates those problems and, consequently, creates a broader horizon of possibilities for the future (Engeström 2001). The expansive learning process can be characterised as a collective, conscious, usually enabled learning process that benefits different types of interventions and reflective material (Engeström 2004). The pursuit of newness is central to this process that aims at collectively exploring rather radical, discontinuous and qualitative changes in the activity—the kind of activity that does not yet exist.

Contribution to Innovation Debate

What is brought to innovation by knowing-in and knowing-in-between perspectives is an idea of learning that implies a process or action in ongoing interaction whereby the subject, object, context and knowledge are meshed in an indistinguishable way. Learning manifests itself through emergent, concrete and relational actions leading by unconscious and conscious attempts to produce and reproduce practices (Brown and Duguid 1991, 2001; Wenger 2000) or to order and negotiate in a nexus of practices (Gherardi 2000; Nicolini et al. 2012). This provides an additional idea of innovation far from the assumption of the inventor able to bring onto the market and effectively control the knowledge progress of a specific scientific field (Cooper 1994). A planned and actor-focused conception of innovation could be overcome in favour of a perspective of "innovating" where attention is paid to how learning and innovation are accomplished in practice and enacted in a daily work setting.

The expansive learning process instead refers to a profound and dynamic development process, to a transformative learning that radically broadens the shared object of activity by systemic changes in the whole activity system. This implies that, if a broadened object of activity is achieved, novel tools are also needed as multilevel and integrated constellations, as well as novel kinds of rules and distributions of labour (Engeström 2004). Thus, in this way, an expansive learning process may serve as a seed for comprehensive, discontinuous, qualitative changes in activity and practices and explain the process of innovating as a learning process.

5.4.2 Social Aspects of Knowing and Learning

The knowing-in approach focuses on the way that individuals interpret or make sense of their knowledge and experience in the social and material worlds (Brown and Duguid 1991). In this sense, the social dimension is related to an idea of learning as the epistemic link with the world that makes sense only if related to the assumption of a collective subject (Cook and Brown 1999; Brown and Duguid 1991, 2001). The social nature of learning is seen not only in the sense of Lave and Wenger's (1991) process of building community identity by participation. It is mainly an aspect of individuals doing and knowing in interaction with others and with the object and activity mediating among subjects (Brown and Duguid 1991, 2001). What individuals learn inevitably reflects the social context in which they learn it and in which they put it into practice. Social practices constrain and guide human knowing and the context of the actions; without social practice, knowing does not exist (Wenger 2000). Individual cognition and the agent role are still considered to be a basic unit of learning, but the focus is more on how learning is socially enacted in a particular setting evolving around practices. Common to the knowing-in scholars is the view that learning mainly involves understanding the dynamic interaction between the agent, knowing and practices. In this sense, learning is much more than an incremental knowledge gained from performing ordinary activities; it is a dynamic two-way relationship between individuals and the social and material context in which they participate (Brown and Duguid 1991).

In the discourse of knowing-in-between authors, the social aspect takes the lead in learning understanding that is described as stemming from the ongoing conversation among practitioners on what makes a practice good or how it needs to be changed (Gherardi 2008). This approach proposes the idea that knowing and learning are not only based on social interactions but also on discursive social behaviour, which gives rise to social order. This social order involves plurality and diversity, human and non-human actants and instances of negotiating and resolving differences (Nicolini et al. 2003). Learning is understood as being mediated by comparison among the perspectives of all the co-participants in practices that also involve symbols, meanings, artefacts and any other mediated objects (Gherardi and Nicolini 2000, 2002; Nicolini 2011; Nicolini et al. 2012). The understanding of learning underpins the interest of knowing-in-between authors in the power and ordering actions (Nicolini 2011). The knowing-in-between approach well illustrates the collective achievement that allows practitioners to evaluate the various performances of their working practices and in so doing to contest or negotiate a new order of actions, i.e. new practices. What becomes apparent is that learning is sustained and created in multiple interactions and social practices that always proceed through collective mediated and coordinated actions.

The expansive learning approach (Blackler et al. 2000; Engeström 1987, 2004, 2007) stresses that human capacities develop and abilities emerge in collaboration with others in the co-construct of social reality. The theory takes into account the social structures and systems dynamics and focuses on the emancipation of thought and the creation of new, collective skills, activity and practices. It is described as possible that, in certain potential learning episodes, learners unite with specific situational solutions and ideas of general systemic change. Thus, the personal and the collective, the immediate and the future oriented can merge, and the actors can move between cognition, actions and activity, and among past, present and envisioned futures, and transform the activity system (Engeström 2007).

Moreover, learning among different activity systems is needed today that further emphasises the importance of social aspects of learning and knowing. Engeström (2004, 2007) calls this "co-configuration" (see also Victor and Boynton 1998) that requires continuous collaborative learning and the creation of solutions according to customers' and users' needs. This requires flexible "knotworking" (viz. distributed collaborative expertise in pursuit of a task) that takes place in heterogeneous patchworks, unnoticed and noticeable actions, in which no single actor has the sole, fixed authority or ready-made rules.

Contribution to Innovation Debate

When it comes to innovation, to the dynamic perspective of change in practice, knowing and learning, the practice-based learning approaches see that human capacities develop and abilities emerge in collaboration with others in the co-construct of social reality. In this way, they shed more light on the innovations from the perspective that they cannot be thought of as the result of a "single act" of "a single actor" but more as the collective activities and common capabilities of multiple actors and activity systems operating in a novel way. Also, these perspectives contribute better to an understanding of innovation as a process that does not consist of a simple distribution of activities and works among different competent actors in a settled and fixed context. The emergence of activities derives from interaction among multiple actors which unfolds step by step, and innovation is constituted as a collective activity of innovating where the distribution of activity and tasks has to be constantly and collectively reproduced in a context.

5.4.3 Attention to the Unorderly

Although much of the idea of practices has built on the common idea of routine and standards (Schatzki 2005), the knowing-in approach offers a useful route into the conceptualisation of change as an endogenous process of practising. This approach has developed the notion of knowing and learning, which assumes that individual performances of practices play a key role in both producing and reproducing practices where the changes (unorderly) in practices are one of the possibilities produced through the variation of practice enactment (Brown and Duguid 1991, 2001). The distinction between practices as a canonical (or espoused) and a non-canonical (divergent from conventions) task runs throughout Brown and Duguid's (1991) arguments to clarify the link between working (doing) and innovating (the emerging of something new). Practice is said to be a mode that is relatively stable in time and a socially recognised way of doing in interaction so that they always remains attentive to emerging events (Wenger 1998, 2000). So, while work practices often follow the prescribed lines of canonical communities, learning is more likely to emerge across the fault lines of working activities. Learning is seen as a sort of bridge between working and innovating, i.e. learning refers to development that emerges in practical accomplishment and is deeply implicated in the series of canonical and non-canonical practices that take place in performing working activities (Brown and Duguid 1991, 2001).

In contradiction to the long-standing emphasis on stability and routine, knowing-in-between scholars start with the position that changes are a constituent element of practices and are continually produced (Nicolini 2009, 2011). They emphasise the fact that learning emerges by collectively knowledgeable actions that forge relations and connections among the available resources and constraints present in a social interconnected context (Gherardi 2009a, b; Nicolini 2011). The view of knowing based on a social nexus of practices is concerned with the socially negotiated character of learning that develops as an unstable provisional accomplishment in the becoming of practices (Bjørkeng et al. 2009; Nicolini 2009, 2011).

In the expansive learning approach (Engeström 2004, 2007), internal and external contradictions and tensions between the current way of acting and the needs for the future work as the initiator of change and development—i.e. expansive learning. The motivation behind expansive learning is thus often the conscious and collective attempts to expand the object of activity in order to better meet future needs in a process that is not exactly known beforehand but triggered by conflicts and uncertainties (Engeström 1987, 2004, 2007). This also means that work organisations do not act in isolation but with constant interaction with the environment that also creates the possibilities and needs for change. Moreover, the organisations currently need novel ways to cope with more rapidly changing environment, and thus, they need to learn novel ways of "how-to-learn". To solve this, an unorderly and time-consuming learning process is needed where the activity system resolved its internal and external contradictions by constructing and implementing a qualitatively new way of functioning for itself. In this process, people collectively come to recognise tensions, reflect on their origins, discuss alternatives and collectively refashion an activity system around the reconceived object of activity (Blackler et al. 2000; Blackler and Regan 2009). In this process, contradictions between the internal ways of acting meet the external needs.

Contribution to Innovation Debate

If we think of innovations, usually they are needed to respond to the changing needs of the market, to change the logic of a market or even to create novel markets. Thus, there are tensions (Engeström 1987) or "something that is going to happen" (Brown and Duguid 2001; Gherardi and Nicolini 2002) that trigger the need to change the existing ways of acting (Nicolini et al. 2003) and the system dynamics of the activity systems (Engeström 1987). Then, how and with what kind of processes can organisations respond to these needs?

Both knowing-in and knowing-in-between approaches start from the assumption that in practice-based processes, there is a common practical context with working activities to be executed, and the alignment and order—emerging around both sustained and contested practices—have to be (re)established continuously (i.e. common problems to be solved). Expansive learning approach describes the fact that, typically, innovation requires constructing and implementing a qualitatively new way of functioning for the activity system(s) in a process that is not exactly known beforehand but requires learning back and forth between the different development phases (Engeström 2007).

Altogether, the unorderly is present in the learning approaches in two ways: First, the tensions and contradictions in the current way of working and towards the environment provide the motive for the learning. Second, the process of learning is very much unorderly. Although some pre-existing phases can be recognised, learning back and forth between different phases is recognised, referring also to innovating as an emergent, uncertain and recursive phenomenon (Toivonen and Tuominen 2009; Sundbo and Toivonen 2011).

5.4.4 Knowledge as Situated in a Spatio-temporal Context

The knowing-in approach builds on the assumption of an essential relation between individual and context explanations of learning and on the notion that context and contextualised entities constitute one another (Cook and Yanow 1993). Learning is located in the context in which it is performed, and it is specific to particular contexts. To understand particular knowing and learning in practice, it is necessary to look beyond the observable social or human relations and to look at broader social patterning (Cook and Brown 1999). The context is much more than simple containers of work activity; it solicits actions and provides resources for actions. It becomes apparent that, in different contexts, some practices are more dominant than others and that knowing and learning are significant for the objects and institution (norm, rules, meanings, schemas) in which events are located.

In the knowing-in-between approach, the actors always undertake their action within multiple historical and cultural settings (Nicolini 2011). The dimension of the provisional and of the historically situated dimension is combined in the everyday knowing and learning of actors through a dynamic articulation and negotiation on what has to be done, how to do it and with whom to build the future action or to discern and resolve specific problems (Gherardi 2008). The sociocultural factors influence how these skills are manifested in a particular circumstance, but they also constantly evolve according to the participants involved in it, as they know how to move in space and time with different practices, materials, rules, etc.

In expansive learning, it is stressed that knowledge and knowing are not separable from the knowers. Hence, although a certain physical and material world exists and can embody knowledge, it cannot be known in one single way. Furthermore, knowledge and learning do not exist independently of the actors, but are constructed in people's interpretations and activity (Engeström 1987; Vygotsky 1978). "Activity" is defined as systemic by its nature and comprises the individual "action" and the collective "interaction". Thus, for knowledge and learning to have a meaning, they must always stem and belong to a specific space and time.

Contribution to Innovation Debate

All of the three learning approaches studied stress that knowledge and learning are understood as local, contextual processes that arise and are situated in the cultural-historical context, time and work activities. In terms of innovations, this causes us to wonder whether some innovations and their processes are the same. For example, it is argued (Orlikowski 2002) that for "best practices" to be meaningful, they have to be created in the local contexts, as a common effort of the actors, and that they can only be supported by developing people's capacity to enact in their own particular, local situations with "useful" rather than "best" practices (Orlikowski 2002). Consequently, we could assume that the creation of innovations is greatly embedded in the social, situated work activities, and they can accumulate from everyday actions (Ellström 2010; Melkas and Harmaakorpi 2012; Orlikowski 2002).

This fuller understanding of the context is necessary as innovation especially in service contexts increasingly involves activities addressing complex interactions between human actors (producers, users and other actors), outputs (services), tools (languages, objects) and other devices (infrastructures) with increasing space and time divided among the actors. The understanding of the new way of practising and changes in practices, which can be an innovation, requires the need to see what the changes are built on and to question how the newness is related with an old and situated way of doing (Nicolini 2011). Acceptable new knowledge and practice in one context may be quite inappropriate in another, as competence could be different and the goals for performance may be quite distinct. These factors influence not only the activities that actors engage in innovation but also what is taken to be the necessary competences to be addressed (Swan et al. 2002, 2007). Hence, our discussion suggests that innovation and its processes are situated in sociocultural contexts, and for this some sort of collective competence enactment is required in order to make the particular innovation recognised and salient (Melkas and Harmaakorpi 2012; Nooteboom 2012).

5.4.5 Tools, Artefacts and Objects

Knowing-in scholars, in line with social theory scholars (Knorr-Cetina 1997; Lave and Wenger 1991), stress that one of the main characteristics of sociality is that it is widely mediated by objects and material artefacts (Knorr-Cetina 1997). The view of knowing as socio-material doing implies that materiality is integral to learning (Brown and Duguid 1991), as the social and the material are constitutively entangled in everyday life (Orlikowski 2002, 2006). Although the material dimension of objects matters, such materiality emerges when it enters into situated practices that are activated by social interaction. What is stressed is not the static sense of things (Orlikowski 2002), as an object can be acted on, and people can be in interaction with it. Yanow (2006) also refers to the artefact as a constitutive element of practices and that the term covers both material and immaterial phenomena; it may be language, acts and objects, and the category also includes values, beliefs and feelings. So, any artefact, as well as other material elements, technology and tools, are also seen as an achievement of learning as they allow the link between the past, present and future of social practicing or of acting at a distance (Orlikowski 2002). Learning is construed as an integral part of sociocultural and material interaction, in which various forms, of using and creating tools, objects and other material products, are seen as a relevant part of the doing work.

Knowing-in-between scholars address the topical role of object and artefact by stressing their capacity to serve as bridges between intersecting social and cultural worlds (Gherardi 2012a). The term "boundary object" has been used by Carlile (2002, 2004) to illuminate how an object can function as a constrainer or facilitator for knowing and any learning in and between interconnected communities of practices. He refers to the boundary objects as mechanisms acting at the interface of different knowledge domains and as an "integrating device" through which knowledge is transformed and collective learning can be achieved. The bridging function not only includes the linking together of practitioners and their activity but needs to include the aspect of transforming the knowledge and ideas being transferred, and at best it allows the widening of cognitive, cultural and temporal distances between different sociocultural settings. Star (2010) stresses how boundary objects can have a very different meaning in different communities. Similarly, Knorr-Cetina (2001) also points out the characteristics of object as to be open-question generating and complex, in order to open up further opportunities for learning and changes.

The important key concepts of expansive learning (Engeström 1987, 2004, 2007) relate to the view that human action is object oriented and that those actions take place in a sociotechnical activity system, in a specific historical situation, as communities with certain material and conceptual tools, rules and division on labour. The object of activity does not refer to a material object but "is heterogeneous and internally contradictory, yet enduring, constantly reproduced purpose of a collective activity systems that motivates and defines the horizon of possible goals and actions" (Leont'ev 1978). For example, the object of activity for a doctor can be the treatment of outcomes instead of just diagnosing the symptoms. When it comes to materiality, people aim to reach, reflect and refashion the object of the work with and through the use of certain tools that can be material (an artefact) or more abstract, such as conceptual signs or tools. In this way, material or conceptual tools and signs mediate intellectual and practical operations, reflection and learning that are understood as epistemic, cultural and collaborative activity. Also, when reaching or refashioning the object of work, real-time interpretation and synthesising of information between multiple parties are needed, and this also requires the creation of new dialogical and reflective tools, artefacts, collaboratively constructed rules and infrastructures (Engeström 2004, 2007).

Contribution to Innovation Debate

Altogether, the three practice-based learning approaches represent the kind of worldview that artefacts can embody knowledge but that the material and physical world cannot be known in a single way (Gherardi 2009a). Rather, it is in a constant state of construction and refashioning by people's interpretations and actions (Brown and Duguid 2001; Engeström 1987; Orlikowski 2002; Swan et al. 2007; Vygotsky 1978). According to these learning approaches, material or conceptual tools and artefacts mediate and embody intellectual and practical operations, reflection, learning and innovating that are understood as epistemic, cultural and collaborative activity. Virkkunen and Ahonen (2011) importantly state that the mediators do not, however, determine individuals' actions, but provide building blocks for them to creatively form actions and collective activity in order to reach their objective in a situation.

To summarise, objects and tools provide a link between the past and the present (i.e. knowing-in), a bridge or a way to mediate between different contexts (i.e. knowing-in-between) or something to shape and direct activity towards (expansive learning).

5.5 Summary and Discussion of the Learning Features to Contribute to the Innovation Debate

In this chapter, we aim to summarise and discuss the counterparts of a broad and novel service-based view of innovation as social human action and process of learning. We have moved from the three practice-based learning theories we named as knowing-in, knowing-in-between and expansive learning and have examined how they theoretically approach the processes on learning (i.e. changes and transformation in ongoing practices of groups of people in sociocultural and historically developed organisational contexts at workplaces, by exploring their key concepts).

In greater detail, three main research questions have guided our theoretical contribution. The first question concerns the analysis of the key concepts that characterise practice-based theories on learning. To address this question, we utilised and further elaborated the five common features in practice-based learning studies recognised by Nicolini et al. (2003) (see also Svabo 2009). The framework proved useful to scrutinise and compare the three approaches and provided a well-founded guide to further development.

The concepts we opened up in depth were practices, action, activities, learning process, knowledge, knowing, time and space, tools, artefacts and object of activity. In particular, our results foster and also renew the common features in practice-based studies of learning (see Table 5.2).

To summarise the process dimension, i.e. "where" learning occurs, the learning theories explained how learning emerges as an ongoing accomplishment in practices, nexus of practices and activity systems in interaction with the social and physical world. The social aspects of learning stress the argument that human capacities develop and abilities emerge in collaboration with others in the co-construct of social reality, i.e. "how" learning occurs. Furthermore, the unorderly issue provides evidence on the possibilities of some differences in the "why" learning unfolds. A more emergent and unconscious idea of learning comes from both the knowing approaches in comparison with a prominent conscious and object-oriented view of the learning mechanisms behind expansive learning theory. The spatio-temporal issues detect the role of context and situatedness of learning, and the reason is needed why a fuller understanding of how each element of context has a place in relation to the others. In a way, it discusses issues of "who" and "in which context" learning occurs. Finally, the artefacts and materiality take centre stage in the tales of practices and learning, i.e. the role of objects and tools. In practice discourse, the object, tools and artefacts can embody knowledge, but the material and physical world can be known only in a social, unique and collective way.

However, in the search for a more robust understanding of practice-based learning approaches, we find it useful also to ponder the question of "what" learning can produce as an outcome. We consider that this element could be the "sixth" clarifying element to add to the five key elements of Nicolini et al. (2003) (see also Svabo 2009). Yet, some differences emerge among the approaches on this point. In the knowing-in perspective, the focus is on how practice is produced and reproduced within the community of its practitioners; the knowing-in-between illuminates the collective achievement that allows practitioners to evaluate the varying performance of their working practices and in so doing to contest or negotiate a "new order of actions", i.e. new practices. Expansive learning then focuses on the reconceived object of activity as well as activity systems and its elements such as new reconceived tools.

The second research question relates to the contribution of key practice-based learning concepts to the process of creating service innovation. We synthesise the

Table 5.2 Key learning elements and their contribution to service-based view of innovation

Approach and key elements	Key concepts in practice-based learning	Contribution to a broad service-based view of innovation and its processes		
		Knowing-in	Knowing-in-between	Expansive learning
1. Process dimension: where learning occurs	Practice	Nexus of practices	Activity systems	Focus on innovation that is accomplished in practice, nexus or practices and activity system enacted in a daily working activity
2. Social aspects: how learning unfolds	Emerges through relationship between individuals and the social and material context	Is sustained and created in multiple interactions proceeding through collective mediated and coordinated actions	Emerges in collaboration with others in the co-construct of social reality. It is dynamic, profound and conscious	Innovation is constituted as a collective (social) action where the distribution of activity and tasks has to be constantly and collectively reproduced in a context
3. Attention to the unorderly; why learning emerges	Unconscious and conscious attempt to acquire and produce knowledge	Unconscious and conscious attempt in organising resources in a nexus of practices	Triggered by contradictions and uncertainties in the current activity, conscious attempt to expand the object of activity to better meet the needs for the future	Contradictions give the motive for learning and innovation. This implies that innovation is an emergent, uncertain and recursive phenomenon
4. Knowledge situated in a spatio-temporal context: who learns and in which context	Individual in communities or among them; learning is located in a context	Communities of communities: individual/collective subjectivity within multiple historical and cultural settings	Individuals and communities learn in activity systems or between them and learning belongs to and evolves in a specific space and time	Innovation is situated in socio-cultural contexts, and some sort of collective competence enactment is required to make particular innovation recognised and salient
5. Material and artefactual aspects: the role of objects and tools	Provide a link between the past and the present	A bridge or a way to mediate between different contexts	Something to shape and direct the activity towards	Material or conceptual tools and artefacts mediate and embody intellectual and practical

(continued)

Table 5.2 (continued)

Approach and key elements	Key concepts in practice-based learning	Contribution to a broad service-based view of innovation and its processes		
		Knowing-in	Knowing-in-between	Expansive learning
				operations. Innovation is understood as epistemic, cultural and collaborative activity
6. Outcome aspect: what learning produces as an outcome	Production and reproduction of practice	New practices	Reconceived object of activity, novel meaning schemas, tools, activity and practices, as well as reorganised activity systems	Innovations could be perceived as production and transformation of practices in a context

results and contributions under the five key elements recognised by Nicolini et al. (2003; also Svabo 2009) and the added sixth clarifying element of "what" (see Table 5.2).

At first, in the service-based innovation debate, a need is recognised to create greater understanding from the un-linear and messy processes of innovation that have been seen more often to be reminiscent of learning processes (Calborg et al. 2014; Gottfridsson 2010, 2012; Kallio and Lappalainen 2014; Kallio 2015; Payne et al. 2008; Russo-Spena et al. 2014). By opening up the process dimension of the three learning perspectives chosen, we contribute to this debate by giving more light and emphasising theoretically the process ("where"), social ("how"), unorderly ("why"), spatio-temporal ("who" and in which "context") and material ("objects and tools") dimensions of innovations.

Hence, process features of learning contribute to the debate on service innovation by making clear that innovation is accomplished in practice and enacted in a daily working activity setting (Ellström 2010; Gherardi 2008, 2014; Sundbo and Toivonen 2011). Most importantly, learning discussion sees that innovation takes place in a continuum—they are not dichotomic concepts. In other words, change does not happen separately from the activity, as it is also beginning to be understood in the recent service innovation literature (Gottfridsson 2010, 2012; Payne et al. 2008). Depending on the situation, for innovation to take place, it requires a minor or major change in current, contextual and embedded everyday activity (Ellström 2010; Sørensen et al. 2013). In this way, it could be seen that innovation processes take place in the form of a collective learning process aiming for qualitative change and development in practice.

Furthermore, in the innovation debate, there has been increasing attention given to social aspects (Edvardsson and Tronvoll 2013) and novel resource integration in value co-creation (Lusch and Nambisan 2015). From a learning perspective, the main contribution to this aspect of innovation is the co-construction of social reality and context. In terms of expansive learning (Engeström 2004), this means co-constructing the object of activity and the elements of activity systems. In general, it refers to the phenomenon that the context and object are not preordained and given, but they are developed alongside the social learning and innovation process and the development of common capabilities in a work environment (Orlikowski 2002).

In service-based innovation literature, it is noticed that the creation of innovation can be based on contradictions and tensions. However, it is also noticed that innovation can create those contractions and tensions as well as novel logics and markets (e.g. the luxury industry and experiential economy) (Vargo and Lusch 2008). When looking at this phenomenon from the point of view of learning theories, we have found that learning can be based on a need to overcome a contradiction or a tension in a current way of acting and to develop an activity, i.e. the "why". The contradictions thus give the motive for the development in an unorderly process that is emergent and uncertain and that cannot be known exactly beforehand. Similarly, innovation is often seen to require constructing and implementing a qualitatively new way of functioning.

Moreover, in recent innovation theories, there is the beginning of an understanding that innovation is increasingly a contextual and situated phenomenon (Melkas and Harmaakorpi 2012; Orlikowski 2002; Sørensen et al. 2013; Edvardsson and Tronvoll 2013). These newest ideas, at the general level, seem to be reminiscent of the socio-constructivist view on innovation and share rather similar epistemology with practice-based learning studies. However, the learning theories studied go into much greater depth in their arguments. The practice-based learning perspectives stressed that for knowledge, practice, activity or objects to have a meaning, they always stem from and are attached to a specific sociocultural context, space and time. In this way, we argue that innovations are also socially constructed phenomenon, emerging as a result of a social accomplishment situated in specific time and place (Gherardi 2012b; Orlikowski 2002).

In the recent service innovation debate, diffusion and scalability are also topics of current interest, for example, in terms of the service platforms and architecture of participation (Lusch and Nambisan 2015). From the learning studies perspective, they underline the kind of epistemology which states that artefacts and conceptual tools can embody knowledge (Gherardi 2009b), but this "knowledge" does not mean much without the social actors. They have always to be "reinvented" in a local sociohistorical context. However, we have discussed how those tools that are based more on "general principles" can be adapted and reflected as boundary objects (Carlile 2004). The role of these kinds of tools and artefacts is to give communities or activity systems some general principles or structure for collaboration, which becomes more crucial when our societies move towards knowledge and service economies (Calborg et al. 2014; Lundvall et al. 2002). However, they need to be flexible and adaptable according to a specific context; it becomes important how "strictly" one follows those (considered as given) or whether they are commonly co-constructed and negotiated. Table 5.2 summarises our attempt to provide an interpretative framework drawing together relevant concepts from practice-based learning fields and how they contribute to broad, service-based view of innovation and its processes.

Then, finally, our third research question dealt with the issue of what kind of (service) innovation conceptualisation emerges from the practice-based learning approach. If we try to crystallise the key contribution of the learning theories to innovation, we can start formulating the following kind of conceptualisation (Kallio 2015). First of all, in the light of the learning theories studied, innovation should refer both to the process and to an outcome of a changing activity in a daily work setting. The process of innovation is reminiscent of an unorderly learning process that is uncertain regarding its patterns and outcomes. It mostly refers to a multifaceted social process among multiple actors, both within and between organisations, situated in specific cultural-historical contexts. In other words, it seems that innovation tends to develop and emerge in relation with a context and takes place in collective activity mediated by object, signs and tools in the co-construct of social reality.

Usually some sort of collective, situated competence enactment and transformation are required to make an innovation recognised and salient. What is produced as an output often refers to new ways of acting, that is, new practices, competences, resource integration, the received object of activity as well as activity systems and its elements such as conceptual or material tools. However, the novelty of the output is always relative.

5.6 Implications and Further Avenues of Research

Conceptualising the intrinsically reproductive and creative dynamics of the practice-based learning approaches we have studied, it proved helpful to break down the aspects of practices—sometimes used to describe repetitive routine activities—and shed light on the critical process of development and creation of something new in a better way (i.e. the processes of innovation).

As a practical implication, this study, based on different practice-based learning approaches, deepens our understanding of how innovation can take place differently in the collective work activity and how these different learning and innovation mechanisms support each other and are nonmutually exclusive (Ellström 2010; Kallio and Lappalainen 2014; Kallio 2015; Russo-Spena et al. 2014; Sørensen et al. 2013). For example, the knowing-in and the knowing-in-between approaches provide a basis for understanding innovation as something that happens alongside work activity, even without a conscious aim to create an innovation. This makes organisational practitioners and managers more closely acquainted with the fact that innovation cannot always be planned or known beforehand. Typically, it is a question of incremental innovations embedded in a collective, daily work activity. On the other hand, the learning mechanisms behind the expansive learning approach appear to be more normative, and the process is reminiscent of a transformational organisational learning process (Kallio 2015; Russo-Spena et al. 2014). It means that manager needs to have an acquaintance with the aim and motive for learning as well as with conflicts and uncertainties at the basis of changes in the activity. This is aligned with more transformational innovation, though it is a time-consuming process and often needs outside facilitators to guide the process.

Therefore, our study produces new theoretical insights on different learning mechanisms, in order to enhance and facilitate different innovation activities simultaneously within and between organisations in a purposeful manner. When different learning mechanisms and their contributions to alternative innovation processes and outcomes are made visible, this enables the utilisation of resources and innovation potentials of local communities more effectively and efficiently. Thus, as our theoretical and practical conclusion, we state that, by increasing in-depth understanding of different learning mechanisms, we can create better conditions for innovation to take place. The renewal capability is increasingly crucial for competitiveness at every level of service-, network- and knowledge-centric society and might provide the means to build a competitive edge.

As the main contribution and implication of the three different learning streams to the broadened view on innovation and innovating, we found that these learning theories show that the kind of development and learning that can also contribute to innovation is seen to be created through the changes in practical and situated activity. Thus, we can conclude that their emergence can be supported by inspiring human activities, common agency in local contexts and the creation of broader horizons for the potential activity (Ellström 2010; Engeström 2004, 2007; Kallio and Lappalainen 2014; Kallio 2015; Orlikowski 2002; Russo-Spena et al. 2014). The task of the managers then becomes that of enablers of those conditions relevant for learning and innovating. However, the learning process and its development must be collectively created by local actors as beneficiaries. Thus, both the practitioners (as organisational developers, managers and others interested in fostering innovation) and academic scholars could direct a stronger focus on understanding and support the practical efforts of what people do and how they do it and what the consequences are of their doing it. It would therefore seem that, by challenging and supporting people in renewing these everyday patterns, we could also produce the kinds of changes that can be regarded as innovations.

In this way, by focusing on analysing the practice-based studies on learning, this chapter brings new insight into the topical phenomena about complex social mechanisms of learning in order to enhance and facilitate innovation activities within and between organisations in a purposeful manner. The summary in Table 4.2 could be applied as a reference frame when reflecting current activities and exploring alternative ways to innovate within and between organisations in order to enhance and create a competitive edge.

However, we consider this work to be only one step towards better framing innovation with a learning, practice and service mindset. The framework we have debated and expanded offers some highlights that have not been addressed thoroughly so far in the literature. We accordingly call on innovation researchers and practitioners to frame new innovation mindsets, namely, those that take into account the practices and systems of activity situated in cultural context in which actors interact. The call addresses practice- and learning-based approaches to innovation to contribute further both to the theoretical debate and to practice by providing guidelines to organisations and practitioners.

References

Alam, I., & Perry, C. (2002). A customer-oriented new service development process. *Journal of Services Marketing, 16*(6), 515–534.

Bjørkeng, K., Clegg, S., & Pitsis, T. (2009). Becoming (a) practice. *Management Learning, 40*(2), 145–159.

Blackler, F., Crump, N., & McDonald, S. (2000). Organizing processes in complex activity networks. *Organization, 7*(2), 277–300.

Blackler, F., & Regan, S. (2009). Intentionality, agency, change: practice theory and management. *Management Learning, 40*(2), 161–176.

Brown, J. S., & Duguid, P. (1991). Organizational learning and communities-of-practice: Toward a unified view of working, learning, and innovation. *Organization Science, 2*(1), 40–57.

Brown, J. S., & Duguid, P. (2001). Knowledge and organization: A social-practice perspective. *Organization Science, 12*(2), 198–213.

Calborg, C., Kindström, D., & Kowalkowski, C. (2014). The evolution of service innovation research: A critical review and synthesis. *The Service Industries Journal, 34*(5), 373–398.

Carlile, P. R. (2002). A pragmatic view of knowledge and boundaries: Boundary objects in new product development. *Organization Science, 13*(4), 442–455.

Carlile, P. R. (2004). Transferring, translating, and transforming: An integrative framework for managing knowledge across boundaries. *Organization Science, 15*(5), 555–568.

Chesbrough, H. (2011). *Open services innovation: Rethinking your business to grow and compete in a new era.* [Hardcover]. San Francisco: Wiley.

Cook, S. D., & Brown, J. S. (1999). Bridging epistemologies: The generative dance between organizational knowledge and organizational knowing. *Organization Science, 10*(4), 381–400.

Cook, S. D. N., & Yanow, D. (1993). Culture and organizational learning. *Journal of Management Inquiry, 2*(4), 373–390.

Cooper, R. G. (1994). Third-generation new product processes. *Journal of Product Innovation Management, 11*(1), 3–14.

Corradi, G., Gherardi, S., & Verzelloni, L. (2010). Through the practice lens: Where is the bandwagon of practice-based studies heading? *Management Learning, 41*(3), 265–283.

Dougherty, D. (2004). Organizing practices in services: Capturing practice-based knowledge for innovation. *Strategic Organization, 2*(1), 35–64.

Edvardsson, B., & Olsson, J. (1996). Key concepts for new service development. *The Service Industries Journal, 16*(2), 140–164.

Edvardsson, B., & Tronvoll, B. (2013). A new conceptualization of service innovation grounded in SD logic and service system. *International Journal of Quality and Service Sciences, 5*(1), 19–31.

Ellström, P. E. (2010). Practice-based innovation: A learning perspective. *Journal of Workplace Learning, 22*(1/2), 27–40.

Engeström, Y. (1987). *Learning by expanding: An activity theoretical approach to developmental research.* Jyväskylä: Orienta-konsultit.

Engeström, Y. (2001). Expansive learning at work: Toward an activity theoretical reconceptualization. *Journal of Education and Work, 14*(1), 133–156.

Engeström, Y. (2004). New forms of learning in co-configuration work. *Journal of Workplace Learning, 16*(1/2), 11–21.
Engeström, Y. (2007). Enriching the theory of expansive learning: Lessons from journeys towards co-configuration. *Mind, Culture and Activity, 14*(1/2), 23–39.
Engeström, Y., Miettinen, R., & Punamäki, R. L. (1999). *Perspectives on activity theory*. Cambridge: Cambridge University Press.
Engeström, Y., & Sannino, A. (2010). Studies of expansive learning: Foundations, findings and future challenges. *Educational Research Review, 5*(1), 1–24.
Finfgeld, D. L. (2003). Metasynthesis: The state of the art- so far. *Qualitative Health Research, 13* (7), 893–904.
Gherardi, S. (2000). Practice-based theorizing on learning and knowing in organizations. *Organization, 7*(2), 211–223.
Gherardi, S. (2006). *Organizational knowledge. The texture of workplace learning*. Oxford: Blackwell.
Gherardi, S. (2008). Situated knowledge and situated action: What do practice-based studies promise. In D. Barry & H. Hansen (Eds.), *The SAGE handbook of new approaches in management and organization*. London: Sage.
Gherardi, S. (2009a). Knowing and learning in practice-based studies: An introduction. *The Learning Organization, 16*(5), 352–359.
Gherardi, S. (2009b). Introduction the critical power of the practice lens. *Management Learning, 40*(2), 115–128.
Gherardi, S. (2012a). Situated knowledge and situated action: What do practice-based studies promise. In S. Gherardi & A. Strati (Eds.), *Learning and knowing in practice-based studies*. Northampton, MA: Edward Elgar.
Gherardi, S. (2012b). Why do practices change and why do they persist? Models of explanations. In P. Hager, A. Lee, & A. Reich (Eds.), *Practice, learning and change: Practice-theory perspectives on professional learning, professional and practice-based learn*. Australia: Springer.
Gherardi, S. (2014). *How to conduct a practice-based study: Problems and methods*. Northampton, MA: Edward Elgar.
Gherardi, S., & Nicolini, D. (2000). To transfer is to transform: The circulation of safety knowledge. *Organisation, 2*(Special Issue), 329–334.
Gherardi, S., & Nicolini, D. (2002). Learning in a constellation of interconnected practices: Canon or dissonance? *Journal of Management Studies, 39*(4), 419–436.
Gherardi, S., & Strati, A. (2012). *Learning and knowing in practice-based studies*. Northampton, MA: Edward Elgar.
Gottfridsson, P. (2010). Development of personalised services in small business: An iterative learning process. *Managing Service Quality, 20*(4), 388–400.
Gottfridsson, P. (2012). Joint service development – The creations of the prerequisite for the service development. *Managing Service Quality, 22*(1), 21–37.
Greer, C., & Lei, D. (2012). Collaborative innovation with customers: A review of the literature and suggestions for future research. *International Journal of Management Reviews, 14*(1), 63–84.
Kallio, K. (2015). *Collaborative learning with users as an enabler of service innovation*. Doctoral dissertation in VTT science series, Espoo, Finland (VTT science 110).
Kallio, K., & Lappalainen, I. (2014). Learning about service-orientation in KIBS: Understanding the customer as an active actor. *Service Science, 6*(2), 78–91.
Knorr-Cetina, K. (1997). Sociality with objects: Social relations in postsocial knowledge societies. *Theory, Culture and Society, 14*(4), 1–30.
Knorr-Cetina, K. (2001). Objectual practice. In T. R. Schatzki, K. Knorr Cetina, & E. von Savigny (Eds.), *The practice turn in contemporary theory*. New York, NY: Routledge.
Latour, B. (1987). *Science in action. How to follow scientists and engineers through society*. Cambridge, MA: Harvard University Press.

Lave, J., & Wenger, E. (1991). *Situated learning. Legitimate peripheral participation*. Cambridge, MA: University Press.

Leont'ev, A. N. (1978). *Activity, consciousness, and personality*. Englewood Cliffs: Prentice-Hall.

Lundvall, B.-Å. (1992). *National systems of innovation: Towards a theory of innovation and interactive learning*. London: Pinter Publishers.

Lundvall, B.-Å., & Johnson, B. (1994). The learning economy. *Journal of Industry Studies, 1*(2), 23–42.

Lundvall, B., Johnson, B., Andersen, E. S., & Dalum, B. (2002). National systems of production, innovation and competence building. *Research Policy, 31*(2), 213–231.

Lusch, R. F., & Nambisan, S. (2015). Service innovation: A service-dominant (S-D) logic perspective. *Management Information Systems Quarterly, 39*(1), 155–175.

Magnani, L. (2001). *Abduction, reason and science: Processes of discovery and explanation*. New York: Springer.

Mele, C., Colurcio, M., & Russo-Spena, T. (2014). Research traditions of innovation: Goods-dominant logic, the resource-based approach, and service-dominant logic. *Managing Service Quality, 26*(6), 612–642.

Mele, C., & Russo-Spena, T. (2015). Innomediary agency and practices in shaping market innovation. *Industrial Marketing Management, 44*(1), 42–53.

Melkas, H., & Harmaakorpi, V. (2012). *Practice-based innovation: Insights, applications and policy implications*. Berlin, Heidelberg: Springer.

Michel, S., Brown, S. W., & Gallan, A. S. (2008). An expanded and strategic view of discontinuous innovations: Deploying a service-dominant logic. *Journal of the Academy Marketing Science, 36*(1), 54–66.

Miettinen, R. (2002). *National innovation system. Scientific concept or political rhetoric?* Helsinki: Edita.

Nicolini, D. (2009). Zooming in and out: Studying practices by switching theoretical lenses and trailing connections. *Organization Studies, 30*(12), 1391–1418.

Nicolini, D. (2011). Practice as the site of knowing: Insights from the field of telemedicine. *Organization Science, 22*(3), 602–620.

Nicolini, D., Gherardi, S., & Yanow, D. (2003). *Organizational knowledge as practice*. Armonk, NY: ME Sharpe.

Nicolini, D., Mengis, J., & Swan, J. (2012). Understanding the role of objects in cross-disciplinary collaboration. *Organization Science, 23*(3), 612–629.

Nonaka, I., & Takeuchi, H. (1995). *The knowledge-creating company*. Oxford: Oxford University Press.

Nooteboom, B. (2012). A pragmatist theory of innovation. In H. Melkas & V. Harmaakorpi (Eds.), *Practice-based innovation: Insights, applications and policy implications*. Berlin, Heidelberg: Springer.

Ordanini, A., & Parasuraman, A. (2011). Service innovation viewed through a service-dominant logic lens: A conceptual framework and empirical analysis. *Journal of Service Research, 14* (1), 3–23.

Orlikowski, W. J. (2002). Knowing in practice: Enacting a collective capability in distributed organizing. *Organization Science, 13*(3), 249–273.

Orlikowski, W. J. (2006). Material knowing: The scaffolding of human knowledgeability. *European Journal of Information Systems, 15*(5), 460–466.

Payne, A. F., Storbacka, K., & Frow, P. (2008). Managing the co-creation of value. *Journal of Academy Marketing Science, 36*(1), 83–96.

Russo-Spena, T., Kallio, K., Lappalainen, I., & Mele, C. (2014). Knowledge and learning in practice-based studies. In L. Freund & W. Cellary (Eds.), *Advances in human factors and service engineering*. Cracovia: CRC Press.

Russo-Spena, T., & Mele, C. (2012a). Five Co-s' in innovating: A practice-based view. *Journal of Service Management, 23*(4), 527–553.

Russo-Spena, T., & Mele, C. (2012b). Creativity and learning in the practices of service innovation. In J. C. Spohrer & L. E. Freund (Eds.), *Advances in human factors and service engineering*. San Francisco: CRC Press; Taylor & Francis Group.

Schatzki, T. (2005). The sites of organizations. *Organizations Studies, 26*(3), 465–484.

Schatzki, T. R., Knorr-Cetina, K., & Von Savigny, E. (2001). *The practice turn in contemporary theory*. London and New York: Routledge.

Sørensen, F., Sundbo, J., & Mattsson, J. (2013). Organisational conditions for service encounter-based innovation. *Research Policy, 42*(8), 1446–1456.

Star, L. S. (2010). This is not a boundary object: Reflections on the origin of a concept. *Science, Technology and Human Values, 35*(5), 601–617.

Suchman, L. (2000). Organizing alignment: The case of bridge-building. *Organization, 7*(2), 311–327.

Sundbo, J., & Toivonen, M. (2011). *User-based innovation in services*. Cheltenham: Edward Elgar.

Svabo, C. (2009). Materiality in a practice-based approach. *The Learning Organization, 16*(5), 360–370.

Swan, J., Bresnen, M., Newell, S., & Robertson, M. (2007). The object of knowledge: The role of objects in biomedical innovation. *Human Relations, 60*(12), 1809–1837.

Swan, J., & Scarborough, H. (2005). The politics of networked innovation. *Human Relations, 58* (7), 913–943.

Swan, J. A., Scarbrough, H., & Robertson, M. (2002). The construction of communities of practice in the management of innovation. *Management Learning, 33*(4), 477–497.

Syson, F., & Perks, H. (2004). New service development: A network perspective. *Journal of Services Marketing, 18*(4), 255–266.

Toivonen, M. (2010). Different types of innovation processes in services and their organizational implications. In F. Gallouj & F. Djellal (Eds.), *The handbook of innovation and services* (pp. 221–249). Cheltenham: Edward Elgar.

Toivonen, M., & Tuominen, T. (2009). Emergence of innovations in services. *The Service Industries Journal, 29*(7), 887–902.

Vargo, S., & Lusch, R. (2004). Evolving to a new dominant logic for marketing. *Journal of Marketing, 68*(1), 1–17.

Vargo, S., & Lusch, R. (2008). Service-dominant logic: Continue the evolution. *Journal of the Academy of Marketing Science, 36*(1), 1–10.

Vargo, S., Wieland, H., & Archpru Akaka, M. (2015). Innovation through institutionalization: A service ecosystems perspective. *Industrial Marketing Management, 44*(9), 63–72.

Victor, B., & Boynton, A. C. (1998). *Invented here: Maximizing your organization's internal growth and profitability*. Boston: Harvard Business School Press.

Virkkunen, J., & Ahonen, H. (2011). Supporting expansive learning through theoretical-genetic future challenges. *Educational Research Review, 5*(1), 1–24.

Vygotsky, L. S. (1978). *Mind in society: The psychology of higher mental function*. Cambridge: Harvard University Press.

Wenger, E. (1998). *Communities of practice: Learning, meaning, and identity*. Cambridge: Cambridge University Press.

Wenger, E. (2000). Communities of practice and social learning systems. *Organization, 7*(22), 225–246.

Yanow, D. (2000). Seeing organizational learning: A 'cultural' view. *Organization, 7*(2), 247–268.

Yanow, D. (2006). Studying physical artifacts: An interpretive approach. In M. G. Pratt & R. Anat (Eds.), *Artifacts and organizations – Beyond mere symbolism*. Mahwah, NJ: Lawrence Erlbaum.

Zollo, M., & Winter, S. G. (2002). Deliberate learning and the evolution of dynamic capabilities. *Organization Science, 13*(3), 339–351.

Part II
Perspectives: A Practice-Based View

Chapter 6
Innovating as a Texture of Practices

Cristina Mele and Tiziana Russo-Spena

Abstract This chapter focuses on the social ways of doing business innovation in actual conditions of complexity. By applying the concept of practical accomplishment, innovating is seen as "a texture of practices," that is, a set of practices resting on other practices. We use the metaphor of texture to develop a fresh conceptualization of innovating. A texture is made up of a weft and a warp; the warp concerns the vertical threads that run along the length of the yardage and the weft refers to the horizontal threads that run from side to side. In innovating as a texture, the weft represents the practice of co-creating and the warp represents the practice of weaving. The multiple connections in actions that arise at the cross points of warp and weft allow the big picture to emerge. This conceptualization goes beyond the economic stance to emphasize the social–contextual nature of innovating, in which establishing relationships and giving them specific forms within a situated practice activate knowing. The focus shifts to the activities, actors, and resources involved in sociomaterial and cultural practices in which innovating occurs. In this realm, innovating emerges as a continuous process.

6.1 Introduction

> In a global market, where innovation can no longer be approached in the same way and with the same methods used in the past, a radical change in models, methods, instruments must be promoted in order to stimulate the realization of new forms of innovation. (CEO—Loccioni Group)

These words, from the CEO of an Italian company, have been echoed by scholars (Chesbrough 2003; von Hippel 2005) and government (Innovation Union within Horizon 2020 European Commission). Innovation is often recited as a

C. Mele (✉) • T. Russo-Spena
Department of Economics Management Institution, University of Naples Federico II, Naples, Italy
e-mail: cristina.mele@unina.it; tiziana.russospena@unina.it

T. Russo-Spena et al. (eds.), *Innovating in Practice*,
DOI 10.1007/978-3-319-43380-6_6

"mantra" to overcome crises in companies, organizations, and governments. But what does innovation mean?

The traditional linear process of the Stage-Gate model within NPD has been challenged by several authors (Chesbrough 2003; Russo-Spena and Mele 2012). In response, there has been a proliferation of multiple dichotomies to frame the phenomenon, including markets and technologies (Garcia and Calantone 2002), continuous and discontinuous (Tidd et al. 2005), goods and services (Vargo and Lusch 2004, 2011), exploitation and exploration (Geiger and Makri 2006), open and closed (Chesbrough 2003, 2006), firm centric and network centric (Möller and Rajala 2007), and competition and collaboration (Chesbrough 2003). There have also been calls to build new conceptual foundations for innovation because of the belief that previous guidelines are no longer valid (Mele et al. 2014). Scholars assert that new frameworks are needed to account for complex phenomena and overcome traditional dichotomies (Harrison and Kjellberg 2010; Papastathopoulou and Hultink 2012; Vargo et al. 2015).

In this chapter, we move beyond the dichotomy perspective to focus on the social ways of doing business innovation. When we study complexity, we look for patterns or structures that emerge from interactions among elements—"how such changes play out or ... the propagation of change through interconnected behaviour" (Arthur 2013: 11). To detect complexity in innovation, we adopt an epistemology of practices (Corradi et al. 2010; Gherardi 2015). In social sciences, adherents of the practice-based approach advocate the need to understand social phenomena by shifting attention from "nouns to verbs" (Nicolini 2011; Gherardi 2012b). This need has also been recognized in business studies, with authors addressing the opportunity to examine verb–actions rather than noun–entities (Kjellberg and Helgesson 2007; Johnson et al. 2003). In complexity economics, stable end-states (i.e., nouns) are regarded as misleading; the view is that verbs better express "a pattern, a set of emergent phenomena, a set of changes that may induce further changes" (Arthur 2013: 19).

Therefore, we propose moving away from innovation (the noun, meaning creation of a new artifact) to innovating (the verb, meaning the emerging process). The active verbal form better reflects the emergent character of innovation in service ecosystems (Vargo and Lusch 2011; Vargo et al. 2015). By adopting an overarching concept of service innovation as the provision of new benefits for market actors (Mele et al. 2010, 2014; Vargo and Lusch 2011), we question the linear traditional process of innovation (Chesbrough 2003) and ask: How do market actors practice innovation in the age of uncertainty and complexity? This approach is consistent with a postmodern style of thinking "which privileges action, movement, process and emergence rather than nouns, end-states, insulated, discrete social entities" (Chia 1995: 59).

We suggest practices as social phenomena are not fixed and stand-alone activities; they emerge as connections in action involving both human and nonhuman elements interwoven in a texture of interconnected practices (Gherardi 2006, 2012a; Reckwitz 2002; Schatzki 2002, 2005). The study of situated innovating practices allows us to understand that innovation does not develop as a set of prescribed stages but emerges in a continuous motion.

By applying the concept of practical accomplishment and by challenging traditional analytical categories, we propose that innovating can be regarded as a "texture of practices": a set of practices resting on other practices (Gherardi 2012a). This conceptualization allows us to go beyond an economic stance to emphasize the social–contextual nature of innovating in which distributed knowledge is activated by establishing relationships and giving them specific forms within a situated practice. We shift the focus to activities, actors, and resources involved in the sociomaterial and cultural practices in which innovating takes place.

Such an approach is coherent with the idea that theory does not affect "the discovery of theorems of undying generality, but the deep understanding of mechanisms that create these patterns and propagations of change" (Arthur 2013: 19). It addresses the need to develop frameworks that are characterized not by order, determinacy, and stasis but by "contingency, indeterminacy, sense-making, and openness to change," where "agents live in a world where their beliefs and strategies are constantly being 'tested' for survival within an outcome or 'ecology' these beliefs and strategies together create" (Arthur 2013: 5).

The chapter proceeds as follows: First, we address a conceptual background that moves from the notion of innovation as an outcome of sequential activities to the understating of co-creation and practices. Second, we outline our research design. Third, according to our research findings, we present a framework for innovating as a texture of practices. We conclude with discussion and implications.

6.2 Conceptual Background

6.2.1 Innovation as an Outcome of Sequential Activities

The traditional innovation model comes from evolutionary economics (post-positivism). It is a rational process (i.e., a sequence of stages and gates) in which the specific aim is the generation of new knowledge to be applied to produce new outcomes (goods and services) that can be marketed and utilized by customers and consumers (Kogut and Zander 1996). The scarcity of resources within a single company makes it necessary for the company to draw on a network of partners that is built on relationships and collaboration (Ritter and Gemünden 2003). The network may include other companies, research institutes, customers, consumers, and governments. It serves as a source of innovation and, in some cases, of co-production, but the perspective is firm focused; innovation follows a pattern of internal activities and closed processes (Möller and Rajala 2007). However, changes in the global and networked context in the last 10 years have spurred changes in the way that innovation is performed. New ideas about the meaning and development of innovation have emerged.

The concept of open innovation (Chesbrough 2003) has gained great prominence in academia and the business world. The opening of borders allows companies to acquire dispersed knowledge in the global context. Collaboration with external partners is regarded as crucial to value creation; networks possess latent technological and market potential (Chesbrough 2006; Enkel et al. 2009). A particular feature of this new collaboration is the role of the customer as coproducer. The user, newly empowered and directly connected, harnesses new developments in technology (e.g., the Internet, online communities) and brings ideas, knowledge, and resources to company innovation processes. Moreover, community-based innovation (Nambisan 2002; Nambisan and Nambisan 2008) emphasizes the opportunities provided by social platforms and the role of "crowds" (Howe 2008) in feeding innovation and unleashing creativity.

Notwithstanding efforts of scholars to overcome the limits of the traditional innovation process, some shortcomings remain. The epistemological stance of most studies is post-positivist and models are linear and rational. The capacity of a firm and its network to innovate is knowledge based; knowledge is regarded as an object to be accumulated and applied. Market actors are viewed as rational economic agents who have the ability to obtain, apply, and transform knowledge in new artifacts (i.e., a rational, deliberate process).

Innovation scholars address specific problems to be solved: how to tackle the unknown, how to identify problems, how to consider indeterminacy and uncertainty, how to make sense of emergence, how to increase openness to change, and how to address complexity (Edvardsson and Tronvoll 2013; Mele et al. 2014; Sundbo and Toivonen 2011).

6.2.2 Innovation as Co-creation by A2A Collaboration

There have been attempts to overcome the rational-positivistic approach to innovation through the effectuation approach (Read et al. 2009) and service-dominant (S-D) logic (Vargo and Lusch 2004, 2008), which offer new ways of co-creating value (Vargo et al. 2008) within a network of actors who integrate resources for mutual value, transforming external realities into new possibilities (Sarasvathy 2009).

A major contribution of effectuation is its focus on accomplishing innovation through an expanding cycle of resources rather than established goals and planned steps (Sarasvathy 2009). Goals emerge from courses of action and innovation is co-constructed by stakeholders involved in an iterative pattern (Sarasvathy and Dew 2005). This idea of co-construction resembles that of S-D logic in which the service ecosystem perspective emphasizes self-adjustment and implies the existence of a learning process that underlies innovation. Actor-to-actor collaboration is enabled by social practices within which actors foster "meaning-making, actions, and behaviour for mutual gain through service to service exchange" (Lusch and Vargo 2014: 137). Innovation is not merely an output; it is a new service provision, bringing new benefits to the involved actors. It changes or replaces practices and is regarded as the locus of resource integration and value co-creation (Russo-Spena and Mele 2012).

Akaka and Vargo (2014) have highlighted the importance of practices that extend the role of technology as an operant resource for innovation. Lusch and Nambisan (2015) emphasize the role of digital infrastructure in bringing together diverse actors and enabling collaboration in the ecosystem.

Although effectuation and S-D logic offer novel insights, they do not reframe innovation from a fresh perspective. A practice approach can extend these insights.

6.2.3 Innovation as a Knowing in Interconnected Practices

The study of practice has long been part of philosophy and sociology. Now it is being addressed in organization and management research (Gherardi 2012a), using an interpretivist approach.

Practices are shared ways of doing that take place within an ecology of connected elements: actions, individuals, contexts, artifacts, rules, symbols, texts, discourses, prior knowledge, and human and nonhuman actors, brought together in daily activities by processes of ordering, relationships, and performativity (Gherardi 2015). The interaction of the relationships between these elements gives rise to practice as "the outcome of conscious and unconscious, deliberate and emergent operations, processes and events that do not necessarily align in terms of goals, functions, timing, identities, roles, processes, and power relations" (Guzman 2013: 436). Practices are stabilized by repetition and supported by the historical, institutional, and cultural context (Bourdieu 2005; Gherardi 2012a, b; Nicolini et al. 2003).

A fresh understanding of doing and knowing arises from working practices; the epistemology stance changes (Gherardi 2012a) from possession (Cook and Brown 1999) to practice (Gherardi 2001). Knowledge is not a commodity—that is, not a static, embedded capability or stable disposition of actors—but rather is knowing, an ongoing social accomplishment within sociomaterial relationships that form as actors engage the world in practice (Orlikowski 2002, 2007). "The return to practice is part of the movement toward a relational, sociomaterial, or post-humanist epistemology because practice makes it possible to see and to represent a mode of ordering the social in which doing and knowing are not separated and the knowing subject and the known object emerge in their ongoing intra-action" (Gherardi 2015: 6).

By recognizing that practice-connecting relationships reach beyond the boundaries of individual organizations to communities, scholars are moving away from a structural conception of networks (the question of open/close disappears) to focus on interactions and connections in action (Gherardi 2012a, b; Nicolini 2007). Rather than analyzing separations and distinctions, Gherardi (2006, 2012a) proposes the concept of 'texture of practices' to reveal ongoing interactions and connections in actions within interwoven practices.

The perspective of knowing through interconnected actions offers a fresh perspective on how innovation unfolds through working practices (Orlikowski 2002). However, the practice approach to innovation is in its infancy and needs further development at the conceptual level and in the empirical context. Three main issues contribute to the innovation debate.

First, authors view knowing, doing, and innovating as interlinked (Brown and Duguid 2001): "...a doing while inventing the way of doing, that is the creative entanglement of knowing and doing" (Gherardi 2015: 8). Innovation is not simply an ad hoc activity within a discontinuous process, but a part of everyday activity; it becomes interwoven and inseparable from learning and working.

Second, by adopting relational thinking (Østerlund and Carlile 2005) to move away from problematic dichotomies such as subjectivism and objectivism (Giddens 1979), practice theory attributes agency to both humans and nonhumans. Both individuals and sociomaterial relationships are carriers of practices (Orlikowski 2007). Innovation is "the constant refinement of practice within an ecology of sociomaterial relations" (Gherardi 2012a: 218).

Third, practices are seen as a blend of material, mental, social, and cultural resources (Nicolini et al. 2003: 27), in which actions are performed through a network of connections in action (Gherardi 2012a, b). Callon (1999) addresses the emergent interconnected aspect of innovation:

> Innovation is by definition an emergent phenomenon based on gradually putting into place interactions that link agents, knowledge, and goods that were previously unconnected, and that are slowly put in a relationship of interdependence [...]. What marks innovation is the alchemy of combining heterogeneous ingredients: it is a process that crosses institutions, forging complex and unusual relations, the market, law, science, and technology. (Callon 1999: 2)

By applying the concept of texture to these three issues, we can offer a fresh conceptualization of innovation as "innovating in practice."

6.3 Research Design

To understand the social way of carrying out business innovation in actual conditions of complexity, we carried out an empirical investigation of "innovating in practice" from 2010 to 2014.

In our project, we assumed a practice point of view (Gherardi 2012a; Schau et al. 2009) and investigated practices as the unit of analysis because they "are loci—spatial and temporal—in which working, organizing, innovating ... occurs" (Gherardi 2012b: 2). Practices are not simply empirical objects; they also represent an epistemology close to relational materialism. The difference exists in "the attribution to practice of a realist ontology (that objectifies practices as primary units) and social constructionist conception that does not distinguish between the production of knowledge and construction of the object of knowledge (between ontology and epistemology)" (Gherardi 2012a: 77). In this view, practices are regarded not as practical phenomena defined by elements within boundaries but as sociomaterial relationships between humans and nonhumans that are shaped through interconnections. To understand the texture produced by these connections, we must follow the pattern of its warp and weft: "The researcher thus puts him/herself in the place of the weaver to follow how the threads are woven and knotted together to form a thicker or thinner texture" (Gherardi 2012b: 78).

The methodological principle of "follow the practices" implies that a researcher observes "a situated practice and moves up from it to the institutional order or moves down from it to the individual-in-situation" (Gherardi and Nicolini 2006, p. xviii). The aim is to trace the connections among actors and institutions, the contexts in which the connections take place, and the resources that are used (Nicolini et al. 2003; Gherardi 2012a, b).

At the beginning of our investigation, we focused on the practices of companies engaged in web-based innovation to account for the specific role of technology in affecting sociomaterial practices (Orlikowski 2007; Russo-Spena and Mele 2012). We used company websites to represent the encounter space (Payne et al. 2008) and the Internet to represent the enabler of contacts and interactions (Verona et al. 2006). Six months later, we identified other information-rich case studies (Piekkari et al. 2010) by using keywords such as "open innovation," "crowdsourcing," and "web-based innovation." In this way, we expanded our first collection of cases by adding new evidence identified according to a "snowball" process. Our results were formed from a judgment sample of 51 case studies (see Table 6.1), grouped into three categories: (1) suppliers of new solutions, (2) intermediaries of innovations—third party agents connecting searchers and providers of idea and solutions—and (3) online communities. We wrote a case report for each category, using a netnography approach and following the suggestions of Kozinets (2002) on obtaining data about actors in their online performance of innovating practices. We augmented the data collected with information from additional documents, case reports, and other related materials (Richardson 2010). In line with Carlile (2002), we studied innovation practices through the analysis of what actors do, which activities they perform and resources they use, and how they interact with other market actors, integrate resources to reach their aims, and give sense to their actions and relations. Our aim was to observe, understand, and report the ways in which practices are accomplished and describe the territories of actions, meanings, knowing, and artifacts within which they unfold.

6.4 Innovating as a Texture of Practices

The metaphor of texture is useful for developing a fresh conceptualization of innovating. In weaving, texture is composed of a weft and a warp; weft is the term for the thread or yarn that is drawn through the warp yarns to create cloth. Warp referes to the vertical threads that run along the length of the yardage. Weft concerns the horizontal threads that run from side to side. In innovating as

Table 6.1 Case studies: website and/or platform

Companies		Intermediaries	Community/ blog
Articulate	NIKEiD	100openNineSigma	Fiat
Blizzard Entertainment	Open Xerox	10EQS	IdeaConnection
BMW Co-creation Lab (until 2013)	P&G Connect + Develop	Idea Bounty	InnoCentive
Cisco	Plantronics	IdeaConnection	Innoget
Dell IdeaStorm	Samsung Digital Native (until 2013)	Ideo Labs	Lego
Electrolux Design Lab	Siemens	InnoCentive	Mulino Bianco
Fiat Mio project (until 2010)	Spreadshirt	Innoget	Nestlé
L'Oreal	Swarovski	Innovation Exchange	Nike
Lego Mindstorms	Threadless	OpenIDEO	OpenIDEO
Lumia Beta Apps	Volkswagen (2010)	yet2.com	L'Oreal
MyStarbucksIdea		YourEncore	Swarovski
Microsoft		Zooppa	Threadless
Nel Mulino cheVorrei			Volkswagen
Nestlé			YourEncore
			Zooppa

a texture of practices, the "weft" is the practice of co-creating and the "warp" is the practice of weaving. Innovating is the overall "picture" that emerges from the intertwining of these practices. By "zooming out" from the cross points of warp and weft, we see the texture, that is, the weaving together of interconnections in action; by "zooming in," we see single practices and their minutiae (e.g., actors and resources).

6.4.1 *The Weft: Practices of Co-creating*

Market actors discover new ways of innovating together by participating more effectively in intricate networking activities. A new vocabulary arises that speaks of collaboration, cooperation, community, coordination, and coproduction. Companies cooperate, intermediaries pool resources, consumers collaborate, and web-based communities create contexts that spur innovation and bring it to market. An endless series of relationships coproduce solutions and co-create value. They are fostered by coordination and communication in a collaborative context. Innovation becomes a "doing together."

The suffix "Co" signals the "togetherness" of innovating practices. We can identify two main co-creating practices that unfold in the ongoing process of creating something new: (1) co-raising an idea and (2) co-fulfilling a solution.

The first practice refers to all the actions and resources actors perform and use to generate and evaluate ideas, that is, how they gain understanding from insights and sense possibilities for increasing value. The second practice refers to the actions and resources needed to move from idea to outcome, that is, how to bridge the gap between needs and solutions. These two practices can be divided into five

Table 6.2 "Co" practices

Co-raising	Co-ideating	Collaborating to generate ideas, sharing insights, commenting on proposals, socializing
	Co-evaluating	Cooperating to appraise proposals, vote on ideas
Co-fulfilling	Co-designing	Collaborating to develop "pictures" into concrete solutions
	Co-testing	Working together to improve solution prototypes
	Co-launching	Collaborating to launch solutions in the market

co-creating practices (see Table 6.2): (1) co-ideating, (2) co-evaluating, (3) co-designing, (4) co-testing, and (5) co-launching.

These "co" practices are not phases of the innovation process to be performed sequentially; they are configurations of specific practices of innovating in which actors interact, collaborate, integrate, and use resources to accomplish collective actions and give those actions sense and meaning. They are sociomaterial practices that use web-based technologies to create open spaces that overcome time and space constraints, in which new actors can propose ideas, solve problems, and generate new solutions. Each "co" can be deployed independently, wherever collaborative innovation flourishes.

Co-ideating

Insights and creativity are the keys to innovation potential. Co-ideating develops from a wide network of actors generating and sharing new ideas. It is not easy for companies to predict where the next ingenious idea will come from, but they can increase their odds of finding new business opportunities and success by extending their connections.

To change past ways of doing, companies deliberately design websites to encourage market actors to connect and share knowledge and ideas:

> P&G Connect + Develop is a platform we created for opening up product and methodology problems to outside resources in the hopes of coming to better solutions. With the world being so highly connected, there is huge opportunity for P&G to learn from, and teach others, innovative techniques and processes. It allows companies to draw on the greater pool of ideas to solve problems. (Source: interview to P&G manager; June 2013)

Actors who participate in co-ideating include companies and their customers—who propose new ways of satisfying needs (expressed or unexpressed)—and as experts and professionals. Increasingly, intermediaries take an active role in generating and shaping ideas. They prompt creativity and innovation through collective action:

> The innovation marketplace challenges the highly disaggregated context of innovation due to the global dispersion of resources and knowledge by enabling and making clear the most valuable connections. It confronts the lack of transparency due to information sensitivity,

IP protection and management-created barriers, as well as enabling R&D staff to recognise and look beyond their own strictly closed network. (Source: www.Ninesigma.com; accessed May 2012)

BMW wanted concepts and ideas to enhance the urban driving experience in cities of the future so it turned to the crowd. It teamed up with the crowdsourced vehicle design company Local Motors and its community of 30,000 designers, engineers, and enthusiasts. The BMW Urban Driving Experience wasn't a vehicle design competition, more a search for ideas to improve the driving experience of premium vehicles in the megacities of the future. Approximately 3500 design boards were sent in support of 400 concepts, and though the entries were unique, a number of common underlying themes emerged including gamification, car sharing, and doing good. (Source: www.ideaconnect.com; accessed May 2012)

Co-evaluating

Evaluation of an idea proceeds through a series of actions such as appraisal of proposals, commenting, and voting. Communities and social networks play a crucial role in fostering participation and establishing connections between various market actors (e.g., consumers, experts, and fans). Clues flow through the web of connections and are evaluated and selected in a collective way. Multiple actors intervene in choosing ideas to be implemented:

The Threadless website has specific sections through which users and fans participate in evaluating the most popular ideas. It is possible to propose illustrations and design to put on T-shirts, to vote and comment on the ideas already submitted, to describe the reason for the vote, and to suggest improvements to the proposed idea. (www.threadless.com; accessed June 2012 and May 2013)

Through websites, actors can also track the status of an idea as it moves through the innovation life cycle (under review, reviewed, coming soon, launched, or rejected). Timelines and methods for evaluation are settled in advance, and the final decision to move the idea into the market emerges as the result of collective conversation on common issues:

An effective process that we have experimented with many of our customers is to evaluate ideas in rounds. Beginning with an easy task, we apply filters that support customers in determining which ideas to respond to first. We use tools such as votes, views, comments, to determine which ideas are in most need of response, which ones are potential winners, and which can easily be taken out of the mix. Ideas that have broad support from our participants through votes or volunteering show promise for implementation and are good candidates to move to internal stages of evaluation, while ideas that draw minimal engagement are deferred for future consideration. (Source: interview to Digital Media Manger; January 2011)

Co-designing

Cooperation is crucial for bridging the gap between identified ideas and market solutions. In co-designing practices, the roles of designers are shuffled. Users are engaged in the process and assigned the role of "expert and competent practitioner" according to their abilities, experience, and background knowledge. On the basis of interest, passion, and effort, users connect with organizations and other actors and define their roles and contributions to the advancement and evolution of ideas:

> The designer is an integrator, who unleashes his/her creative potential by providing interactive tools: I'm speaking about the ability of designer to build on their own or ideas from others. His toolbox is now made much more of open source software and open source objects, documentation, sandbox, and user loop feedback. (Source: www.pgcommunity.com; accessed March 2013)

Innovative projects convey the design abilities of people around the world, leveraging emotional and experiential connections with hobbyists and experts. They are not only business affairs but also social ways of taking part in processes in which designing is part of enjoying and having fun:

> In 2010, Fiat launched the Fiat Mio Project to have people working together to create a car for the future. The futuristic concept car based on the ideas of thousands of people around the world was developed by firms who documented and showed the entire process on the web through the 'Making of Fiat Mio' blog. More than 17,000 participants from around the world submitted more than 11,000 ideas. During the design process, people were able to take part in many decisions like car colours and doors configuration. (Source: case reports on Fiat and Ideaconnection)
>
> Spreadshirt is your creative platform for custom T-shirts and custom clothing. You can explore our easy to use T-shirt Designer and make your own T-shirt or other personalized clothing products to spread your unique message. Design your own T-shirt by customizing your own text, design, or image on a variety of custom products, like hoodies, hats, or aprons. Spreadshirt has over 100 unique products for all your T-shirt printing and T-shirt design needs. (Source: www.spreadshirt.it; accessed June 2013)

Co-designing practices take place through actions sustained by learning connections and knowledge-based networking. Web design contexts are complemented by the use of Internet features such as wikis, blogs, aggregators, and forums that enable participants to be share documents and other materials, competencies, and knowledge with others:

> IdeaStorm was created to give a direct voice to our customers and an avenue to have online 'brainstorm' sessions to allow you the customer to share ideas and collaborate with one another and Dell. Dell added 'Storm Sessions' where Dell posts a specific topic and asks customers to submit ideas. These Storm sessions will only be open for a limited time, therefore making Storm Session discussions targeted, relevant, and time bound. In 2012, Dell added idea "Extensions" where an idea poster can promote a comment made on their idea to be considered a part of it. These "Extensions" enable ideas to evolve over time through collaboration. (Source: www.ideastorm.com; accessed September 2012 and May 2013)

Co-testing

In co-testing practices, actors improve prototypes of product and services and test their marketability. Users express specific perceptions and propose improvements to help refine products and services by and improve the user experience:

> Xbox Preview Program has been launched by Microsoft to test new Xbox One software. The testers use the Dashboard app that is the place for preview participants to give feedback about the program, get the latest news, change console enrollment settings, and report problems. Xbox has also designed a forum in which actors can leave their feedback and receive support from official Xbox Support staff and Xbox Community Ambassadors. (Source: interview to Microsoft software developer; May 2013)

Crowdsourcing platforms connect users and experts to test pre-commercialized applications. They often include discussion boards and user experience surveys to test preliminary concepts according to the actual needs and experiences of users:

> Our consumer is active, alert and aware, to contribute their opinion to improve products for himself, for his/her friends and for his/her family. Within the community we create a space for participation where consumers can know and provide the information needed to be involved in the projects; he/she participates in product testing, then judges and sends suggestions for improvement. In focus groups and forums in place or online, thanks to a virtual room, he/she exchanges opinions, comments, photos and movies. (Source: interview to Lumia Beta Apps Community Manager; January 2014)

Companies benefit from user testing and feedback. The tested technologies are usually early or incremental releases that have not gone through the rigorous testing and validation generally conducted by product testing groups. Market actors as problem-solvers are challenged to use their creativity and knowledge for detailed projects and prototypes that can be integrated into the innovation process:

> Through the slogan "play test, give the opinion, get the free software," Microsoft invites people to XBOX playtest: People create an account but the participation is on invitation. Microsoft contacts registered users when there is a study that is a good fit for them, but beta users can invite their friends and other experts they know to participate. (Source: www.innocentive.com; accessed February 2014)

Co-launching

Co-launching practices are similar to the patterns of actions and resources companies use to co-create value with users and sustain the success of new products and services in the market. There is an inherent challenge in overcoming initial diffidence toward unknown and often beta or provisory versions of product and services; companies manage content information and rely on mass appeals to push users to try new offerings. Co-launching is closely related to co-testing by its scope of making the product or service ready for the market:

> In the case of Threadless, the users are invited to vote on design attractiveness and the highest scoring designs voted are ultimately produced. Each design is evaluated by an average of 1500 users. In addition, customers' work is used more in the launch phase and they take over some market risk, too. Customers assume responsibility for advertising and photographing for catalogues and find new customers. (Source: www.threadless.com; accessed March 2012)

User-generated content companies are emerging as simpler, less expensive ways to create product awareness during launch and postlaunch activities. Companies use communication contests to excite users about new products, by linking the possibility of winning something to participation in social web launch events:

> In the run-up to the launch of their latest game, 'Heroes of the Storm,' Blizzard Entertainment implemented an all-encompassing digital countdown campaign featuring a YouTube teaser video, social hub, and live-streamed celebration event. The participants sign-up using Facebook, Twitter or Google+, and can upload a photo where they see themselves getting warped into the Nexus character. Best entries are being showcased on the website's social hub, along with other #NexusCalling user-generated content conversations. Plus, entrants have the chance to be featured in the live-stream celebrations. (Source: interview to Blizzard Entertainment communication manager; January 2014)
>
> To encourage users dialogue or the launch of Windows 10, Microsoft created a captivating rich media page integrating social tools such as Pinterest, Facebook, and Twitter. As part of their social content strategy, these tools provide a social hub enriched with persuasive and reloadable user-generated content, so users could experience the new software and encourage advocacy. (Source: microsoft.com; accessed February 2014)

6.4.2 The Warp: Practices of Weaving

In the previous section, we described practices of co-creating that unfold as an ongoing process of creating something new. In this section, we contend that to understand and represent the process of innovating, we must also appreciate the wider connectedness of practices. We identify two main practices of weaving: networking and knowing.

Networking practices are ways of connecting heterogeneous actors and resources. Companies search for relevant partners and knowledge bases. Through interaction and collaboration, a knowledgeable community develops. Being inside or outside a firm's boundaries is not as relevant as establishing connections as a precondition to action.

Through networking action, knowing practices emerge; actors question the what, why, and how of current and emerging knowledge. Knowing unfolds from the context of its production and is anchored in the sociomaterial resources of that context.

The two practices can be further divided into four practices that weave the texture (see Table 6.3): (1) engaging, (2) exploring, (3) exploiting, and (4) orchestrating. Each practice mediates the interactions and relationships among actors and supports their activities and resource integration.

Engaging

The practices of engaging refer to the ways actors become involved as part of communities and networks. Through the web context, different groups of actors connect to generate and evaluate ideas, develop designs, and test and launch new solutions. To facilitate ongoing dialogue and communication, talented and

Table 6.3 Practices of weaving

Networking	Engaging	Actors and resources becoming connected and socialized in an expanding network
	Orchestrating	Framing the contexts and conditions to mobilize and strengthen relationships between actors and resources
Knowing	Exploring	Generating, articulating, and experimenting "learning and doing" within the connections of actors and resources
	Exploiting	Discussing, appropriating, and disseminating "learning and doing" within the connections of actors and resources

enthusiastic actors are solicited to collaborate on new projects and become acquainted with opportunities for connections in co-creation actions. Distinct company and intermediary roles emerge as bridges are built to connect actors and assign new tasks for cooperation:

> Our challenge is not only to build infrastructure but also to design effective innovation experiences, so that actors stay engaged, discover opportunities, and achieve the desired outcomes. We develop new digital platforms and build knowledge-sharing systems for managing information and data across organizations and other actors, all of which we center around the needs of core users. (Source: interview to Ideo Labs manager; April 2012)
>
> The Innoget.com website is the tool through which our clients connect with companies, research institutions, and innovation professionals worldwide. In Innoget.com, organizations seeking innovation meet innovation-generating organizations. Our clients capture external knowledge (via the publication of Technology Requests) and promote the knowledge generated within their organizations. Making the knowledge flow, we foster innovative technologies, products, methods, procedures, and services. (Source: www.innoget.com; accessed May 2012)

Actors are solicited by sharing contexts such as workshops, playrooms, and open spaces. Challenging projects match their expertise, competencies, and shared views on problem and tasks.

> The Lego Mindstorms community is conceived to bring together minds and creative thinkers in a very professional community. Here professionals conduct ongoing conversations with the company and other experts to get suggestions, clarifications, or details about projects they are interested in developing. (Source: mindstorms.lego.com; accessed May 2013)

In expanding the range of possibilities, companies and intermediaries continuously look beyond their peripheral vision. The web context operates like an enlarged focus group in which creative minds are invited to participate in informal ways. This format allows for a large number of users who have diverse backgrounds and views:

> Nokia's Ideas Project is an online community for everybody from all around the world to brainstorm. Facebook, Ovi, and Twitter links allow users to be in real-time connection with companies, developers, experts, NGOs, and all who can help in the realization of ideas. (Source: inventwithnokia.nokia.com, accessed February 2013)

These social and community spaces give interested and talented actors the freedom to interact on the basis of their resources and expertise and according to their own terms; actors communicate with each other, share ideas or documents, and conduct conversations on new product concepts and fully developed projects. By socializing, connected users transform their separate understandings about ideas and projects into a more coherent view of new solutions:

> We are moving from being a fully connected society to a hyper-connected one where people act and feel as global citizens and participants of a virtual technology platform. The members of our platform are characterized by having digital literacy, using online social networks, having multitasking capabilities, socializing and learning on the net, consuming and producing digital information, and imagining and visualizing while communicating online. (Source: www.mindstorms.lego.com; Lego Mindstorms Digital Manager published interview; accessed May 2013)

Exploring

Exploring practices consist of applying, leveraging, and extending knowing-based resources and activities through knowledgeable communities of actors. Experts, fan users, and other actors prove their ability in a practical context of experimentation by generating fresh ideas and recognizing new opportunities. Most creative actions are defined in problem-finding contexts and open-ended structures within the web nexus as a stage for creative solutions:

> Solutions to the challenges innovation faces today can no longer rest solely in the hands of an elite few. These challenges require collaboration, imagination, and action at an unprecedented scale. What would the innovation look like if everyone saw the problems in the world as opportunities? What if everyone believed in his or her creative abilities and had a means of productively tapping into them? We know the need to develop, nurture, and facilitate the skills, behaviors, and mind-sets of people as future design thinkers. (Source: interview to Innocentive manager; March 2013)

A wide range of resources and tools are used to enable actors to articulate specific aims, metrics, and tests for collaborative work. Actors are invested with compelling activities—documenting, writing, drawing, showing, and modeling. By explicating and formalizing the connections between projects, ideas, and solutions, knowing emerges from doing. Through these activities, the "co" practices can advance and innovation can unfold:

> At the P&G Connect + Develop platform, any project takes place as an iterative process. It starts by a project brief in which actors elicit general scope, objective, and success criteria and then involves partners to work on ideas or concepts and integrate it into the co-design or co-test. To design activities, the participants are required to work with detailed projects and prototype that have already been reduced to practice. (Source: www.pgconnectdevelop.com, September 2012)
>
> By participating in the Lego Mindstorms NXT project, users work on an authentic new project, involving skills, knowledge, and competencies often at the level of those used by professionals and experts in the design field. They work with specific knowledge artifacts to visualize, store, and improve knowledge and creativity expression. At the heart of user's activities there is a device known as the Robotics Command System, an autonomous microcomputer with an infrared link that can execute user-created code sent from a PC. Users can combine and recombine various blocks of code to bring those creations to

> life and can post the photos and the description of their projects. Moreover, 'Mindstormers' have the possibility to write blogs that describe their prototypes. (Source: www.mindstorms.lego.com; project manager published interview; accessed May 2013)

Activities such as codification, documentation, and communication feed knowing practices by enabling knowledgeable actors to participate in cooperative efforts to create functional solutions or test market potential. All concepts that are generated are shareable, remixable, and reusable by a knowledgeable community. In this way, users not only contribute original ideas, designs, and pictures but also experiment with ideas, objects, or knowledge created by other actors. Such ideas, designs, and thoughts play a prominent role in interaction and provide useful groundwork for other users' creativity:

> Bringing new ideas to life is an essential part of what we do. We invite our users to work on tools they can share with others. The pictures, designs, and proof prototypes are those tools we ask users to use in order to challenge their skills. They are early proofs of concepts and usually they are far from pretty, and they do not work. These tools are mainly ways of helping users to start to work on [an] idea and explore, learn, and think in a collective way. They're prototypes—not finished products. Most of our prototypes are not let out into the world. Our aim is to develop stuff that we can share, and help users to unlock and foster their creativity. (Source: www.labs.ideo.com; accessed September 2012)

Exploiting

Exploiting practices consist of combining current and integrating and transforming knowing-based resources and activities. By moving across a range of contexts, actors discover how the knowledge and technologies that are available in each context can be used, combined, and refined in new ways.

In exploiting, actors discuss and evaluate ideas and artifacts created by others. In this way, the most promising ideas and concepts are selected from ideas that inspire the community. Actors cogenerate new solutions by inspiring ideas, designs, and concepts of others and making associations among existing ideas and projects to create new concepts. By discussing which ideas or concepts are most critical, actors' attentions and perceptions converge toward a specific class of issues; by identifying priorities, they determine the criteria used to judge the solutions:

> Our discussion section works as s a sensing platform to acquire knowledge on how people think differently. Each discussion starts with a short demand and different issues are presented. Users are invited to write a comment and to provide their opinions on ideas from others. They hold different suggestions and ask users how they could work together are our way to work. New ideas require shedding 'either/or' thinking and seeing the whole as inclusive of opposition and open to a third (or fourth or fifth) solution. The practice of discussing allows us to live with and resolve paradox and contradiction. (Source: www.ninesigma.com; accessed June 2013)

Through appropriating actions, actors make associations and reinterpret existing ideas in new contexts to create their own novel commercial concepts. This way of doing emphasizes the highly connective nature of innovation and the role of association in bridging various actors' contexts to exploit design innovation and boost the creation of new concepts and original designs:

> To value the meaning and quality of the original source material is what we have tried to encourage. In The Fiat Mio Project we adopted the tool of Commons creative licenses. This method allowed users to get, diffuse, and combine the ideas developed by the Fiat team, as well as to make available to whoever used the technological knowledge generated by Fiat Mio. (Source: https://www.ideaconnection.com/open-innovation-success, Fiat Mio project manager published interview; May 2011)

The success of new ideas also depends on earning acceptance from a wider audience. Acceptance occurs when effective networking and disseminating activity persuades the greatest share of users to use the new solution. Through the use of various online collective tools (e.g., user-generated communication contests, web-enabled word of mouth, viral messaging, video sharing, virtual communities, and common license), actors establish themselves as sounding boards to promote the diffusion of ideas and innovation projects across different contexts and fields. Disseminating actions inspire others to use new solutions; users become "evangelists" who preach and spread the appreciated features of newness. The actors attract others, thus forming a critical mass of actors who have a vested interest in the success of the innovation:

> For its Aggregated Services Router (ASR) launch, Cisco executed the introduction entirely online by leveraging social media. The company used video to educate its audience and engage them in a gaming world. In addition, the firm assembled videos and images in a widget format, encouraging video blogging and the spreading of information. Plantronics, for the launch of its new product, CS70N, identified brand evangelists through a survey. Once selected, they were given access to online tools to help facilitate their conversations about the product and its dissemination. (Source: communities.cisco.com, accessed May 2014)

Orchestrating

The practices of orchestrating capture the influential role of networking in shaping how actors collaborate and share their resources in action. Knowledgeable actors need to know how to access, interact, and participate in the community. Orchestrating frames contexts and conditions to improve the flow of actions within the practices of engaging, exploring, and exploiting. It also includes actions to improve the quality of interactions and create effective stimuli for strengthening sense of community and social relationships.

Companies and intermediaries arrange and execute roles, resources, and rules. They clearly state the risks and benefits of actor engagement and establish incentives and property rights. They support effective coalitions, align actors, and contribute to the sharing of resources and integration of knowledge:

> The best way to prevent incomplete or underdeveloped innovation as well as prevent conflicts is to be as specific as possible about what you are looking for and our users could expect when launching a contests. Establish a strong context and use the project description along with a Kindling Post to state what our goals are and what factors you'll

need to evaluate and select ideas. Being transparent about our evaluation criteria helps us ensure that participants propose ideas that are relevant to the challenge and include all the information they are looking for. (Source: www.electroluxdesignlab.com; accessed October 2011)

The Design Quotient (DQ) is a measure of your contributions to OpenIDEO. It corresponds to how active you are in the inspiration, conception, and evaluation phases of a challenge. It also measures your collaboration, increasing every time you comment or build on other peoples' inspirations and concepts. (Source: www.openideo.com; accessed July 2012)

Intermediaries enable the framing of innovation networks by improving actors' responsiveness and ensuring their adaptability in mutually beneficial interactions. Furthermore, they generate the necessary roles, resources, and rules to facilitate the effective alignment and linkage of actors by overcoming technological, social, and cultural differences:

> We provide deal-facilitation services to ensure that value is generated for both the parties helping to close transactions and assist parties. Our focus is on accelerating deals between partners mainly when larger solution-seekers and small problem-solvers are involved. We also support firms in moving introductions through initial logistics, potentially complicated NDA/IP issues, technical evaluation, economic discussions, and final negotiations. (Source: www.Yet2.com; accessed September 2012)

Actors' interactions are also enabled when users maintain their focus, energy, and commitment to innovation-based relationship. Mobilizing actors and helping them to have confidence in projects or innovation foster their abilities to face collaborative innovation requirements. In the attempt to make market actors inspired and allow connections to be established and developed over time, companies make actors aware of the value of social relationships and sustain their participation in an open and connected innovation context:

> We work at fostering the creativity of users who have not been closed in their background but who have been encouraged to join their ideas and experiences in a collaborative effort. We know that this effort requires not only a simply transfer of information or other form of material knowledge. We strive to encourage our partners to work with something that others can understand and share and then transfer to the innovation activities of others (internally and externally). This requires more attention to the way our users dialogue and to the tools and artifacts (documents, protocols, datasets) they use to interact even when we know that everyone has his/her goal to accomplish. (Source: interview to Lego Mindstorms community manager; May 2013)

6.5 Discussion and Implications

In this chapter, we developed a fresh conceptualization of innovation by adopting a practice-based approach. By framing the phenomenon as a verb rather than a noun, we addressed "innovating" as a texture of practices—a complex interconnectedness of actors, actions, contexts, and resources. The texture metaphor allows us to move away from traditional models based on multiple phases of innovation within

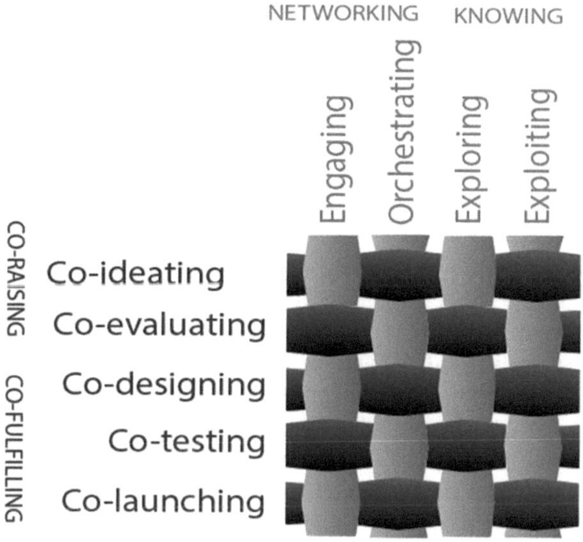

Fig. 6.1 Innovating as a texture of practices

absolute dichotomies (such as open–closed or market–technology) toward complex patterns composed of interwoven elements.

Innovating is the overall picture that emerges from the intertwining of the practices of co-creating (the weft) and the practices of weaving (the warp). Such practices emerge from the pattern of connections in actions between multiple actors; they reflect how actors constitute the context and are affected by it, as the result of these connections (see Fig. 6.1).

Actors discover new ways of knowing and doing together. A new vocabulary arises that stresses "togetherness" and speaks of collaboration, cooperation, coordination, and community. Companies cooperate, intermediaries pool resources, consumers collaborate, and web communities create enabling contexts.

In such a framework, we identify two core co-creating practices: co-raising ideas and co-fulfilling solutions. The first practice relates to how actors gain understanding from insights to increase value. The second relates to actions and resources needed to move from ideas to solutions. Co-creating practices can be divided into (1) co-generating, (2) co-evaluating, (3) co-designing, (4) co-testing, and (5) -co-launching. Two core practices of weaving sustain the co-creation process: networking and knowing. Networking practices are ways of connecting heterogeneous actors and resources. Through interacting and collaborating, a knowledgeable community develops and knowing practices emerge; actors question the what, why, and how of current and emerging knowledge. These two core practices are divided into (1) engaging, (2) exploring, (3) exploiting, and (4) orchestrating. Each practice mediates the interactions and relationships among actors and supports their activities and resource integration.

Innovating thus arises from these nine practices, nested one within the other. By "zooming out" from the cross points of warp and weft, we see the texture, that is, the weaving together of connections in action; by "zooming in," we see layered and interwoven patterns of minutiae (e.g., actors, resources). In this approach, we can see practices as "in time and in becoming."

This thick texture is shaped by human and nonhuman connections. Practices are socially and materially performed, that is, they are constantly co-constructed by the actions of participating actors in sociomaterial interactions that support relationships in alignment of aims, meanings, and activities. Practitioners interact with technology and its material and symbolic features. Information and communications technology and the material setting provide the relational infrastructure to support the performance of practice (Nicolini et al. 2003). In the entanglement of the social and the technological, a structuring process for innovating emerges and becomes stabilized and, in some ways, institutionalized (Gherardi 2012a, b).

Innovating is no longer just a business issue; it is a shared knowing and doing of society (Gherardi 2006). It involves a great variety of actors—companies, competitors, venture capitalists, research centers, universities, users, design institutes, and intermediaries—who collaborate in a shared context by developing and connecting activities, resources, interests, languages, and objects. It is a complex phenomenon that comes from deliberate activities and emerging opportunities in working practices produced on a daily basis.

6.5.1 Implications for Scholars

The proposed framework of "innovating as a texture of practices" aims to foster a debate about practice in innovation studies. The practice-based approach contributes epistemologically and methodologically (Gherardi 2012a; Raelin 2007) to the theorizing process:

> Turning to practice does not mean becoming more engaged or making social science more relevant to the practical concerns of practitioners; rather, it means developing a type of theory (called theoria) that is "about proceeding from within an activity, making its 'grammar' explicit, opening new possibilities for action, and informing mindful, caring, and wise conduct [Eikeland and Nicolini (2011: 169) in Gherardi (2012a: 4)]."

In line with Bourdieu (2005), we advocate the practice approach as the locus for the dialectics of opus operatum (the result of practices) and modus operandi (the mode of practices), stressing the creative dimension of the practical and the collective existence of human beings (see also Fuchs 2003). We move beyond dichotomist arguments to suggest that practices are at the core of innovating. This view allows us to grasp the phenomenon of complexity and understand it not only in terms of NPD and NSD, (i.e., stand-alone activities) but also in terms of its reproduction of sociomaterial relationships and of society. In this way, scholars can grasp how sociomaterial relationships affect—and are affected by—innovating.

The very reason for studying practices is to study innovation in its making. By focusing on practices, we can detect innovation in the process of being practiced and in the relationship between a situated practicing and the broader texture of practices (Gherardi 2006). Our attention is thus focused on becoming: "Phenomena are not treated as entities, as accomplished events, but as enactments—unfolding processes involving actors making choices interactively, in inescapably local conditions, by drawing on broader rules and resources" (Tsoukas and Chia 2002: 577).

When relational thinking (Østerlund and Carlile 2005) is applied, the "dichotomies dilemma" disappears, allowing us to adopt a more holistic view of a phenomenon both planned and emergent. Many types of multiplicity converge, including pluralities of values, pluralities of rationalities, multiple discourses, unequal power relations among stakeholders, and asymmetries of information.

Further studies could deepen knowledge of how a practice-based conceptualization contributes to complexity theorizing, according to contingency, indeterminacy, and sense-making (Arthur 2013). The practice conception holds an ordering principle due to the institutionalization of activities and ways of doing which are sustained by both material and social relations. At the same time, it is disordering principle as it is transitory and unsteady because practices hold a logic that is contingent and in a becoming: thus they not only help to reduce uncertainty but introduce indeterminacy (Gherardi 2012a, 2015).

As part of the theorizing effort, scholars also need to develop a practice-based vocabulary (Nicolini et al. 2003) for innovating: "New vocabularies populate the world differently: they not only change existing ideas, they introduce new and alternative values. [Scholars should develop] new linguistic and conceptual repertoires ... to free us from the prevailing notions" (Nicolini et al. 2003: 7). This requires cross-fertilization with studies on knowing, learning, and doing.

6.5.2 Implications for Managers

We should not regard the economy as a given but should grasp its performative character and understand how it is shaped by a continuously changing set of technological innovations, institutions, and arrangements. This understanding gives rise to three managerial implications.

First, faced with such a complex "becoming," managers should develop new perspectives on innovation and business. The process of innovation is not linear and sequential, but is an intricate and endless series of connections in action, in which there are ongoing interactions between networks of actors that share, comment, communicate, draw, learn, and discuss. Each interaction and connection can be seen as a co-creating opportunity. Companies should consider more clearly the full options of co-creation activities and be involved in designing and responding to co-creation initiatives.

From this perspective, managers should exchange their understanding of innovation as a new outcome for a revised A2A view that regards innovating as a set of collaborative practices (Vargo et al. 2015). They should create opportunities and contests to enable collaborative ways of innovating and fostering the integration of multiple resources of the service ecosystem.

Similarly, companies should shift perspectives, from the planned-and-governed approach of traditional management studies to the concept of innovating as ongoing actions that can be mediated and situated within a context of interconnected practices. Innovating arises from a mixture of emergent, adapted, and adopted practices that are refined over time by interaction within sociomaterial, interconnected activities. Thus, a key challenge for managers is to recognize and represent practices to empower practitioners. They should "zoom out," beyond minutiae, to see the bigger, textured picture of relationships and connections.

Second, managers must also grasp the interconnections between humans and nonhumans. This requires the ability to think and act collectively and cooperatively. Fostering the practice of engagement, manager can bridge disconnected worlds and build new innovation opportunities for market actors. Specific actions can enable connections, solicit talented and interested people, foster socialization, and expand innovation networks.

By promoting the integration of ideas, tools, images, and languages, managers can foster practices of exploring and exploiting. They can promote market innovation and focus on methods and tools to foster co-generation of ideas, transform ideas into solutions, and experiment with alternatives.

Third, companies should open up the business landscape to find out how social, cultural, material, and contextual connections in action affect and are affected by the evolution of social arrangements and institutional structures that support new ways of innovating.

New methods and tools can be useful to the orchestrating practice. For example, such methods and tools can be used to frame better contexts and conditions for improving the efficacy and efficiency of actors' interactions, resource exchanges, and integration practices. For this reason, it is important to increase the commitment of market actors and foster the development of a collective sense of co-created market innovation.

References

Akaka, M. A., & Vargo, S. L. (2014). Technology as an operant resource in service (eco) systems. *Information Systems and e-Business Management, 12*(3), 367–384.

Arthur, W. B. (2013). *Complexity economics: A different framework for economic thought*. Oxford: Oxford University Press.

Bourdieu, P. (2005). *The social structures of the economy*. Cambridge: Polity Press.

Brown, J. S., & Duguid, P. (2001). Knowledge and organization: A social-practice perspective. *Organization Science, 12*(2), 198–213.

Callon, M. (1999). Le reseau comme forme émergente et comme modalité de coordination. In M. Callon, P. Cohendet, J.-M. Dalle, F. Eymard-Duvernay, D. Foray, & E. Schenk (Eds.), *Reseau et Coordination*. Paris: Economica.

Carlile, P. R. (2002). A pragmatic view of knowledge and boundaries: Boundary objects in new product development. *Organization Science, 13*(4), 442–455.

Chesbrough, H. (2003). The logic of open innovation: Managing intellectual property. *California Management Review, 45*(3), 33–58.

Chesbrough, H. W. (2006). *Open innovation: The new imperative for creating and profiting from technology*. Boston, MA: Harvard Business Press.

Chia, R. (1995). From modern to postmodern organizational analysis. *Organization Studies, 16*(4), 579–604.

Cook, S. D., & Brown, J. S. (1999). Bridging epistemologies: The generative dance between organizational knowledge and organizational knowing. *Organization Science, 10*(4), 381–400.

Corradi, G., Gherardi, S., & Verzelloni, L. (2010). Through the practice lens: Where is the bandwagon of practice-based studies heading? *Management Learning, 41*(3), 265–283.

Edvardsson, B., & Tronvoll, B. (2013). A new conceptualization of service innovation grounded in SD logic and service systems. *International Journal of Quality and Service Sciences, 5*(1), 19–31.

Eikeland, O., & Nicolini, D. (2011). Turning practically: Broadening the horizon. *Journal of Organizational Change Management, 24*(2), 164–174.

Enkel, E., Gassmann, O., & Chesbrough, H. (2009). Open R&D and open innovation: Exploring the phenomenon. *R&D Management, 39*(4), 311–316.

Fuchs, C. (2003). Some implications of Pierre Bourdieu's works for a theory of social self organization. *European Journal of Social Theory, 6*(4), 387–408.

Garcia, R., & Calantone, R. (2002). A critical look at technological innovation typology and innovativeness terminology: A literature review. *Journal of Product Innovation Management, 19*(2), 110–132.

Geiger, S. W., & Makri, M. (2006). Exploration and exploitation innovation processes: The role of organizational slack in R & D intensive firms. *The Journal of High Technology Management Research, 17*(1), 97–108.

Gherardi, S. (2001). From organizational learning to practice-based knowing. *Human Relations, 54*(1), 131–139.

Gherardi, S. (2006). *Organizational knowledge: The texture of workplace learning*. Oxford: Blackwell.

Gherardi, S. (2012a). *How to conduct a practice-based study: Problems and methods*. Cheltenham: Edward Elgar.

Gherardi, S. (2012b). Why do practices change and why do they persist? Models of explanations. In P. Hager, A. Lee, & A. Reich (Eds.), *Practice, learning and change: Practice-theory perspectives on professional learning, professional and practice-based learn*. Australia: Springer.

Gherardi, S. (2015). To start practice theorizing anew: The contribution of the concepts of agencement and formativeness. *Organization*, on line first November 2015.

Gherardi, S., & Nicolini, D. (2006). *Organizational knowledge: The texture of workplace learning*. Malden, MA: Blackwell Publishing.

Giddens, A. (1979). *Central problems in social theory: Action, structure, and contradiction in social analysis*. Berkeley: University of California Press.

Guzman, G. (2013). The grey textures of practice and knowledge: Review and framework. *European Business Review, 25*(5), 429–452.

Harrison, D., & Kjellberg, H. (2010). Segmenting a market in the making: Industrial market segmentation as construction. *Industrial Marketing Management, 39*(5), 784–792.

Howe, J. (2008). *Crowdsourcing: How the power of the crowd is driving the future of business*. New York, NY: Random House.

Johnson, G., Melin, L., & Whittington, R. (2003). Micro strategy and strategizing: Towards an activity based view. *Journal of Management Studies, 40*(1), 3–22.

Kjellberg, H., & Helgesson, C. F. (2007). On the nature of markets and their practices. *Marketing Theory, 7*(2), 137–162.

Kogut, B., & Zander, U. (1996). What firms do? Coordination, identity, and learning. *Organization Science, 7*(5), 502–518.
Kozinets, R. V. (2002). The field behind the screen: Using netnography for marketing research in online communities. *Journal of Marketing Research, 39*(1), 61–72.
Lusch, R. F., & Nambisan, S. (2015). Service innovation: A service-dominant logic perspective. *MIS Quarterly, 39*(1), 155–175.
Lusch, R. F., & Vargo, S. L. (2014). *Service-dominant logic: Premises, perspectives, possibilities.* Cambridge: Cambridge University Press.
Mele, C., Colurcio, M., & Russo-Spena, T. (2014). Research traditions of innovation: Goods-dominant logic, the resource-based approach, and service-dominant logic. *Managing Service Quality, 24*(6), 612–642.
Mele, C., Russo-Spena, T., & Colurcio, M. (2010). Co-creating value innovation through resource integration. *International Journal of Quality and Service Sciences, 2*(1), 60–78.
Möller, K., & Rajala, A. (2007). Rise of strategic nets – New modes of value creation. *Industrial Marketing Management, 36*(7), 895–908.
Nambisan, S. (2002). Designing virtual customer environments for new product development: Toward a theory. *Academy of Management Review, 27*(3), 392–413.
Nambisan, S., & Nambisan, P. (2008). How to profit from a better virtual customer environment. *MIT Sloan Management Review, 49*(3), 53–61.
Nicolini, D. (2007). Studying visual practices in construction. *Building Research and Information, 35*(5), 576–580.
Nicolini, D. (2011). Practice as the site of knowing: Insights from the field of telemedicine. *Organization Science, 22*(3), 602–620.
Nicolini, D., Gherardi, S., & Yanow, D. (2003). *Knowing in organizations: A practice-based approach.* ME Sharpe.
Orlikowski, W. J. (2002). Knowing in practice: Enacting a collective capability in distributed organizing. *Organization Science, 13*(3), 249–273.
Orlikowski, W. J. (2007). Sociomaterial practices: Exploring technology at work. *Organization Studies, 28*(9), 1435–1448.
Østerlund, C., & Carlile, P. (2005). Relations in practice: Sorting through practice theories on knowledge sharing in complex organizations. *The Information Society, 21*(2), 91–107.
Papastathopoulou, P., & Hultink, E. J. (2012). New service development: An analysis of 27 years of research. *Journal of Product Innovation Management, 29*(5), 705–714.
Payne, A. F., Storbacka, K., & Frow, P. (2008). Managing the co-creation of value. *Journal of the Academy of Marketing Science, 36*(1), 83–96.
Piekkari, R., Plakoyiannaki, E., & Welch, C. (2010). 'Good' case research in industrial marketing: Insights from research practice. *Industrial Marketing Management, 39*(1), 109–117.
Raelin, J. A. (2007). Toward an epistemology of practice. *Academy of Management Learning and Education, 6*(4), 495–519.
Reckwitz, A. (2002). Toward a theory of social practices a development in culturalist theorizing. *European Journal of Social Theory, 5*(2), 243–263.
Read, S., Dew, N., Sarasvathy, S. D., Song, M., & Wiltbank, R. (2009). Marketing under uncertainty: The logic of an effectual approach. *Journal of Marketing, 73*(3), 1–18.
Richardson, I. (2010). *Writing. A method of inquiry* (2nd ed.). New York, NY: Sage.
Ritter, T., & Gemünden, H. G. (2003). Network competence: Its impact on innovation success and its antecedents. *Journal of Business Research, 56*(9), 745–755.
Russo-Spena, T., & Mele, C. (2012). Five Co-s in innovating: A practice-based view. *Journal of Service Management, 23*(4), 527–553.
Sarasvathy, S. D. (2009). *Effectuation: Elements of entrepreneurial expertise.* Cheltenham: Edward Elgar.
Sarasvathy, S. D., & Dew, N. (2005). New market creation through transformation. *Journal of Evolutionary Economics, 15*(5), 533–565.

Schatzki, T. (2002). *The site of the social: A philosophical exploration of the constitution of social life and change*. University Park, PA: Penn State Press.
Schatzki, T. R. (2005). Peripheral vision the sites of organizations. *Organization Studies, 26*(3), 465–484.
Schau, H. J., Muñiz, A. M., Jr., & Arnould, E. J. (2009). How brand community practices create value. *Journal of Marketing, 73*(5), 30–51.
Sundbo, J., & Toivonen, M. (2011). *User-based innovation in services*. Cheltenham: Edward Elgar.
Tidd, J., Bessant, J., & Pavitt, K. (2005). *Managing innovation: Integrating technological, market and organizational change*. Chichester: Wiley.
Tsoukas, H., & Chia, R. (2002). On organizational becoming: Rethinking organizational change. *Organization Science, 13*(5), 567–582.
Vargo, S. L., & Lusch, R. F. (2004). Evolving to a new dominant logic for marketing. *Journal of Marketing, 68*(1), 1–17.
Vargo, S. L., & Lusch, R. F. (2008). Service-dominant logic: Continuing the evolution. *Journal of the Academy of Marketing Science, 36*(1), 1–10.
Vargo, S. L., & Lusch, R. F. (2011). It's all B2B... beyond: Toward a systems perspective of the market. *Industrial Marketing Management, 40*(2), 181–187.
Vargo, S. L., Maglio, P. P., & Akaka, M. A. (2008). On value and value co-creation: A service systems and service logic perspective. *European Management Journal, 26*(3), 145–152.
Vargo, S. L., Wieland, H., & Akaka, M. A. (2015). Innovation through institutionalization: A service ecosystems perspective. *Industrial Marketing Management, 44*, 63–72.
Verona, G., Prandelli, E., & Sawhney, M. (2006). Innovation and virtual environments: Towards virtual knowledge brokers. *Organization Studies, 27*(6), 765–788.
von Hippel, E. (2005). *Democratizing innovation*. Cambridge, MA: MIT Press.

Chapter 7
Innovating as an Interaction of Practices: The Case of Customers' Use of a New Website

Stefano Pace, Antonella Carù, and Bernard Cova

Abstract While innovation is the final outcome of a company-based process, 'innovating' is innovation in action, that is, it takes place within the normal consumer practices surrounding a new product or service. By engaging with a new product, consumers shape innovation. Thus, consumers provide the final, essential stage in a company's innovation. This chapter will focus on the actual use of new web services by customers. Through a mixed method approach based on observation and introspection, we analysed the practices of Internet users that unfold when facing a website that is new to them and addressed the question of which practices customers adopt in order to allow an innovation to enter their daily practices. We argue that the use of a new service is an interaction between the consumer practices required by the design of the new service (such as pressing buttons, following a detailed sequence of options, etc.) and the regular daily practices of the customers (such as opening a link, putting keywords in a search engine, etc.). The success of this interaction determines the success of the innovation, while its failure requires an adjustment to the innovation design.

7.1 From Cocreative Projects of Innovation to Consumer 'Innovating'

Innovation today is so complex that companies cannot achieve it without the support of consumers, who now play a vital role in the cocreation of innovations (Prahalad and Ramaswamy 2004; Vargo and Lusch 2004; von Hippel 2005). The accepted perspective on how companies and consumers innovate together assumes that this innovation cocreation takes place within the boundaries of a well-defined

S. Pace (✉) • B. Cova
Department of Marketing, Kedge Business School, Marseille, France
e-mail: stefano.pace@kedgebs.com; bernard.cova@kedgebs.com

A. Carù
Department of Marketing, Bocconi University, Milan, Italy
e-mail: antonella.caru@unibocconi.it

T. Russo-Spena et al. (eds.), *Innovating in Practice*,
DOI 10.1007/978-3-319-43380-6_7

project. These cocreative projects are initiated by companies or by customers themselves, but in both cases, they refer to something that lies outside of daily routines. In the case of the former, a company may invite customers to contribute their creativity to the development of a product design, like in the case of Threadless, a platform where consumers can participate in the company's activity by submitting their own designs for T-shirts (Fuchs et al. 2010; Schreier et al. 2012). As for customer-initiated innovative projects, customers may start autonomous communities, where they design new models of products they are interested in. For example, fans of sneakers who also have design expertise may define new styles of sport shoes in an online community (NikeTalk) based upon a company's existing styles and models (Füller et al. 2007).

The present chapter takes a different stance from the isolated innovation process mentioned above. The research argues that innovation increasingly takes place within normal consumer practices. By engaging with new products or services, consumers shape the meaning of innovation. Therefore, the practices of consumers are a key element for companies to understand when developing innovations. It is in the consumers' practices, which are arguably the final stage of the development of an innovation, that the action of 'innovating' occurs. Some of the existing literature in consumer research introduces this view. Akrich et al. (2002a, b) deal with the idea of innovation 'in the making'. According to their theory, innovation is 'patiently constructed; it is more and more a result of a collective activity, and no longer the monopoly of an inspired and dedicated individual' (Akrich et al. 2002a: 189). Innovation stems from a complex nexus of different actors. While end users are relevant, Akrich and colleagues note that market research may be at odds with the correct understanding of customers, because 'customer is king, but of an empire whose boundaries are poorly defined and whose laws are vague. It is an enigmatic entity' (Akrich et al. 2002a: 200). In order to address the enigmatic nature of customers, it is necessary to enter their world and study them while they are using and shaping the innovation. Rather than focusing on the anticipatory knowledge about customers provided by surveys and lab experiments, innovation should be rooted in the actual practices of customers.

Shove and Pantzar (2005) describe how consumer usage led to the development of Nordic walking as a new practice and cite it as an example of why innovation should be 'understood less in terms of the "diffusion" of innovations than in terms of their "reinvention" by those who are to adopt it as it stands' (Cochoy 2015: 143). However, cases like that of Nordic walking are already based on practices (i.e. walking), and thus the interaction with new products is less evident. More generally, the notions of prosumers (Ritzer 2014) or working consumers (Cova and Dalli 2009) describe consumers who produce value in association with the company. However, innovation is not the core objective of these contributions. Whenever innovation is at stake, the emergent view is that of innovation as a separate process from consumption, a process that is explicitly aimed at generating something new that will be offered to the market at a later time (as in the case of the project-based innovation explained above). The argument put forth in this chapter is that the normal use of a new product informs innovation, that is, the practice of interacting with the new product is 'innovating'.

In terms of what to offer, products and services are bundles of so many innovative features that it is difficult for the company to define which ones will really make the difference in the market. When it comes to the market side, any product offered enters people's real lives, and this opens them up to a multitude of possible, alternative uses. Each of these alternatives can lead to further innovation of the product or service. For instance, early mobile phones were used by consumers to write text, which is a somewhat unexpected use of a device that has been designed to make voice calls. It is this customer practice that is the source of current text-based apps such as Twitter and WhatsApp.

Individuals encountering an innovation cannot be considered passive consumers, i.e. subjects who will adopt the innovation according to the company's intention for its use. This passivity paints the user as a sort of 'on/off' agent, who either adopts the innovation exactly as it is prescribed by the company or not at all. On the contrary, consumers are real people who embed products and services into their daily practices (Arnould and Thompson 2005; Holt 1995). As long as a market offer can be incorporated into those practices, it can enjoy market success. Consumers are cultural agents who enhance innovation, rather than absorb it. An innovation is successful if it becomes integrated into the practices of individuals. This act of embedding an innovation into common practices is the responsibility of the consumers, who creatively bend innovation to suit their lifestyles.

The ability of consumers to change the essence of innovation applies to big and small innovations alike (Troilo and Vicari 2008). An example of the former is Second Life, an online innovation that had its peak some years ago, when it was depicted as the future of the web and even a threat to offline reality. While Second Life remains an interesting virtual environment, it failed to change the daily practice of users in the way that other web-based innovations, such as Facebook or Pinterest, have done. This relative failure may be due in part to the fact that Second Life offered an alternative and parallel universe, while more successful innovations fit into regular daily practices and enhance them.

Successful innovation enhances life without disrupting it. Small innovations also have to go through the cultural vetting of consumers. For instance, Facebook's 'poke' function was intended to be a way for users to express a friendly sentiment towards another user in a simple and direct way (i.e. a virtual hug) or a way to get the attention of another. It was similar to the Facebook 'like', only addressed to a person rather than a post. However, this simple option was not widely adopted by users likely because it did not fit with the emerging practices of Facebook users, who were responding more to content than to profiles.

The traditional view of innovation sees consumers as merely adopting new products (Bass 1969; Gatignon and Robertson 1985). On the contrary, the adoption of innovations is essentially a cultural act through which consumers make sense of the new product (Arnould 1989). This sense-making is a source of innovation in itself owing to the ability of consumers to collaborate with the company and express their skills (Cova and Dalli 2009). By observing how the practices of users adapt to the new product or service, the company can get a real sense of the possibilities of what it has produced and can complete the innovation process by streamlining the product to function most effectively according to the way in which it is regularly used by consumers. Practices represent consumer culture and are the observable expression of the 'innovating' process.

In summary, we can call 'innovating' the set of consumer practices through which users embed a new product or service in their normal consumer practices, interact with it and enact the innovation that is present in the new product. Innovating practices actualise the potential innovation of the new product or service.

7.2 'Innovating' and Consumer Practices

We can observe the act of 'innovating' in consumer practices. Practices are the set of actions, understanding and final aims that form human behaviour (Reckwitz 2002; Schatzki 1996; Schau et al. 2009). Practices are 'routinized ways in which bodies are moved, objects are handled, subjects are treated, things are described and the world is understood' (Reckwitz 2002: 250). Practices express the implicit knowledge of individuals through their overt and observable actions. Travelling with a car, for instance, is a practice involving a set of explicit rules (the traffic code), implicit understandings to interpret those rules (driving safely, respecting other road users), actions (all the bodily movements to drive the car) and the final aims of their emotional engagement (commuting to work, reaching a holiday destination). Consumption stems from practices (Warde 2005), because practices dictate what desires a consumer can have and which products or services can satisfy those desires. Analogously, we may posit that innovation stems from practices. By adopting certain types of practices, the user unleashes the potential for the innovation of new services.

In practice theory, the role of objects and technology is almost at the same level as individuals. Objects are part of the practices in the same way that their users are. The consumer agency is present together with the 'object agency', that is the set of possibilities that is inscribed into an inanimate object. In 'practice theory, objects are necessary components of many practices—just as indispensable as bodily and mental activities. Carrying out a practice very often means using particular things in a certain way' (Reckwitz 2002: 252). Other streams of research—notably, the actor-network theory (Latour 2005)—acknowledge the importance of objects as *actants*, i.e. as entities that able to perform an activity. Consequently, in addition to having a focus on individual intentions, feelings and thoughts, a researcher should also pay attention to the technical elements forming the new service, such as the options available to the user, their layout on the screen and the technical aspects of the service.

Practices are particularly useful to understand the customers' use of services and innovative services in particular. Services require a direct involvement of the customer, entailing a blurred boundary between the action of the user and the service function that is used. Innovative services present users with new options and functionalities. By using those options, customers become enmeshed with the innovation and actualise innovation.

Observing practices is the preferred way to consider 'innovating' because it combines observing the actions of the users and the features of the object used. On a website, the pressing of the button and the button itself are inherently linked. Instead of investigating the intentions or attitudes of the customer (i.e. her mental state), we can consider the practices of the customer, not as an expression of meaning but as having meaning in themselves. Enquiring into the thought processes of the customer does not provide a true understanding of the purpose of the action but rather addresses a component of the practice itself. As the methodology section of this chapter will demonstrate, the observation of the actions and the investigation of the inner state of the user are both necessary to gauge the practice.

The idea that the concept of 'innovating' is replacing that of innovation seems to have proof in some recent innovations. Consider the launch of the much anticipated Apple Watch, and whether it displays the attributes of 'innovating'. For the first time in the history of Apple, the company seems to have launched a concept rather than a finished product. Consider the differences in the presentation of the iPod and the Apple Watch. In the case of the old iPod launched by Steve Jobs, each and every significant feature of the new product was dutifully described by Jobs: the number of songs, the function of the click wheel, the battery duration and so on, whereas the Apple Watch has relatively vague features, its distribution has been planned for some months after its presentation, and no one can really say what the most relevant application of this new product will be (see Fig. 7.1).

The concept of releasing an unfinished product may be a sign (though not conclusive proof) that an innovative and complex lifestyle product like the Apple Watch needs to be embedded in consumer practices in order to be truly defined. The innovation will embody the practices of Apple consumers, including how they envision using the future Apple Watch, and their fantasies and criticisms regarding the product. In this scenario, the company (in this case, Apple) is essentially the factory implementing the innovations developed by consumer practices.

Another example of innovating practices can be found in Google Glass, an innovative platform with many potential uses (Pace 2013). Google Glass can be a device to store memories of extraordinary experiences (extreme sports) or special moments (a birthday), a new way to share emotions with people from distant places (while visiting an exotic place) and an always-on device allowing access to the web at any time.

Users of Google Glass do not limit its usage to the definition of its main functions and benefits; they establish the nature of this product. Is Google Glass an electronic gadget that can be used in everyday life or is it a product for special occasions, such as holidays? Which meaning prevails will be determined by users through their practices. The Google Glass project has been discontinued by Apple, and the product does not have a wide distribution. Therefore, most of the consumer elaboration takes form in user-generated content, such as YouTube videos, where

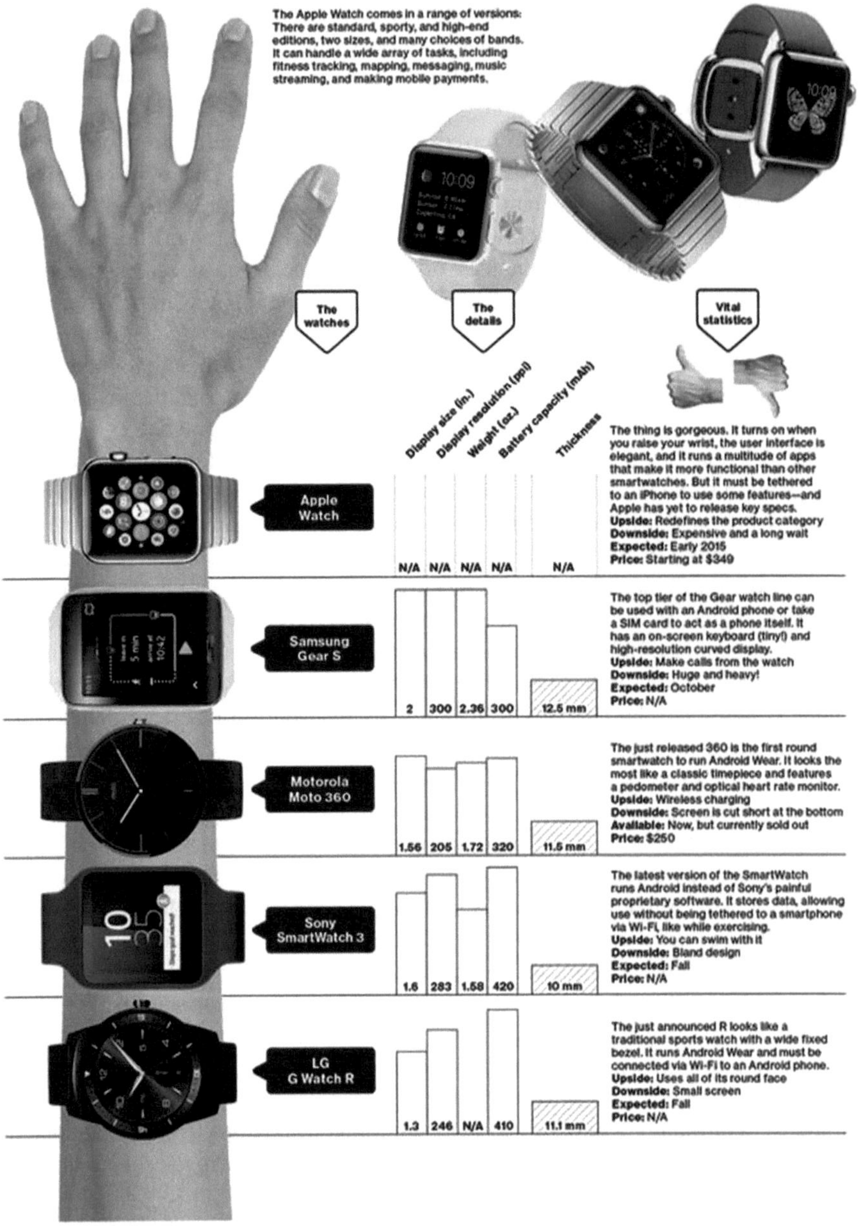

Fig. 7.1 The features of the Apple Watch compared to other devices: generic characteristics to foster 'innovating' (source of the image: www.bloomberg.com)

Fig. 7.2 'Glassing' as a form of 'innovating' (source: image from the Instagram/Twitter account of Barry Bonds)

the possible benefits and risks of this innovation are explored. Consumers invent scenarios where Google Glass is used in common practices, such as walking on a street, being in a relationship with someone or driving a car (Pace 2013). These videos replicate the innovating process of embedding a product in user practices and explore the possible innovations conveyed by the product (see Fig. 7.2).

The concept of 'innovating' is called forth (though in evocative terms) in a case linked to Google Glass. The baseball celebrity Barry Bonds published on his Instagram profile an image of himself calmly drinking some coffee, looking outside and wearing the Google Glass. What is interesting about the image is the slogan that reads, 'I am Glassing'. This slogan shows that innovation is not a noun anymore, it is a verb. It is innovating. Innovating is about shaping innovations through daily practices like drinking coffee in the morning. Bonds' simple gesture defines Google Glass as a daily companion, not a special device; in other words, it becomes an intimate object for introspective moments, rather than a glamorous technological gadget.

The rest of this chapter focuses on investigating which practices users adopt to navigate through a new website and how they make sense of it. Answering the question of how consumers make sense of a new website allows the company to imbue its innovation with meaning and, ultimately, to complete it.

7.3 Methodology

The present empirical study stems from a research project that was intended to study pleasure in online browsing, in particular, in exploring a website that was new to the user. We focus on the online platform Google Earth, a new service that all the participants were using for the first time in the research study. The informants were asked to use Google Earth for 30–40 min while the researcher observed them, noticing their overall behaviour and interaction with the service and collecting the verbalisation of thoughts and sensations. During the observation, from time to time, researchers asked questions regarding the ongoing experience to deepen their understanding of interesting stages in the use of the platform.

After using Google Earth, informants were asked to write an introspective report of their experience. Consumers who participated in our experiment were asked to write, within 24 hours, a narrative expressing their internal state with regard to their experience with the new website (Holbrook and Schindler 2003). Each introspective narrative drafted by consumers was coupled with a short report written by the researcher on the main elements emerging from ethnographic observations.

The objective of the study was to understand how consumers manage their experience with a new website and how they are able to mobilise the elements that the new website makes available to them, using a mixed methodology (Carù et al. 2014) that combined observation and introspection (Carù and Cova 2008). The nature of practices has driven the epistemology of this research and subsequent methodology. As seen above, practices are composed of overt actions and latent unobservable understandings. Both the observation of external actions and introspective accounts of the inner states of the informants are thus necessary to grasp the unfolding of practices. On the one hand, the mere observation of the actions of the informants (such as clicking on a link, scrolling through a page, etc.) would miss the mental references used by the subjects to conduct a given action. For instance, clicking on the name of a city can be conceived by the individual as a way to access some information about that city or a simple attempt to see what can happen upon clicking it. The two cases show different understandings and aims of the action of clicking. On the other hand, focusing exclusively on the words of the subjects to explore their inner states would risk an over-theorisation of the actions by the subjects. In addition, many actions would not have upfront meaning for the subjects themselves, due to the novelty of the explored service. The verbalisation of these actions would not be easy. Observations would compensate for 'people's inabilities to report fully on the complex interweaving of culturally significant behaviours' (Arnould and Wallendorf 1994: 487).

The nature of practices calls for a *flattening* of the epistemology (Bajde 2013), i.e. to remain as close as possible to the context of the study, with the assumption that neither the theoretical background of the researcher nor the observed facts would generate by themselves knowledge about the phenomenon under study. The interactions between the informant and the researcher together with the interactions between different methods (introspective accounts and observations) offer the final understating of the phenomenon. A person browsing an online service is an assemblage of actions, movements, intentions, elements appearing on the screen and computers and servers reacting to the requests of the users. The epistemology supporting the study of this practice requires the acknowledgement of the assemblage, because the 'making of knowledge requires varied deployment of heterogeneous assemblages of humans and nonhumans' (Bajde 2013: 230).

To conduct the research, the following steps were taken: (1) we chose the type of online service to study, i.e. Google Earth, (2) we selected the people involved in the research project, (3) we designed the data collection methods, and (4) we chose the data analysis methods.

1. We were motivated to select Google Earth as the new service to analyse due to the richness of the system and its novelty for the informants. Unlike regular websites with standard options and ways of navigation, Google Earth stimulates rich experiences and a variety of possible approaches. It can be a way to get a sense of exotic places or to plan a working trip. Its use can be hedonic or strictly functional. Google Earth is also similar to common offline maps, but, at the same time, it augments the experience of a normal map. This position between common and extraordinary objects triggers individual research of new ways to use a map and stimulates the innovating practices through which users define the innovative power of Google Earth.
2. To select the informants to analyse, we focused on their familiarity with the web. In order to adopt the new website, consumers should exhibit some form of competence. Consumer research differentiates between two dimensions of competency, familiarity and expertise (Alba and Hutchinson 1987, 2000), which are explained as follows: 'Whereas familiarity accumulates naturally from any product-related encounter (whether its effects benefits are beneficial or not), expertise is measured relative to a performance criterion and implies increased ability' (Alba and Hutchinson 2000: 123). In this study, we essentially assess consumers' level of competence on the basis of their familiarity with online environments. This choice is linked to the problem of calibrating competences (Alba and Hutchinson 2000), a problem that is well known in the field of online research (Pillai and Hofacker 2007). The difficulty with correctly evaluating consumers' competences by means of auto-assessment leads us to concentrate our attention on familiarity. Indeed, Söderlund (2002: 871) argues that 'while consumer familiarity increases, customer expertise is likely to increase'. Consequently, familiarity is regarded as a basic element of consumer competence. When familiarity is high, expertise increases and as a result, so does the overall level of competence. We evaluate familiarity at two levels: general familiarity with the web and specific familiarity with the online environment under investigation. Both results are classified as either high, medium or low. With respect to familiarity with the web, we asked consumers to provide a self-assessment of their web use frequency and the variety of sites visited. A filtering question was included to determine the participants' varying levels of competence. With

respect to familiarity with the website, we asked consumers if they had already used the site that the researcher asked them to browse.

The study is based on a rather varied sample group in terms of general familiarity, ranging from novice (9 subjects), to average (29) and to highly familiar (17) consumers. Conversely, the composition of the sample with regard to familiarity with a specific site was less varied, with the vast majority of consumers being unfamiliar with the site. In this study, we focus on users that experienced Google Earth for the first time. The group was geographically split between large- and medium-sized cities and comprised 55 French and Italian consumers.

3. The data collection included introspective reports by the informants and observations of the researcher while the subjects used Google Earth.

   Narratives permeate consumption practices and, in general, consumption experiences (Chronis 2008). People use stories to make sense of their lives and to build their identities (Shankar et al. 2001). Currently, recording consumers' introspective narratives is considered the best way to access their subjective experience (Belk 2006; Patterson et al. 2008). Starting with the postulate that, 'although agreeing that we can never know anything about a subjective experience as such', it is possible to search for an intermediate solution 'by arguing that we can at least study the reports that are made' (Richardson 1999: 470). The method chosen to account for each consumer's subjective experience is therefore the analysis of introspective narratives. Despite similarities to the category that Wallendorf and Brucks defined as guided introspection (1993), retrospective introspections of this kind differ in that the individual writes his/her narrative without the researcher being present. This assumes a preparation phase during which the individuals involved receive instructions that the researchers have previously tested on themselves (Richardson 1999). In our research, the experiences largely unfolded in these consumers' natural environments—in front of a computer screen at home or at work. Consumers agreed to browse a website that the researcher had chosen for a period of 30–40 min. They also agreed to be observed and to write an introspective report. This helps to bring a double method into play that not only features observations that generate 'small stories' but also introspection that generates 'big stories' (Bamberg 2006). Small stories are composed of ordinary conversational exchanges and the communication of ongoing events. Big stories are made up of life stories and autobiographical accounts.

   Observation brings the specific experiential context into play, allowing a merger of the consumer and researcher horizons (Thompson 1997). Owing to the presence of the researcher as the informants navigated Google Earth, we were able to catch incidents that occur naturally and 'that stimulate real-time interpretive insights and launch systematic analysis of additional data' (Arnould and Wallendorf 1994: 485).

4. Given the complexity of the problem, we developed a mixed method approach to the analysis. The analysis of the introspective reports took a conventional approach, in which the assessment was based upon a number of predefined

conceptual categories. The use of codes (in our case, some were predefined, while others emerged in the course of the analysis) helped us to structure the content (Arnould and Wallendorf 1994; Celsi et al. 1993). For the observations, we employed a more interpretive approach. The reason for this dual analysis lays in the complexity of the material, which included introspective accounts and observations.

7.4 Findings

The empirical findings can classify the practices enacted by consumers into three main emerging themes: calibrating, framing and activating. Calibrating refers to the initial stages of the use of the new service, when users adjust their level of confidence. In this stage, users try to impose their prior ideas on what Google Earth should be (for instance, they might try to categorise it as either a map or a search engine). The use of the new service relies more on the inner thinking of the user and less on the actual options offered by the service.

Framing is the stage when the user selects a familiar activity or point of reference to make sense of the new service. In this stage, the actual possibilities provided by the innovation start to unleash themselves. For instance, simulating flying over unknown places with Google Earth's aerial view could seem like an innovative option for users.

Activating refers to the peak moment, when users embed the innovation into their practices. It is the moment where the user and the software seem to come together and become one entity. Seen from the perspective of subjective inner states, it resembles the moment of flow that has been well documented in the literature of online experiences. However, practice theory includes the object and suggests a broader view, one where the user and the innovation entwine. The success of innovative products does not reside in sense satisfaction or ease of use but rather in this moment, when the new service expresses its full potential as brought out by the user. Innovating, then, is about an empowered individual and an actualised technology.

7.4.1 *Calibrating*

Many people approaching Google Earth overestimated their abilities in this type of environment. The users exhibited a miscalibration in the correspondence between confidence and accuracy of knowledge (Pillai and Hofacker 2007). This miscalibration may affect the use of the website, because 'overconfident individuals would face bottlenecks during their navigation and wouldn't know how to remedy the situation since they are unaware of their ignorance' (Pillai and Hofacker 2007: 264).

Individuals who maintain that they have a good knowledge of the web show some tendency to consider their general knowledge as a benchmark for any website that they may visit. This leads to overconfidence and to deadlocks during navigation, which results in frustration and the possible adoption of different cognitive strategies. The most common is a defence strategy to save face, in which the individual does not regard his/her knowledge as faulty or incomplete. This strategy is common when facing difficulties using web-based services (Meuter et al. 2000). Individuals prefer to blame the website or other factors unrelated to his/her knowledge when they are unable to use the website effectively. This strategy can be seen in the case of one study participant, Myriam, who approached Google Earth like a normal search engine (instead of an interactive map), inserting the name of the place she wanted to visit, which did not produce the results she expected. Myriam explained:

> I disliked not being able to see, for example, Austria or some place with mountains, because it did not understand the name of the town or place I was asking for. So, I closed the program without being able to use the 3D function. What a bore! I think you have to insert the original language; but how do I know how to write things in German if I do not speak German!! Frustrating... (source: Myriam's interview).

Later in her navigation, she is disappointed that the images are not in real time, an expectation based on the structure of other online services, such as webcams or chat:

> The only disappointment is that it would have been fantastic if the images had been in real time, so that you could also see the people... we can imagine why ... (Myriam's interview).

In another instance, the participant did not type the name of the city correctly when trying to locate it on the Google Earth map and blamed the website for searching for incorrect locations:

> At the beginning, it gave me some trouble because, as there were other cities called Seville, it automatically searched for these (Myriam's interview).

The same participant claims:

> A satisfactory experience, even if I have to say that once you finally understand how it functions, or better, how it wants us to write the name of the towns, you have to feel happy, because it takes you two hours to work it out and it is pretty frustrating (Myriam).

By discovering that his/her knowledge is insufficient, a consumer hypothetically has an opportunity to start exploring the options of the website. Overconfident individuals do not undertake such exploration, seeming to prefer defensive strategies. These individuals show some unwillingness to start an investigation to compensate for the gap in their knowledge.

Conversely, underconfident consumers show some tendency towards discovery and investigation:

> I immediately clicked on a random point of the planet and, in a few seconds, I was in Miami like in a tele-transport. I started to use the toolbar (zoom, rotate the map, etc.) until I got to

> see the shapes of buildings, bridges; I felt a nice sensation because I could see any place and move everywhere (Alessandro).

7.4.2 Framing

Users tend to compare the experience of using a website to some familiar activity or point of reference. They select another online or offline environment as a frame of reference to use in order to gain familiarity with the new website. For instance, first-time users of Google Earth would use known places to search for unknown places. In the following excerpt, a new place (Helsinki) is compared to a known one (Oslo):

> In any case, it was entertaining to see how places that I had visited, e.g., Helsinki, are not as I had perceived them. I thought it was very wild and green, but from what I could see it seemed very built up like Oslo, which I thought were both typical northern European cities, very green (Myriam).

Participants also used tools that were common in their experience, though not necessarily common to the majority of the population. For instance, geographical coordinates are a common element in offline maps, and the following subject uses them to navigate Google Earth:

> I almost immediately felt at ease: the graphics helped me a lot. I was not ready to use the highlighted co-ordinates to change position, but seeing them on the screen was reassuring; it was as if I was moving in a territory that I knew how to control (Armando).

When the website provides tools to put the user in control, the feeling of being at ease emerges, indicating that clear tools help users to enjoy a website and encourage them to explore it.

The activity through which the user explores the new website does not necessarily have to be concrete. Internal simulations wherein the user imagines how they might navigate the website can also provide an experience of the site. Consider, for instance, this excerpt from a narrative regarding Google Earth:

> I sat down and he launched a magnificent site with a beautiful blue planet. I quickly saw that I could travel the world without leaving my chair. Getting aerial views without any airsickness was fantastic. After a few minutes, I started to click on the places I wanted to go (Armando).

In this excerpt, the user expresses an imaginary activity (flying over the planet), but does experience what flying around might feel like. He assimilates the memory of aerial views that one can see in other contexts, such as photographic reports, in order to have the above-mentioned experience. He then applies his imaginary travel to search for known places on Earth.

The past can also be an anchor for the subject:

> And I liked the idea of being able to go and see what was happening in a place where I knew my friend was... I was astonished! I looked at the monitor as happy as a child on Christmas Day. I looked at the images, I lingered and remembered events linked to the places that I

> was now observing from above. Amused and fascinated, I commented on what I saw (Armando).

This anchoring to normality provides the user with a sense of 'home' regarding the innovation.

The application of familiar practices may lead to unsuccessful attempts to make sense of the new environment. For instance, if Google Earth is framed as software that is similar to Word applications, it may lose its novelty for the user and create some false expectations:

> My attention moved to the "bulky" toolbars that struck me for the high number of options and raised a positive sensation in me. Apparently, they were like the toolbar in Microsoft Word Art and thus I thought that I could have had fun with the different available tool. [. . .] At the same time, I felt irritated when I could not do what I wanted, or when the computer seemed to act by itself or it did not understand my intentions (Claudia).

The same may happen for common buttons that one finds online (like the 'download' button) if its use is associated with negative experiences like getting software viruses:

> Seeing "download" on the left made me suspicious: I don't usually download anything via the network, particularly.exe files (Armando).

Common practices can thus lead to making sense of the innovation or failing to do so, depending on the subjective experiences of the user.

7.4.3 Activating

Once users get past the early stages of becoming familiar with the innovation through normal practices, they can then start to truly understand the innovation. The innovation then enters the practice of the consumer, who in turn expands his range of competency and moves from normal practices (searching for known places, replicating daily routines) to innovative ones:

> Little by little I got to know the dashboard at the bottom of the screen and discovered how to change the perspective. I could go down to ground level and add the third dimension (Armando).

When users engage with the full potential of the website, they express enjoyment and may even enter a state of flow:

> I cannot take my eyes off the screen, eager to find familiar places with a smile on my lips (Caroline).

This enthusiasm may encourage further discoveries of the website's features by enabling users to go beyond their concerns and to be lost in the online space.

A sense of limitless experience may emerge once the website's novelty is absorbed by the user:

> After several minutes of hazardous manipulations, the earth turned in front of my eyes, it came closer, then went away, it enlarged and let me foresee countless possible experiences (Roger).

Users move between boredom and excitement. In the first stage, the innovation does not get past the normality of common practices. In the second stage, the user overcomes her expected limitations, and the innovation gains meaning. In the third stage, the individual apprehends the novelty, activates both the options of the system and his new competencies, a sense of play emerges, and the user begins to enjoy the new features and the exploration of new corners of the online environment.

The three stages examined here show that users find their way through innovation in different ways. Still, regardless of the starting point and the path followed, the adoption can be full and complete. For instance, a user can employ Google Earth only as a more detailed map as compared to regular paper maps. Then, that user can discover that zooming (which is a typical action to enjoy more details) implies movement and that this movement can also be used to go elsewhere. This may lead the user to explore other options of the platform, such as heading elsewhere with the help of a compass. The reference frame of a map can introduce the user to other frames of references, such as Google Earth as a flying machine. The continuous interaction between the system and the user is a cocreation of the final experience of innovation.

7.5 Discussion and Conclusions

The participants investigated in this study progressed through two primary phases as they explored the new online environment:

1. Application of common frames of references and normal practices to the new environment. For instance, a user can frame Google Earth as a landscape that one sees while flying above the earth's surface, while another user can think about it as a different form of search engine for geographical places.
2. If the previous stage is successful, the individual is able to further explore the options and features of the new environment. It is in this second phase that the previous frames and practices are challenged, and the user achieves a greater sense of the innovation.

These two phases suggest an interaction between normal practices and new practices (see Fig. 7.3).

It is through these interactions that the 'innovating' process takes place. Individuals adopt their own approaches in using the innovation, and through these approaches, the innovation obtains its real purpose. The company offering the innovation may look at what the recurring practices and frameworks adopted by the users are, and how the subjects move from those to the adoption of different

Fig. 7.3 Interaction between normal practices and new practices

options. By observing the user, the company can gain a greater understanding of the potential of its innovation. For the company, this observation is not just a way to improve its innovation or to have ideas, but it is a method for making sense of what it has offered to the market. The practices of users embody the true essence of an innovation, because it is only through interaction with the new environment that the user can shape the innovation meaning.

Play represents a central element in this process. Many informants in this study, once they overcome a possible sense of boredom, start to play with the innovation; they push its limits by recombining its options and searching for a sense of enjoyment that goes beyond it use. Consumers can apply a ludic agency (Kozinets et al. 2004) over the environment, combining respect for the rules and slight transgression from them. This combination of going beyond the limits without breaking them pushes the user and the environment beyond their ascribed limits. Play helps discover the range of opportunities provided by the innovation. The sense of play is one of the typical behaviours online, where thousands of users may aggregate out of a mere sense of curiosity around an initiative and generate an outcome based on collective intelligence. For instance, some people enjoy the task of classifying Mars craters in the Clickworkers project by NASA. This task is done for its own sake as part of a game. The final result is the marking and classification of the Mars surface, which is an endeavour that neither a single scientist nor a set of computers can achieve. Innovative services and products can get the same treatment online if they engage the playful side of consumers.

In a few years, we could live in the world of 'makers' (Anderson 2012) where everyone will have tools (like 3D printers) to manufacture certain goods and express personal creativity. This final convergence between companies and customers is currently in a state of transition, with companies and customers converging around 'innovating' as a common ground to develop new products and services.

Our research also presents some managerial implications. The first managerial implication is that companies can shape innovation such that some degree of serendipity is left to the user. Users can play with new services, if those services allow for some flexibility. This serendipitous exploration can lead to the discovery of practices that the company can absorb for future advancement of the innovation.

A further implication concerns the managerial mindset for innovation. Today, many managers and innovators rely on the idea that Big Data contains the secret of success for innovation. Any tiny novelty introduced in a digital service can be studied by designing experiments and collecting data about alternative variants of the innovation, from the background colour of a website to the position of a banner and from the size of a smartphone display to the weight of a smartwatch. A further step in the Big Data mindset has to do with the phenomenon of the quantified self, in which data are automatically generated by the body of the consumer. Big Data can convey rich and actionable knowledge that managers can use for their decisions. However, the search for big and even 'bigger' data may present some risks. A purely numeric approach, dealing with consumers as digital profiles and not as people, may hide the insights that managers can draw from the introspections of users and by observing them. This contact with the user on the other side of the screen does not necessarily entail the deployment of a research team (though it would be ideal to catch the subtle meanings of the user experience). Some consumers are eager to share their experiences as they are unfolding. For instance, 'haul videos' are popular videos where a consumer shows the opening of purchased items, commenting on characteristics and other aspects of the consumption experience as it happens. A similar example is in the video game context, where advanced users comment on their gaming session while they play, so that other users can enjoy and learn from it. These phenomena suggest that managers and innovators have a rich repository of first-hand innovating practices that can inform their decisions.

References

Akrich, M., Callon, M., & Latour, B. (2002a). The key to success in innovation. Part I: The art of interessement. *International Journal of Innovation Management, 6*(2), 187–206.

Akrich, M., Callon, M., & Latour, B. (2002b). The key to success in innovation. Part I: The art of choosing good spokespersons. *International Journal of Innovation Management, 6*(2), 207–225.

Alba, J. W., & Hutchinson, J. W. (1987). Dimensions of consumer expertise. *Journal of Consumer Research, 13*(4), 411–454.

Alba, J. W., & Hutchinson, J. W. (2000). Knowledge calibration: What consumers know and what they think they know. *Journal of Consumer Research, 27*(2), 123–156.

Anderson, C. (2012). *Makers: The new industrial revolution*. New York: Crown Business.

Arnould, E. J. (1989). Toward a broadened theory of preference formation and the diffusion of innovations: Cases from Zinder Province, Niger Republic. *Journal of Consumer Research, 16*(2), 239–267.

Arnould, E. J., & Thompson, C. J. (2005). Consumer culture theory (CCT): Twenty years of research. *Journal of Consumer Research, 31*(4), 868–882.

Arnould, E., & Wallendorf, M. (1994). Market-oriented ethnography: Interpretation building and marketing strategy formulation. *Journal of Marketing Research, 31*(4), 484–504.

Bajde, D. (2013). Consumer culture theory (re)visits actor-network theory: Flattening consumption studies. *Marketing Theory, 13*(2), 227–242.
Bamberg, M. (2006). Stories: Big or small: Why do we care? *Narrative Inquiry, 16*(1), 139–147.
Bass, F. M. (1969). A new product growth for model consumer durables. *Management Science, 15* (5), 215–228.
Belk, R. (Ed.). (2006). *Handbook of qualitative research methods in marketing*. Cheltenham: Edward Elgar.
Carù, A., & Cova, B. (2008). Small versus big stories in framing consumption experiences. *Qualitative Market Research: An International Journal, 11*(2), 166–176.
Carù, A., Cova, B., & Pace, S. (2014). Combining observation and introspection in practice: A contextualized account of the evolution of consumer studies. *Management Decision, 52*(4), 777–793.
Celsi, R. L., Rose, R. L., & Leigh, T. W. (1993). An exploration of high-risk leisure consumption through skydiving. *Journal of Consumer Research, 20*(1), 1–23.
Chronis, A. (2008). Co-constructing the narrative experience: Staging and consuming the American civil war at Gettysburg. *Journal of Marketing Management, 24*(1-2), 5–27.
Cochoy, F. (2015). Consumers at work, or curiosity at play? Revisiting the prosumption/value. Cocreation debate with smartphones and two-dimensional bar codes. *Marketing Theory, 15*(2), 133–153.
Cova, B., & Dalli, D. (2009). Working consumers: The next step in marketing theory? *Marketing Theory, 9*(3), 315–339.
Fuchs, C., Prandelli, E., & Schreier, M. (2010). The psychological effects of empowerment strategies on consumers' product demand. *Journal of Marketing, 74*(1), 65–79.
Füller, J., Jawecki, G., & Mühlbacher, H. (2007). Innovation creation by online basketball communities. *Journal of Business Research, 60*(1), 60–71.
Gatignon, H., & Robertson, T. S. (1985). A propositional inventory for new diffusion research. *Journal of Consumer Research, 11*(4), 849–867.
Holbrook, M. B., & Schindler, R. M. (2003). Nostalgic bonding: Exploring the role of nostalgia in the consumption experience. *Journal of Consumer Behaviour, 3*(2), 107–127.
Holt, D. B. (1995). How consumers consume: A typology of consumption practices. *Journal of Consumer Research, 22*(1), 1–16.
Kozinets, R. V., Sherry, J. F., Storm, D., Duhachek, A., Nuttavuthisit, K., & Deberry-Spence, B. (2004). Ludic agency and retail spectacle. *Journal of Consumer Research, 31*(3), 658–672.
Latour, B. (2005). *Reassembling the social – An introduction to actor-network-theory*. Oxford: Oxford University Press.
Meuter, M., Ostrom, A. J., Roundtree, R. I., & Bitner, M. J. (2000). Self-service technologies: Understanding customer satisfaction with technology-based service encounters. *Journal of Marketing, 64*(3), 50–64.
Pace, S. (2013). Looking at innovation through CCT glasses: Consumer culture theory and Google glass innovation. *Journal of Innovation Management, 1*(1), 38–54.
Patterson, A., Hodgson, J., & Shi, J. (2008). Chronicles of 'customer experience': The downfall of Lewis' foretold. *Journal of Marketing Management, 24*(1-2), 29–46.
Pillai, K. G., & Hofacker, C. (2007). Calibration of consumer knowledge of the web. *International Journal of Research in Marketing, 24*(3), 254–267.
Prahalad, C. K., & Ramaswamy, V. (2004). *The future of competition: Co-creating unique value with customers*. Boston: Harvard Business Press.
Reckwitz, A. (2002). Toward a theory of social practices: A development in culturalist theorizing. *European Journal of Social Theory, 5*(2), 243–263.
Richardson, A. (1999). Subjective experience: Its conceptual status, method of investigation, and psychological significance. *The Journal of Psychology, 133*(5), 469–485.
Ritzer, G. (2014). Prosumption: Evolution, revolution, or eternal return of the same? *Journal of Consumer Culture, 14*(1), 3–24.

Schatzki, T. R. (1996). *Social practices: A Wittgensteinian approach to human activity and the social*. New York: Cambridge University Press.
Schau, H. J., Muñiz, A. M., & Arnould, E. J. (2009). How brand community practices create value. *Journal of Marketing, 73*(5), 30–51.
Schreier, M., Dahl, D., & Fuchs, C. (2012). The innovation effect of user design: Exploring consumers' innovation perceptions of firms selling products designed by users. *Journal of Marketing, 76*(5), 18–32.
Shankar, A., Elliott, R., & Goulding, C. (2001). Understanding consumption: Contributions from a narrative perspective. *Journal of Marketing Management, 17*(3-4), 429–453.
Shove, E., & Pantzar, M. (2005). Consumers, producers and practices: Understanding the invention and reinvention of Nordic walking. *Journal of Consumer Culture, 5*(1), 43–64.
Söderlund, M. (2002). Customer familiarity and its effects on satisfaction and behavioral intentions. *Psychology and Marketing, 19*(10), 861–880.
Thompson, C. J. (1997). Interpreting consumers: A hermeneutical framework for deriving marketing insights from the texts of consumers consumption stories. *Journal of Marketing Research, 34*(November), 438–455.
Troilo, G., & Vicari, S. (2008). The role of unexpected markets events in market creation strategies. In A. Carù & K. Tollin (Eds.), *Strategic market creation*. Hoboken, NJ: Wiley.
Vargo, S. L., & Lusch, R. F. (2004). Evolving to a new dominant logic for marketing. *Journal of Marketing, 68*(1), 1–17.
Von Hippel, E. (2005). *Democratizing innovation*. Cambridge: The MIT Press.
Wallendorf, M., & Brucks, M. (1993). Introspection in consumer research: implementation and implications. *Journal of Consumer Research, 20*(3), 339–359.
Warde, A. (2005). Consumption and theories of practice. *Journal of Consumer Culture, 5*(2), 131–153.

Chapter 8
Practices of Service Innovation Diffusion

Daniela Corsaro, Roberta Sebastiani, and Cristina Mele

Abstract Service innovation has often been conceptualized as a linear multistage development process within a main innovator system. In this logic, innovation is communicated through certain channels over time, and adopters play a key role in the process of innovation diffusion. This study moves away from this conceptualization and proposes service innovation diffusion (SID) as an emergent process of co-construction and sense-making by companies, customers and other partners. Such actors enact practices by interacting and integrating resources in a social, economic and cultural context. In particular, thanks to the analysis of ten case studies, we identify specific practices of SID in which actors make sense of and share meanings about innovation, both individually and collectively. These practices enable confidence in service innovation, creation of shared meanings among actors and leveraging of actors' competencies. The study also highlights the dual role of users, who can simultaneously act as adopters and rejectors.

8.1 Introduction

The diffusion of innovation is "the process by which an innovation is communicated through certain channels over time among members of a social system" (Rogers 1995: 5). This definition is representative of the mainstream conceptualization of diffusion in innovation literature in the last 50 years. Diffusion research is indeed a particular type of communication research, with a linear stage model of development, involving different categories of adopters (Mahajan et al. 1990;

D. Corsaro (✉) • R. Sebastiani
Department of Economics and Business Management, Catholic University of The Scared Heart, Milan, Italy
e-mail: daniela.corsaro@unicatt.it; roberta.sebastiani@unicatt.it

C. Mele
Department of Economics, Management, Institution, University of Napoli Federico II, Naples, Italy
e-mail: cristina.mele@unina.it

T. Russo-Spena et al. (eds.), *Innovating in Practice*,
DOI 10.1007/978-3-319-43380-6_8

Rogers 2010). The main focus of extant studies has been on the elements that foster or slow the adoption process, especially the role of technology. Recently, some scholars have begun to address a wider view of innovation diffusion to move beyond the individual as principal change agent in the communication process to include the social context and the learning conditions (MacVaugh and Schiavone 2010; Wenger 1998).

The need for a fresh conceptualization of innovation diffusion is linked to overcoming innovation as a linear multistage development process within a main innovator system (Mele et al. 2014). Service research has suggested that service innovation does not result from a single economic agent but from a complex process in which several actors interact and integrate resources (Edvardsson and Tronvoll 2013; Lusch and Nambisan 2015). Viewing service innovation as an evolutionary, non-linear, contextual process of multiple actors (Mele et al. 2014; Möller 2010) shifts the interest from the output of innovation to the social mechanisms that influence its development and diffusion. In this regard, it is important to understand how parties align perceptions and expectations, exchange knowledge and share meanings in practising service innovation.

In a similar way, studies adopting a practice-based approach have conceptualized innovation not just as the outcome of deliberate activities introducing discontinuities but as produced on a daily basis as "a continuous process situated in work practices" (Gherardi 2012a: 219). Innovation seems to emerge from the interaction among agents, knowledge and other symbolic and material resources "that were previously unconnected, and that are slowly put in a relationship of interdependence" (Callon 1999 cit. in Gherardi 2012b: 224).

By taking into consideration criticisms of the traditional view of diffusion research and building on recent conceptualizations of service innovation and the practice-based approach, this chapter aims to offer a fresh understanding of innovation diffusion. We carried out ten practice-based case studies on actors engaged in innovation processes. Epistemologically, relational thinking is adopted (Østerlund and Carlile 2005) through which three sociomaterial practices of service innovation diffusion (SID) emerge: (1) enabling confidence, (2) creating shared meanings and (3) leveraging actors' competencies.

Innovation diffusion is framed not as a (final) stage after other steps but as innovating in practice; that is, it is seen not only as a matter of communication and adoption but as a process of co-construction and shared understanding among a multiplicity of actors. Innovation emerges as a collective phenomenon that goes beyond simply producing new artefacts to creating new practices, meanings and values (Mele et al. 2014; Russo-Spena and Mele 2016) with companies, customers and other partners, performing a process of co-cocreation and sense-making in a social, economic and cultural context (Edvardsson et al. 2010).

The remainder of this chapter is composed as follows. Next, the authors review the literature on innovation diffusion and service research and on the practice-based approach. Then they describe the methodology of the research and present the findings. The chapter ends with conclusions, implications and suggestions for further research.

8.2 Theoretical Background

8.2.1 Innovation Diffusion and Service Research

Diffusion research is a wide interdisciplinary area ranging from education to sociology, economics and information systems (Mahajan and Peterson 1985), whereas the mainstream conceptualization is given by Rogers' (1962, 2010) theory of diffusion of innovation (DOI), which describes the adoption or non-adoption of new technologies. The DOI is depicted as a special type of communication "in that the messages are concerned with new ideas" (Rogers 1995: 5). The linear process starts with personal knowledge about new technology and proceeds through four other steps—persuasion, decision, implementation and confirmation—which also address conditions and obstacles to the final outcome, i.e. the adoption or rejection of the new technology (Rogers 2010).

Most studies on DOI have focused on technology adoption related to new products/services. Recently, however, some scholars have gone beyond the mere technological standpoint and the communication issue. MacVaugh and Schiavone (2010: 198) stated that "innovation diffusion is affected by technological, social and learning conditions while operating in the contextual domain of the individual, community or market/industry". Individuals' decisions are linked to a process of information exchange and sense-making to include social communities' acceptance. It is not the utility but the social context that influences the use of new technology. A process of social contagion emerges: "diffusion is not only the means by which innovations become useful by being spread throughout a population, it is also an intrinsic part of the innovation process, as learning, imitation, and feedback effects which arise during the spread of a new technology enhance the original innovation" (Hall 2006: 460).

Furthermore, two main contributions motivate a fresh understanding of innovation not as linear and multistage but as a collective effort.

Firstly, scholars have introduced a network perspective in service innovation (Agarwal and Selen 2009; Ordanini and Maglio 2009; Voss and Hsuan 2009). The creation of new proposals is increasingly the result of interaction between different players within inter-organizational collaborations (Vanhaverbeke and Cloodt 2006). Innovation does not occur in isolation but increasingly within a network setting, which includes the interplay of different and committed actors, where the innovation activities of the firm must take into account the complementary innovation activities of other actors (Perks and Jeffery 2006). Companies present higher levels of interdependence, generating a complex network of vertical and horizontal relations, formal or not, which result in a greater external collaboration for innovation (Tidd et al. 2001). Innovation is generated by the resources that a company is able to develop internally and through the external actors' access to other resources (Snehota and Hakansson 1995). Through exploiting external relationships, firms realize the co-development of innovative products and services (Gulati and Gargiulo 1999). The internal/external divide at the basis of traditional stage gate or funnel models blurs in favour of a network-to-network conceptualization of the innovative process. The network perspective opens the linear model to external contributions, which lead to the emergence of the concept of service network (Henneberg et al. 2013). Innovation networks include all actors involved in the generation, commercialization and diffusion of new and improved products, processes and services. Different actors can share knowledge resources in order to generate new ideas and bring them to market (Chesbrough 2003; Hagedoorn 2002).

The second insight for a wider conceptualization of service innovation is that scholars have argued that effective service innovation requires the integration of the company and its partners' capabilities with customers' needs (Dougherty 2004). Such ambition involves understanding how the complex social processes of introducing new services interact with using the services from a customer perspective (Leonard-Barton 1998). The focus moves from the output (the new service) to the process coming out of actors' relationships and within a set of practices that overcome industry borders (Bessant and Tidd 2007). Innovation is enabled collectively in the context of relational interaction (Ballantyne and Varey 2006; Grönroos 2010), where it is difficult to draw the line between inputs and outputs. The relational context, in turn, is not a given entity; it is shaped by individuals interacting and guided by a set of value drivers that vary over time, depending on the space–time context of the subjects. The competitive environment becomes a "context" instead of a structure.

Exploring heterogeneity in the capabilities and skills of individuals is an important source of innovation and learning (Cohen and Levinthal 1990). When a person is in contact with different groups of people, it increases the likelihood of making a concrete contribution to the resolution of problems (Perry-Smith and Shalley 2003). The complementarity between these people's knowledge and skills represents a stimulus to the development of new ideas and renders their diffusion more effective.

The conceptualization of service innovation as a complex social process means that the diffusion does not happen only at an individual level; rather, one should look at the different actors involved in the process, which, in turn, implies the application of practices at collective level.

8.2.2 A Practice-Based Approach to Innovation

Central to the practice perspective is acknowledgement of the social, historical and structural contexts in which actions take place (Corradi et al. 2010; Erden et al. 2014; Nicolini et al. 2003; Russo-Spena and Mele 2012). A practice is conceived as a way of doing in which "human and non-human agents constantly interact and where knowing, doing and innovating interweave" (Gherardi 2012a: 228–229).

Theories of practice adopt a "post-humanist" epistemology (Gherardi 2015) with relational thinking based on interdependencies between subject and object, person and world, and networks and society (Østerlund and Carlile 2005). By going beyond the traditional dualisms (i.e. action/structure, human/non-human, etc.), these theories address not the individual actions as building blocks of the social but social practices with a plurality of artefacts, actors and institutions (Gherardi 2001). The practice lens joins the individual and collective dimensions and the human and technological elements and conceptualizes innovation as the product of connections (Gherardi 2012a). It develops through working, knowing and learning: it is not a separate activity but a continuous process in which old and new situated knowledge coexist in a constant tension. Innovation comes from deliberate, as well as unforeseen, activities, following an array of ideas and creating multiple connections in action within an endlessly changing process (Gherardi 2012a, b; Nicolini 2012). Russo-Spena and Mele (2016) recently addressed innovating as a set of collective practices performed by a knowledgeable community: an emergent process arising through negotiation and alignment among actors, resources, activities and aims.

The practice-based approach advocates the social, contextual and process-oriented views of innovation. Innovation is a participative social process (Corradi et al. 2010) or a "cocreation process" (Russo-Spena and Mele 2012), which does not result from a single company's processes but rather is generated in practical constellations (Schatzki et al. 2001) in which actions are performed and resources are used and integrated by multiple actors (Mele and Russo-Spena 2015).

Methodologically, the focus of extant research has been on describing the sociomaterial mechanisms that enable innovation to develop in time and space and through multiple heterogeneous actors (Nicolini 2012). Gherardi (2010: 505) proposed three levels of analysis (Table 8.1) to address the idea that "a stabilized way of doing things becomes a practice when it is institutionalized and made normatively accountable both for its practitioners and for those who view it from outside".

The first level is "practices from outside". The investigation is on the pattern of activities, which are socially recognized and normatively sustained: "The recursiveness of practices is the element that enables both practitioners and researchers to recognize a practice as practice" (Gherardi 2010: 505). A practice is thus a way of doing, which is socially recognized as an institutionalized action: it is the actors' shared understanding that permits the replication of activities.

The second level is "practices from inside". Scholars have adopted the perspective of the activity, "with its temporality and processuality, as well as the emergent and negotiated order of the action being done" (Gherardi 2010: 505). Innovation practices are systematic actions carried out by actors using resources in their value search (Mele and Russo-Spena 2015). They are practical accomplishments that emerge from the actors' ability to use and integrate their capabilities and competences in the complex net of interactions and relationships among people and material artefacts (Gherardi 2015).

The third level is related to the social outcomes of individual practices, in terms of their influence in creating and recreating social structures and institutions. Scholars have focused on practices in their "doing" of society (Bourdieu 1977). A practice involves the production of social relations: "Practices are not only

Table 8.1 A practice-based analytical framework

Levels of analysis	Definition of practice	Research question
Practice from outside	A pattern of activities that are socially recognized and normatively sustained	What is the "object" of the practice?
Practice from inside	A knowledgeable collective action	How is knowing-in-practice enacted?
Practice as social outcomes	The social is the effect of interconnected practices and their circuit of reproduction	What does doing a practice do?

Source: Gherardi (2010: 507)

recurrent patterns of action (level of production) but also recurrent patterns of socially sustained action (production and reproduction). What people produce in their situated practices is not only work, but also the (re)production of society" (Rouse 2002: 16).

8.3 Research Methodology

In studying SID, the present work follows suggestions by scholars to move the unit of analysis from dyadic relationships to practices across networks of actors (Schau et al. 2009): "the methodological principle of 'follow the practices' acquires concrete meaning when the researcher observes a situated practice and moves up from it to the institutional order or moves down from it to the individual-in-situation. Or in other words, when she/he explores 'a connective web'" (Gherardi and Nicolini 2006: xviii).

The practice-based approach advocates a contextual and process-oriented view of innovation (Gherardi 2012b). Researchers can study the process of innovating by examining developed and shared systems of practices. Focusing on practices enables analysis of the social connections among individuals, collectives, organizations, institutions and the social contexts in which these connections are formed. As Barnes (2001: 17) affirmed, "practice is all there is to study and describe".

We chose ten case studies of SID (Table 8.2) and focused on how practices develop according to the levels addressed by Gherardi (2010), that is, "from outside" and "from inside".

The investigation was based on analysis of archival records and interviews with key informants in the company's service network such as CEOs, service development personnel, sales managers, service managers, business partners and customers. A total of 62 interviews were conducted, with each lasting approximately from 40 min to 2 h. The interviews were tape-recorded and transcribed. The interviews aimed to let the idea of practices be reconstructed.

The researchers analysed archival materials in order to gather additional background information and to substantiate the interview data. In addition, it adopted an

Table 8.2 The case studies

Firm	Type of business	Number of interviews
Aggiustami.it	Search engine that compares body shops' prices for car repair	3 (CEO, 2 customers)
Bravosolution.it	Provider of supply management excellence, delivered through software, professional services and category expertise	7 (CEO, marketing manager, account director, 2 suppliers, 2 customers)
Diet-to-Go	Diet delivery service that provides nutritionally balanced, freshly prepared, real food for weight loss and a healthier lifestyle	5 (CEO, franchising manager, 3 customers)
Innovuum	Open and collaborative innovation platform	6 (CEO, 3 managers, 2 customers)
Just Eat	Online takeaway ordering service	9 (CEO, marketing manager, sales manager, 6 customers)
Neronote	E-store of made-to-measure, personalized, made in Italy shirts	7 (CEO, marketing manager, web manager, contact centre manager, 3 customers)
Sfogliaviaggi	Tour operator specializing in accessible and disability travel and tours	4 (CEO, communication manager, 2 customers)
Smartika	Social lending	7 (CEO, 2 managers, 4 customers)
Tabbid.com	Social network and online marketplace for odd jobbers	5 (general director, marketing manager, 3 customers)
Zalando	E-commerce company specialized in selling shoes, clothing and other fashion and lifestyle products online	4 (country manager Italy, 3 customers)

expert judgement technique (Coffey and Atkinson 1996) to select interviewees who could provide insights (Adams and Schvaneveldt 1985).

Data collection and data analysis were carried out by the research team to achieve complementary insights and enhanced confidence in the findings (Eisenhardt 1989). The data analysis was based on detailed case study write-ups for each case. The narrative approach was then used to present the findings (Czarniawska 2004; Hummel 1991). Every narrative offered a plausible story for a particular outcome and the connected sequences of events, thereby reducing equivocality (Weick 1995).

8.4 Practices of Service Innovation Diffusion

In the analysed case studies, SID is not simply a process of communication but is spread through a process of sense-making and social construction by a multiplicity of actors in their relations with resources and contexts.

By adopting the lens of the practice from outside, the *object* of SID is not only to communicate the service innovation to foster its adoption; in addition, actors interact and integrate resources to co-construct a new service provision for multiple beneficiaries. The diffusion process is not linear nor always intended, but the innovation emerges from a set of interactions in which actors have the opportunity to enable confidence and increase support to each other with the specific aim of creating a new service provision. Service innovation may undergo a lengthy period of negotiation among potential adopters, in which it is discussed, contested and reframed. New practices arise when actors share knowledge and ideas in order to continuously refine and improve the new proposal. The diffusion happens jointly with development of the new solution. Thus, the success or failure of a service innovation is related to how it is co-constructed by the actors and the stories that are told in its development and diffusion.

Going deeper into the activities being performed (i.e. from inside), three sets of actions and relations emerge as specific practices of SID in which actors make sense and share meanings about the newness, both individually and collectively:

- *Enabling confidence* in the service innovation concerning how actors are trusted to interact and integrate resources
- *Creating shared meanings* among actors with the aim of grounding service innovation in sense-giving and sense-making
- *Leveraging actors' competencies* to manage and support resource integration by widening the network

8.4.1 Enabling Confidence

Actors promote service innovation by letting users and other partners become confident in the new value proposal. Such assurance acts as a glue for the relationships at the heart of innovation. By interacting, actors develop the confidence that they can apply their abilities to develop a service innovation and to succeed together.

An interesting example in this sense is represented by Tabbid.com, which is a platform that connects people who are looking to outsource their daily activities when they have a small budget. Tabbid helps people to devote their free time to the things they love most with the help of the community. Benefits for the publishers of announcements include possibilities to enjoy more free time to be used in other things, to publish job adverts free of charge and to choose what price to pay for a specific job. At the same time, "Tabbiders", who make their tasks available, can provide "network services" based on strong social and mobile technologies and geolocation to engage people around them. Tabbiders who are reliable and serious will be most often requested by users. The Tabbiders' quality is determined by the community. However, it has been quite difficult to spread the new service idea, mostly because potential users tend not to trust people they do not know and are reluctant to allow them to enter their houses. To overcome this problem, the company decided to introduce a new practice: rather than just filling out a form to describe what they are able to do, Tabbiders now have to send a video presenting themselves and guaranteeing that they are able to do a specific job. In addition, users can write reviews on Tabbiders, and these are visible to all other users.

The socialization practice enabled the development of such service; persisting in increasing people's confidence made it possible to overcome barriers and innovate.

In the Sfogliaviaggi case, actors faced several challenges in delivering a new service for all-inclusive accessible travel and tours. One of the main issues was the lack of specialized professionals who are capable of meeting the requests of tourists with disabilities, especially during guided tours in museums and exhibitions.

> I realized that I had to create part of the network necessary to deliver the service. I contacted several organizations in this field and during some meetings, I explained my project to create an all-inclusive cultural and leisure experience for disabled tourists. At the beginning, they rejected my proposal since there were no professionals in their organizations able to provide these types of services. Then I decided to design a training programme focused on the new specific competencies and I presented it to the organizations previously contacted. Some of them adhered to the programme and this allowed them to enter in the network. (Sfogliaviaggi CEO)

In this case, increasing awareness fostered confidence on the service innovation within the professional community, allowing new ventures to target disabled tourists. Moreover, it determined the emergence of a culture of disability in the tourism industry, also thanks to word-of-mouth promotion. However, realizing the needs of this specific segment was not enough to push the development of the new offering, as collaboration from more players was also needed. In this sense, training activities have been key to provide real opportunities to cocreate value collectively.

Actors improve confidence and trust by enabling interaction with customers and partners through information and communications technology (ICT). This implies continuous proactive collaboration and learning with customers in order to involve them cognitively, affectively and behaviourally in service innovation, as in the practice of Just Eat (an online takeaway food delivery service). The online platform allows customers to search for their favourite local takeaway restaurants and place orders online, choosing from pick-up or delivery options. It is interesting to observe that in developing the Italian market, Just Eat's sales force faced a problem with Chinese restaurants, which did not trust the company and did not want to join the service. To manage this problem, the Italian CEO had the following idea:

> We asked a Chinese restaurant who was already our customer to make a video, in Chinese language, in which he invites colleagues to listen [to] what our people have to say as reliable [sources]. In practice, the salesman enters the Chinese restaurant with an iPad in front of his face, he does not say anything ... [he just shows] the video! (CEO Italy)

Involving existing customers in Just Eat's communication with Chinese restaurants promoted the latter's acceptance of the new service innovation and allowed the former to increase the Chinese presence in the company portfolio. In this sense, listening to the voice of existing customers has shown to be a very effective way to generate trust and confidence.

Similarly, in the case of Zalando (which is an e-commerce platform for fashion products), potential partners did not trust the initiative and the company was not able to immediately develop a collaborative network. In order to build actors' confidence in the new way of buying, Zalando developed the concept of "accessible fashion for everyone" and designed its website with enhanced usability to let the buying experience become intuitive and immediate.

> We want to ensure that the platform is shaped [based on the] user: we know what he/she wants, what ... articles [he/she] returns [to], we know even where he/she points his/her mouse around the screen. We can build an experience [to tailor] use. This is the case for example of the 'Image Recognition System' with which Zalando helps users to find shoes and clothing seen in the real world. We went over the idea of the classic bar code. Just a photo, to start recognition, the application can see if the same product is in the catalogue or if there are similar products. The consumer wants to have immediate answers, and technology today can give [this to] him/her. (Country Manager Italy)

Zalando "broke the rules" by being the first online fashion company to advertise on television. This new approach supported the credibility of the concept of "luxury for everyone", not just for consumers but also suppliers, distributors and opinion leaders.

8.4.2 Creating Shared Meanings

Service innovation is a collective practice performed by actors' integrating their service provision interpretations from dispersed, individual viewpoints. The process of understanding and meaning construction comes from interaction based on social, linguistic and cultural resources, norms and rules. In particular, through connection actors are able to subjectively and intersubjectively interpret the innovation they experience and to derive shared meanings from this in order to overcome constraints and difficulties.

In the case of BravoSolution, the BravoSolution Education Network was developed to support the alignment of actors involved in the development of the e-sourcing platform. This is a social learning environment that is available online to BravoSolution customers, business partners and employees and contains both free and premium content used to build procurement and functional skills in procurement professionals. The purpose is to give, especially to customers, the skills, knowledge and resources needed to succeed as procurement professionals by using BravoSolution applications and providing suggestions to advance the service innovation over time. Rolling out the new practice was not easy, as it faced resistance from both customers and business partners in relation to adopting the new solution.

> When they proposed to adopt e-sourcing to support the [procurement] activity, I was really doubtful and reluctant to change the [procurement] routines we had in the firm. I asked the Bravo Solution representative: I do not understand if the advantages connected to this new solution are more than the costs I have to sustain to adopt these new practices. He suggested for me to access the forum in order to [address] these issues with other colleagues and ask for explanations. (BravoSolution customer)

Conversation through the above-mentioned forum plays a crucial role in maintaining alignment between BravoSolution and its customers and business partners. The general forum for procurement professionals was developed after the initial service launch as a place for individuals to ask other procurement professionals questions on e-sourcing and exchange e-sourcing best practices. In this way, a shared understanding was generated with respect to how the new way of managing procurement could provide increased benefits for the actors involved.

In fostering service innovation, companies often face difficulties with the institutional way of doing things, as they face constraints from national associations and government. In this case it is necessary to move into the service ecosystem and act at a higher level to promote a shared understanding and new meaning of the benefits involved in the innovation, not only for suppliers and customers but also for the multiple other actors involved.

In the case of Aggiustami.it (an online platform to find the best auto body shops in a given area and collect estimates for repairing vehicles), the user can request and compare quotes from the comfort of their own home, without wasting time visiting the various body shops in their city. In addition, potential users can read reviews that other customers have published on the website. At the heart of this new approach was a need for customers to fix their cars at more competitive prices and increase their contractual power over body shops. This was not the first time this idea for changing behaviour was launched into the market, as another actor had attempted to do so in the past; however, this previous attempt was a total failure as the national body shop association obstructed the changes. The main problems were the low cultural level of body shop owners, the fact that they are quite adverse to change and the issue that they did not understand how the service worked. Aggiustami.it's strategy was to introduce very simple, direct language to communicate with body shop owners, making it very clear how the "value pie" would be shared between the body shops themselves, the customers and the online intermediaries. In this case, interaction clearly became a means for creating a shared understanding among the actors. Sharing the same language allowed the actors to coordinate meanings around the new service and thus increased its acceptance; thousands of body shops are now part of Aggiustami.it's network. By practising interaction over time, these meanings consolidated, allowing for further development of the service and for its diffusion.

The Smartika (social lending context) platform was launched in 2008 under the name Zopa. At the beginning of 2009, it had a community of 5000 people and total issued loans of 7 million euros. The Bank of Italy then ceased Zopa's activities and rejected the initiative, as it lacked understanding of the conditions in which it was legal to lend money to people. Zopa's CEO struggled to convince the authorities of the legality of the company's project (thanks also to the application in Italy of European norms on institutional payments). Three years later, in 2012, the project started up again under the new name, Smartika. The company name refers to a smart way of doing things, emphasizing the advantage of service innovation adoption.

> Through this name we wanted to communicate the message of being open, fair and balanced, offering benefits and opportunities to all actors. (Smartika CEO)

Smartika aimed to develop a new, positive meaning around the idea of social lending between individuals based on the creation of a community in which applicants (those applying for a loan) and providers (those who invest their money by lending it to others) can interact directly with each other without resorting to intermediaries. This process offers better conditions for both: lower rates for those who get the loans and higher interest rates for moneylenders. The result was "relocation of the Zopa community" of 45,000 members, who had remained loyal over time, to Smartika.

8.4.3 Leveraging Actors' Competencies

Successfully implementing a new, attractive offering is increasingly related to the company's ability to interact and cooperate with all the related network actors, including customers, in order to better grasp how to develop the new solution in line with users' needs. It is necessary to leverage actors' competencies to enable a new approach concerning the diffusion of innovative ideas and proposals.

Actors in the innovation network engage in dialogue and interaction during service innovation design, production, delivery and consumption to leverage each other's competencies. This aspect is evident in the case of Neronote (the e-store for made-to-measure, personalized, made in Italy shirts). The prerequisite for its success is that the customers have the competencies to perform their role. In developing the innovative business model, the company's management quickly realized that customers needed to be guided carefully to integrate their resources and processes with those of Neronote.

> We have created in the website a step-by-step guide that provides all the information that customers need to customize a shirt, design his own shirt, take measurements on the body or on another shirt, choose a standard size and take care of the Neronote shirt. After the introduction phase of the project, some of the potential customers reacted to our innovation, writing in the website forum and telling the contact centre's staff that they would not be available to take part in the process of creating a bespoke shirt online since they believed they were not able to do it correctly. (Neronote marketing manager)

Starting from these comments, the company decided to try and better understand the key issues to be solved and to prepare a set of tools that could be easily used and shared with other customers. The goal was to help customers to express the competences they have, but which would have remained hidden if not adequately stimulated. Since that moment, the implementation practices have been continuously performed on the basis of the non-stop feedback from Neronote's customers.

In the case of Diet-to-Go (home delivery service of healthy meals), a new practice emerged through the company's aim of changing the traditional view of the diet as punitive. Actors foster the idea of food as a pleasure (which is especially pertinent in the Italian culture), even for people who want to lose weight; these customers can do so while having a great experience. Every morning, the customer receives, at his home or office, a well-designed Diet-to-Go bag containing all the just-cooked meals and snacks needed for the entire day. The new diet concept was not defined nor imposed by the entrepreneur, who had no previous experience in the industry at all. The new meaning emerged as the result of collaboration between different players, which favoured the company's acceptance by customers.

> Nutritionists help to study diet programmes to [enable customers to] lose weight progressively but without any health damage. Top chefs use food to experiment [with] new tastes and the presentation of meals. Kitchens grant the overall quality of the output and synchronize [with] each other to generate meals with the same characteristics throughout the country. Delivery men assure in-time deliveries and a positive interaction with consumers. A network of franchisees coordinate the service provision, maintain the relationships with consumers and also develop[ing] the market locally. (Diet-to-Go founder)

In the Diet-to-Go network, each actor makes its competences and ideas available to the others, and this allowed for the creation of a new concept of dieting. In this sense, the new concept was less planned or imposed as it emerged more as an effect of interaction. The customers are also part of the process; through different communication channels, they interact with the other players using various touch points and thus contribute to developing the service innovation, as well as to its acceptance. However, customers also rejected a specific part of the service idea, and this spurred further improvement.

> Consuming Diet-to-Go's meals at the office was too expensive and [the lunch was] too large for office time. In reaction, consumers started to share one bag between two persons. After realizing this adaptation, the company developed a unique dish for lunch breaks while at the office, with great success. (customer's quote)

Also in the case of Innovuum, innovation required cultural changes that needed to be explained and shared. In Italy, there are often a number of small- and medium-sized enterprises that actually possess great entrepreneurial skills, flair and creativity, but have few resources to support their ideas. Without innovation, the possibilities for growth are few. The new practices developed by Innovuum act on what the company calls the "innovation matching". A specific tool has been introduced to enable this matching: the Innovation Marketplace. This is the first Italian virtual notice board for searching for and offering innovation.

In order to match demand and supply of innovation, it is necessary that each knows how to track down the other and for the other to become "visible". These two conditions are often hampered by a difficulty in the language of generic demand searches and presentations full of supply technicalities. In this context, Innovuum set up a new practice to provide support for the development of innovation through a real contribution to the realization of new projects and interesting collaborations between companies.

> You can find new opportunities to innovate products, renew production processes and access to new markets. The platform contains a portfolio of innovations in demand and

> supply that enable companies to develop their own ideas and undertake innovative projects with other firms and institutions. It facilitates the contact between the demand and supply of innovation through (1) mapping a constant supply of innovation; (2) the management and organization of events and activities of matching; and (3) the refinement of services devoted to business and targeted to the matching of supply/demand of innovation. (Innovuum manager)

Innovuum, therefore, also represents the renewed role of intermediation, which is not only about connecting disconnected actors and integrating technologies but generating more and more deals through "soft" coordination of heterogeneous cultures and visions so that value can be effectively cocreated.

8.5 Conclusion

This chapter moves away from the traditional conceptualization of innovation diffusion as a linear multistage development process within a main innovator system and where adopters play a key role (Rogers 2010). In this study, SID is framed more as an emergent process of co-construction and sense-making by companies, customers and other partners. Such actors carry out practices by interacting and integrating resources in a social, economic and cultural context (Gherardi 2015). Through participation, they negotiate meanings and aim to shape actions and relationships, thus cocreating value (Lusch and Vargo 2014).

In considering the pattern of activities and relations, we can frame the "practice from outside" (Gherardi 2010), that is, as the actors' way of developing and diffusing service innovation. By adopting the lens of "practice from inside", we went deeper and our study detected three specific sets of actions and relations as outcomes. Such a view allowed us to frame SID as connections in action, in which actors make sense and share meanings about the newness, both individually and collectively (see Fig. 8.1). The three actions and outcomes are as follows:

- *Enabling confidence* in service innovation concerns how actors come to trust interacting with each other and integrating resources. Actors encourage service innovation by letting users and other partners become confident in new value propositions. However, this assurance is not predetermined ex ante but is developed in interactive patterns; while interacting, actors increase the confidence that they can apply their abilities to foster a service innovation and to succeed together. Without this confidence, coordination is highly complicated or impossible.
- *Creating shared meanings* among actors entails grounding service innovation in sense-giving and sense-making. Managers face difficulties in integrating huge pieces of scattered individual cognitive visions of the various actors. This becomes even more relevant when there are different actors involved in the innovation process and they all have heterogeneous views. Through interactive patterns, service innovation becomes a collective practice performed by actors

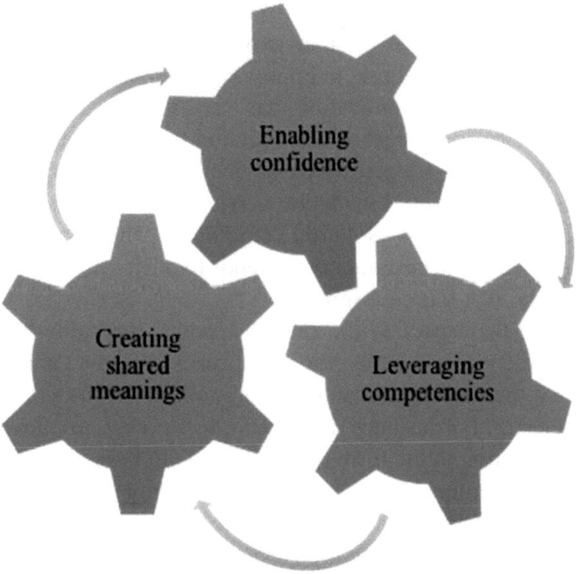

Fig. 8.1 The practices of service innovation diffusion

integrating their service provision interpretations from dispersed pieces of individual viewpoints. In this way, shared and overlapped views are created; these stabilize and thus support the organization of activities and the combination and integration of resources.

- *Leveraging actors' competencies* to manage and support resource integration by widening the network. Customer competencies in performing their role have been undervalued for a long time. However, innovation increasingly passes through an innovative business model in which the customer is highly central with respect to the contribution they can provide to the new value proposition. Of course, the customer is one of multiple actors with whom companies interact. Service innovation arises from the integration and leveraging of these actors' competencies and processes.

These three SID practices delineated above are carried out by actors using resources in their value search (Mele and Russo-Spena 2015). They are practical accomplishments that emerge from the actors' ability to participate in actions with the needed competencies in the complex network of relationships (Gherardi 2001). The practices should not be seen as distinct but rather as deeply interrelated; enabling confidence fosters the creation of shared meaning, which in turn is a requisite for leveraging actors' competencies. Together, they support the diffusion of service innovation. Such diffusion influences the creation and recreation of social structures and institutions.

This work offers a second contribution to the literature, as it shows that SID results from the interaction between actors who, by acting either as adopters or rejecters, play a constructive role by influencing how service innovation is dynamically developed and spread. As adopters, they shape the practices and thus allow the process of innovation to be developed successfully. As rejecters, they assume a critical role in improving the innovative service content. The roles actors play in service innovation may vary with regard not only to the intensity of their participation in resource integration but also to the approach (either favourable or critical) they adopt towards service innovation. For such reasons, companies benefit from facilitating interaction between rejecters and adopters, as highlighted in the case of the BravoSolution community. The role of the adopter or rejecter is not limited to the customer but extends to other players involved in the service network, as the Just Eat case shows.

With respect to the previous point, the study not only shows that there are different roles involved in the process of developing service innovation but also that the same actor can play both roles and thus simultaneously act as adopter and rejecter. For instance, in the Diet-to-Go case, employees in companies refused the innovation in the form of the Diet-to-Go bags, while at the same time adopting the idea of unique meals and, even more importantly, contributing to the evolution of this idea. In the Aggiustami.it case, the national body shops association initially opposed the innovation presented by the company, but later strongly promoted the initiative. It is therefore evident that in the development of complex innovative solutions, users and providers should not be seen as distinct entities; rather, each actor could behave simultaneously as a provider who brings resources to be integrated and as users who make use of the resulting combination to solve their specific problem (Cantù et al. 2012). This makes service innovation a process which is very much context-dependent.

A third contribution comes from our focus on the purpose of collaboration in creating a shared vision that goes beyond any particular party (Chrislip and Larson 1994); the development of a shared understanding of problems and solutions greatly facilitates communication and further learning (Brown and Duguid 1998; Powell et al. 1996). Confidence and shared meanings are essential in setting priorities and coordinating activities to accomplish the objectives of each party (Mohr et al. 1996). They also contribute to the operations needed to implement an innovation, as the integration of resources among multiple actors requires communication and coordination (Maglio and Spohrer 2013). In the Tabbid case, for instance, actors exchange reciprocal services because they share a common idea about mutual support and trust each other's competences. Similarly, in the Innovuum case, innovation could only happen thanks to a cultural change that occurred at system level.

A key challenge of SID resides in the actors' ability to interweave the social and ICT nets in order to foster practices for open innovation as they enable users' knowledge and creativity. Collaborative users who create and learn from (and with) others in the workplace of innovation (the web setting) call for managing a new kind of web of interactions aimed at co-constructing service innovations. The importance of ICT is evident in the Zalando case, which has the biggest online marketing department in Europe. Zalando developed a system that is able to adapt in real time to changing customer needs and user experience. In addition, Neronote has been particularly effective in developing an e-platform that is able to provide to customers all the information needed to customize shirts.

8.6 Implications and Further Research

Social practices are situational, recurrent, materially bounded and shared forms of social activities that produce and structure life in organizations. Such practices have been shown to be increasingly important for the development of service innovation. Indeed, most innovations are a mixture of emergent adapted and adopted practices, which become sharpened over time via interaction within social interconnected activities, leading to different implications at theoretical level.

First of all, rather than relying on linear descriptions of work flows composed of actions, researchers should adopt a systemic view and seek to understand how actors interact, integrate resources and develop relationships within their social context. This implies the ability to capture contextual knowledge and develop an appropriate arrangement and disarrangement of the practices and constituents that are the net of actions, people, contextual artefacts and tools.

Specific research questions could then be considered based on the three practices identified in this study, as follows:

Enabling confidence: How do actors foster development of the idea of service innovation? How can interaction be shaped to transform rejecters into adopters?

Creating shared meanings among actors: How can interaction support the alignment of actors involved in the service innovation? How can diverging views of service provision be negotiated? What is the role of values and institutions in creating shared meaning among actors, both as adopters and rejecters?

Leveraging actors' competencies: How can actors be engaged? How do actors' values and competences enable or hinder the practice of SID? Which resources and tools can foster coordination among adopters, companies and other actors?

In summary, adopting or rejecting an innovation requires not simply making a decision or taking an action but the performance of different service innovation practices.

This study also has some implications for managerial practice. Overall, managers should see the process of service innovation not as a company's set of activities but as a process of co-construction and sense-making by a network of actors. They should realize that service innovation is a collective effort. In addition, by analysing the role of actors as adopters and rejecters of service innovation, managers can design strategies to engage adopters in diffusing service innovation. They should take an active role in adopters' practices and at the same time interact with rejecters to understand possible weaknesses of the process. Finally, managers can help actors develop the confidence and persistence needed to overcome obstacles, complete various tasks and therefore innovate.

References

Adams, G. R., & Schvaneveldt, J. D. (1985). *Understanding research methods*. New York, NY: Longman.
Agarwal, R., & Selen, W. (2009). Dynamic capability building, in service value networks for achieving service innovation. *Decision Sciences, 40*(3), 431–475.
Ballantyne, D., & Varey, R. J. (2006). Relationship marketing and the "new" service-dominant logic of marketing: Are they rival logics? Conference paper accepted for ICRM, Leipzig Germany, 13th–16th September 2006.
Barnes, B. (2001). Practice as collective action. In T. Schatzki, K. Knorr-Cetina, & E. von Savigny (Eds.), *The practice turn in contemporary theory* (pp. 17–28). London: Routledge.
Bessant, J., & Tidd, J. (2007). *Innovation and entrepreneurship*. England: Wiley.
Bourdieu, P. (1977). *Outline of a theory of practice*. Trans. Richard Nice. Cambridge: Cambridge University Press.
Brown, J. S., & Duguid, P. (1998). Organizing knowledge. *California Management Review, 40*(1), 90–112.
Callon, M. (1999). Actor network theory – The market test. *The Sociological Review, 47*(S1), 181–195.
Cantù, C., Corsaro, D., & Snehota, I. (2012). Roles of actors in combining resources into complex solutions. *Journal of Business Research, 65*(2), 139–150.
Chesbrough, H. (2003). The logic of open innovation: Managing intellectual property. *California Management Review, 45*(3), 33–58.
Chrislip, D., & Larson, C. (1994). *Collaborative leadership*. San Francisco: Jossey-Bass.
Coffey, A., & Atkinson, P. (1996). *Making sense of qualitative data*. London: Sage.
Cohen, W. M., & Levinthal, D. A. (1990). Absorptive capacity: A new perspective on learning and innovation. *Administrative Science Quarterly, 35*, 128–152.
Corradi, G., Gherardi, S., & Verzelloni, L. (2010). Through the practice lens: Where is the bandwagon of practice-based studies heading? *Management Learning, 41*(3), 265–283.
Czarniawska, B. (2004). *Narratives in social science research*. London: Sage.
Dougherty, D. (2004). Organizing practices in services: Capturing practice-based knowledge for innovation. *Strategic Organization, 2*(1), 35–64.
Edvardsson, B., Gustafsson, A., Kristensson, K., & Witell, L. (2010). Service innovation and customer co-development. In P. Maglio, A. A. Kieliszewski, & J. C. Spohrer (Eds.), *Handbook of service science*. New York: Springer.
Edvardsson, B., & Tronvoll, B. (2013). A new conceptualization of service innovation grounded in SD logic and service systems. *International Journal of Quality and Service Sciences, 5*(1), 19–31.
Eisenhardt, K. M. (1989). Building theories from case study research. *Academy of Management Review, 14*(4), 532–550.
Erden, Z., Schneider, A., & von Krogh, G. (2014). The multifaceted nature of social practices: A review of the perspectives on practice-based theory building about organizations. *European Management Journal, 32*(5), 712–722.
Gherardi, S. (2001). From organizational learning to practice-based knowing. *Human Relations, 54*(1), 131–139.
Gherardi, S. (2010). Telemedicine: A practice-based approach to technology. *Human Relations, 63*(4), 501–524.
Gherardi, S. (2012a). Why do practices change and why do they persist? Models of explanations. In P. Hager, A. Lee, & A. Reich (Eds.), *Practice, learning and change* (pp. 217–231). Netherlands: Springer.
Gherardi, S. (2012b). *How to conduct a practice-based study: Problems and methods*. Cheltenham: Edward Elgar.
Gherardi, S. (2015). How the turn to practice may contribute to working life studies. *Nordic Journal of Working Life Studies, 5*(3a), 13–25.

Gherardi, S., & Nicolini, D. (2006). *Organizational knowledge: The texture of workplace learning*. Malden, MA: Blackwell.
Grönroos, C. (2010). Adopting a service logic in manufacturing: Conceptual foundation and metrics for mutual value creation. *Journal of Service Management, 21*(5), 564–590.
Gulati, R., & Gargiulo, M. (1999). Where do interorganizational networks come from? *American Journal of Sociology, 104*(5), 1439–1493.
Hagedoorn, J. (2002). Inter-firm R&D partnerships: An overview of major trends and patterns since 1960. *Research Policy, 31*(4), 477–492.
Hall, B. H. (2006). Innovation and diffusion. In J. Fagerberg, D. C. Mowery, & R. Nelson (Eds.), *The Oxford handbook of innovation* (pp. 459–484). New York: Oxford University Press.
Henneberg, S. C., Gruber, T., & Naudé, P. (2013). Services networks: Concept and research agenda. *Industrial Marketing Management, 42*(1), 3–8.
Hummel, R. (1991). Stories managers tell: Why they are valid as science. *Public Administration Review, 51*, 31–41.
Leonard-Barton, D. A. (1998). *Wellsprings of knowledge: Building and sustaining the sources of innovation*. Boston: Harvard Business School Press.
Lusch, R. F., & Nambisan, S. (2015). Service innovation: A service-dominant logic perspective. *MIS Quarterly, 39*(1), 155–175.
Lusch, R. F., & Vargo, S. L. (2014). *Service-dominant logic: Premises, perspectives, possibilities*. Cambridge: Cambridge University Press.
MacVaugh, J., & Schiavone, F. (2010). Limits to the diffusion of innovation: A literature review and integrative model. *European Journal of Innovation Management, 13*(2), 197–221.
Maglio, P. P., & Spohrer, J. (2013). A service science perspective on business model innovation. *Industrial Marketing Management, 42*(5), 665–670.
Mahajan, V., Muller, E., & Srivastava, R. K. (1990). Determination of adopter categories by using innovation diffusion models. *Journal of Marketing Research, 27*(1), 37–50.
Mahajan, V., & Peterson, R. A. (1985). *Models for innovation diffusion. Quantitative Applications in the social science* (Vol. 48). Beverly Hills, CA: Sage.
Mele, C., Colurcio, M., & Russo-Spena, T. (2014). Research traditions of innovation: Goods-dominant logic, the resource-based approach, and service-dominant logic. *Managing Service Quality, 24*(6), 612–642.
Mele, C., & Russo-Spena, T. (2015). Innomediary agency and practices in shaping market innovation. *Industrial Marketing Management, 44*, 42–53.
Mohr, J. J., Fisher, R. J., & Nevin, J. R. (1996). Collaborative communication in interfirm relationships: Moderating effects of integration and control. *The Journal of Marketing, 60*, 103–115.
Möller, K. (2010). Sense-making and agenda construction in emerging business networks—How to direct radical innovation. *Industrial Marketing Management, 39*(3), 361–371.
Nicolini, D. (2012). *Practice theory, work, and organization: An introduction*. Oxford: Oxford University Press.
Nicolini, D., Gherardi, S., & Yanow, D. (2003). *Knowing in organizations: A practice-based approach*. New York: ME Sharpe.
Ordanini, A., & Maglio, P. P. (2009). Market orientation, internal process, and external network: A qualitative comparative analysis of key decisional alternatives in the new service development. *Decision Sciences, 40*(3), 601–625.
Østerlund, C., & Carlile, P. (2005). Relations in practice: Sorting through practice theories on knowledge sharing in complex organizations. *The Information Society, 21*(2), 91–107.
Perks, H., & Jeffery, R. (2006). Global network configuration for innovation: A study of international fibre innovation. *R&D Management, 36*(1), 67–83.
Perry-Smith, J. E., & Shalley, C. E. (2003). The social side of creativity: A static and dynamic social network perspective. *Academy of Management Review, 28*(1), 89–106.
Powell, W. W., Kaput, K. R., & Smith-Doer, L. (1996). Interactional collaboration and the locus of innovation: Networks of learning in biotechnology. *Administrative Science Quarterly, 41*, 116–145.
Rogers, E. M. (1962). *Diffusion of Innovations*. New York: Free Press.

Rogers, E. M. (1995). Diffusion of Innovations: Modifications of a model for telecommunications. *Die Diffusion von Innovationen in der Telekommunikation, 17*, 25–38.
Rogers, E. M. (2010). *Diffusion of innovations*. New York: Simon and Schuster.
Rouse, J. (2002). *How scientific practices matter*. Chicago, IL: University of Chicago Press.
Russo-Spena, T., & Mele, C. (2012). "Five Co-s" in innovating: A practice-based view. *Journal of Service Management, 23*(4), 527–553.
Russo-Spena, T., & Mele, C. (2016). Knowing and innovating in practices through networking. In M. Toivonen (Ed.), *Service innovation: Novel ways of creating value in actor systems*. Springer.
Schatzki, T. R., Knorr-Cetina, K., & von Savigny, E. (2001). *The practice turn in contemporary theory*. London: Routledge.
Schau, H. J., Muñiz, A. M., & Arnould, E. J. (2009). How brand community practices create value. *Journal of Marketing, 73*(5), 30–51.
Snehota, I., & Hakansson, H. (1995). *Developing relationships in business networks*. London: Routledge.
Tidd, J., Bessant, J., & Pavitt, K. (2001). *Managing innovation: Integrating technological, organizational and market change*. New York: Wiley.
Vanhaverbeke, W., & Cloodt, M. (2006). Open innovation in value networks. In H. Chesbrough, W. Vanhaverbeke, & J. West (Eds.), *Open innovation: Researching a new paradigm* (pp. 258–281). Oxford: Oxford University Press.
Voss, C. A., & Hsuan, J. (2009). Service architecture and modularity. *Decision Sciences, 40*(3), 541–569.
Weick, K. (1995). *Sensemaking in organizations*. Thousand Oaks, CA: Sage.
Wenger, E. (1998). *Communities of practice: Learning, meaning, and identity*. Cambridge: Cambridge University Press.

Chapter 9
Innovation with Effectuation: An Opportunity for the Public Sector

Kirsti Mäensivu, Marja Toivonen, and Karo Tammela

Abstract This chapter focuses on new innovation models in the public sector. It analyses the specificities of the public sector—institutionalism and professionalism—as the context of innovation. On the other hand, it argues that entrepreneurial innovation processes combined with the perspectives of learning and networking are very well applicable—not only in the private companies but in public organizations as well. Based on this view, the effectual approach, closely linked to service-dominant logic (SDL), is examined as a theoretical construct and tested in practice. An 'effectual-type' process, in which a city administration collaborates with citizens groups to foster the mutual expansion of resources, is described as a case and the characteristics showing analytical generalizability are identified.

9.1 Introduction

In recent decades, innovation has been considered crucial for fostering economic growth and welfare. The background of this emphasis is the rapid rate of change and the significance of knowledge in present societies. In the public sector, a specific need for innovation emerges from the challenge of providing services at times of increasingly severe problems and scanty financial resources.

Simultaneously, a strong tradition of innovation is missing in the public context: innovations have been mainly incremental and isomorphic, representing continuous improvement (Albury 2005). Their aim has often been to ensure stability. Path

K. Mäensivu (✉)
Freelance Scientist, Valtakatu 32 AS 10, 96200 Rovaniemi, Finland
e-mail: kirsti.maensivu@kolumbus.fi

M. Toivonen
VTT Technical Research Centre of Finland, Helsinki, Finland
e-mail: marja.toivonen@vtt.fi

K. Tammela
Aalto University School of Business, Helsingfors, Finland
e-mail: karo.tammela@aalto.fi

T. Russo-Spena et al. (eds.), *Innovating in Practice*,
DOI 10.1007/978-3-319-43380-6_9

dependence (Aagaard 2012) and homogeneity have strengthened the 'status quo' in public institutions (DiMaggio and Powell 1983). The role of innovation has been promoted by the proponents of the new public management (NPM), which is the dominant public paradigm in many Western countries. The mission of this paradigm is to improve efficiency by marketization, private-style management, entrepreneurship, performance standards and output/outcome control (Hood 1991; Pollitt 1993; Tummers 2013). While public services have traditionally been authoritative, rule based and supply driven, aiming to ensure equity (Torfing and Triantafillou 2013), NPM highlights deregulated and demand-driven services aiming to ensure efficiency and user satisfaction.

Recently, a new framework of network governance (NPG) has been emerging parallel with NPM. It focuses on inter-organizational cooperation (Osborne et al. 2012; Sørensen and Torfing 2011). NPG assumes that working across organizational, professional and political boundaries will enable more efficient and effective services and policy implementation. In this framework, innovations refer to both transformational changes and continuous improvement (Hartley 2005). They may occur at central or local levels in the perpetually changing context. NPG includes new tools to empower and engage stakeholders in public problem solving and service production (Torfing and Triantafillou 2013). This 'post-NPM' seems to be more about working together in a pragmatic and intelligent way than about formalized collaboration across organizational, institutional and public-private boundaries (Christensen 2012; Sørensen and Torfing 2011, 2012).

The present chapter aims to provide additional insights into the practical ways in which the NPM and NPG can be integrated in the context of public innovation. At a more specific level, we examine a process model that would foster this integration. A long-lasting dispute in innovation literature has been between the STI model (science and technology-based innovation) and the DUI model (learning by doing, using and interacting). Our starting point is the latter—a reason being that it takes into account important aspects raised by NPG. The DUI model highlights competence building in networks, interaction with users and empowerment of employees (Lundvall 2007). It reflects the broadening view on the nature of innovation, and it is near to the ideas of open innovation (Chesbrough 2006) and innovation democracy (von Hippel 2005). In managerially oriented research, it is close to the studies on user-based and employee-driven innovation (Kesting and Ulhøi 2010; Langergaard 2011; Sundbo and Toivonen 2011), which have aimed to concretize the coevolving of innovation with everyday activities.

Demands for an innovative stance create a strong pressure on public institutions to find new entrepreneurially oriented ways of transforming services. Here, the public sector will undergo a paradox: entrepreneurship is 'a science' of turbulence and change. It does not highlight continuity and does not exist in static conditions (Arend and Chen 2012; Chiles et al. 2010), typical of the public sector.

We examine the approach of effectuation as an answer to this challenge (Sarasvathy 2008). The effectual approach has close linkages to the basic thoughts of service-dominant logic (SDL) developed by Vargo and Lusch (2004, 2008). It also relates to the public sector applications of SDL by Osborne et al. (2012) and has similarities with the views on institutional work by Lawrence and Suddaby (2006). These linkages are important for the examination of the public sector as an institutional context for innovations and entrepreneurship. We focus on several characteristics of institutions: the role of the agency, institutional work and the role of professionals in transforming institutional practices.

From now on, the chapter has been structured as follows. In the second section, we analyse the theoretical literature, focusing on the two topics that are most relevant for our study: the public sector as an innovation context and the approach of effectuation as an innovation model. In the third section, we present our empirical research: we first introduce the case and our methodology and thereafter analyse the results. The fourth section includes our concluding discussion.

9.2 Theoretical Background

9.2.1 Public Sector as a Context for Innovations

This literature-based subsection examines the public sector as an innovation environment. We start by analysing the central characteristics of institutions and then continue to the characteristics of actors. Finally, we focus on the issues of transformation and practices and on the role of the professionals as change makers.

Characteristics of Institutions

The main characteristics influencing innovation in the public sector are its political and institutional contexts (Seo and Creed 2002; Windrum 2008). The trend to more entrepreneurial and innovative ways to produce services can be seen as a policy change in itself and may as such create a big challenge to the traditional institutions in the public sector.

As an institution, the public sector is defined according to rules, norms and beliefs which describe the reality of the institution/organization (Hoffman 1999; Garud et al. 2007), reduce uncertainty and direct the behaviour and beliefs of the actors (DiMaggio and Powell 1983, 1991; Dorado 2005). Institutional elements are regulative, normative and cognitive. Rationality plays an important role in the creation of formal organizations (Meyer and Rowan 1977) and provides, together with cultural-cognitive elements and resources, stability and meaning for the organization and its social life (Scott 1995).

Friedland and Alford (1991) have introduced the concept 'institutional logics' to describe the various practices and beliefs of institutions. Thornton et al. (2012: 132) have suggested that institutional logic provides 'frames of reference that conditions actors' choices for sense making, the vocabulary they use to motivate action and their sense of self and identity'. Thus, a change of institutional logics is a core phenomenon when a public organization becomes more entrepreneurial and innovative.

In addition to rules and logics, the interrelatedness of institutions may impact on how various practices can be transformed. Interrelatedness refers to the relationships between different inter-institutional orders: family, community, religion, state, market, profession and corporation (Thornton et al. 2012). It demonstrates the partial autonomy of actors and organizations from other actors and organizations. The local community has a remarkable influence on how local institutions are enabled to change and transform their practices. The community embodies local understandings, norms and rules that become mental models upon which individuals and organizations define problems and situations (Marquis et al. 2007).

Together with the interrelatedness, the complexity and multiplicity of institutional logics determine the possibilities to transform institutional practices. Institutional complexity describes situations in which divergent prescriptions from multiple institutional logics collide (Greenwood et al. 1996, 2011). A multiplicity of logics is in play in any particular context, and its composition is constructed, not given (Thornton et al. 2012). This composition defines the agency as multiple and dynamic, and it frames the institutional work in the everyday practice of individuals coping with the institutional complexities of their work.

Who Is Enabled to Transform Institutional Practices?

Defining factors in the transformation of institutions are (1) agency, (2) resources and resource mobilization, (3) individual and institutional opportunities and (4) multiplicity, heterogeneity and the level of institutionalization of the organization (Dorado 2005). Actors are only partially autonomous from the institutional structure (Seo and Creed 2002). A crucial issue is whether the institutional scripts and rules are taken for granted (DiMaggio and Powell 1983). Another issue is how the actors are able to change the institution (Lounsbury and Crumley 2007). This is called 'the paradox of embedded agency', and it is linked to the concept of institutional entrepreneurship (Garud et al. 2007). New institutions arise when organized actors with sufficient resources see an opportunity to realize their highly valued interests (DiMaggio 1988). Actors reproduce and maintain practices within the existing institutional logics, but they have capacity to innovate and mobilize resources to transform institutions (Scott 2008; Thornton et al. 2012).

The nature of agency has been theorized from various standpoints. Agency can be habitual (Giddens 1984), strategic (DiMaggio 1988) and sense-making (Weick 1995). All three forms are simultaneously involved in agency, but one dominates. According to Emirbayer and Mische (1998), social actions (including innovation) are influenced by the combination of temporally rooted orientations. These orientations may inform actors from the past (habitual orientation), the future (capacity to imagine alternatives) and the present (capacity to conceptualize past habits and future projects within the contingencies of the moment) (ibid.).

Referring to Emirbayer and Mische (1998), Dorado (2005) suggests that institutional actors can adopt three different kinds of profiles in changing an institution: entrepreneurship, convening and partaking. Smets and Jarzabkowski (2013) apply a categorization into iterative, projective and practical—evaluative dimensions. The projective dimension supports future change and entrepreneurial and creative behaviour (Battilana et al. 2009; Lawrence et al. 2009). In the dialectical framework, the human agency is conceptualized as praxis and positioned as the essential mediating mechanism that links the institutional embedded actor, contradictions and change. According to Seo and Creed (2012: 240) 'the possibility of institutional change is rooted in the aptitude and opportunity for praxis'.

Individual actors have a remarkable role in changing and creating institutional practices. In spite of that, the nature of institutional entrepreneurship is not individual but collective. It is a group-bounded activity: groups motivate, inspire and enable the engagement (Dorado 2013). This raises the question about the conditions in which socially embedded unreflective actors become conscious of the social arrangements in which their interests are unmet, mobilize other similarly situated actors and take collective action for change (Seo and Creed 2002). The starting point for change is the extent to which groups are dissatisfied with the way their interests are accommodated (Greenwood and Hinings 1996).

How Institutions Will Be Transformed

Lawrence and Suddaby (2006) have introduced the term 'institutional work' to describe how actors build, sustain and change institutions. The concept aims to focus attention, not on the social structures but on the processes, i.e. move from the institution in itself to the purposive action. Institutional work can be described with three characteristics: (1) institutional actors are reflexive, goal oriented and capable, (2) actors' actions are the centre of institutional dynamics, and (3) the structure, agency and their interrelations are captured (Battilana et al. 2009).

Institutional work is an umbrella concept (Hwang and Colyvas 2011). It highlights the question of how actors become motivated and enabled to change the taken-for-granted practices and norms of the institution they are involved in (Smets and Jarzabkowski 2013; Seo and Creed 2002). Decision making, sense-making and collective mobilization are the key elements of institutional work. They link together the dynamics of practices and identities (Thornton et al. 2012).

The concept of institutional work has pushed the institutional theory towards practice-based views. According to Lounsbury and Crumley (2007), practice should be understood as a kind of institution in itself, including material and cognitive activities that are evolved and shaped by the broader cultural framework. Here, activities are interpreted in the same vein as Jarzabkowski (2005) and Engeström (1987): they are actions and interaction between actors as they perform their daily duties and roles. Practice refers to activity patterns that are infused with broader meanings and provide tools for ordering social life and activity. Practice is members' phenomenon—something that actors draw upon, monitor and orient to in real-time interactions (Llewellyn and Spence 2009).

These views make it plausible to suggest that the concept of practice integrates innovation theories and institutional theories. It also provides conceptual tools to understand innovation processes in the real life context of the public sector. The recent discussion about the 'practice turn' aims to connect the micro-level of individual activities with the meso-level of organizations and with the macro-level of the broader organizational (and institutional) field (Suddaby et al. 2013).

Currie et al. (2012) link the concept of institutional work with the sociology of professions (Abbott 1988) to highlight the role of professions as change agents. They analyse institutional work especially in a situation where the professional dominance is threatened. Three issues emerge here. First, how different types of institutional work interact and cross the activities of creating or maintaining institutions. Second, how the social position or status of actors frames the institutional work they engage in (intra- and inter-professionally). Third, how the institutional work of 'theorizing' by professional elites shows its significance. These questions lay grounds for a broader discussion about professions as change agents.

Professionals as Carriers of Practices

In the institutional theory, professionals are seen as key agents of social change (DiMaggio and Powell 1983, 1991). Professionals consist of several groups: (1) service professionals, including old and traditional professions and new 'weaker professions' (Etzioni 1969), (2) professionals seeking this status (Llewellyn 2001) and (3) professionals managing services (Noordegraaf 2011).

Professionalism used to be one of the most trust-producing, formal public structures (Misztal 2002). Correspondingly, professionals have been considered to be 'the anchors of the order, someone trusted and respected, individuals given a status, autonomy and social elevation' (Dent and Whitehead 2002: 8). Public services have been developed by the strong input of professionals whose status has been legitimized by scientific knowledge. Professionals have had the monopoly on the knowledge, social status and autonomy to make decisions and to choose the right way to act according to the profession's moral and cultural code. Professionals are also seen as bearers of important social values (Tummers 2013).

Suddaby and Viale (2011) identify four essential dynamics through which professionals reconfigure institutions and organizational fields. First, professionals use their expertise and legitimacy to challenge the incumbent order and to define a new, open and uncontested space. Second, they use their social capital and skills to populate the field with new actors and new identities. Third, professionals introduce new rules and standards that recreate the boundaries of the field. Fourth, they manage the use and reproduction of social capital within a field. Professionals aim to establish professional control (Freidson 2001) and occupational closure (Abbott 1988) to govern themselves and to block the outside interventions.

Within the new governmental context (NPM and NPG), professionals are expected to be entrepreneurial, creative and efficient lifelong learners and team workers (Dent and Whitehead 2002; Tummers 2013). According to Du Gay (2007), the principle of the new governmental logic is to create a new professional and managerial subjectivity (Halford and Leonard 2002). These requirements may be in contradiction with the traditional values and rules of professionals and may force them to make sense of the change in relation to their practices and to the individual and professional identity (Emison 2010). From the viewpoint of the employees, new entrepreneurial expectations connected with the new managerial work appear as an identity regulation.

The contradiction is realized when knowledge, being the main property on which the professionals base their decisions and choose the means to act, is challenged by the demand to cooperate with other professions and citizens. From the service point of view, the question is of the ability of professionals to share and transform their knowledge (Vargo and Lusch 2008).

New practices may threaten the power and status of elite professionals in particular (Currie et al. 2012). The profession as an institution demands stability, but innovation yields change (Emison 2010). The expected changes do not take place without resistance (Wasserman and Frenkel 2011). An additional challenge is that professionals often have difficulties to identify the policy programmes they are expected to implement (Tummers et al. 2009). It has even been argued that as a consequence of NPM, the relationships between politicians and professionals have moved from trust-based to a low-trust relationship (Windrum 2008). The reactions of professionals to new demands are discussed from many perspectives in literature: NPM and resistance, managerial governance, the non-spread of innovations and professional self-regulation in a changing architecture of governance (Dent and Whitehead 2002; Ferlie et al. 2005; Tummers 2009).

Professionalization is a continuing process. It is one important base that determines the isomorphic organizational change (DiMaggio and Powell 1983). Professionals transform services, but at the same time they transform themselves and their jurisdiction (Suddaby and Greenwood 2005; Suddaby and Viale 2011). Professionals are inhabitants of both the institution in question and the institution of their profession. In addition, they are citizens and inhabitants of their community. Professionals create and maintain these three environments/institutions but also disrupt them; simultaneously, they maintain and negotiate their roles and identities (Noordegraaf 2011). They have to solve their relationships with the institution they work for, with their own and other professional groups and with citizens and the community (Evetts 2012; Noordegraaf 2007).

In our study, the concept of performance is linked to professional work. 'Professional performance' (Dent and Whitehead 2002) includes two dimensions. The first dimension highlights the knowledge based on which performance is measured and legitimized (Lyotard 1984). As stated earlier, the emerging discourse has delegitimized the traditional narrative professional knowledge. Professional accounts are no longer sufficient as such, but must be measured and inspected against external criteria or targets of performance (Dent and Whitehead 2002). The second dimension focuses on the performed character of identity and on the implications it has for agency, resistance and subjectivity (Butler 1990). Performance, in this sense, comes into existence by language and practices (deeds) and gives voice to the actors' knowledge about the world (Dent and Whitehead 2002).

9.2.2 *Effectuation as a Process with Expanding Cycles of Resources*

The approach of effectuation has its background in theories of decision making in situations of uncertainty, where decision makers are unsure about the consequences and preferences (Sarasvathy 2008). These theories have also been important for the approach of service-dominant logic (SDL) in its foundational conceptualization. An important source of insights has been the empirical knowledge on the behaviour of successful entrepreneurs. According to the developers of the effectual view, entrepreneurship is inextricably intertwined with uncertainty, i.e. entrepreneurial expertise means expertise in uncertainty (Read et al. 2009). Entrepreneurs operate through three logics (Sarasvathy and Dew 2005): (1) the logic of identity, (2) the logic of action and (3) the logic of commitment to stakeholders and value-creation. Each logic has complementary decision criteria: identity (who you are), action/knowledge (what you know) and commitment/network (whom you know).

Effectuation suggests the replacement of predictive thinking with a means-oriented approach to tackle uncertainties. It highlights that the cooperative shaping of the situation, rather than a competitive scramble for (predicted to be) valuable resources, drives the development. Available resources form the starting point; goals emerge in the course of action. Any given resource can be made more or less valuable and more or less capable of producing long-term advantages: thus, what people do with resources matters. This approach clearly differs from a linear process, which starts from the identification of an initial opportunity, sets a goal and aims to achieve it in a preselected context (ibid.). Figure 9.1 presents the basic ideas of the effectual process [a modified version from Read et al. (2009)].

Expanding cycles of resources characterize the activities in effectuation. The resources are cocreated through commitments within a network of partners, customers and other stakeholders. Also the process of stakeholder acquisition is iterative and reflects the basic idea of expanding resources. Stakeholder relationships are usually created directly, one step at a time. A result is that they are able to generate rich first-hand knowledge related to the effort and will quickly have a sense of whether the activities have real promise (Sarasvathy 2008).

An essential part of iterative processes is adaptive trial and error. As predictive information does not support decision making in the best possible way in an uncertain operative environment, more reasonable is reliance on strategies that enable direct control, cocreation and transformation of situations towards positive outcomes. Quickly realized small successes and failures help avoid the risk that

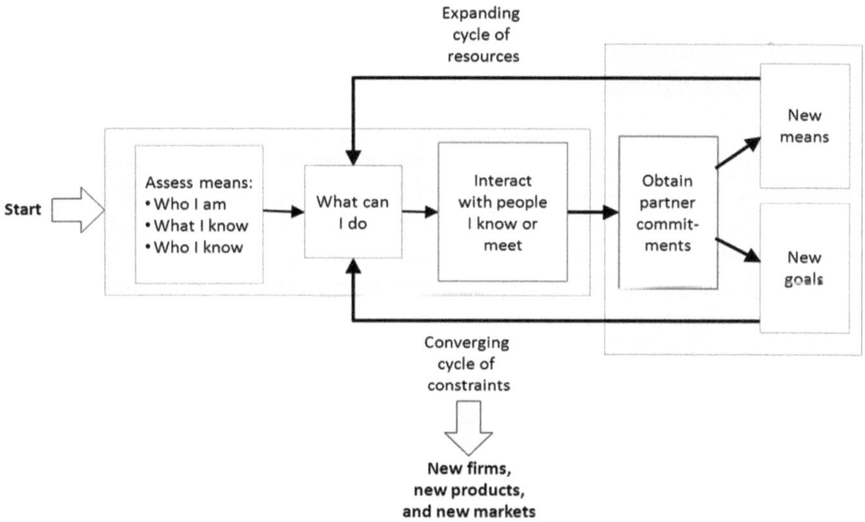

Fig. 9.1 The main idea of the effectual process (Read et al. 2009; modified)

some action puts the entire effort in jeopardy. Preparedness for considering changes in value propositions should be actively embraced, even if it necessitates a change in the strategy (Sarasvathy and Kotha 2001).

Even though setting the goal and proceeding towards it in systematic steps is not an efficient way to tackle the unknown, an alternative approach must include enough structure to support the utilization of resources and to foster collaborative creativity. This can be achieved via framing the problem in hand comprehensively: using a framework or schema within which specific decisions and their linkages to other decisions can be contextualized. The ability to group problems into fundamental categories and relate them to other problems results in knowledge architectures that link multiple decisions in the task domain over time, with feedback and interpretation—not isolated decisions (Read et al. 2009).

The approach of effectuation has not been explicitly linked to the theories of innovation. However, this kind of linkage would be highly beneficial. We see connections between the effectuation model and the process model of practice creation by Lounsbury and Crumley (2007), the change model by Dorado (2005), the dialectical perspective to institutional change by Seo and Creed (2002) and the process model of legitimacy by Drori and Honig (2013). We also argue that the main principles of effectuation are applicable, not only in the market context but also in the public sector, where the need for innovation is as urgent as it is in private companies. Thus, we illustrate with our case study, how a public innovation process could be understood as an effectual process.

9.3 Empirical Research

9.3.1 The Case and the Methods

Our case study describes a two-phase development project of municipal services for children and youth in a middle-sized Finnish city; the project was part of wider development efforts of the city in this service area. The city uses a procurer-provider model, i.e. it has separate departments for these functions in the services for children and youth (and also for some other services). The need for the project emerged after the procurer organization had launched a new service strategy. This strategy focused on service provision that would (1) promote the wellbeing of children, youth and families in a holistic way and (2) foster collaboration with citizens and different service providers, including cross sector allocation of responsibilities. Instead of the earlier, administrative division into day care, schools, club activities, etc., these services were collected together to form an integrated whole in the suburbs of the city.

In this situation, the traditional way to forecast service needs and to produce or procure services on that basis was insufficient. The procurer organization needed deeper understanding on how to build an integrated service system and to base this system on the user and stakeholder values (Osborne et al. 2012). The situation showed typical features of uncertainty: the decision makers were ambiguous regarding the preferences and consequences (Read et al. 2009).

In addition to this renewal, the city wanted to encourage citizens' initiatives regarding the development of services and empower employees to engage in experimental development. The procurer organization of services for children and youth concretized these aims in a project whose purpose was to foster small innovative experiments (mini-pilots) among citizens, companies, third sector organizations and city employees. The goal was to increase the knowledge concerning (1) the service needs of citizens and (2) the willingness of various actors to participate in the cooperative development of services.

As the mini-pilots were successful, the procurer organization wanted to take a further step based on the results achieved. At this stage, a more systematic mechanism was sought: the aim was to provide—not only 'nice add-ons' to the service repertoire of the city—but to foster a more regular bottom-up input that could influence the 'official service system'. For this purpose, another more specifically defined pilot was planned together with researchers (the authors of this chapter). This pilot tested the collaboration between citizens and grassroots employees in workshops that were based on the ideas of effectuation.

Thus, our study includes phase I analysing the mini-pilots and phase II analysing the experience from the workshops. The research approach is a qualitative case study which—despite its weaknesses in generalizability—enables an insightful examination of phenomena that evolve and change (Gephart 2004). The methods used include data acquisition via document analysis, interviews, observation and action research-type participation in workshops. All three authors of this chapter have participated in the collection of empirical material. The role of researchers has varied from an outside analyst in the document analysis of phase I to participatory activity in the workshops of phase II; observations and interviews in phase I represent an intermediate form between these extremes. In each of the methodological approaches, guidelines given in the relevant literature have been applied in order to diminish the impact of subjectivity and avoid other typical biases of qualitative research (Kvale 1996; Stake 2000). The implementation of individual methods is described in more detail in the following subsections.

Phase I: Mini-Pilots

Our study started with the follow-up and analysis of the mini-pilots which had an important pioneering role in the adoption of a new open culture in the city administration. In order to be accepted as a mini-pilot, the applicants had to fulfil two requirements. First, there had to be at least two actors representing different actor groups. Second, the idea included had to be new, beneficial and replicable for children and young people. Regarding the contents, there were no specific restrictions because the aim was to encourage the emergence of a variety of mini-pilots. The actors were allowed to define the problem themselves and then implement the solution in the way they found most suitable. A project manager (agent) was hired to coordinate the project and a steering group was established to ensure its realization. The financial support for each mini-pilot was 500 €.

The city compiled various data about the mini-pilots. This data was in the form of (1) case reports that covered 103 mini-pilots, (2) a short summary including basic quantitative and qualitative descriptions of the mini-pilots and (3) documents which included research reports, web pages and articles concerning the wider development in which the mini-pilots were part. We used this data as the starting point in our study. In addition, we observed some workshops held in the context of mini-pilots. We also carried out 17 face-to-face interviews among the mini-pilots actors: citizens, employees and managers from different levels of the city organization and people from non-profit organizations.

In the analysis of the material, we first compiled the data acquired from different sources (city documents, observation notes and interview notes). Thereafter we carried out the data analysis. We did not use a coding tool but carried out several rounds of analysis to derive meanings from data and to reduce the amount of data (Huberman and Miles 1994). While reading the documents and interview notes, we uncovered the most common and typical themes (open coding) and classified and structured them. We were particularly interested in who were participating in the mini-pilots and what were the topics tackled and the innovations emerged. Further, we analysed how the topics and innovations were related to the core of the services, what kinds of resources were mobilized and how the mini-pilots were experienced by different actors and stakeholders. Finally, we linked our observations to the theoretical views on institutional work and effectuation and made conclusions in a dialog between theory and empirical findings.

Phase II: Effectual Workshops

The positive results of the mini-pilots encouraged the procurement organization of services for children and youth to broaden its experimental approach towards the core of the daily life of citizens and employees. At this stage, the above-mentioned organizational transformation towards holistic and cooperative services was just implemented. This transformation included new, wider collaboration at the suburbs level and a new management/leadership model. Support was needed because the new organization required innovativeness from the employees and emphasized collaboration, both internally and externally. Thus, deviating from the first phase, the second phase was deliberately used to facilitate the transfer towards the new organization of services in local suburbs of the city.

The idea was to collect service initiatives from the grassroots level and to combine them with strategic administrative aims in a multidirectional dialog. The focus was on school and day-care services in two suburbs. The effectual approach shown in Fig. 9.1 was the explicit starting point. The approach was tested in four workshops: the first one consisted of citizens and grassroots employees (25 participants), the second one of school and day-care managers (4 participants), the third one of the administrative managers for children and youth services (3 participants) and the fourth one of all these stakeholder groups (15 participants). Due to the experiential nature of the pilot, the participants were invited individually—there was no open invitation.

Workshops at three different levels were organized in cooperation with the managers of the service provider for children and young people. The topics for the workshops were set according to the effectual model: who we are, what we have/know to develop further, who we need to cooperate with and what concrete deeds we are able and willing to take up. The first-level workshops took place in two different neighbourhoods. The second-level workshops engaged middle management from both of these neighbourhoods, and the third-level workshop engaged strategic management of the service provider.

The researchers documented the discussions and circulated the summaries of the discussions results from one level to the next. Finally, all the participants of the workshops gathered together in a common workshop. From the research point of view, the final workshop was used to validate and supplement the data gathered in earlier workshops. The analysis of data followed a similar process as in phase I. Workshop notes were structured according to three effectual topics: (1) institutional conditions to enable innovations (the logics of commitment, (2) agency (the logics of identity) and (3) focus of innovations initiated (the logics of action).

9.4 Results

9.4.1 *Mini-Pilots: Features of Resource Integration and Effectuation*

Via the mini-pilots, the city authority aimed to find new knowledge and other resources for solving the existing problems and also succeeded in this aim. The results reveal that the actors participating in the mini-pilots understood and shared the view of the city about the basic problems and challenges that made the development project necessary. The data collected by the city itself provided important basic information about the success of mini-pilots:

- The composition of actors in mini-pilots was rich. It included individual citizens, companies, public and non-profit organizations. The most initiative actors to carry out a mini-pilot were schools, enterprises and non-profit organizations whose shares in total number of mini-pilots were 19 %, 15 % and 17 %, respectively. The shares of other active participants were parent associations 7 %, youth organizations 7 %, kindergartens 5 %, senior associations and elderly houses 5 %,and sport clubs 4 %.
- The versatile contents of the mini-pilots distributed are as follows: cooperation with enterprises 19 %, social clubs for children and families 17 %, enrichment of the curriculum in schools 17 %, social events 13 %, citizens as active participants in local cooperation 12 % and cross generation cooperation 11 %.
- Innovative resources mobilized in the mini-pilots were characterized as local cooperation 21 %, 'a school opening its doors' 17 %, cooperation with companies 17 %, citizens organizing a service or an activity 11 %, cross generation cooperation 9 % and service development/innovations 9 %.

In the interviews and observations, the main finding was that the core service processes like basic curriculum/teaching in the classroom or actual care in the kindergarten remained untouched. All in all, resources were mobilized and integrated with the core of services in four different ways: (1) to enrich the core activity of the service, (2) to support or supplement the core activity of the institutions, (3) to develop more radical boundary breaking activities and (4) to strengthen the viability and cooperation within the broader community. Most mini-pilots were aimed to enrich and support or supplement the core of the services and communal resources. Two or three mini-pilots (3 %) could be defined as radical boundary breaking activities, in which the actors were able to transform also the core of service processes. In Fig. 9.2 we present an example on how the resources were integrated from the viewpoint of one actor (school). We also illustrate what kind of institutional work—decision making, sense-making or collective mobilization (Thornton et al. 2012) were needed in order to enable mini-pilots.

An illustrative, and also innovative, example of the mini-pilots was various activities aiming to increase dialog between the generations: to promote the seniors' interaction with children and youth. Both generations benefited each other with

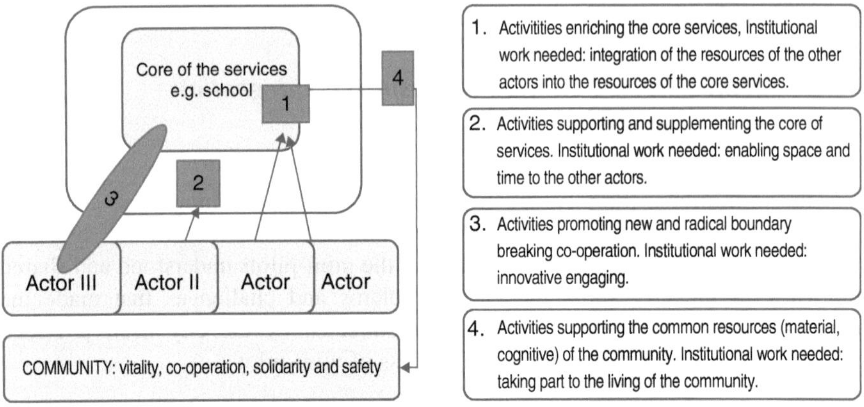

Fig. 9.2 The focus of the mini-pilots activity in an exemplary case (school)

their own competences, and the younger generation learned in a natural way about life in former times. Another innovative example deals with the activities carried out by an individual: a young unemployed man. He organized activities for his friends whom the employment officials had been unable to reach and engage. In this example, the roles of professionals and a lay person were exchanged, the professionals staying in the 'back office' and supporting the layman.

The interviews revealed that in the mini-pilots the role of an actor was experienced to be stronger than the roles of a consumer, user or citizen. Thus mini-pilot actors can be described as 'authorized with the given resources'. Professionals took part in the mini-pilots as facilitators, but they were rarely in an active role. Managers more often adopted an entrepreneurial way of acting and, as they also possessed resources, played the role of a gatekeeper.

From the viewpoint of effectual theory, it is interesting that the mini-pilots project showed several features that are in line with it. Figure 9.3 summarizes these features and concretizes the problems, actors, means and goals with illustrative examples. The figure shows that small-scale initiatives—even single events and individual services—may include a seed for the emergence of much broader goals of securing and enriching services and preventing social exclusion.

On the one hand, the experience of mini-pilots shows the willingness of actors to widen the traditional role of public institutions (e.g. schools) and cooperation between different actors. In this context, we were able to verify a finding by Battilana (2006): the lower positions in organizations and communities favour engagement in experimental development, which might initiate radical changes in existent practices as well as creation of new ones. On the other hand, the interviews indicated that the need for being engaged is sometimes a stronger motive to participate in different kinds of activities than the actual service needs. From this viewpoint, the contribution of the mini-pilots was primarily social and communal rather than professional.

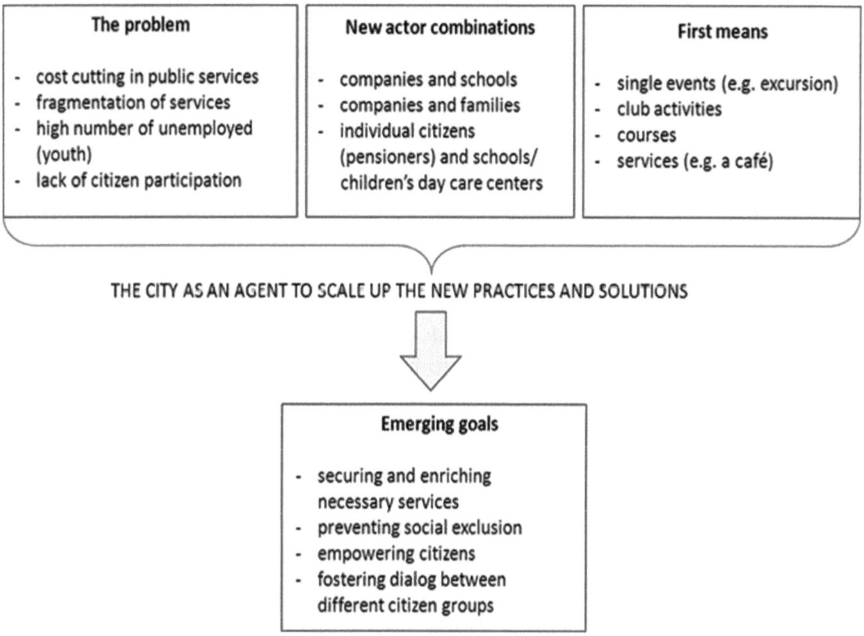

Fig. 9.3 Effectuation-based features of the mini-pilots

9.4.2 Effectual Workshops: The Logics of Commitment, Identity and Action

As mentioned above, we structured the results of the second phase of our study on the basis of the effectual logics, exploring if and how the characteristics of the effectual logics were identifiable in the discussions of the workshops. In the following, we analyse these logics one by one: the logics of commitment, the logics of identity and the logics of action. We link these logics to the issues of institutional conditions, agency and focus, respectively.

Conditions (the logics of commitment): The elements that seemed to promote the commitment of the actors to take part in the service development were the strategies of the city and the experienced cohesiveness of the neighbourhood. A strategy which highlights cooperation was welcomed by the actors and was experienced as a common resource. Also the slogan applied in one neighbourhood 'everyone is permitted to innovate' was seen to provide individual and collective wellbeing and relief in the work. On the other hand, the new managerial arrangements in the wider suburbs were not yet well understood and created some confusion about the activities that were desirable. New arrangements were not experienced as resources at the grassroots level. At the same time, the strategic management group expected initiatives to support the transformation.

There was also some difference in the commitment to the neighbourhood in the two suburbs. In one of them, actors described them to be very autonomous and possessing a strong local identity ‘we are like a small village’. In the other neighbourhood, actors described their suburb as less coherent expecting external support like ‘who in the city could help us and do something for our suburb’.

Agency (the logics of identity): The actors in the first-level workshops were citizens and grassroots employees. They represented simultaneously several different roles: personal experiences and knowledge as parents, professionals, active citizens and even as politicians (Osborne et al. 2012; Sarasvathy and Dew 2005). The multiplicity was experienced as an extra resource. This result resonates with the notion of the multidimensional nature of agency (Smets and Jarzabkowski 2013).

Iniativeness in taking the role of innovators was found to flourish in a situation which is characterized by easiness and freedom. On the other hand, advice to promote cooperation was desired: how to find partners and how to build a network. Thus, both cognitive and emotional efforts seemed necessary for engagement in innovation (Foss and Lorenzen 2009; Lawrence et al. 2013). From the managerial viewpoint, this means that arousing and maintaining the reflexive awareness and casting a glance at the future will be a critical question in fostering bottom-up innovations in the context of traditional institutions.

Focus (the logics of action): In the workshops, we could identify four levels of initiatives: (1) individual-level initiatives, (2) workplace and neighbourhood-level initiatives, (3) suburb-level initiatives and (4) city-level initiatives. The discussions in the workshops concerned mostly the first and second levels. On these levels, the actors seem to have a sense of possessing resources and autonomy to adhere to the issues ‘I just take a call and tell my idea’. Leadership in the own workplace was experienced as an important resource to encourage innovation. This was expressed by the slogan ‘leading from the centre with minimal regulations and pressures’. This resonates with the notion that small groups promote entrepreneurship (Dorado 2013).

Issues from the third level (suburb) were identified by the actors, but the alternatives and resources to handle with them did not take shape by them. This may be explained by the novelty of the new arrangements in the suburb. Sense-making and sense-giving were desired from leaders and managers to make the new arrangements understandable. There were also some issues concerning the broader cross sector and city-level practices and policy (the fourth level). In these cases resources and coordination were expected from the even higher managerial level, ‘from the city’.

9.5 Conclusions and Implications

9.5.1 A Theoretical Model to Foster Grassroots Innovative Activities

Our results revealed that innovative attitudes may be quite general in a public context: in our study, it was quite easy to encourage citizens and employees to be initiative. However, the activities emerged were small-scale trials. They were performance-based activities, not institutionalized practices (a set of activities).

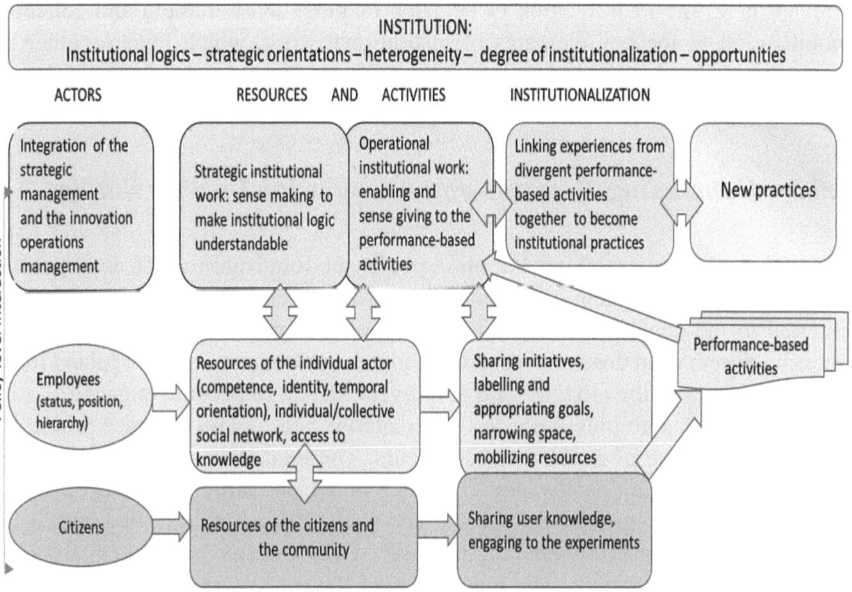

Fig. 9.4 Institutionalization of the performance-based activities

They were local and pragmatic focusing on the tiny issues that improve the daily life of inhabitants and provide small variations to the extant service processes. Even though tiny variations are essential to promote changes in the public sector (Aagaard 2012), the impact would be stronger if the grassroots initiatives could be transformed into established institutional practices. In order to establish lower-level initiatives as institutional practices, the institutional work of strategic and operational management is crucial. This means continuing interactions between the micro-, meso- and macro-levels (Suddaby et al. 2013). Based on our results, we have built an ideal/theoretical process model (Fig. 9.4) that illustrates how such a transformation (from tiny experiments to institutional practices) could take place and gain legitimacy among citizens, employees and stakeholders.

Our theoretical model highlights that both on the individual level and on the levels of workplaces and neighbourhoods, actors need a sense of sufficient autonomy to be able to mobilize resources and engage in innovations. In addition, experiments or innovations are not enough, but entrepreneurial efforts have to gain legitimacy to be successful (Drori and Honig 2013; Lounsbury and Glynn 2001). In order to become established as institutional practices, the various experimental grassroots-level activities should be (1) supported by sense-making by the strategic and operational management and (2) linked together and analysed by sense-giving with actors and strategic and operational management. These views are in line with the emphasis of earlier literature on the projective dimension that supports future change and entrepreneurial and creative behaviour (Battilana et al. 2009; Lawrence et al. 2009). Also the ideas of Thornton et al. (2012) are relevant here: the identification of decision making, sense-making and collective mobilization as the key elements of institutional work, which links together the dynamics of practices and identities.

9.5.2 Concluding Remarks and Avenues for Further Studies

This chapter has analysed innovation in public services from the effectual point of view. The innovative activities that we found in our empirical study were mainly external to the core services, aiming to support, enrich and supplement them. A question arises: what does this result tell about the characteristics of expected future services and about the citizens' and employees' desire to become engaged?

The scale of activities describes the material and cognitive space where the actors have a sense of enabled entrepreneurs. The focus and scale may be understood in terms of a 'bandwagon process', in which participants negotiate the possible issue of cooperation. Fuglsang and Eide (2012) highlight that on a bandwagon the different activities of different actors become structured around a common field of meaning. The boundaries of the process may be mental, social or physical (Montgomery and Oliver 2007). Boundary objects will be found by labelling (identifying the meaning of the issue in question), by tackling the issue collectively to resonate with the community and by narrowing the conceptual and technical space to make cooperation possible (Corradi and Verzelloni 2010; Fujimura 1992; Star 1989).

Lounsbury and Crumley (2007) have raised the question of how the performativity-driven variations in activities can spur field-wide efforts to establish new innovations as a practice. We have formulated an ideal model (see Fig. 9.4) to describe the development of performance-based activities, from the individual and group levels to the level of established institutional practices. We assume that it is the task of innovation management to formulate the central questions when enhancing public innovations. In order to promote innovations that start as small-scale activities, the institution has to find actors who (1) are able to recognize the tiny elements of innovations, (2) to feed these (possibly delicate) elements and (3) to carry on their implementation systematically.

In the first experiment of our study, the role of the agent (hired) was central for the operational management of innovation—this role can be interpreted in terms of meta-governance. The agent was able to use institutional tools like the structure and design of the organization, the political and discursive framework, process management and direct participation (Sørensen and Torfing 2011). The steering group was responsible for the strategic management. In the second experiment, both the strategic and operational managers were engaged in the effort.

Our results confirm several views presented in earlier literature. They are in line with the view of Arend and Chen (2012) about the importance of strategic management in entrepreneurship and innovation and with the view of Sørensen and Torfing (2012) about the importance of leadership. The central role of middle managers for the continuation of social initiatives was clearly visible and confirms the view of Sharma and Good (2013) about the role of this group in maintaining the hybridity of profit and social logics. Our results also highlight the significance of innovative culture in the fostering of a systematic way of developing services and point out the integration of operational service delivery and strategic service orientation (Teboul 2006). Osborne et al. (2012) summarize most of these factors in their emphasis on the simultaneous implementation of strategic orientation, marketing, co-production and operations management.

To become established as institutional practices, innovative activities need to be interpreted against the institutional goals, logics and expectations. In our case, these activities focused mainly on the boundary levels and produced intangible outputs that were social, organizational and individual in nature (Hitt et al. 2011; Osborne et al. 2012). The actors reported increasing wellbeing, joy and communal trust as the outcomes of the activities. A question arises whether these results are in harmony with the expectations, goals and logics of the institution. What is the added value they provide to the knowledge base of the strategic and operational management? In our case, the city was satisfied with the results and has continued to conduct mini-pilots (1) as activities in some suburbs and (2) as a model building activity for systematic and experience-based service development.

One more question to be raised based on our results concerns the increasing complexity of the institutions and the inter-institutional orders that is a consequence of many organizational changes. The emerging complexity raises questions about how to handle it: how it changes the innovation context. Actors may experience the changing environment as hazy and inchoate. They may not be able to imagine alternatives and opportunities nor to identify the resources to be used or to find the access to knowledge. Therefore, actors need an adequate organizational structure and support from the managers whose work is to interpret the divergent logics of complex organizations. It is important that the perspective of institutional dynamics and complexity is taken into account and linked to the issues of the actors in their concrete daily work. This is a task for both the managers who create the innovation enabling institutions and for the researches who analyse these institutions (Smets and Jarzabkowski 2013).

Our study contributes to the effectual model by highlighting two elements that are particularly relevant when this model is applied in the public context. First, public innovations are collective activities of partially autonomous actors, implying that effectuation has to be analysed as a collective activity, too. Second, effectuation reached in the individual or group-level innovations becomes more general via institutional work. This means that also strategic and operational management processes should follow the effectual logic. Actually, in the public context the first question of effectuation 'what we want to be' should be supplemented with a question 'what the institution wants to be' (Kodeih and Greenwood 2014).

An important conclusion based on our study is that the effectual approach provides an innovation process model whose main principles are applicable, not only in the market context but also in the public sector. Even though the resource expansion in our case is still going on, we have been able to make some initial observations that are crucial and whose validity in other contexts could be tested in further studies. One such observation is the importance of improvisation, courage and tolerance towards failures as factors that favour the emergence of novelties. Another area for further research is the applicability of service-dominant logic (SDL) in the public sector. Our example illustrated the actions of citizens (users) as innovators and the role of public professionals (providers) as facilitators in the innovation process—a 'work division' that is in line with the premises of SDL (Vargo and Lusch 2004, 2008). Also the institutionalization of the new practices is a topic in which our case provides interesting material for future research.

References

Aagaard, P. (2012). Drivers and barriers of public innovation in crime prevention. *The Innovation Journal: The Public Innovation Journal, 17*(1), 2–17.

Abbott, A. (1988). *The system of profession: An essay on the division of labor*. London and Chicago: Chicago University Press.

Albury, D. (2005). Fostering innovation in public services. *Public Money and Management, 25*(1), 51–56.

Arend, R. J., & Chen, Y. (2012). Entrepreneurship as dynamic, complex, disequilibrious: A focus that benefits strategic organization. *Strategic Organization, 10*(1), 85–95.

Battilana, J. (2006). Agency and institutions: The enabling role of individuals' social position. *Organization, 13*(5), 653–676.

Battilana, J., Leca, B., & Boxenbaum, E. (2009). How actors change Institutions: Towards a theory of institutional entrepreneurship. *Academy of Management Annals, 3*(1), 65–107.

Butler, J. (1990). *Gender trouble: Feminism and the supervision of identity*. New York: Routledge.

Chesbrough, H. (2006). *Open business models: How to thrive in the new innovation landscape*. Boston, MA: Harvard Business School Press.

Chiles, T. H., Tuggle, C. S., McMullen, J. S., Bierman, L., & Greening, D. W. (2010). Dynamic creation: Extending the radical Austrian approach to entrepreneurship. *Organization Studies, 31*(1), 7–46.

Christensen, T. (2012). Post-NPM and changing public governance. *Meiji Journal of Political Science and Economics, 1*, 1–11.

Corradi, G. G., & Verzelloni, L. (2010). Through the practice lens: Where is the bandwagon of practice-based studies heading? *Management Learning, 41*(3), 265–283.

Currie, C., Lockett, A., Finn, R., Martin, G., & Waring, J. (2012). Institutional work to maintain professional power: Recreating the model of medical professionalism. *Organization Studies, 33*(7), 937–962.

Dent, M., & Whitehead, S. (2002). Introduction: Configuring the "new" professional. In M. Dent & S. Whitehead (Eds.), *Managing professional identities* (pp. 1–16). New York: Routledge.

DiMaggio, P. J. (1988). Interests and agency in institutional theory. In L.-G. Zucker (Ed.), *Institutional patterns and organizations* (pp. 3–21). Cambridge: Ballinger.

DiMaggio, P. J., & Powell, W. W. (1983). The iron cage revisited: Institutional isomorphism and collective rationality in organizational fields. *American Sociological Review, 48*(2), 147–160.

DiMaggio, P. J., & Powell, W. W. (1991). Introduction. In P. J. DiMaggio & W. W. Powell (Eds.), *The new institutionalism in organizational analysis* (pp. 1–40). Chicago: University of Chicago Press.

Dorado, S. (2005). Institutional entrepreneurship, partaking and convening. *Organization Studies, 26*(3), 386–414.

Dorado, S. (2013). Small groups as context for institutional entrepreneurship: An exploration of the emergence of commercial microfinance in Bolivia. *Organization Studies, 34*(4), 533–577.
Drori, I., & Honig, B. (2013). A process model of internal and external legitimacy. *Organization Studies, 34*(3), 345–376.
Du Gay, P. (2007). *Organizing identity: Persons and organizations after theory*. London: Sage.
Emirbayer, M., & Mische, A. (1998). What is agency? *American Journal of Sociology, 103*(4), 962–1023.
Emison, G. A. (2010). Ethics of innovation for public service professionals. *The Innovation Journal: The Public Sector Innovation Journal, 15*(3), 2–10.
Engeström, Y. (1987). *Learning by expanding: An activity-theoretical approach to developmental research*. Helsinki: Orienta-Konsultit.
Etzioni, A. (Ed.). (1969). *The semi-professions and their organization: Teachers, nurses, social workers*. New York: Free Press.
Evetts, J. (2012). Professionalism in turbulent times: Changes, challenges and opportunities. *Propel International Conference*, Stirling 9–11 May 2012.
Ferlie, E., Fitzgerald, L., Wood, M., & Hawkins, C. (2005). The nonspread of innovations: The mediating role of professionals. *Academy of Management Journal, 48*(1), 117–134.
Foss, N., & Lorenzen, M. (2009). Towards an understanding of cognitive coordination: Theoretical developments and empirical illustrations. *Organization Studies, 30*(11), 1201–1226.
Freidson, E. (2001). *Professionalism: The third logic*. Cambridge: Polity.
Friedland, R., & Alford, R. (1991). Bringing society back in: Symbols, practices and institutional contradictions. In P. J. DiMaggio & W. W. Powell (Eds.), *The new institutionalism in organizational analysis* (pp. 232–263). Chicago: University of Chicago Press.
Fuglsang, L., & Eide, D. (2012). The experience turn as "bandwagon": Understanding network formation and innovation as practice. *European Urban and Regional Studies, 20*(4), 417–434.
Fujimura, J. (1992). Crafting science: Standardised packages, boundary objects and translation. In A. Pickering (Ed.), *Science as practice and culture* (pp. 168–211). Chicago, IL: University of Chicago Press.
Garud, R., Hardy, C., & Maguire, S. (2007). Institutional entrepreneurship as embedded agency: An introduction to the special issue. *Organization Studies, 28*(7), 957–969.
Gephart, R. (2004). What is qualitative research and why is it important? *Academy of Management Journal, 47*(4), 454–462.
Giddens, A. (1984). *The constitution of society: Outline of the theory of structuration*. Berkeley and Los Angeles: University of California Press.
Greenwood, R., & Hinings, C. R. (1996). Understanding radical organizational change: Bringing together the old and the new institutionalism. *The Academy of Management Review, 21*(4), 1022–1054.
Greenwood, R., Raynard, M., Kodeih, F., Micelotta, E. R., & Lounsbury, M. (2011). Institutional complexity and organizational responses. *The Academy of Management Annals, 5*(1), 317–371.
Halford, S., & Leonard, P. (2002). New identities? Professionalism, managerialism and the construction of self. In M. Exworthy & S. Halford (Eds.), *Professionals and the new managerialism in the public sector* (pp. 102–120). Buckingham: Open University Press.
Hartley, J. (2005). Innovation in governance and public services: Past and present. *Public Money and Management, 25*(1), 27–34.
Hitt, M. A., Ireland, D., Sirmon, D. G., & Trahms, C. A. (2011). Strategic entrepreneurship: Creating value for individuals, organizations and society. *Academy of Management Perspectives, 25*(2), 57–75.
Hoffman, A. J. (1999). Institutional evolution and change: Environmentalism and the US chemical industry. *Academy of Management Journal, 42*(4), 351–371.
Hood, C. (1991). A public management for all seasons. *Public Administration, 69*(1), 93–109.
Huberman, A. M., & Miles, M. B. (1994). Data management and analysis methods. In N. K. Denzin & Y. S. Lincoln (Eds.), *Handbook of qualitative research* (pp. 428–444). Thousand Oaks, CA: Sage.

Hwang, H., & Colyvas, J. A. (2011). Problematizing actors and institutional work. *Journal of Management Inquiry, 20*(1), 62–66.

Jarzabkowski, P. (2005). *Strategy as practice: An activity-based approach.* London: Sage.

Kesting, P., & Ulhøi, J. P. (2010). Employee-driven innovation: Extending the license to foster innovation. *Management Decision, 48*(1), 65–84.

Kodeih, F., & Greenwood, R. (2014). Responding to institutional complexity: The role of identity. *Organization Studies, 35*(1), 7–39.

Kvale, S. (1996). *InterViews. An introduction to qualitative research interviewing.* London: Sage.

Langergaard, L. (2011). Understanding of "users" and innovation in a public sector context. In J. Sundbo & M. Toivonen (Eds.), *User-based innovation in services* (pp. 203–226). Cheltenham and Northampton: Edward Elgar.

Lawrence, T. B., Leca, B., & Zilber, T. B. (2013). Institutional work: Current research, new directions and overlooked issues. *Organization Studies, 34*(8), 1023–1033.

Lawrence, T. B., & Suddaby, R. (2006). Institutions and institutional work. In S. R. Clegg, C. Hardy, T. B. Lawrence, & W. R. Nord (Eds.), *Handbook of organization studies* (pp. 215–254). London: Sage.

Lawrence, T. B., Suddaby, R., & Leca, B. (2009). Introduction: Theorizing and studying institutional work. In R. Suddaby, B. Leca, & T. B. Lawrence (Eds.), *Institutional work: Actors and agency in institutional studies of organizations* (pp. 1–27). Cambridge: Cambridge University Press.

Llewellyn, S. (2001). Two-way windows: Clinicians as medical managers. *Organization Studies, 22*(4), 593–623.

Llewellyn, N., & Spence, L. (2009). Practice as a members' phenomenon. *Organization Studies, 30*(12), 1419–1439.

Lounsbury, M., & Crumley, E. T. (2007). New practice creation: An institutional perspective on innovation. *Organization Studies, 28*(7), 993–1012.

Lounsbury, M., & Glynn, M. A. (2001). Cultural entrepreneurship: Stories, legitimacy, and the acquisition of resources. *Strategic Management Journal, 22*(6/7), 545–564.

Lundvall, B.-Å. (2007). *Innovation system research: Where it came from and where it might go.* Working Paper Series, 2007-01. The Global Network for Economics of Learning, Innovation, and Competence Building System (GLOBELICS).

Lyotard, J.-F. (1984). *The postmodern condition: A report on knowledge.* Minneapolis: University of Minnesota Press.

Marquis, C., Glynn, M. A., & Gerald, D. (2007). Community isomorphism and corporate social action. *Academy of Management Journal, 32*(3), 925–945.

Meyer, J. W., & Rowan, B. (1977). Institutionalized organizations: Formal structure as a myth and ceremony. *American Journal of Sociology, 83*(2), 340–363.

Misztal, B. A. (2002). Trusting the professional: A managerial discourse for uncertain times. In M. Dent & S. Whitehead (Eds.), *Managing professional identities: Knowledge, performativity and the "new" professional* (pp. 19–37). London: Sage.

Montgomery, K., & Oliver, A. L. (2007). A fresh look at how professions take shape: Dual-directed networking dynamics and social boundaries. *Organization Studies, 28*(5), 661–687.

Noordegraaf, M. (2007). From "pure" to "hybrid" professionalism: Present-day professionalism in ambiguous public domains. *Administration and Society, 39*(6), 761–785.

Noordegraaf, M. (2011). Risky business: How professionals and professional fields (must) deal with organizational issues. *Organization Studies, 32*(10), 1349–1371.

Osborne, S. P., Zoe, R., & Nasi, G. (2012). A new theory for public service management? Toward a (public) service-dominant approach. *American Review of Public Administration, 43*(2), 135–158.

Pollitt, C. (1993). *Managerialism and the public services.* Oxford: Blackwell.

Read, S., Dew, N., Sarasvathy, S. D., Song, M., & Wiltbank, R. (2009). Marketing under uncertainty: The logic of an effectual approach. *Journal of Marketing, 73*(May), 1–18.

Sarasvathy, S. D. (2008). *Effectuation: Elements of entrepreneurial expertise*. Cheltenham and Northampton: Edward Elgar.
Sarasvathy, S. D., & Dew, N. (2005). Entrepreneurial logics for a technology of foolishness. *Scandinavian Journal of Management, 21*(4), 385–406.
Sarasvathy, S. D., & Kotha, S. (2001). Dealing with Knightian uncertainty in the new economy: The real networks case. In J. Butler (Ed.), *Research on management and entrepreneurship* (Vol. 1, pp. 31–62). Greenwich, CT: IAP Inc.
Scott, R. (1995). *Institutions and organizations*. Thousand Oaks: Sage.
Scott, W. R. (2008). Lords of the dance: Professionals as institutional agents. *Organization Studies, 29*(2), 219–238.
Seo, M. G., & Creed, D. W. E. (2002). Institutional contradictions praxis and institutional change: A dialectical perspective. *Academy of Management Review, 27*(2), 222–248.
Sharma, G., & Good, D. (2013). The work of middle managers: Sensemaking and sensegiving for creating positive social change. *The Journal of Applied Behavioural Science, 49*(1), 95–122.
Smets, M., & Jarzabkowski, P. (2013). Reconstructing institutional complexity in practice: A relational model of institutional work and complexity. *Human Relations, 66*(10), 1279–1309.
Sørensen, E., & Torfing, J. (2011). *Enhancing social innovation by rethinking collaboration, leadership and public governance*. Nesta Social Frontiers: The next edge of social innovation research.
Sørensen, E., & Torfing, J. (2012). Introduction: Collaborative innovation in the public sector. *The Innovation Journal: The Public Sector Innovation Journal, 17*(1), 1–14.
Stake, R. E. (2000). Case studies. In N. K. Denzin & Y. S. Lincoln (Eds.), *Handbook of qualitative research* (2nd ed.). London: Sage.
Star, S. L. (1989). The structure of ill-structured solutions: Boundary objects and heterogeneous distributed problem solving. In M. N. Huhns & L. Gasser (Eds.), *Distributed artificial intelligence* (pp. 37–54). London: Pitman.
Suddaby, R., & Greenwood, R. (2005). Rhetorical strategies of legitimacy. *Administrative Science Quarterly, 50*(1), 35–67.
Suddaby, R., Seidl, D., & Lê, J. K. (2013). Strategy-as-practice meets neo-institutional theory. *Strategic Organization, 11*(3), 329–344.
Suddaby, R., & Viale, T. (2011). Professionals and field-level change: Institutional work and the professional project. *Current Sociology, 59*(4), 423–442.
Sundbo, J., & Toivonen, M. (2011). Introduction. In J. Sundbo & M. Toivonen (Eds.), *User-based innovation in services*. Cheltenham and Northampton: Edward Elgar.
Teboul, J. (2006). *Service is front stage: Positioning services for value advantage*. Basingstoke: Palgrave.
Thornton, P. H., Ocasio, W., & Lounsbury, M. (2012). *The institutional logics perspective: A new approach to culture, structure and process*. Oxford: Oxford University Press.
Torfing, J., & Triantafillou, P. (2013). What's in name? Grasping new public governance as a political-administrative system. *International Review of Public Administration, 18*(2), 9–25.
Tummers, L. (2009). *Policy alienation and the power of professionals: Confronting new policies*. Cheltenham and Northampton: Edward Elgar.
Tummers, L. (2013). Connecting public administration and change management literature: The effects of policy alienation on resistance to change. *11th Public Management Research Conference*, Madison, Wisconsin.
Tummers, L., Bekkers, V., & Steijn, B. (2009). Policy alienation of public professionals: Application in a new public management context. *Public Management Review, 11*(5), 685–706.
Vargo, S., & Lusch, R. (2004). Evolving to a new dominant logic for marketing. *Journal of Marketing, 68*(1), 1–17.
Vargo, S., & Lusch, R. (2008). Service-dominant logic: Continuing the evolution. *Journal of the Academy of Marketing Science, 36*(1), 1–10.
von Hippel, E. (2005). *Democratizing innovation*. Cambridge, MA: MIT Press.

Wasserman, V., & Frenkel, M. (2011). Organizational aesthetics: Caught between identity regulation and culture jamming. *Organization Science, 22*(2), 503–521.
Weick, K. E. (1995). *Sensemaking in organizations.* Thousand Oaks: Sage.
Windrum, P. (2008). Introduction. In P. Windrum & P. Koch (Eds.), *Innovation in public sector services: Entrepreneurship, creativity and management* (pp. 3–20). Cheltenham: Edward Elgar.

Part III
Perspectives: Conceptual Merging

Chapter 10
Framing the New Social–Service Innovation Mindset

Tiziana Russo-Spena, Maria Colurcio, and Monia Melia

Abstract This chapter addresses a proposal for an integrated perspective on social–service innovation. In particular, research effort is devoted to developing a framework that captures the challenges of new social issues within the contemporary landscape of service innovation. The research interest is twofold: (1) to understand conceptually how social and service innovation perspectives can be combined into a more integrated approach and (2) to provide a social–service innovation framework within which to understand how social innovation can be built onto concepts and practices of service innovation logic. Building on insights and categories from the existing literature on social innovation and service innovation, the chapter identifies five issues that are integral to the proposed social–service innovation framework. The proposed framework aims to add to the efforts of scholars in developing new ideas, concepts and tools that more deeply explicate the complex and multifaceted nature, structure and processes of social innovation. By combining the main social and service dimensions, it addresses the opportunity for furthering research discussion and practical implications concerning a new social innovation mindset.

T. Russo-Spena (✉)
Department of Economics Management Institutions, University of Naples Federico II, Napoli, Italy
e-mail: russospe@unina.it

M. Colurcio • M. Melia
Department of Legal, Historical, Economic and Social Sciences, University Magna Graecia of Catanzaro, Catanzaro, Italy
e-mail: mariacolurcio@unicz.it; monia.melia@unicz.it

T. Russo-Spena et al. (eds.), *Innovating in Practice*,
DOI 10.1007/978-3-319-43380-6_10

10.1 Introduction

Over the past 20 years, both politics and governments have begun moving towards new ideas about social and economic issues. The Europe 2020 Strategy makes clear that the wealth and wellbeing of society will depend far more on smart, sustainable and inclusive economic growth.

Under these pressures, research communities have deployed a range of concepts to address new challenges related to the economy. Thus, social issues and sustainability have become popular topics, spanning different research domains of politics and economics (Drucker 1994), strategy and management (Carroll 1999; Porter and Kramer 2006, 2011), technology and information and communications technologies (ICT; Hasan and Meloche 2013) and marketing and service (Belz and Peattie 2009; Edvardsson and Enquist 2009).

In terms of innovation, researchers have shown that notable conceptualizations and the social innovation literature have grown significantly as the result of entrepreneurship efforts (Thompson and MacMillan 2010; Yunus et al. 2010), socio-economic studies (Dawson and Daniel 2010; Hämäläinen and Heiskala 2007; Pol and Ville 2009) and business scholars' research (Prahalad and Hammond 2002; Prahalad and Hart 2002).

Beyond the plurality of definitions flourishing with new concepts, at its core social innovation stands out as the antagonist to approaches that have their roots in the business domain (Mulgan 2006). Social innovation pertained more to the social needs of the poorest segment in society and public, and non-profit organizations took the lead in these processes (Dawson and Daniel 2010; Hämäläinen and Heiskala 2007). When they were considered, private companies were viewed as boundary actors of sorts that did not consider profit as their first goal (Dawson and Daniel 2010; Pol and Ville 2009).

More recently, increasing debate in the business and academic literature has begun to question whether "business" and "social" sectors can still be treated as distinct and conflicting forms of innovation (Franz et al. 2012; Rubalcaba et al. 2013).

Some empirical evidence has proven societal challenges to be of particular interest to companies, not as charitable favours to society but because they open up to present tremendous business opportunities (Porter and Kramer 2006, 2011). Many multinational corporations (i.e. Danone, Unilever, Johnson & Johnson) have recently launched social business projects with the explicit intention of making them levers of strategic renewal. Moreover, new organizations such as for-benefit corporations (Sabeti 2011) are emerging as new forms of enterprise driven by businesses that are motivated by social aims.

The existence of links between social innovation and business orientation has also recently been proposed by some social innovation scholars (Jacobsen and Jostmeier 2012; Osburg and Schmidpeter 2013) in their attempts to tackle the new challenges of modern and global economies.

Moreover, the question of linking social and business issues appears to be gaining increasing interest in the research agendas of service scholars. Ostrom and colleagues (Ostrom et al. 2010, 2015) have stressed the critical role of service in transforming consumers' lives and improving collective and societal wellbeing, and Vargo and Lusch's service dominant logic (2004) has produced fervent debates newly devoted to service innovation concepts. More than simply dealing with technology or goods, service innovation is stated as a primary approach for how firms can better serve others (Edvardsson and Tronvoll 2013; Michel et al. 2008; Ordanini and Parasuraman 2011; Vargo et al. 2015).

Notwithstanding advances in academic and empirical debates, the links between social innovation and service innovation remain unclear, and disciplinary boundaries widely dominate the respective social and business analyses. The time may have come for social innovation and service innovation to begin talking to one another more often.

This chapter addresses the proposal for an integrated perspective on social–service innovation. The research effort is devoted to developing a framework that captures the challenges of new social issues within the contemporary landscape of service innovation. The research interest is twofold: (1) to understand conceptually how the perspectives of social innovation and service innovation can be combined into a more integrated approach and (2) to provide a social–service innovation framework within which to understand how social innovation can be built onto concepts and practices of service innovation logic. Building on insights and categories from the existing literature on social innovation and service innovation, the chapter identifies five issues that are integral to the proposal for a social–service innovation framework. The proposed framework aims to add to the efforts of scholars in developing new ideas, concepts and tools that more deeply explicate the complex and multifaceted nature, structure and processes of social innovation. By combining the main social and service dimensions, the framework addresses the opportunity for furthering research discussion and practical implications concerning a new social innovation mindset.

The rest of the chapter is structured as follows. Section 2 clarifies the conceptual nature of the discussion and motivates the research design of the study with case studies as examples. Section 3 presents the analysis of relevant literature on social innovation and service innovation. In this way, it serves as a helpful basis to our reasoning about the overlapping aspects of social and service issues. Our proposed social–service conceptual framework is presented in Sect. 4. Section 5 clarifies and empirically illustrates the key elements of the framework. Section 6 discusses and further details how the framework logic of social–service innovation unfolds. Finally, Sect. 7 presents the main implications of the current study and discusses further research.

10.2 Research Process

The chapter is conceptual and is based on a focused literature review (Tranfield et al. 2003) that consisted of three main steps.

In the first step, we made clear the specific literature interest of a qualitative study by deciding on the relevant topics and inclusion criteria for the studies to be considered. The data were selected from the ISI Thomson Web of Knowledge by searching for articles, proceedings and book chapters with titles, abstracts or keywords containing the terms "social innovation/s", "service/s innovation" and the combination of both of these terms (pertinence criterion). To ensure the focus of the analysis, we checked publications in the three main research categories: business, management and operation research. Specifically, regarding the literature on social innovation, we found 97 contributions published between 1999 and 2015 (June 30): 57 articles, 32 proceedings papers and eight book chapters. The query "service innovation" highlighted 481 contributions published between 1996 and 2015 (June 30): 54 % articles, 40 % proceedings papers and 6 % book chapters. The search for contributions that jointly investigated service and social innovation ("social innovation" + "service innovation") provided only three contributions. These contributions showed significant interdisciplinary approaches ranging from and intersecting such categories as operations research and environmental sciences, computer science and ecology and business economics.

The second step in the research was directed at an in-depth review of the literature analyses for both disciplines. To grasp meanings and connections among elements in the different literature contributions, a list of key topics were developed. The reading of articles and the analysis of words and content in documents was used to do so (Tesch 1990). The topics included common issues in both types of literature such as defining innovation, forms of innovation as well as the process dimension and outcomes of innovation. All of these issues were scrutinized in-depth within both types of literature. The analysis was not a one-off process but emerged rather as a cyclical process of searching and questioning insights and key topics. This step facilitated navigating among the different literatures and helped to (1) make clear differences between innovation concepts and (2) highlight how the debate regarding these concepts was evolving. This analysis provided a first basis for questioning whether the distinction between social and service domains of innovation was still valid and useful for depicting the complexity of new market and social phenomena.

The final step involved both theoretical and empirical work. In this phase, a critical reading of the contributions was conducted to scrutinize how a set of concepts from both types of literature could be linked into a broader, more useful and integrated approach. Building on insights and categories from previous stages, a theoretically based framework was derived to illustrate the social–service innovation conceptualization. The main components of the proposed conceptual framework were defined by an iterative process that enriched theoretical arguments via a context-specific analysis (Flyvbjerg 2006). The purpose of doing so was to empirically illustrate and further extend the concepts at the core of the proposed conceptual framework. Examples involved four service companies: two from the profit sector and two from the non-profit sector. The companies were selected on the basis of their theoretical suitability, accomplished by the outstanding approaches they had undertaken towards social- and service-orientated innovation. The first non-profit example was Altromercato Consortium, the first Fair Trade central, who was expanding its social innovation strategy to embrace service and business issue. Coop Italia as an example proved useful to the research intent, as it represented the largest food retailer in Italy striving to increase its commitment to social issues. On a similar basis, Banca Popolare Etica from the non-profit sector and Humanitas Research Hospital as a private service company provided additional data on how social–service innovation mindsets work in real social business contexts.

Combining social and service innovation is an area that is still limited in terms of existing research; therefore, the examples in this study can help to generate more concrete and context-specific knowledge to advance this area of research (Flyvbjerg 2006).

10.3 Theoretical Boundaries of Social Innovation and Service Innovation

This section defines the theoretical boundaries of the chapter by explaining the foundations of previous studies on social innovation and service innovation as they relate to this study. This provides preliminary insights into some commonalities between social and service innovation and lays the groundwork for an integrative perspective that will be discussed further below.

10.3.1 Social Innovation: Main Concepts and Evolutions

The review conducted revealed that social innovation as a field of research has gained great interest from scholars during the past 20 years, thus transforming this field from anecdotal and policy-based evidence into a stand-alone emerging academic area of investigation (Mulgan 2006). However, although the concept gained popularity in the late 1980s (Drucker 1987), the boundaries of social innovation research have not yet been completely defined; as for any emergent topic, the literature remains fragmented, disconnected and scattered among a wide variety of fields (Cajaiba-Santana 2014).

To date, few attempts can be presented from scholars who have provided guidance in the incipient research endeavours on what social innovation means and the scope of changes it produces (Howaldt and Schwarz 2010; Mulgan et al. 2007). First, Hubert (2010, 2013)[1] has provided a categorization by distinguishing between social, societal and systemic innovation and according to the different scopes that each type of innovation strives to address. In particular, social innovation is stated as the type of innovation that is directed towards vulnerable groups in society; societal innovation is defined as societal challenges directed towards society as a whole; and systemic innovation is described as sources and outcomes of wellbeing (Hubert 2010). While the differentiation into social, societal and systemic innovation seems useful, the definitions provided for them seem to narrow down the debate on these terms, thus avoiding making clear on what aspects of each scholars have focused.

Moving from the previous categorization, we went towards an in-depth analysis of progressions in the literature in order to better contribute in this sense. The following section provides a synthesis of the social innovation debate; it begins with more traditional studies, then moves on to deal with studies attempting to open up social debate about business and economic implications and concludes with new conceptualizations of social innovation within a broader perspective.

Social Demand Approach

An initial line of study—which we named the social demand approach—has been distinguished in the efforts of scholars whose focus has been strictly on social instances of innovation to provide benefits to the public or to society; these instances have traditionally not been addressed by the market or governments (Mulgan 2006; Hämäläinen and Heiskala 2007).

Within this perspective, social innovation emerged as a distinct and complementary form of technology and business innovation (Hubert 2010). Instead of satisfying private issues (such as gains for entrepreneurs, investors and ordinary non-disadvantaged consumers), social innovation is stated as concerning complex issues that respond to pressing social demands that involve vulnerable groups of society (Mulgan et al. 2007).

Nevertheless, social innovation includes various aspects that address local solutions to specific problems such as employment, poverty, housing, health care and benefiting society (MacCallum et al. 2009). A definition provided by a Stanford social innovation research team serves as a good reference point: social innovation is a "novel solution to a social problem that is more effective, efficient, sustainable,

[1]These three levels are clearly identified and make up a topical line of argument and analysis in the so-called BEPA Report (2010). This work, carried out by Hubert (2010) and other researchers who participated in the BEPA Workshop on Social Innovation, is one of the most detailed studies on the subject of social innovation and provides a thorough description of the key issues of social innovation, on the basis of which a majority of social authors have developed their own definitions of social innovation.

or just than existing solutions and for which the value created accrues primarily to society as a whole rather than private individuals" (Phills et al. 2008: 39).

The newness that social innovation conveys was stated as generally arising in the form of new ideas such as new products, services and models, new, creative and imaginative community initiatives or innovative activities and services (Mulgan 2006; Mulgan et al. 2007). Often, social innovation was defined as strictly connected with process and service innovation. However, social innovation and service innovation remained distinct phenomena because social innovation had as a primary characteristic to address broader social scopes (Mulgan et al. 2007), to prove effects that are longer term (Howaldt and Schwarz 2010) and to be social both in its ends (new output) and its means (i.e. the way in which it is realized), as well as in the potential it created for society (Hubert 2010).

The type of value that innovation is expected to deliver is less tied to issues concerning profit, as noted by Pol and Ville (2009: 884): "social innovation resolves societal changes and it regards new ideas improving quality or quantity of life, not necessarily linked with economic profits; the ultimate end of social innovation is to help create better futures".

This had important implications not only in terms of goals and beneficiaries but also in terms of primary actors as agents of innovation. Different scholars claimed social innovation as predominantly being developed and diffused through organizations whose primary purposes were social (Mulgan 2006). These scholars tended to limit social innovation domains to public and non-profit organizations, implicitly or explicitly, and excluded firms and other organizations. Although scholars made valiant efforts to broaden prevailing conceptions of social innovation to those of social entrepreneurship and social enterprise (Thompson and MacMillan 2010), the trade-offs between social and business innovation remained preeminent. The responsibility to take care of society and prevent negative impacts in the long term was the only challenge with which social innovators had to deal (Yunus et al. 2010). The managers and partners of social companies operated without seeking to add pecuniary or other material gains to their assets and profits, when they occurred, which were accepted as principal, because investment in the enterprise was required (Hoogendoorn et al. 2010).

Societal Strategic Approach

A second line of studies—which we named the societal strategic approach—places social innovation into a broader strategic discourse. Innovation is not only seen as a way of responding to social problems but as a tool for addressing global challenges (including instances such as climate change, pollution, etc.) and to ensure economic performance under a new integrated paradigm that aligns social value activities with strategic economic activities (Porter and Kramer 2006). This line of studies spanned a wide range of different fields, and although different in their respective research interests, they nonetheless illustrated some synergy and informed us about (1) the closer relation between social and business innovation—in this case, new concepts of corporate social responsibility (CSR) innovation and corporate social innovation were established (Porter and Kramer 2006; Kanter 1999); (2) the broad focus on societal needs, problems, wealth, etc. (for the Base of the Pyramid innovation concept, see Prahalad and Hammond 2002 and Prahalad and Hart 2002); and (3) the various sectors and organizations where social innovation can start and occur (Porter and Kramer 2011).

The novelty of these studies relied not only on the broadest needs of society that social innovation was stated to address but primarily on the fact that they significantly blurred the boundaries between social and economic factors. The emphasis is on the strategic role of social innovation and the close link that it created between business and society, even if the subordinate role of social issues to business logic was highlighted. Indeed, Porter and Kramer (2011) referred to the innovative activities of exploiting social issues as an opportunity to identify unmet needs and to develop solutions that create new markets (Kanter 1999; Prahalad and Hart 2002). For companies, social frameworks were seen as needed in order to complete more traditional technological and economic innovations; in this way, companies were able to reach systemic synergies, productivity growth, increasing returns and steadily growing incomes (Kanter 1999). Strategic social issues were often viewed as being at the core of large or new corporations accepting social responsibility and fully integrating it into their business models.

The newness of social innovation included new products and services, as well as new markets and new business models (Ellis 2010; Eyring et al. 2011), and social innovation tended to be seen more as a "pattern-breaking" process involving the responsibility of multiple actors to become effective (Ellis 2010; Porter and Kramer 2011). Even if this perspective was business oriented, collaboration in the network with other businesses, individual non-profits and government sectors was deemed necessary for innovation to develop effectively and expand, thus bringing real societal advantage.

This view also suggests the idea that profit for companies and the health of the communities in which they operate are mutually dependent. Porter and Kramer (2011) coined the term "shared value" to capture the more comprehensive concept of value-creating activities in which social and economic goals are addressed in a balanced manner.

Multiple Approach

Although definitions tend to fall under the label of social innovation, more recent contributions have put forward the idea of social innovation as a complex and interdisciplinary concept that crosses sectors and disciplinary boundaries (Rubalcaba et al. 2013). We grouped these studies under the perspective named the "multiple approach", and we identified this approach as mainly concerned with social innovation as the primary means of encouraging and directing social transformation at both public and private levels (Nicholls and Murdock 2012). As a means of social transformation and far removed from an instrumental understanding of previous perspectives, social innovation is in this instance mentioned predominantly in a qualitative context. First, rather than a material and merely quantitative increase in innovation, the focus is on the systemic and processual dimensions of social innovation. Neumeier (2012) stated that social innovation involved not simply the material medium of a new technical artefact simply because it implied a nested service, product or process impact; in a similar vein, Caulier-Grice et al. (2012) and Hochgerner (2013) emphasize the creation of new social systems as the main aim of social innovation.

In addition, the newness of social innovation, even when associated with a form of service innovation, is stated to comprise more than the traditional service relationship content (Franz et al. 2012; Rubalcaba et al. 2013). On this point, the literature refers strongly to new practices and capacity-building dynamics as key elements through which the process dimension of social innovation was created. According to Viñals (2013: 4), social innovation involved design and implementation processes, as well as a process for disseminating new social practices and policies. Similarly, Cajaiba-Santana (2014: 3) defines social innovation as "the collective creation of newly legitimized social practices aimed at creating social change". Hochgerner (2013) affirms that social innovation cannot be built on the basis of established practices and necessarily involves a renewal of resources, practices, values and schemas with the power to transform society and its practices.

All of the above definitions portray the essence of social innovations as innovation that works towards systemic social change, including changes in values, power, beliefs and practices. They focus on the "multidimensional" aspect of the innovation process, meaning that organizations, technologies, services and multiple networks and partner relationships have to be developed simultaneously. How technological and economic innovation can be combined with changes in attitudes and values, strategies and policies, organizational structure and processes is also important, as are the responsibilities and tasks of different actors involved, both at individual and institutional levels (including businesses, policies and regulatory issues). Many authors strongly agree that social innovation cannot simply be reduced to the contribution of a single actor, no matter how extraordinary or grand it might be (Neumeier 2012; Osburg and Schmidpeter 2013; Rubalcaba et al. 2013).

Researchers looked at interrelations between actors with an emphasis on collective creation for social innovation and its networking implications. Moreover, social innovation included the idea of collaboration between multiple empowered agents who act together in the development, implementation and diffusion of new ideas and practices to promote social transformation (Cajaiba-Santana 2014; Neumeier 2012; Rubalcaba et al. 2013). The empowerment of agents is a key complementary aspect of social innovation, as "social" was described as reshaping society in the direction of a more participative arena where empowerment represents the key source of wellbeing (Hochgerner 2013). There is multiple evidence that social innovation will become of growing importance not only concerning social integration, equal opportunities and dealing with the greenhouse effects but also with regard to preserving and expanding the innovative capacity of companies and societies (Franz et al. 2012).

This dimension was crucial in the depiction of social innovation outcomes through the multidimensional construct of social value (Franz et al. 2012) as interconnected with a collective level of benefits or wellbeing.

10.3.2 Service Innovation: Main Concepts and Evolutions

Service innovation also represents an area of investigation (Maglio and Spohrer 2008) that has focused on the interests of researchers over the past decade (Carlborg et al. 2014) and that has gained relevance and significance in different fields (strategy, marketing, operations, economics, information systems) where new approaches and contributions are continuously emerging (Lusch and Nambisan 2015). The state of the art therefore depicts a body of scholarly research that is heterogeneously characterized (Ostrom et al. 2010, 2015). However, some contributions have provided critical reviews of the literature that clarify the concept of service innovation and the evolution of its scope and ambitions.

Particularly, a recent systematic review of literature by Carlborg et al. (2014) has identified an evolutionary pattern of service innovation clustered into three distinct phases and separated by content and perspective: (1) formation phase, (2) maturity phase and (3) multidimensional phase. The formation phase—extended from 1986 until 2006—included seminal contributions that ran parallel to services marketing research (Edvardsson and Olsson 1996), which aimed to empower the concept of service and called for theories that were more directly applicable to services (Gallou and Djellal 2010). The common denominator among studies was to dignify service innovation as a distinct research area, rather than to define the concept per se (demarcation perspective). The focus was on a conceptualization of different innovation models that supports firms "to generate ideas and formalize them as concrete offering" (Carlborg et al. 2014: 376). The maturity phase had a short time extension (2001–05), despite representing a moment of strong growth, in terms of both quantity and advancing scientific research. The focus was in this instance still on observing the process of designing innovation, but managerial and organizational factors came to the fore, together with more established engineering and technological issues of innovation. Customer and provider–customer interactions were the main factors leading the NSD debate. Research increased significantly as a result of pressure to provide an answer to managerial and organizational issues such as (1) promoting customer participation (Gallou and Djellal 2010) and (2) organizing interactions and managing organizational structure (Toivonen 2010). The concept of service innovation definitively did away with the strictly technological framework typical of earlier studies (Sundbo and Toivonen 2011). The final phase—the multidimensional phase—is characterized by the peak of publications on the topic. In this phase, service innovation is established as a more solid research area that attracts and stimulates multidisciplinary approaches. With respect to the previous phases, the research strives and calls for a more expansive and integrative approach to service innovation, embracing more general issues such as strategy and policy. The main topics were (1) the higher level of service innovation management issues related to the firm's overall strategy and marketing strategy and (2) the contributions to general discussions about the nature and theory of what service innovation was, as well as its specific characteristics. Service innovation has developed into a pervasive concept that offers a new theoretical framework that is not limited to services but is instead an interpretive lens for better understanding the mechanisms and dynamics of value processes (Kindström 2010).

Multi-actor and Context Focus: The SDL Approach

The analysis of more recent literature provides us the opportunity to integrate the review of Carlborg et al. (2014) with an additional phase of research on service innovation, which is currently ongoing and can be referred to as phase 4—multi-actors and social centric perspective (2011–15).

Studies and research grouped in this phase are affected by the coeval emerging service dominant logic (SDL; Vargo and Lusch 2004), which emphasizes service as a broadening concept of all exchange and a transcending concept on which all of economic science should be built (Vargo and Lusch 2011). Due to its infancy, contributions that adhere to this perspective are few; however, they provide a completely renewed type of service innovation debate by addressing many new concepts.

First, these scholars have claimed an idea of innovation as processes that overcome the traditional products and services divide (Vargo and Lusch 2004, 2011) and that are defined exclusively in terms of a service for the sake of service provision (Lusch and Nambisan 2015; Vargo et al. 2015). Moreover, SDL focuses on the perspective of the beneficiary, for whom a new service is seen as a new resource or competence for his/her action in the unique experience of that benefits (Michel et al. 2008; Vargo and Lusch 2011). Additionally, the relevance of collaborative and social dimensions has been developed in these studies with wider implications for service innovation. Vargo and Lusch (2011) emphasize that economic actors use resources to access additional resources through service-for-service exchanges; the authors coined the term "service ecosystem" to depict the complex and dynamic dimensions of service-for-service networks and markets.

In a service ecosystem, innovation emerges as an open link process based on interactive and iterative activities connecting firms, users and organizations (Lusch and Nambisan 2015; Mele et al. 2014; Vargo et al. 2015). In the service approach, scholars recognized service innovation as inherently value and network centred sourced by changes that occur as a result of the way in which multiple actors integrate resources (Lusch and Nambisan 2015; Mele et al. 2014; Vargo et al. 2015) to cocreate mutual value. Accordingly, service innovation is a new way, experienced by an actor in a given context, to integrate and use resources for capturing value, as Lusch and Nambisan (2015: 161) explain: "service innovation consists of re-bundling of diverse resources that create novel resources that are beneficial (i.e., value experiencing) to some actors in a given context".

Edvardsson and Tronvoll (2013) put forward the social aspect of service innovation on the basis of the social structure and social system concepts. They moved from the idea that social structures affect social activities and interactions and constitute the foundation for service innovation. As a consequence, service innovation takes place in a social context and it has been defined as "changes in structure that stem from either a new configuration of resources or a new set of schemas and that result in new practices that are valuable for the actors in a specific context" (Edvardsson and Tronvoll 2013: 27).

In similar vein, Vargo et al. (2015) advance the service dominant logic of innovation by directing their attention to social mechanisms (social norms, rules, values, meanings and practices; i.e. institutions) and their role in the service ecosystem and innovation (Vargo and Lusch 2011). They provide a more dynamic view of social and organizational service ecosystem structures within which innovation is created by explicating a recursive relationship between actors' interactions and institutions. Institutions influence interactions that contribute to the creation and evaluation of value among multiple actors, even those that result in the rejection of value propositions (Skålen et al. 2015; Vargo et al. 2015). Thus, institutions become central to innovation that is broadly conceptualized as "the cocreation or collaborative recombination of practices that provide novel solutions for new or existing problems" (i.e. the at least partial disruption of existing institutions) (Vargo et al. 2015: 70). The maintenance, disruption and change of institutions fully concern the process of innovation, and they lead the distinction between technology innovation—that is, new value propositions that are influenced by existing institutions—and market innovation, considered as the institutionalization of new practices (i.e. new solutions) (Vargo et al. 2015).

The service dominant perspective addresses the social nature of service innovation and opens up some interesting insights on how service innovation can better align current new global social issues.

10.4 Crossing Social Innovation and Service Innovation: A Proposal of Conceptual Framework

The brief literature review strengthens the idea that a distinction between social and service innovation is restrictive and not useful for depicting the complexity of new markets and social phenomena. As is demonstrated above, the literatures on social innovation and service innovation have oscillated between a wide set of definitions and concepts, and recent social studies as well as service studies are developing according to multidisciplinary and interdisciplinary research approaches (Rubalcaba et al. 2012, 2013). Furthermore, several authors in the field of social innovation in many ways cross service research boundaries that share common

Table 10.1 A comparison between social innovation and service innovation

	Context	Process	Output	Outcome
Social innovation (multiple approach)	Innovation occurs at the interfaces between sectors and involves actors from across sectors	Is produced by large numbers of people working on collective projects. Is distributed in systems where innovation and initiative are dispersed, distributed and local	It can be identified by new resources, as well as the type of relationships and practices created with and between their beneficiaries	Individual and collective wellbeing is obtainable only by mutual dependence and social value
Service innovation (actors and social centric perspective)	Innovation nests both service and tangible goods into an integrated overarching service view	Depends on open processes and is carried out within networks. Relational processes influence service innovation processes	Involves changes in structure that stem from either a new configuration of resources or a new set of schemas that result in new practices that are valuable for the actors in a specific context	Actors re-bundle and integrate the available resources in a social system to create something new that is beneficial (i.e. value experiencing) to all actors in a given context

interests involving the public sector, local and regional development and more general public policy issues. Similarly, some service scholars have begun to conceptualize service perspectives in social contexts (Edvardsson and Tronvoll 2013; Vargo et al. 2015) by focusing on values and schemas that influence business and market behaviours.

By following the recent evolution of social innovation and service innovation literatures, it is possible to discern several elements that are common to the thinking of the social and service authors we reviewed. As Table 10.1 shows, the contexts of social innovation and service innovation have started to cut across and occur in all sectors; additionally, the process dimensions of both types of innovation studies are characterized by an open and collaborative nature involving a wide range of actors, and innovation is experiential, as it is created "with" and "by" users and not delivered "to" and "for" them. Other similarities include the idea that the newness of both social and service innovations has to be identified in terms of new resources and new practices that are valuable to the actors in their specific context. Finally, the notion is common that individual and collective value is closely related to superior and wider-sought outcomes that are obtainable by mutual dependence and value cocreation.

Based on these arguments, in this chapter we argue that the debate linking social and service innovation has gained momentum, thus opening up debate about a new social innovation perspective in which social and service issues are being combined.

Addressing social challenges by means of innovation suggests the need for a new framework that captures the complex nature of social innovation and the multiplicity of its social, service and business drivers.

Arguments about closeness have also been proposed in recent work by Rubalcaba et al. (2013), while other social authors (Gallou and Djellal 2010) support such arguments through the emphasis they place on the multifaceted nature of social innovation. However, we find that the whole story has not yet been told. An understanding of the mechanisms of social innovation remains encapsulated within a tight social mindset with strictly designated actors, objectives and means.

In the view that we intend developing here, any adequate theoretical understanding of social innovation has to embrace both of the aspects of the social perspective and the service perspective, and the proposal of an integrated framework where social and service domains are more closely connected is attempted.

The framework that we want to propose moves forward in terms of questioning the specific service concept that will be useful for developing a new social–service innovation conceptualization. Specifically, our interest in this chapter focuses on opening up both the "newness" and the "process" dimensions of social–service innovation, that is, to provide an in-depth description of what is being innovated (the newness of social–service innovation) and the actions and mechanisms at the core of this newness (the process of social–service innovation).

10.5 Key Elements of the Social–Service Innovation Mindset: Case Illustrations

In this section, we discuss the proposal of a conceptual framework for understanding social–service innovation. The focus is on the key elements of social–service innovation as we have conceptually identified them from the literature and here theoretically and empirically present in more depth. The framework we propose establishes five basic elements: (1) new social–service resource offering, (2) new social–service experience, (3) new social–service practices, (4) resource integration in a multi-actor context and (5) collective learning. The first three elements detail aspect of the newness of social–service innovation; the final two are employed to deepen the process dimensions. Some empirical illustrations based on case studies are also presented to detail the content of each dimension and to provide an indication of the usefulness of the proposed conceptual framework.

10.5.1 New Social–Service Resource Offering

The social–service resource offering is the first key element of the framework. It captures companies' efforts to develop a new resource offering that moves beyond social and business divides and that enters crosswise into business. Over the decades, the paths of social and business environments have become increasingly convergent and new processes, technologies and services (Gallou and Djellal 2010; Rubalcaba et al. 2012, 2013) are emerging as a crossing point according to an integrated approach.

Moreover, the fundamental change taking place within the concept of service—that is, the application of competencies for the benefit of others (Vargo and Lusch 2004)—provides an extension of the social aspects of innovation that are potentially included in any type of innovation. If innovation concerns how firms can better serve (Lusch and Nambisan 2015; Michel et al. 2008; Vargo et al. 2015), we can view social–service innovation as the way in which firms can better serve society, that is, viewing social–service innovation as the application of new competencies for the benefit of others. In this sense, social–service offering transcends the social and business divide and reflect a shift towards the provision of any type of resource newness—product, service, process, technology or their combination—the effect of which is devoted to making the customer better off by creating a better society.

The cases in Exhibit 10.1 clarify the different paths that some companies are undertaking towards the proposition of a new resource offering that integrates market and social pay-offs and balances economic and social value.

Exhibit 10.1: New Social–Service Resources Offering

Altromercato Consortium is a top-selling, non-profit organization that distributes Fair Trade products in Italy. Over the past 10 years, the consortium has undergone radical and continuous innovation in terms of doing social business, specifically along three dimensions: new products and brands, new communication strategies and new social–service projects. New products and brands (Natyr, Auteurs du Monde) contribute to the company's complete offering in order to better address the expectations of consumers of the Fair Trade brand; new communication strategies with the integration of new media and social networks enhance the knowledge of Fair Trade values and allow for engaging different consumer groups; new business projects (i.e. an eco-district-building project) increase the diffusion of Fair Trade values within new businesses. Moreover, Altromercato has significantly reshaped its retail chain. In addition to the "Bottega"—small shops that are reminiscent of traditional markets with a clear social scope—Altromercato has also opened new stores: a large store in Treviso, designed more as a big retail store, and two outlets in Milan and Turin. The aim is to promote the diffusion of Fair Trade values to all types of consumers.

Coop Italia is a consumer cooperative leader in the Italian large-scale retail distribution market. During the past few years, Coop has revolutionized its core business by taking on a more social dimension. Specifically, it has

(continued)

Exhibit 10.1 (continued)

progressively expanded its offering in order to meet the new social needs of consumers. To this end, Coop has introduced two new private-label products, "ViviVerde" and "Solidal", pertaining to organic and Fair Trade approaches in order to favour and promote environmental and social values. In addition, it has promoted several activities and projects for environmental protection and ethical respect, for instance, animal wellbeing, energy saving, workplace safety and anti-child labour practices. Coop is in fact changing the concept of its stores into that of eco-stores, which are designed to meet higher standards of social and environmental performance.

Banca Popolare Etica is the first bank to be inspired by financial ethics. The mission of Banca is to meet the needs of entrepreneurs and investors. The bank is inspired by the values of sustainability and social development. However, created to finance projects exclusively within the third sector, it has nonetheless also extended its services to the private sector (citizens, families and firms), thereby pursuing both social and economic objectives. It has enhanced its service offering by also providing internet banking for firms and families/single individuals. Today, Banca is a bank for all purposes: it is possible either to open a bank account or apply for a mortgage—the only condition is that the customer must provide the purpose for which he/she wants to acquire a service, which is in line with the bank's social value responsibilities.

Humanitas Research Hospital is a leading Italian private hospital group. Since being established, it has always been recognized as a hospital with an excellent and advanced medical team and services. Recently, to better answer the needs of more informed and active users (patients and their families), Humanitas has innovated itself by adopting a more social dimension. It has extended its mission of providing education and assessment that are aimed at creating a social and collective culture that promotes health within a wider community context. It also expanded its offering through the provision of new social services such as ad hoc conventions for accommodation, financial services (hospitality for economically disadvantaged patients and their families) and psychological support (mutual self-help for patients and their families).

The examples above illustrate that social–service innovation can not only be understood as an offering that integrates product and process technologies for the needs of a specific social or market target, it is also something that occurs at the cross-linking of different activities, technologies and services aimed at addressing the more complex resource demands of actors and individuals interested in improving the quality of their lives by improving the wellbeing of society.

10.5.2 New Social–Service Experience

Social–service innovation is not only limited to the consideration of new resource offering, even if it is intended to be in line with the overarching service concept (Vargo and Lusch 2004). What makes social–service innovation a distinct typology is also its foundation on service innovation, viewed as a process that simultaneously considers providers and users. The perspective is that of the user for whom a new service is a new resource or competence that can be used for whatever specific action they need to achieve (or experience; Vargo et al. 2015). Thus, the social–service innovation concept assumes that innovations offered from within a production context need to be activated in the use context in order to address the social scope for which they have been created. The service–social divide is bridged by the view that it is the beneficiary—that is, the person who experiences innovation—who evaluates that innovation within his/her social context (Edvardsson and Tronvoll 2013). Innovation is experienced in a context of practices that the beneficiary undertakes in connection with his/her networks. This view invites a better understanding of how customers experience innovation as well as the context of this contribution. Thus, the social–service innovation is always linked to the beneficiary's experience within a social context. The following illustrations (Exhibit 10.2) clearly identify this (see Exhibit 10.2).

Exhibit 10.2: New Social–Service Experience

The **Altromercato** business model widely recognizes the role of social and community contexts at the core of its service strategy. Shops are conceived as places where consumers can interact and socialize with others (consumers, employees, volunteers) to become aware of Fair Trade values and to act as agents in the diffusion of social and sustainability values. Initiatives for the creation of a sustainable market, Fair Trade educational sessions or collective readings of books are some examples of activities co-produced by Altromercato and its customers. The creation of social–service experience is also pursued by Altromercato through the use of social networking and Altromercato communities, in which consumers take an active role by working with the company to disseminate information and Fair Trade values through ad hoc co-produced communication campaigns and events.

Coop is conceived in a multiple interaction context, in which consumers, employees, non-profit organizations and other firms work together according to social–service logic. Among all of these actors, consumers play a strategic role, as they are viewed not only as "buyers" but as a link that encourages the development of relationships in the social context. The online context, such as a Facebook page, YouTube channel and company website, allows consumers to engage in dialogue, to gain insights and make suggestions and to share their experience with Coop employees, partners and with all the other

(continued)

Exhibit 10.2 (continued)

members of this social context. By sharing their experiences, consumers make their knowledge, feelings and actions known to other members, thereby enhancing their engagement with companies at the wider social level.

Banca Popolare Etica activates different interactions with customers and with the other beneficiaries of his social context. Through the use of innovative communication tools such as social networks (Facebook page and blog), it brings together many people who are able to share information and exchange experience. Additionally, a YouTube channel is a useful tool that can be used by the company for the development of its service strategy. Consumers can share their experience pertaining to ethical finance and promote the diffusion of positive experience, thereby contributing to creating trust for those who use the services for the first time.

Humanitas enables the creation of social–service experience by encouraging several actors to play active roles in the process of diagnosis, therapy, counselling and wellness. Users are considered not only as recipients of new care processes or beneficiaries of activities but as active parts of these processes. Specifically, patients are conceived as collaborators that share their knowledge, experiences and emotions. Ad hoc events, support groups and conferences are delivered by the company to encourage patients to take part in dialogue with specialists and other patients. Patient participation as testimonials in training sessions and prevention days as well as the creation of social initiatives by patients (such as the creation of ex-smoker communities) allows for making available to others healing practices that favour the creation of new social–service experiential contexts. A company online community and YouTube channel are aimed at patients; moreover, new contexts for sharing resources and for cocreating a new trusted relationship with professionals, other patients, medical students and patients' relatives have also been created.

The above cases illustrate that social–service experience is created when services are used by customers in a social context. The beneficiaries remain socialized when and by using services, and the new services are developed by the companies to increase interaction with and among beneficiaries. Consumers actively participate in order to cocreate experience in a social context. Aspects of cocreation in multiple user contexts promote the possibility of social innovation where organizations, users and their relevant networks play a role in the customer's experience and innovation. Hence, it becomes more evident that in social–service innovation, it is no longer only providers that make social offers to consumers through their own agency; instead, beneficiaries become active themselves, and without their involvement, the value of social–service innovation is not created or diffused.

10.5.3 Resource Integration in a Multi-actor Context

The third element of our framework concerns the process dimension that we define as resource integration within a multi-actor context. The framework that we propose encapsulates the service research idea that social–service innovation is not only about what is new but also about the mechanisms and process through which innovation unfolds.

Here, discovering new ways of creating innovation is conceived mainly as a process that includes the contribution of any actor participating in the intended social–service aims (Vargo and Lusch 2011). A social–service innovation is pursued as an open process in which companies, consumers, suppliers, the public or non-profit organizations and other institutions mobilize and integrate their resources (skill, competence, information; Edvardsson and Tronvoll 2013) and share the responsibility (Rubalcaba et al. 2013) for creating novel resources.

Even if companies are stated as being primary actors, at least regarding their business responsibilities, the ways in which multiple actors are mobilized and resources integrated into a coherent new proposal become a key factor for effectively responding to both business and social challenges (see Exhibit 10.3).

Exhibit 10.3: Resource Integration in a Multi-actor Context
Altromercato encourages the interaction, sharing and integration of resources in networks, both online and offline. All actors are solicited to enact, share and integrate their resources. The company's website serves as a useful tool for interaction, dialogue and engaging suppliers. These actors are called on to talk about what it means to work as suppliers in a Fair Trade context, to explain their role and tasks and to share their resources and competencies with the others. Within its stores, Altromercato engages employees and volunteers to share their experience and knowledge with other actors related to Altromercato and Fair Trade issues by responding to comments and questions to disseminate information further. The central role of interactions and integration emerges also through events such as the "sustainable market", where employees, consumers, volunteers, citizens and public institutions are called on to make available to other actors their skills and knowledge on social, sustainability and responsibility issues.

Coop has become a platform where a range of professionals can collaborate and interact on specific issues, enabling the generation of new innovation within everyday community practices. A range of different actors from medical (paediatricians, nutritionists, dieticians, etc.), social (voluntary association, ONG, international humanitarian organizations, cultural and scientific institutions), environmental (environmental associations) and public (local government) fields are called on to work together to develop projects and activities. Some of these projects yield innovative results regarding the

(continued)

Exhibit 10.3 (continued)

improvement of resource consumption, the use and reuse of unsold products or other waste materials and new nutritional education programs for children.

Banca Popolare Etica has conceived resource network in which different actors engage to share, combine and integrate their resources, thus becoming fundamental members of the company's existence. Thus, employees, financial experts, customers (and potential customers) and social and environmental activists (and/or associations) interact, engage in dialogue and integrate resources for the fulfilment of "special projects". Some of these projects concern innovation in energy saving, microcredit, social cooperation and a fair economy. In particular, the online platforms of Banca Popolare Etica work as elective contexts that enable different actors to remain engaged with one another.

Humanitas has posited collaboration and actor participation in, for example, the inspiring principle of its managerial philosophy. The company is conceived as a collaborative network in which patients, doctors, researchers, students, specialists and other leading hospitals worldwide and research centres (such as the Universities of Lovanio and New York, the Madrid Biotechnology Centre, the Queen Mary School of Medicine in England and the Pasteur Institute of Paris) can participate in the innovation processes. This network is conceptualized as a platform for resource access and integration. Stakeholders are called on to integrate and/or re-bundle new and existing resources in order to innovate clinical processes and conceptualize services and activities that satisfy patient needs in new ways.

It is worth noting from the above cases how actors are not simply providers or users of resources according to a computational process; rather, the resource integration process implies the incorporation of all actors' resources into the processes of all actors involved and in accordance with common expectations, needs and capabilities (Vargo et al. 2015). Through these processes, each actor becomes a co-innovator and takes an active role in social–service innovation networks (Mele et al. 2014; Russo-Spena and Mele 2012).

10.5.4 New Social–Service Practices

From the social–service perspective, innovation is not only intended as service provision in an experienced service network.

The social literature first introduced the idea of innovation viewed as the result of wider activities devoted to satisfying human needs through the significant transformation of social relations. Social scholars have long considered innovation as the creation of new social practices (Cajaiba-Santana 2014). The service literature proves useful in clarifying practices as not simply synonymous with good actions or best routines; the idea of practices recalls the consideration of subjects and their interactions, values, schemas and the contexts in which they occur (Edvardsson and Tronvoll 2013; Vargo et al. 2015). Practices emphasize the role of the sociocultural context of innovation and the need to look at social–service innovation as something that implies change in the way actors effect actions and primarily changes concerning how actors think and behave regarding these actions. In this sense, social–service innovation emerges as legitimized actions that are aimed at social changes that revolve around the creation and diffusion of new competencies, values and schemas (new practices) within service networks.

Exhibit 10.4 provides evidence of companies' efforts to solicit new sets of values and schemas to promote new ways of thinking and actions regarding problems and opportunities.

Exhibit 10.4: New Social–Service Practices

The mission of **Altromercato** is focused on the promotion and diffusion of new practices in an ethical context. The eco-district, a co-housing neighbourhood project that encourages residents to share services, produce energy together and minimize waste, is an example of the company's network commitment to configuring a new way of conceiving responsible consumption. Eco-districts are conceived as relational loci in which different actors such as Altromercato, residents, owners and architectural experts work together and collaborate in an ongoing effort to align different interests, promote sense making and (re)configure in a collective way a new idea of what an urban city should be.

Coop enables the creation of new practices that favour relationships and the integration of resources among consumers, local government, associations, etc. The promotion and support of the "EcoCourt" project (a European project through which real communities, families and condominiums joined together with virtual actors to adopt good solutions for saving energy, water and waste by reducing household expenses) have allowed the development of innovative behaviours that have changed the schemas and practices of citizens for the sake of saving resources.

Banca Popolare Etica has over recent years significantly changed its modus operandi to activate and support activities for improving society's quality of life. It has engaged several actors such as cultural associations, universities, firms and non-profit organizations to create a "new economy" and realize actions that are more appropriate for community development. It supports special projects on energy and Fair Trade that combine economics, finance, law, reciprocity, evolution and ethics in an attempt to overcome the dichotomous approach between profit and non-profit organizations; this

(continued)

Exhibit 10.4 (continued)

approach includes free and professional activities that have economic and social value.

Humanitas can be viewed as a complex network in which patients, physicians, researchers, institutions, patient associations and other hospitals collaborate to create new health practices that have been conceived according to a more social and service-oriented logic. Campaigns and prevention days are some examples of activities promoted by Humanitas and realized with the assistance of others to promote healthy lifestyles, prevention and early diagnosis and to define personalized paths of prevention. In these contexts, different actors (and specifically, patients) are encouraged to cocreate new practices in a spontaneous manner and choose the most suitable approach for their personal health problems and lifestyles.

As the above examples show, the companies have undertaken processes that involve reflection with various actors, not only on the nature of innovations for benefiting customers (or at least their relevant contexts) but also about how innovation has an impact on society as a whole. The focus is more on social mechanisms such as meaning, values, beliefs and rules (Edvardsson and Tronvoll 2013; Vargo et al. 2015) and their enactment, sharing and integration and how this influences the creation of new practices as emergent outcomes of social–service innovation activities.

10.5.5 Collective Learning Process

When the creation of new practices is expected as an intended outcome of social–service innovation, the processes that enable new practices (i.e. the innovations) for diffusion and stabilization become additional elements to be addressed.

As recent service literature shows, the development and diffusion of innovation imply institutional work (Vargo et al. 2015) that relies on the enactment of practices and the determination of meaning from beneficiaries, companies and other actors. All of these actors take an active role in promoting conditions in which new practices can be incorporated into existing institutional fields (existing systems of rules, meanings and values) or provide a foundation for changes in existing systems; both mechanisms contribute to the acceptance or rejection of innovation.

The way in which different values and schemas are enacted and arranged in shared practices calls into question users' and actors' competence-building processes. Social–service innovation implies a collective learning process through which each actor develops more competent participation aimed at innovation. This process concerns how companies, customers and other actors learn more about each other's expectations and needs, adjust their knowledge and actions to each other's behaviours and values and develop their competencies about what innovation means and what makes a good innovation or how it may need to be changed. Collective learning therefore becomes a basis and a condition for developing common ways of thinking and acting and thus sustains the service network in the process of transforming society (see Exhibit 10.5).

Exhibit 10.5: Collective Learning

Altromercato promotes collective learning processes, organizing interactive moments in which consumers, employees, volunteers and others become more aware of their roles, values and commitment. Actors are encouraged to take collective responsibility in sharing knowledge and Fair Trade values. For example, the "Fiera 4 Passi", an annual promotional event, is conceived not only as a traditional exhibition but as a working and relationship context that allows more actors to engage in dialogue about what a good innovation should be and to encourage its creation. The "sustainability moments" are special appointments through which those present (consumers, employees, volunteers) can discuss innovative projects or initiatives. Some of these projects are then selected and financed by companies.

Coop promotes the engagement of all actors in innovation activities in order to favour the development of new competencies and the empowerment of all actors. Laboratories, as training environments on specific issues, as well as events and projects, enable the exchange and acquisition of knowledge, the sharing of experience and the development of new capabilities. All of these activities are designed, developed and implemented with the active involvement of all actors to increase feelings of enjoyment and empowerment among them. Moreover, they favour the development of knowledge and the sharing of common values.

Through the experience of laboratories, together with businesses, academia and third-sector organizations, **Banca Popolare Etica** encourages different actors to think together and has developed a concrete alternative to the "depersonalization of economics and finance". The opening initiative was launched for the first time in 2012 in an event named "Laboratories on New Economy". This event is replicated every year in different Italian cities to provide a context in which people can meet, share, learn, study and explore the issues and experiences of new economies. These events are not only educational opportunities that can affect daily life and/or future perspectives but tools through which it is possible to understand the needs of people and that can be used to create support for sustaining and institutionalizing social transformation.

Humanitas collective learning processes are based on the empowerment of the patient, who is viewed as an active and skilled agent who collaborates

(continued)

Exhibit 10.5 (continued)

with medical staff for the creation and diffusion of new health practices. Humanitas organizes several events that are conceptualized as training opportunities for students and families. Through interactive games, these events disseminate and promote healthy and natural collaboration. These new interactive approaches for conceptualizing health favour the sharing of knowledge and the emergence of new practices that are able to redefine the entire concept of health care.

The cases above illustrate that companies engage actors in a range of roles in joint activities through which common understandings and meanings emerge. Actors develop common interpretations of what innovation should be, and through interaction with companies and other actors, the individual actor's capacities and abilities are stimulated through the collective construction of innovation. To produce the intended effects, all actors have to support social–service innovation by developing their competencies and understandings in a shared domain of values and interests. The conscious and aware participation of actors is necessary to stimulate changes in thinking and behaving. The emphasis in this context is on the social actions by which actors and groups collectively develop their skills and competencies. Any change requires actions of learning to challenge the status quo and support the implementation of new values, norms and behaviours. The emergence of collective learning facilitates sense making and creates support for sustaining and institutionalizing social transformation. In this way, the learning process opens up to innovation opportunities, and any actor can become an agent of change in society.

10.6 Discussion and Contributions to the Social Innovation Debate

This chapter links the debate on social innovation to the discourse on service innovation, following the emergence of politics and businesses as they relate to new concepts about what social innovation should be (Rubalcaba et al. 2013; Schmidpeter 2013).

In the context of what some service scholars have recently defined as the "servitization of innovation" (Carlborg et al. 2014) for counteracting the traditional basis of pure technological and material innovations (Lusch and Nambisan 2015; Vargo et al. 2015), this chapter argues in favour of a service approach to social innovation. The service innovation perspective provides a framework for putting forward recent advances in the social innovation debates aimed at rendering salient the need for an integrated approach (Rubalcaba et al. 2013).

The focus is on "key elements" of social innovation and how they can be conceptualized according to an integrated social–service mindset. By theoretically and empirically dealing with the social and service innovation literature, a social–service innovation framework has been established. The framework includes five key elements that contribute to detailing both the newness (newness dimension) and the how (process dimension) of social–service innovation.

Specifically regarding what social–service innovation is, the framework identified three main interrelated outcomes: (1) new social–service resource offering, (2) new social–service experience and (3) new social–service practices. These outcomes are connected and influence one another; they can be seen as different levels at which we can observe social–service innovation by focusing on core concepts and the inclusive aspects of social–service conceptualization. In addition, our framework also focuses on two additional elements: (1) resource integration in a multi-actor context and (2) collective learning. These elements detail the process dimensions of social–service innovation; that is, the mechanisms that enable the promise of new resource offering are effectively conveyed as revolving around the betterment of customers and society as a whole.

All of the above five elements are integral to a social–service framework and put the focus on the processes needed to complement the outcome dimensions of innovations.

An in-depth discussion of how these elements work together provides us with a structure to further detail how the framework logic of social–service innovation unfolds and allows for the consideration of further concepts (see Fig. 10.1).

First, focusing on the first level of the framework, the "what" dimension, the emphasis is on firms' standpoints, and social–service innovation is seen as the creation of *new social–service resource offering* for addressing human needs that are relevant at both a business and a social level. The integration of social and service aspects is a factor in new economies. After all, by adding social aspects to

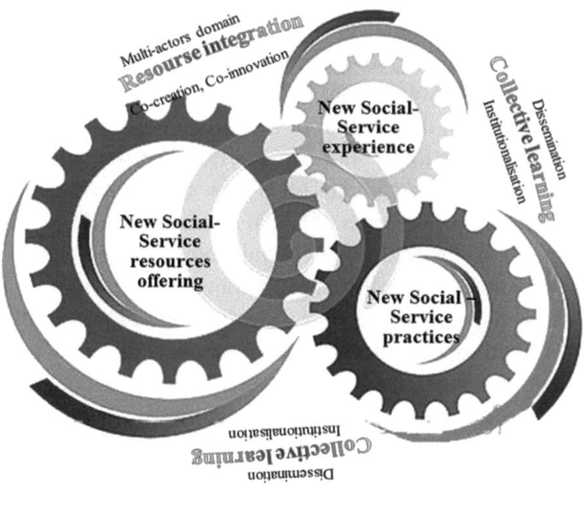

Fig. 10.1 Social–service innovation framework

business services or by increasing the service logic of social business efforts, companies are increasing their innovation efforts towards the more complex instances of individuals interested in how they live well together with others. In this sense, social–service innovation, more than simply being new products or new services, is a new resource that sustains a service ecosystem for societal benefits. This contributes further to the idea that social–service innovation cuts across the business context and takes place in all sectors (Moulaert et al. 2013; Painter-Morland 2013).

When the context is that of a firm interacting with a customer, the outcome of social–service innovation deals primarily with creating *social–service experience*. This means that the newness of social–service innovation takes the form of new or improved resources and contexts that are at the disposal of beneficiaries' new experience and cocreation activities. Social–service innovation assumes that resources are offered from a provisional context to be activated in and by the context employed. In social–service innovation, beneficiaries become active for themselves in order to cocreate an experience that suits them. The need here is to look at social–service innovation that is also linked to beneficiaries' experience, as well as cocreation activities in their wider social contexts.

With regard to both the innovation outcomes that the framework outlines, an initial consideration for the process dimension comes into focus. Specifically, we propose a view of social–service innovation that accounts for *resource integration in multi-actor contexts* as the underlying driver linking new social–service resource offerings and customers' social experience. In line with service studies (Lusch and Nambisan 2015; Vargo et al. 2015), social–service innovation involves a wide range of actors working together to integrate resources. In this way, the importance of technologies or other in-house company resources becomes apparent to the role of interconnected links and relationships that companies and other actors are able to mobilize around new outcomes. On the basis of this representation of the social–service process, innovation emerges as a change in the way resources are combined or re-bundled, which in turn gives rise to several modes of innovation, that is, radical, incremental and re-combinative. Thus, social–service innovation can be defined according to several modes; therefore, whether innovation happens with or without technology breakthroughs is not a crucial issue. This becomes more prominent in the context of resource-constrained environments, in which much of social innovation is stated to emerge (Westley and Antadze 2010). It also provides evidence and arguments for a new social–service perspective that challenges the traditional weaknesses of social business. Social–service innovation overcomes the limitations of traditional social companies in terms of speeding and scaling up innovation for its diffusion; it transforms businesses and their networks in a true alternative to the social agency of public or non-profit institutions.

When we consider the broader aspects of social–service innovation newness, the creation of new practices is identified as a complementary outcome of social–service innovation. Social innovation is stated to be systemic in nature, as it includes changes in values, beliefs and practices (Cajaiba-Santana 2014; Rubalcaba et al. 2013). Thus, the focus of social–service innovation is on the production and transformation of practices in a social context of experience. The matter is not only an innovation that deals with new resources and experience; innovation also affects how beneficiaries and other actors behave and the way in which they make sense when they use (experience) new resources and create value. Social–service innovation assumes that actors, in their respective roles (providers, users etc.), are ready for new activities and interactions and that in doing so, they also change expectations and behaviours. The focus here is on the actions of users, as well as on all actors and their use and integration of new resources, and the ways in which they transform society by collectively changing the basis of established practices.

Social–service innovation has been accepted and stabilized not only due to aspects of innovation but also because innovation itself has to be one of its intended consequences (Vargo et al. 2015). The current study also highlights the need for looking at the collective learning process at the core of the social–service innovation framework. Social–service innovation implies a joint or social learning process in which different agents exchange and acquire knowledge, share their perspectives and experiences and collectively build meanings and expectations about what innovation should be. The emphasis is on social actions by which actors collectively develop their competencies and enact, share and integrate meanings, values, beliefs and rules within and across the organizations and groups in which they participate. The emergence of collective learning facilitates sense making among actors' experience and develops a common way of acting that influences the creation and diffusion of new practices.

As Fig. 10.1 shows, the collective learning process takes together all three elements depicting the content of social–service innovation newness. Through this process, the new social–service experience of actors becomes part of new practices, which are then diffused, accepted and integrated by others over time. In turn, collective learning produces a kind of collective enactment that sustains new practices and contributes to making new social–service resource offering recognized and salient.

As a result, the framework of social–service innovation does not come to the fore as simply a list of key elements. Rather, it provides a synthesis of interrelated concepts that sets the boundaries of social–service innovation and promotes a scheme for interpreting it. Its foundational elements allow us to *define social–service innovation* as *the provision of new or improved resources that act as a transformative force for society when they are experienced and activated by multiple actors in their social contexts; social–service innovation relies on the learning abilities of actors to integrate, transform and share resources and schemas collectively; collective enactment influences new practices to be created, accepted and integrated and sustains social–service innovation as a transformative force.*

10.7 Conclusions and Practical Implications

This study proposes a framework for addressing the conceptualization of social–service innovation.

The discussion is useful for breaking down aspects of what it means to be innovative in the context of a social mindset and how firms can improve their ability in tackling new challenges within society.

First, managers have to become acquainted with the notion that societal development always has business and social aspects and that most innovation opportunities should consequently link these various dimensions to encourage and integrate the concept of social–service innovation.

The concept of social–service innovation allows managers to improve their understanding of what social innovation involves. It delimitates social–service innovation boundaries as characterized by three main interdependent outcomes and two related process dimensions. The point for managers is therefore not simply to provide better or improved resources following the logic to benefit someone. The systemic nature of social–service innovation becomes a focal point. The issue concerns not only the creation of a collaborative network of actors for the pursuit of actors' primary goals. The organizations involved take the lead in terms of social and business changes if they are more proficient at interactively shaping and collaboratively developing values, standards and rules that coalesce the interests of wider social, economic and institutional actors.

The above aspects also have implications for policy. It becomes more salient that within a social–service mindset, companies have power in the same way that public institutions have power. The business environment can become influential in the same way that public institutions in many parts of society are and can become more proactive in response to popular or special interests in order to change society. In this way, the business environment can become an integral part of a broader system of economic, social, cultural and technological importance. The prominence of various private and public actors and their influence depend on their ability to shape a new social business innovation context.

Thus, the areas for dialogue raised in this chapter include additional questions regarding the different modes for organizing social–service innovation, its appropriation regimes and the evaluation of its impacts. It also provides a useful framework for thinking about different types of financing, cultural and social support that innovators and innovations need in order to grow. A better understanding of social innovation in the light of service innovation is likely to help reduce even further the hidden innovation gap within our economies and enable us to advance towards a new and comprehensive innovation paradigm.

Acknowledgements The authors are grateful to two anonymous reviewers. The authors also acknowledge the useful comments received from the seminar participants at the University Magna Graecia of Catanzaro. Finally, the authors thank Claudia Casbarra for her research assistance in collecting data for the illustrations.

References

Belz, F. M., & Peattie, P. (2009). *Sustainability marketing. A global perspective*. Chichester: Wiley.

Cajaiba-Santana, G. (2014). Social innovation: Moving the field forward. A conceptual framework. *Technological Forecasting and Social Change, 82*(2), 42–51.

Carlborg, P., Kindstrom, D., & Kowalkowski, C. (2014). The evolution of service innovation research: A critical review and synthesis. *The Service Industries Journal, 34*(5), 373–398.

Carroll, A. B. (1999). Corporate social responsibility: Evolution of a definitional construct. *Business and Society, 38*(3), 268–295.

Caulier-Grice, J., Davies, A. P., & Norman, W. (2012). Defining social innovation. A deliverable of the project. In *The theoretical, empirical and policy foundations for building social innovation in Europe (TEPSIE), European Commission – 7th Framework Programme*. Brussels: European Commission, DG Research.

Dawson, P., & Daniel, L. (2010). Understanding social innovation: A provisional framework. *International Journal of Technology Management, 51*(1), 9–21.

Drucker, P. F. (1987). Social innovation—management's new dimension. *Long Range Planning, 20*, 29–34.

Drucker, P. F. (1994). *Post-capitalist society*. New York: HarperBusiness.

Edvardsson, B., & Enquist, B. (2009). *Values-based service for sustainable business: Lessons from IKEA*. London: Routledge.

Edvardsson, B., & Olsson, J. (1996). Key concepts for new service development. *Service Industries Journal, 16*(2), 140–164.

Edvardsson, B., & Tronvoll, B. (2013). A new conceptualization of service innovation grounded in SD logic and service system. *International Journal of Quality and Service Sciences, 5*(1), 19–31.

Ellis, T. (2010). *The new pioneers: Sustainable business success through social innovation and social entrepreneurship*. London: Wiley.

Eyring, M. J., Johnson, M. W., & Nair, H. (2011). New business models in emerging markets. *Harvard Business Review, 89*(½), 88–95.

Flyvbjerg, B. (2006). Five misunderstandings about case-study research. *Qualitative Inquiry, 12* (2), 219–245.

Franz, H.-W., Hochgerner, J., & Howaldt, J. (2012). *Challenge social innovation. Potentials for business, social entrepreneurship, welfare and civil society*. Berlin: Springer.

Gallou, F., & Djellal, F. (2010). *The handbook of innovation and services: A multidisciplinary perspective*. Cheltenham: Edward Elgar.

Hämäläinen, T. J., & Heiskala, R. (2007). *Social innovations, institutional change and economic performance: Making sense of structural adjustment processes in industrial sectors, regions, and societies*. Cheltenham: Edward Elgar.

Hasan, H., & Meloche, J. (2013). Innovative ICT-mediated activities for people, profit and planet. *European Journal of Innovation Management, 16*(3), 335–354.

Hochgerner, J. (2013). *Social innovation. In encyclopedia of creativity, invention, innovation and entrepreneurship*. New York: Springer.

Hoogendoorn, B., Pennings, E., & Thurik, R. (2010). *What do we know about social entrepreneurship: An analysis of empirical research*. Amsterdam: ERIM.

Howaldt, J., & Schwarz, M. (2010). *Social innovation: Concepts, research fields and international trends*. Dortmund: Sozialforschungsstelle Dortmund.

Hubert, A. (2010). Empowering people, driving change: Social innovation in the European Union, the Bureau of European Policy Advisers. Report available at http://ec.europa.eu/bepa/pdf/publications_pdf/social_innovation.pdf

Hubert, A. (2013). *Social policies and innovation policies in the EU current agenda*. Bureau of European Policy advisors (BEPA), Granada, 29 May 2013.

Jacobsen, H., & Jostmeier, M. (2012). What is social about service innovation? Contributions of research on social innovation to understanding service innovation. In H.-W. Franz, J. Hochgerner, & J. Howaldt (Eds.), *Challenge social innovation. Potentials for business, social entrepreneurship, welfare and civil society* (pp. 107–118). Berlin: Springer.

Kanter, R. M. (1999). From spare change to real change: The social sector as beta site for business innovation. *Harvard Business Review, 77*, 122–132.

Kindström, D. (2010). Towards a service-based business model—Key aspects for future competitive advantage. *European Management Journal, 28*(6), 479–490.

Lusch, R. F., & Nambisan, S. (2015). Service innovation: A service-dominant logic perspective. *MIS Quarterly, 39*(1), 155–175.

MacCallum, D., Moulaert, F., Hillier, J., & Vicari, S. (2009). *Social innovation and territorial development*. Aldershot: Ashgate.

Maglio, P. P., & Spohrer, J. (2008). Fundamentals of service science. *Journal of the Academy of Marketing Science, 36*(1), 18–20.

Mele, C., Colurcio, M., & Russo-Spena, T. (2014). Research traditions of innovation: Goods-dominant logic, the resource-based approach, and service-dominant logic. *Managing Service Quality, 26*(4), 612–642.

Michel, S., Brown, S. W., & Gallan, A. S. (2008). An expanded and strategic view of discontinuous innovations: Deploying a service-dominant logic. *Journal of the Academy Marketing Science, 36*, 54–66.

Moulaert, F., MacCallum, D., Mehmood, A., & Hamdouch, A. (2013). General introduction: The return of social innovation as a scientific concept and a social practice. In F. Moulaert, D. MacCallum, A. Mehmood, & A. Hamdouch (Eds.), *The international handbook on social innovation*. Cheltenham: Edward Elgar.

Mulgan, G. (2006). The process of social innovation. *Innovations, 1*(2), 145–162.

Mulgan, G., Tucker, S., Ali, R., & Sanders, B. (2007). *Social innovation: What it is, why it matters, how it can be accelerated*. Oxford: Skoll Centre for Social Entrepreneurship.

Neumeier, S. (2012). Why do social innovations in rural development matter and should they be considered more seriously in rural development research? – Proposal for a stronger focus on social innovations. Rural development research. *Sociologia Ruralis, 52*(1), 48–69.

Nicholls, A., & Murdock, A. (2012). *Social innovation: Blurring boundaries to reconfigure markets*. Hampshire: Palgrame, MacMillan.

Ordanini, A., & Parasuraman, A. (2011). Service innovation viewed through a service-dominant logic lens: A conceptual framework and empirical analysis. *Journal of Service Research, 14*, 3–23.

Osburg, T., & Schmidpeter, R. (2013). *Social innovation solutions for a sustainable future*. Berlin: Springer.

Ostrom, A. L., Bitner, M. J., Brown, S. W., Burkhard, K. A., Goul, M., Smith-Daniels, V., Demirkan, H., & Rabinovich, E. (2010). Moving forward and making a difference: Research priorities for the science of service. *Journal of Service Research, 13*(1), 4–36.

Ostrom, A. L., Parasuraman, A., Bowen, D., Patrício, L., & Voss, C. (2015). Service research priorities in a rapidly changing context. *Journal of Service Research, 18*(2), 127–159.

Painter-Morland, M. (2013). *The role of business in society. Social innovation solutions for a sustainable future*. Heidelberg: Springer.

Phills, J., Deiglmeier, K., & Miller, D. (2008). Rediscovering social innovation. *Stanford Social Innovation Review, 6*(4), 34–43.

Pol, E., & Ville, S. (2009). Social innovation: Buzz word or enduring term? *Journal of Socio-Economics, 38*(6), 878–885.

Porter, M. E., & Kramer, M. R. (2006). Strategy and society: The link between competitive advantage and corporate social responsibility. *Harvard Business Review, 84*, 78–92.

Porter, M. E., & Kramer, R. M. (2011). Creating Shared Value. *Harvard Business Review, 89*(1/2), 62–77.

Prahalad, C. K., & Hammond, A. (2002). Serving the world's poor, profitably. *Harvard Business Review, 80*(9), 4–11.
Prahalad, C. K., & Hart, S. (2002). The fortune at the bottom of the pyramid. *Strategy and Business, 26*, 54–67.
Rubalcaba, L., Di Meglio, G., & Gallego, J. (2013). Public-private Innovation networks and social innovation in the service economy. In C. R. Viñals & C. Parra Rodriguez (Eds.), *Social innovation: New forms of organisation in knowledge-based societies*. London: Routledge.
Rubalcaba, L., Michel, S., Sundbo, J., Brown, S. W., & Reynoso, J. (2012). Shaping, organizing, and rethinking service innovation: A multidimensional framework. *Journal of Service Management, 23*(5), 696–715.
Russo-Spena, T., & Mele, C. (2012). Five Co-s' in innovating: A practice-based vie. *Journal of Service Management, 23*(4), 527–553.
Sabeti, H. (2011). The for-benefit enterprise. *Harvard Business Review, 11*, 1–7.
Schmidpeter, R. (2013). *Social innovation: A new concept for a sustainable future? Social innovation solutions for a sustainable future*. Berlin: Springer.
Skålen, P., Gummerus, J., Koskull, C., & Magnusson, P. (2015). Exploring value propositions and service innovation: A service-dominant logic study. *Journal of the Academy of Marketing Science, 43*(2), 137–158.
Sundbo, J., & Toivonen, M. (2011). *User-based innovation in Services*. Northampton, MA: Edward Elgar.
Tesch, R. (1990). *Qualitative research: Analysis types and software tools*. Bristol, PA: Falmer.
Thompson, J. D., & MacMillan, I. C. (2010). Business models: Creating new markets and societal wealth. *Long Range Planning, 43*(2/3), 291–307.
Toivonen, M. (2010). Different types of innovation processes in services and their organizational implications. In F. Gallouj & F. Djellal (Eds.), *The handbook of innovation and services* (pp. 221–249). Cheltenham: Edward Elgar.
Tranfield, D., Denyer, D., & Smart, P. (2003). Towards a methodology for developing evidence-informed management knowledge by means of systematic review. *British Journal of Management, 14*, 207–222.
Vargo, S. L., & Lusch, R. F. (2004). Evolving to a new dominant logic for marketing. *Journal of Marketing, 68*(1), 1–17.
Vargo, S. L., & Lusch, R. F. (2011). It's all B2B and beyond...: Toward a systems perspective of the market. *Industrial Marketing Management, 40*(2), 181–187.
Vargo, S. L., Wieland, H., & Akaka, M. A. (2015). Innovation through institutionalization: A service ecosystems perspective. *Industrial Marketing Management, 44*, 63–72.
Viñals, C. R. (2013). Introduction to social innovation as a new form of organisation in knowledge-based societies. In C. R. Viñals & C. Parra Rodriguez (Eds.), *Social innovation: New forms of organisation in knowledge-based societies* (pp. 3–15). London: Rutledge.
Westley, F., & Antadze, N. (2010). Making a difference: Strategies for scaling social innovation for greater impact. *The Innovation Journal: The Public Sector Innovation Journal, 15*(2), 3–20.
Yunus, M., Moingeon, B., & Lehmann-Ortega, L. (2010). Building social business models: Lessons from the grameen experience. *Long Range Planning, 43*, 308–325.

Chapter 11
Conceptualizing Resource Integration to Advance Service Innovation

Maria Colurcio, Angela Caridà, and Bo Edvardsson

Abstract The chapter provides a literature review of the resource integration (RI) phenomenon through a bibliometric analysis including 1250 ISI listed journal articles. The main contribution of the study is an overview and analysis of how RI has been portrayed in marketing and management research and the six most important RI dimensions or themes. Furthermore, the analysis highlights two main research areas and approaches on RI: the traditional school of good-dominant logic (GD logic) and the recent school of service-dominant logic (SD logic).

11.1 Introduction

In the last decade, the sharing and the integration of resources have been emphasized both in managerial practice and in literature, although with various approaches and terminology, in different areas of management and marketing research. The book *Co-business* of Turiera and Cross (2013) showed the positive business effects of matching and integration of resources by providing 50 examples of business collaborations between diverse actors (competitors, company/customer, supplier/company, companies of different industries) who combined resources (technologies, applications, services) that resulted in improved solution's value.

In the field of strategic management studies, the term co-opetition coined by Brandenburger and Nalebuff (2011) encloses the essence of cooperation, even between competitors, as the authors stress that "in today's business environment, most companies can achieve more success in a dynamic industry than they ever could working alone [...] business, unlike war, is not a winner takes it all

M. Colurcio (✉) • A. Caridà
Department of Legal, Historical, Economic and Social Sciences, University Magna Græcia of Catanzaro, Catanzaro, Italy
e-mail: mariacolurcio@unicz.it; angela.carida@unicz.it

B. Edvardsson
CTF, Karlstad University, Karlstad, Sweden
e-mail: bo.edvardsson@kau.se

T. Russo-Spena et al. (eds.), *Innovating in Practice*,
DOI 10.1007/978-3-319-43380-6_11

proposition. The objective is to maximize the return on investment regardless of how well or how poorly other people or other companies perform." However, studies on collaboration and sharing of resources in this area date back to the end of the 1980s (Hamel et al. 1989).

In the field of marketing, the latest contributions on service-dominant logic (SD logic) consider resource integration (RI) as the precondition of both the service exchange and the value co-creation process (Vargo and Lusch 2008). Authors implicitly agree on an open view of innovation (Chesbrough 2006) which is based on the actors' abilities to mobilize their own resources and thus to become co-innovators (Mele et al. 2010).

Hence, the theme of integration of resources between diverse actors for the co-creation of successful innovation recurs frequently, but a focused literature review revealed that so far published research that considers these issues is very scarce.

The overview shows that RI is not understood as a theoretical concept but rather as a way of sharing and exchanging of resources (mainly knowledge), and the way the term RI is used depends on the aim of the research and on the scholarly background (marketing, management, operations). Two main streams of research can be identified (Skålén et al. 2015): research into good-dominant logic (GD logic) and research into SD logic (Vargo and Lusch 2004, 2008, 2011). GD logic research narrows the matter mainly to the availability of resources (typology and value), which determine the competitive advantage of the firm, and neglects the links to the value creation process.

SD logic studies stress interconnection between successful (service) innovation and practices for integrating resources into attractive value propositions (Skålén et al. 2015) and provide meaningful insights for unifying innovation framework (Löbler and Lusch 2014) in which innovation is based on the novel integration of resources and capabilities to support value co-creation (Ngo and O'Cass 2009). In spite of the relevance of the topic, the contributions which mean RI according to an SD logic doesn't provide a clear definition of RI focus more on the relationship between RI and the value co-creation process instead on RI and the processes of innovation.

The aim of this chapter is to contribute to the debate in marketing and service research by providing an overview of research in the area of RI.

To fulfill the aims, we conducted a systematic review of literature in order to summarize research on RI issues and to discuss the implications for innovation research. The remainder of this chapter is organized as follows. First, the methodology used to review the literature is illustrated. Second, the concept of RI is analyzed in the light both of traditional and consolidated research in management studies (GD logic) and of recent contributions in marketing studies (SD logic). Third, the main contribution on RI as a concept "per se" is outlined. Finally, areas for future research are discussed.

11.2 Methodology

We decided to do a systematic review of RI literature in the fields of marketing and management. Figure 11.1 summarizes the procedure for the selection of the articles we analyzed. To carry out the study, we collected data from ISI Web of Science, as it is one of the most important bibliographic databases (Cobo et al. 2011; Sakata et al. 2013).

In order to define the research domain, we searched for scientific journal articles using the term "resource integration" and narrowing the investigation to management and business categories for the entire period for which databases provide online coverage. The database we built consists of 1250 articles that have been published from 1985 (year of publication of the first paper on the topic) to 2015. We selected only articles excluding other types of documents (e.g., letters, editorials, reviews, etc.) since the articles best reflect the production of the original research (Benavides-Velasco et al. 2013).

To frame the conceptual structure of the topic and to contribute to the advancing as well as to the consolidation of this discipline (Benavides-Velasco et al. 2013), we

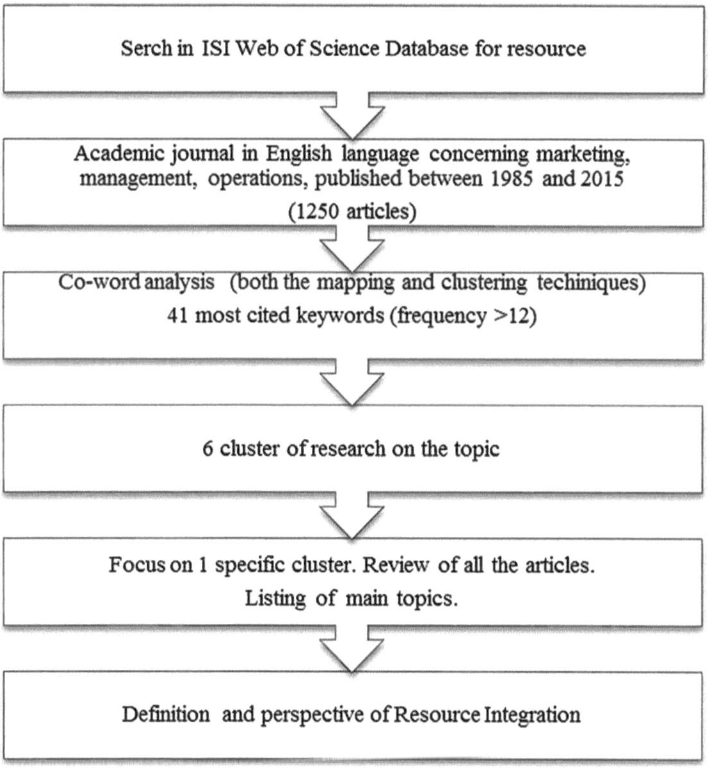

Fig. 11.1 Procedure for the selection of the articles analyzed

performed co-word analysis (CWA) (Benavides-Velasco et al. 2013; Callon et al. 1991; Muñoz-Leiva et al. 2012). CWA is well recognized as a powerful technique that offers a significant approach to knowledge discovery (He 1999); indeed, it allows identifying and displaying structural and dynamic aspects of scientific research, showing linkages among subjects in a field, and tracing the emerging research areas (Bhattacharya and Basu 1998; Börner et al. 2003; Callon et al. 1991; Cooper 1982).

To perform the co-word analysis, we first extracted the author-provided keywords from the entire dataset, and second, we selected them, on the basis of their frequency of occurrences, to create co-word pairs. Because keywords were ranked on the basis of their maximum occurrence, they are the most important concepts which informed researchers (Bhattacharya and Basu 1998) on the investigated topics.

We collected a total of 2649 unique keywords from 1250 articles. Some keywords were revised and standardized, because they were expressed in different form but had the same meaning (Zhu and Guan 2013). We included a total of 2554 keyword in the final dataset. Finally, we built our research sample by including only the 41 most cited keywords with frequency $>$12. The selection of the first 41 keywords satisfies two criteria: (1) the selection of important keywords on the basis of their frequency distribution and (2) the number of keywords suggested, arriving at interesting results.[1]

We combined the mapping and the clustering techniques (Waltman et al. 2010) to better outline the scientific production of RI and to map the nature of the relationships between its subfields. Indeed, the combined use of these techniques allows researchers to depict: (1) the main topics or the main research fields within a certain scientific domain, (2) how these topics or these fields relate to each other, and (3) how a certain scientific domain develops over time.

According to the network approach, we describe each word as a node connected by the relation of co-occurrence (the ties or the links) to another node, whereas through the clustering technique (Van Eck and Waltman 2010) we found strong associations and divisions between the several subfields of the RI stream of research (Muñoz-Leiva et al. 2012). Furthermore we built a bibliometric distance-based map (e.g., bubble chart)[2] to visually show the associations. The distance-based map is based on a co-occurrence matrix and reflects the strength of the relation between the items: smaller distances indicate a stronger relation. Specifically, we adopted the

[1]On the point, Persson et al. (2009) claimed that around 40–50 keywords make still quite nice maps, whereas a maximum is around 80–90 keywords.

[2]Although the science mapping analysis can be performed using generic software for social network analysis (Cobo et al. 2011; Börner et al. 2003), we used the free software Bibexcel (http://www8.umu.se/inforsk/Bibexcel/), as it has been specifically developed to manage data from different bibliographic sources and to build maps that can be read by software such as Excel, SPSS, VOSviewer, etc. (Cobo et al. 2011). Specifically, we used VOSviewer (version 1.5.4) to create a map based on network data. The map has been created using the VOS mapping technique and the VOS clustering technique (Van Eck and Waltman 2010).

VOS mapping technique (Van Eck and Waltman 2007), as it better structures maps than multidimensional scaling (Van Eck et al. 2010). The VOS mapping technique—which can be viewed as an alternative mapping technique to the multidimensional scaling—requires a similarity matrix as input and uses the association strength as a similarity measure: "the higher the similarity between two items, the higher the weight of their squared distance in the summation" (Van Eck and Waltman 2010: 531). The similarity measure can be referred to the proximity index or to the probabilistic affinity index (Van Eck and Waltman 2010).

Finally, to produce consistent results, the VOS mapping technique translates, rotates, and reflects the solution. In such a way, the solution is centered at the origin, and the variance on the horizontal dimension is maximized (principal component analysis) and is reflected in the vertical axis (e.g., if the median of x11,. . .,xn1 is larger than 0) or in the horizontal axis (e.g., if the median of x12,. . .,xn2 is larger than 0).

The review process left 25 scientific articles that have been published from 2008 to 2015. To identify topics related to RI research, we identified three main topics from our reading of the articles: (1) value co-creation, (2) practices, and (3) service innovation. In turn, we systematically categorized each of the 25 articles according to the content of its main topic.

We conducted a qualitative evaluation revising continuously the list of topics to ensure its representativeness.

After the bibliometric analysis, the content analysis, and the topic categorization, we classified the selected 25 articles providing a key of interpretation to understand the progression of research on RI. The analysis allowed us to provide a systematization of RI in literature and to propose some issues to advance research.

11.3 Findings

11.3.1 Resource Integration in Literature: Bibliometric Measures

The analysis of the scientific production highlighted an increasing regular trend with a peak in 2012; contributions come from diverse fields of business such as marketing, management, and operations and show different research interests (e.g., HRM, R&D, innovation, Internet and the interactive communication technologies, competitive advantage, etc.).

The heterogeneity of scientific interest in the topic of RI also emerged from the analysis of the journal on which the articles have been published. Indeed the 1250 papers of the dataset have been published by 186 different journals with an average of less than 7 papers per journal. On the other hand, about 33 % of the journals published more than five papers by producing the whole 76 % of the contributions in the field. Among these, the most productive journals (the first ten journals)

Table 11.1 Most cited keywords

Keywords	Frequencies	Keywords	Frequencies
Resource-based view	70	Competitive advantage	25
Supply chain management	65	Resource integration	23
HRM	63	Knowledge management	20
Integration	43	Network	20
Performance	41	Case study	20
Innovation	39	Outsourcing	18
IT	38	Competencies	18
China	33	Resources	16
Transaction costs	32	Strategy	16
M&A	28	Knowledge	16
Dynamic capabilities	25		

covered the 36 % of the scientific production, while the 48 % of the journals included in our sample published one or two articles on the topic. The analysis of the journals showed that despite the RI topic is rooted in marketing studies, marketing journals are not very important (7.5 % of the sample; productivity rate = 2 6 %), while management journals have published the largest number of publications (productivity rate = 5.7 %).

Also the analysis of author-provided keywords highlighted that the topic of RI attracted interest from different fields. Table 11.1 shows how the well-established themes linked to the strategic management dominate within the contributions on RI.

Shifting beyond the 15 most cited keywords, new ones emerge, such as value co-creation, SD logic, Internet, service industries, sustainable, etc. Furthermore, the analysis shows that RI is a sort of *idiomatic expression* as the 47 % of the keywords are cited just one time: authors refer to the concept in a general sense without addressing precise theoretical concept.

Figure 11.2 shows the bibliometric map (bubble chart). Descriptors[3] (e.g., keywords) of the central concept RI are highlighted as well as information on linkages between each other (Pinto et al. 2014). "Resource-based view" is the central descriptor; it has the greatest number of links with other descriptors, followed by "supply chain management," "innovation," "performance," "HRM," "transaction costs," and "China." The first three significant descriptors belong to different clusters.

The map highlights six different clusters that present research with common descriptors. Each cluster identifies a well-defined research stream which put the first one or another element with respect to others according to the school of thought or the specific field as well as the time of publication. The proximity among clusters indicates a close relationship between their subfields, while clusters far from each other indicate only a weak relationship.

[3]The bubbles' size is straight linked to the number of descriptors occurrences. "The more important an item, the larger its label and its circle" (van Eck and Waltman 2010: 256).

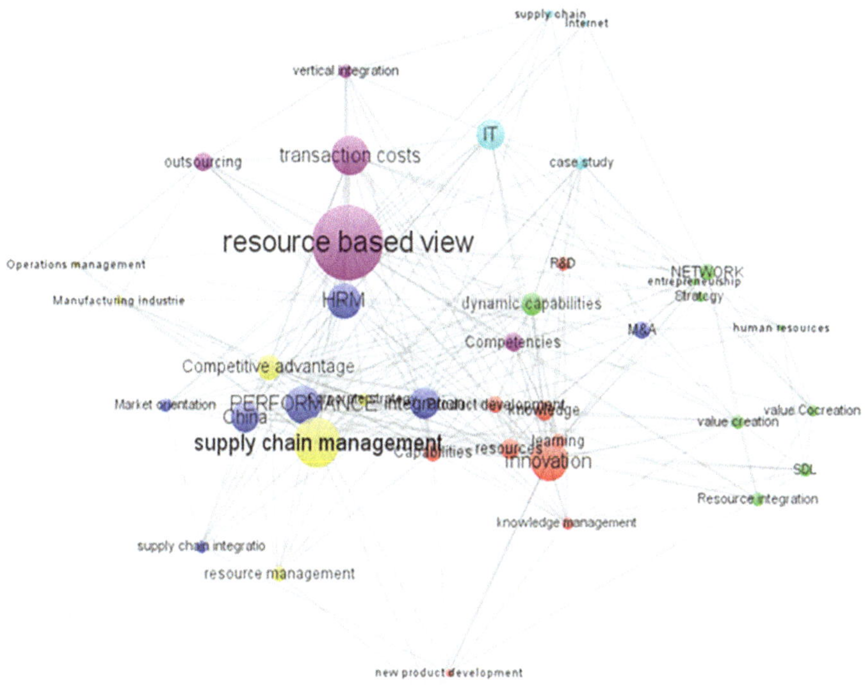

Fig. 11.2 The concept map of RI

The clusters' distribution in the map is heterogeneous, as confirmed by the asymmetrical distribution of the descriptors around the central area (fields between which there are no strong relations are clearly separated from each other).

Interesting aspects on the strength of association of key concepts emerged through the analysis of the (normalized) link strength of the central descriptors and allowed us to compare the co-occurrence frequencies (van Eck and Waltman 2007).

"Resource-based view" shows the strongest relationship with "transaction costs" (3.68), "supply chain" (3.01), and "resources management" (2.32). "Supply chain management" claims the strongest relationship with "resource management" (5.20), "operation management" (4.70), and "cross-functional integration" (4.11). This is a clear expression of a good-dominant logic. "Innovation" has the strongest relationship with "R&D" (4.72), "new product development" (4.02), "knowledge management" (3.76), and "network" (2.87). On the contrary it shows a weak linkage with "resource integration," "SD logic," and "value creation" (1.18) as well as with "value co-creation" (1.43). Of course "resource integration" (green cluster) shows its strongest relationship with "SD logic" (17.36) and "value co-creation" (12.66), as it identifies a specific research domain. Furthermore, it is interesting to note the strong linkage with "new product development" (5.91) in contrast with the weak linkage with "innovation" (1.18).

Table 11.2 Research streams that treat RI

Color	Label	Main content/domain
Red	Open innovation	Technology
Green	Value co-creation	Marketing
Blue	General management	Miscellaneous
Yellow	Supply chain management	Operation
Purple	Strategic management	Strategy
Sky blue	Social/digital marketing	Digital marketing

11.3.2 Resource Integration in Literature: Research Streams

Data analysis and interpretation provided six clusters that identify the main research streams which have treated the issue of RI.

The clusters that have been labeled according to the main content are showed in Table 11.2.

Open innovation focuses on the knowledge sharing as the main condition for the success of innovation and emphasizes the integration of resources as a mechanism to trigger and foster learning processes. This cluster contains mainly contributions from the research area of knowledge management, which are strongly focused on the typology, nature, and process of formation of resources, knowledge, and capabilities.

The three descriptors (1) network, (2) resource integration, and (3) value co-creation depict the research stream *value co-creation* (green cluster), which encloses recent marketing contributions of mainly SD logic scholars who discuss RI as a well-identified concept/process which relates to value co-creation and emphasizes the role of actors and the dynamic of sharing and combining resources. In contrast to the blue and yellow clusters, which are overlapped, the green cluster holds an edge isolated position that marks out differences in the content and in the approach to the topic of RI.

The blue cluster, *general management*, groups diverse cross-field elements and appears as a miscellaneous box enclosing various contributions which treat the issue of RI according to a common and generic meaning without referring to a specific theoretical perspective or to a specific concept. RI addresses the general topic of interfirm cooperation and improvement of firm performance.

Supply chain management identifies contributions from scholars who emphasize the operations' side of relationship and integration according to a supply chain perspective. The focus is on the management of resources rather than on the processes of transformation of resources. Competitive advantage depends on the management of the process of resource exchange and sharing between the actors of the supply chain; the potential of other actors (i.e., consumer) in transforming and integrating resources is still neglected.

Strategic management identifies scholars who focus on the availability of resources (typology and value) which determine competitive advantage of the firm. The process of RI is framed in a transaction cost perspective and refers to practices of outsourcing rather than to a collaborative approach to the business.

Finally, *social media/digital marketing* identifies studies and empirical research (as the descriptor case study points out) which focuses on interactive technologies and stresses the importance of web tools and Internet platforms to enable the cooperation and the integration of resources between actors.

A clear division within the knowledge structure of RI comes to light. The bubble chart (see Fig. 11.2) distinguishes essentially two main blocks: (1) management studies which refer to the GD logic perspective and embrace descriptors belonging to open innovation, supply chain management, general management, and strategic management clusters and (2) marketing studies which refer to the SD logic perspective and encompass descriptors belonging to the edge clusters' value co-creation and social media/digital marketing.

As a further step in the analysis, we created a co-occurrence density map (see Fig. 11.1) that shows the amount of attention researchers paid to the different research topics located in the various areas of a concept map (van Eck and Waltman 2007). The color of each point in the map depends on the density of items at that point: "the larger the number of neighboring items and the smaller the distances between these items and the point of interest, the higher the item density. Similarly, the higher the weights of the neighboring items, the higher the item density" (van Eck and Waltman 2010: 533). The density map provides a quick overview of the important areas within the research field. Colors are assigned by default; red corresponds with the highest item density and blue corresponds with the lowest item density.

The first block of clusters depicts the more traditional and consolidated research trend which focuses on goods logic, whereas the second one depicts more recent trends of marketing studies. Indeed, red color identifies more investigated and well-established research topics, whereas green/yellow colors identify new and emerging research areas. The density map shows three main areas. Each area relates to the three central descriptors of the study: "resource-based view," "supply chain management," and "innovation" that are consolidate domains within the research field we investigated. The boundaries of each domain mark exactly the studies within these fields.

Specifically, the map shows two opposite trends: on the one hand, the clear separation of the "resource-based view" area from the other ones and, on the other hand, the weak separation between the "supply chain management" and "innovation" areas. In this latter case a smaller research domain including the descriptors "integration" and "capabilities" creates a continuum between the two areas (Fig. 11.3).

Other important research areas relate domains, including (1) "dynamic capabilities," (2) "competencies" and "R&D", (3) "IT", and (4) "network," "entrepreneurship," and "strategy," that identify specific and consolidated research areas within the green cluster.

The research area in which "resource integration" is the pivotal descriptor appears lightweight green/yellow colored. It identifies a new emerging domain

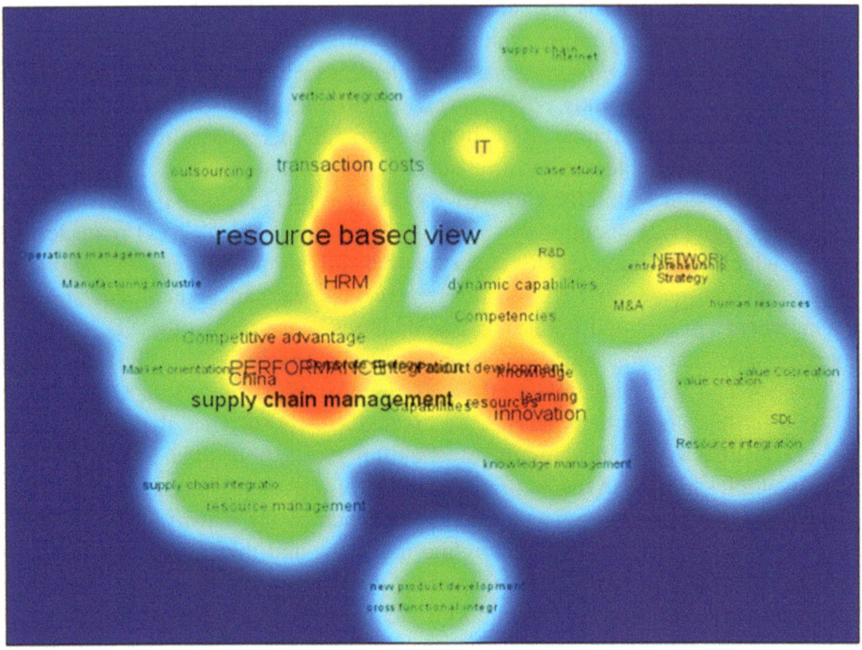

Fig. 11.3 The density map of RI

within the general field of studies on resources and reveals that the label "resource integration" doesn't identify a general field of study but rather it refers to a process to co-create value.

11.4 Resource Integration: An Evolutionary View

Some scholars have called for a more solid theory as the basis for studies depicting RI as a key concept (Kleinaltenkamp et al. 2012; Peters et al. 2014). However, RI is still relatively unexplored and conceptually subordinate to value co-creation.

To better understand RI and provide suggestions for how to advance research, we developed the qualitative analysis and the synthesis of the 25 papers belonging to the cluster "value co-creation" (see appendix). It allows us to critically review the extant research and trace the evolution of RI along the SD logic perspective in the last 10 years (from 2006 to 2015).

Accordingly, RI can be framed through three phases which are referred to different topics and perspectives:

- A: Introduction stage (2006–2008)
- B: Development stage (2009–2013)
- C: Expansive stage (2014–2015)

A: Introduction Stage (2006–2008)

The first stage of research identifies studies which propose new lexicon and labels to support fresh thinking and conceptualization of value co-creation process (Vargo 2008; Vargo and Lusch 2008). The authors highlight differences between customer integration (coproduction, GD logic), as unidirectional and firm-centered process (Moeller 2008), and RI (value co-creation, SD logic), as multidirectional and service beneficiary-centered process. The theoretical discussion remains on an abstract level: authors discuss RI indirectly as a fundamental (pre)condition of both the service provision/exchange (FP1) and the value co-creation process (FP9) (Vargo and Lusch 2008). Indeed, RI is the means through which value is accomplished (Vargo and Lusch 2008). Researchers featuring this stage focus mainly on the role of actors as resource integrator within service economies, the exchange of operant resources as the basis of competitive advantage, and the concept of value as outcome fundamentally derived and determinate in use through the integration and application of resources in a specific context (Vargo 2008; Vargo and Lusch 2008).

In 2008, Vargo and Lusch refine the FP9 broadening the context of value creation to the network and to service systems as configurations of value comprising people, technology, and propositions of value. However, the authors called for further refinement and elaboration of the RI concept and its explicit connection with the interactivity and networking literature—"all social and economic actors are resources integrators"—studies focus more on the dyadic relationship between the firm and customer linking the concept of RI to the customer's participation in a company's value-creating process.

B: Development Stage (2009–2013)

This stage identifies contributions which refine and upgrade the foundational elements of SD logic trying to systematize how value co-creation and RI happen from a theoretical and pragmatic viewpoint.

Studies address the need for a practice-based approach to enrich the "understanding of the actual practical process of RI, and how value stems from this integration" (Korkman et al. 2010: 238). For example, Peters et al. (2009) extend Vargo and Lusch's (2004) SD logic by exploring the relation between value creation and learning practices at the level of the group, organization, and network. The authors link value co-creation to the transformation of individual knowledge (e.g., customers' feedbacks) into organizational learning through the development of learning practice. However not explicitly defined, this transformation implies the process of RI as it requires the processing, interpretation, sharing, and the recreation of individual knowledge at collective level to enact learning practice for value creation. Hibbert et al. (2012) advance SD logic by conceptualizing customer learning in the light of RI and value creation as "any self-initiated or self-directed set of activities in which customers integrate learning resources afforded by organizations and other network actors with their own operant and operand resources with the primary purpose of learning in order to increase their effectiveness in other resource integration processes" (Hibbert et al. 2012: 250). Some authors shift the focus of the analysis from organizational to customers' learning and from customer's value co-creation process to the customer's RI effectiveness, that is, "the customers' proficiency in deploying resources as they engage in value generating processes." Other scholars switch from theory to practice by providing a framework to design and enable RI as interactive customer learning process. The model identifies the characteristics of learning situations, the interactive elements of the learning process, and their links to customers' effectiveness in RI activities. White and Badinelli (2012) define a mathematical model (coproduction RI model) to support the resource integration decision for service processes. The model provides insights into the optimal levels of client involvement and provider commitment in RI, as it captures the effects of the level of client skill and intensity on the efficiency and quality of the service process. Peters (2012) focuses on business-to-business networks to depict the effects of customers' involvement on learning, value co-creation, and innovation.

Whereas the role of the customer in the creation of value has been firmly established by contributions belonging mainly to the first stage of research (Moeller 2008; Prahalad and Ramaswamy 2004; Vargo and Lusch 2008), in the second stage many authors address more explicitly the role of the firm in supporting value co-creation process and RI activities (Gidhagen et al. 2011; Hibbert et al. 2012; Lusch and Webster 2011). Gidhagen et al. (2011) use the orchestrating metaphor to explain the role of firm in the value creation process: "Firm influences value creation through orchestrating interaction—and hence resource integration—by taking the role of an inspirator, a facilitator and an attendant within the value emergence process" (Gidhagen et al. 2011: 404). The authors used the term orchestrating interaction instead of managing, as management is strongly associated with the traditional concept of control that is viewed, from a service logic perspective, as an illogical approach. In the same vein Lusch and Webster (2011) emphasize the role of firm as facilitator in the customers' RI and value co-creation activities: "to be truly customer centric, the firm has to think not about optimizing the firm and its activities but how to support customers in their resource integration and value co-creation activities" (Lusch and Webster 2011: 132). The authors explicitly refer the term RI as antecedent of value creation to the last era marketing in network organizations: "value is created when a customer interacts with the resources and capabilities provided by a relationship with their firm/supplier and other providers of resources" (Lusch and Webster 2011: 132).

Ballantyne et al. (2011) put forward SD logic and refine the FP7 (Vargo and Lusch 2004) which is based on the notion of reciprocal value propositions between focal firm and its counterpart(s) to better reflect the contemporary industrial marketing practice of initiating and responding to the requirements of customers, suppliers, and other stakeholders. Authors refer to RI both as "outcome" and "stage." First, it can be initiated and guided by reciprocal value propositions that are positioned as communication practice with the potential to integrate exchange activities, relationship development, and knowledge renewal. Second, it is one of the three stages, featuring the platform for evaluation of communicative interactions, allowing the linking between the development of reciprocal value propositions and the co-creation and improvement of reciprocal value over time.

Consistent with Ballantyne et al. (2011), Truong et al. (2012) adopt a practice-based approach to explore how reciprocal value propositions are developed (or not) through interaction and RI in the mobile television (TV) sector. In the same vein, Haase and Kleinaltenkamp (2011) adopt a more practical approach to set the value co-creation and the RI process. They focus on the property rights theory to clarify what forms of access to operant and operand resources exist and what economic problems are related to them. RI is "the economic counterpart to the division of labor..." (Haase and Kleinaltenkamp 2011: 151) that regulates and solves collaborative processes to *create value for the whole social system*. Accordingly, it is a mandatory mechanism emerging from the division of labor and the related division of resources and the property rights they imply.

Kleinaltenkamp et al. (2012) explicitly call for deepening the RI according to a practice-based approach: "there is much to learn about the practices of integrating resources and how to design and configure the integration process" (Kleinaltenkamp et al. 2012: 203) and for a general framework to provide a clearer understanding of the role of social and economic factors in RI. Authors' focus is both on the emerging and managing of RI process that "is the means through which resource integrators (actors) co-create phenomenologically determined value" (Kleinaltenkamp et al. 2012: 201) and accordingly requires process(es) and forms of collaboration that are usually voluntary and that are based on actors' clear perception of benefits. Furthermore, this contribution calls for an in-depth analysis of both the role of technology in RI and the role of people interacting with technology. Lusch and Vargo (2012) in their contribution on the forum on Markets and Marketing (FMM) call for the adoption of a more generic, resource-integrating, actor-to-actor (A2A) model of value co-creation. Consistently with these research priorities, McColl-Kennedy et al. (2012) adopt the social practice theory to understand how customers co-create value in the healthcare context. Authors identify five customer value co-creation practices which reflect different roles, activities, and modes of interactions between collaborators within the customer's service network. In line with the network perspective, McColl-Kennedy et al. (2012) introduce different manners of co-creating value which depend on the ability of customers to access and integrate resources from different sources (e.g., market facing, public and private).

A further attempt to practically shape the RI process, and thus to better understand how suppliers integrate resources, and how customers respond to, and participate in the co-creation process, has been done by Hakanen and Jaakkola (2012). Different from previous studies, authors investigated the process of co-creating customer-focused solutions at a business network level between multiple suppliers and their mutual customers. They define the factors (i.e., uniqueness and the clarity of the customer problem to be solved, the customer expectations regarding their role in co-creation process and its value outcomes, the scope and the complementarity of the resources of the suppliers, and their mutual relationships in terms of trust, openness in sharing customer information, etc.) that are relevant for the co-creation of more effective customer-focused integrated solutions. In "value co-creation in solution networks," Jaakkola and Hakanen (2013) provide a new conceptual understanding of value co-creation that emerges from the interplay of actors, resources, and activities in solution business network (ARA model). RI and interaction between suppliers are viewed as the primary characteristic of business and as the main processes affecting the customer perceived value. Jaakkola and Hakanen (2013) identify the potential linkages between the value processes of actors and their wider network. They stress that the type of the resources integrated may be an antecedent to certain benefits and sacrifices. Resources can be positive or negative, as firms or customers may deploy resources that can have a negative effect on value perceptions, leading to value co-destruction. However most part of contributions on the interactive value formation processes, including the RI, frame it as an unproblematic phenomenon linked to positive outcomes (value). Smith (2013) conceptualizes value as a resource gain and addresses the impact of resource gain/loss on the individual's well-being. The author clearly addresses the overlooked concept of value co-destruction, resulting from the accidental or intentional misuse of resources by a system acting in an inappropriate or unexpected manner (Plé and Chumpitaz Cáceres 2010).

Löbler (2013) advances the network perspective to conceptualize the service-dominant networks (SDN). It can be viewed as a resource-integrating phenomenon that performs one or more intended activities:

> "resource integration is the activity that keeps the network together, and everything and everyone can become a resource during the integration process. After the integration process, the SDN disappears as a SDN (integrating resources), though not necessarily as a network of actors (collection of actors) (Löbler 2013: 430) (. . .) the resource integration process might be carried out by a single person, several people, or many people (Löbler 2013: 422) (. . .) every resource integration process is oriented toward or rationalized by one or several intended activities." (Löbler 2013: 423)

Löbler (2013) revises many foundational elements of the SD logic emphasizing the nature, key components, and properties of the SDN construct. The resources' conceptualization and the RI process are the pivotal factors of such a construct.

Moving the focus on complex service systems, Högström and Tronvoll (2012) argue that the understanding of how value is co-created requires the understanding of the complexity of value-creating systems in action by acknowledging multiple types of embeddedness (structural, cultural, political, cognitive). According to this, value-creating system is patterned after not only RI but also political, cultural, and cognitive forces in the embedding social reality. RI is viewed as one of the multiple patterns of value co-creating system. Edvardsson et al. (2012) provide a deeper understanding of what constitutes a service system in practice and how resources are integrated and value is co-created in service systems. Authors link service systems to service practices and social systems and through this to social practices (Giddens 1984) to understand how human actors integrate resources and co-create value; they describe the service systems as constellation a of practices that are embedded in and interacting with social systems. Both the social and service structures act as mediators between social and service practices and on the other social and service systems which subsequently have a major impact on RI, value co-creation, and thus value in context.

C: Expansive Stage (2014–2015)

Studies relating to this stage clearly address service system in action as the fundamental basis for RI and value co-creation. They provide a broader perspective on forces/resources (e.g., technology, institutions), shaping RI processes to create value.

Edvardsson et al. (2014) put forward the actor-centric and systemic nature of RI, as well as the key relevance of coordination and collaboration to the process of RI (Kleinaltenkamp et al. 2012): "resource integration comprises the methods through which resources are integrated and used by actors, so that their potential value is realized in practice. Resource integration consists of cooperative and collaborative processes between actors, leading to experiential outcomes and outputs, as well as mutual behavioral outcomes for all actors involved." In such service ecosystems, institutional logics and institutions—set regulative, normative, and cognitive rules—enable but may also hinder RI. Institutions shape and coordinate actors' roles, activities, and interactions when they integrate resources during value co-creation processes; therefore they are the major components for the recursiveness of RI and the development of service system tout court.

Consistent with the actor-centric and systemic nature of value creation, Paredes et al. (2014) highlight the importance of knowing the resources available to each actor involved in the RI to understand value co-creation process. Firm resources and customer resources can have a synergistic, proportional, or antagonistic nature; therefore it is fundamental to consider both parties (e.g., firm and customers) as co-creation partners overall in highly interactive context such as the e-commerce settings.

In a similar vein, Lempinen and Rajala (2014) emphasize the importance of mutual interests, the power structure, and the knowledge sharing as the main factors affecting value co-creation process overall in a networked environment, as the information technology (IT) services. The authors focus on the central role of the information system (IS) organization in moderating and facilitating the social interaction and the RI process among relevant stakeholders.

Both Paredes et al. (2014) and Lempinen and Rajala (2014) relate RI and value co-creation to IT (context), whereas Lusch and Nambisan (2015), in addition to theemphasis on gradual digitization of social context, depict the emerging and dual role of IT as an integral component of the innovation and as an actor that triggers or initiates the innovation. In line with Kleinaltenkamp et al. (2012), the authors shed light on the role of technology on RI and on RI process too. IT is central to the integration processes as it is both a facilitator-enabler (e.g., operand resource) and an initiator-actor (e.g., operant resource) in continuous service innovation. Innovation is a collaborative process occurring in an actor-to-actor (A2A) network value co-creation that emerges from the recombining of existing resources (Arthur 2009); accordingly, RI is the medium through which value is innovated and co-created and requires mechanisms to support the underlying roles and processes. Some scholars go beyond the concept of RI as a mechanism, raising it toward a meta-theoretical foundation of SD logic (e.g., such definition includes also A2A networks, resource liquefaction, and resource density) that is especially relevant to service innovation.

This contribution complements the SD logic approach with a more macro perspective on RI (meta-theoretical foundation of SD logic) and explicitly addresses the powerful relationship between RI and innovation that existing literature has largely overlooked.

11.5 Discussion and Contribution

The aim of this chapter has been to contribute to the debate in marketing and service research by providing an overview of research in the area of RI.

A bibliometric study (e.g., co-word analysis, mapping, and clustering techniques) and content analysis have been carried out. The co-word analysis included 1250 scientific articles from 1985 to 2015. Six clusters—open innovation, value co-creation, general management, supply chain management, strategic management, and social/digital marketing—have been identified and labeled according to their main content, respectively referred to as technology, marketing, miscellaneous, operation, strategy, and digital marketing.

The knowledge structure of RI is divided in two main blocks (see Fig. 11.1) that cover well-established research topics closed to the GD logic perspective (e.g., open innovation, supply chain management, strategic management, and general management clusters) and new and emerging research topics closed to the service-dominant logic perspective (e.g., value co-creation and social media/digital marketing clusters). The lightweight green/yellow colors and the edge isolated position of the value co-creation cluster in the map (see Fig. 11.2) mark out, respectively, the emerging nature of these studies and the differences in research content related to RI.

Results from the qualitative analysis we conducted on the 25 papers adopting the idiomatic expression RI and belonging to the value co-creation cluster allow us to frame the evolutionary path of RI. They highlight the emergence of different topics and perspectives we synthesize in three main phases: introduction (2006–2008), development (2009–2013), and expansive (2014–2015) stage.

The introduction stage of research relates more to the building of new lexicon to support fresh thinking and conceptualizations of value co-creation process and indirectly of RI. The development stage adopts and calls for a pragmatic viewpoint and refines and upgrades these concepts in the light of the practice-based view to better understand the actual practical process of RI (Korkman et al. 2010: 238) and to support the development of new practices to reproduce it. Finally, the extensive stage provides a broader perspective on resources that shape RI processes. It identifies new potential actors—as institutions and technology—acting both as operand and operant resources for creating and coordinating service innovation system. Furthermore, it reconceptualizes service (eco)system in action not just as the context of analysis of RI but as the main driving force of RI.

These three stages show that although the spectrum of focus areas is broad and varied and different perspectives—ranging from the focal firm to the ecosystem—feature the RI research, two factors, value and the becoming nature of resources, turn out as common to all contributions.

Indeed, the review of RI and the analysis of its linkages to other concepts—e.g., actors' activities and interactions, value, learning, systems, and sharing—that are well recognized as fundamentals of successful innovation allow us to depict RI as the central process of collaborative innovation and to emphasize these elements as the joining links between matters and clusters belonging to GD logic and SD logic.

Innovation is the result of recombining existing resources (Arthur 2009; Lusch and Nambisan 2015; Schumpeter 1934) and thus a resource-based process that arises through business and social interaction (Chesbrough 2003, 2006; Trott and Hartmann 2009). Therefore, we argue that innovation emerges through RI that is context specific, orchestrated collaboratively in actor-driven processes with the intention to co-create value for the involved actors and others. RI is centered on actors' interactive practices that potentially lead to experiential outputs and more important outcome—e.g., innovation/value—for the entire ecosystem.

Accordingly, RI is a practice-based process that dynamically shapes and is shaped by actors acting as resource integrators in service ecosystems.

All actors within these service ecosystems have the potential to become valuable resources and thus to generate value/innovation through the practice of the RI. Our conceptualization of RI emphasizes the dynamic process during which resources become valuable within and through the co-creation process (Edvardsson et al. 2012; Pels et al. 2009; Zimmermann 1951). Resources have no inherent value in themselves. Instead, they possess important potential value, depending on how they are integrated and operated on, in specific contexts with specific intentions (Zimmermann 1951). In this sense, resources need to be understood as *becoming* instead of *existing* in the specific situation, as they are integrated by actors in order to enable value co-creation (Pels et al. 2009).

From this view, we pay attention to different forces—actors' practice, institutions, technology, and service ecosystems as well—shaping social context, driving, and coordinating the actors' integration of available resource, the value co-creation processes, and the actors' value assessment activities. Indeed, attractive value emerging through RI is not an obvious outcome (Smith 2013) as it requires collaboration and the presence of an orchestrating actor that facilitates the entire process of RI from the matching of the actors' available resources to the assessment/interpretation of the co-created value.

This study provides a basis for the ongoing conversation on the conceptualization of RI in service research. First, it puts forward the convergence between the marketing and management perspectives; second, it conceptualizes RI as a process per se that emerges and is shaped by service ecosystems; and third, it clearly addresses the strong link between RI as a process-based practice and innovation.

11.6 Further Research

Future research should focus on the theoretical foundations of what a resource is and how resources are becoming (Zimmermann 1951; Vargo and Lusch 2008) as part of value co-creation and innovation processes.

Edvardsson and Tronvoll (2013) define RI as the involved actors' activities and interactions when resources are integrated and operated upon. This should be focused on in different contexts and service ecosystems such as healthcare, higher education, financial services, social communities, telecom services, and service infusion in manufacturing. The way resources are integrated, the challenges for different engaged actors, as well as RI failures should form an empirical basis for extending the conceptualization of RI in the context of service ecosystems.

Since actors in SD logic are viewed as resource integrators and RI requires process(es) and forms for collaboration, future research should empirically study how and why different forms are selected and how these forms are linked to more or less successful value co-creation practices. Another field for research is how actors' knowledge and skills shape RI across service system boundaries and cultural contexts. It is argued that resources are embedded in social structures and that social capital—understood as a set of potential resources (including both inhibitors and enablers)—is integrated by engaged social and economic actors. How do social capital and social contracts, understood as enablers, guide actors' RI efforts in different practices?

We also suggest research focusing on resource integration across service system boundaries. In practice, actors' resource integration in different practices is often drawing on resources in multiple service systems. Thus service system boundaries become an issue that needs more attention in future research.

Finally we see a need to focus research on the role of RI in service innovation including the scaling up of service innovations. Many innovations are grounded in novel and smarter ways of using existing resources. What do these novel and smarter ways of integrating resources have in common?

Appendix: Overview of RI Concept

Authors	Conceptualization
Ballantyne et al. (2011)	Reciprocal value proposition is a communication practice that brings exchange activities, relationship development, and knowledge renewal closer together Reciprocal value propositions can initiate and guide resource integration activities across a network of firm stakeholders
Edvardsson et al. (2012)	Social and service structures are mediators between social and service practices and on the other social and service systems which subsequently have a major impact on resource integration, value co-creation, and thus value in context

(continued)

Authors	Conceptualization
Edvardsson et al. (2014)	Resource integration comprises the methods through which resources are integrated and used by actors, so that their potential value is realized in practice. It consists of cooperative and collaborative processes between actors, leading to experiential outcomes and outputs, as well as mutual behavioral outcomes for all actors involved
Gidhagen et al. (2011)	Firm influences value creation through orchestrating interaction—and hence resource integration—by taking the role of an inspirator, a facilitator, and an attendant within the value emergence process
Haase and Kleinaltenkamp (2011)	Resource integration is the economic counterpart to the division of labor and as such it is the basic connection point between SD logic and property rights theory division of labor, and the subsequent division of resources and the property rights they imply are the reasons why resource integration is needed to create value for economic actors as well as societal prosperity
Hakanen and Jaakkola (2012)	Solution co-creation refers to the interactive process where actors—the suppliers and their customer—jointly create the solution offering by integrating resources. Co-creation is affected by customer's preferences for participation and value and the degree of competition, clarity of role division. and rapport among the suppliers
Hibbert et al. (2012)	Customer learning resource integration is any self-initiated or self-directed set of activities in which customers integrate learning resources afforded by organizations and other network actors with their own operant and operand resources with the primary purpose of learning in order to increase their effectiveness in other resource integration processes
Högström and Tronvoll (2012)	A value-creating system is patterned after not only resource integration but also political, cultural, and cognitive forces in the embedding social reality
Jaakkola and Alexander (2014)	Through CEB, customers can contribute a range of resources beyond dyadic exchange, contributing to enhanced offerings and value outcomes at a wider service system level. CEB induces broader resource integration
Jaakkola and Hakanen (2013)	Value co-creation occurs from the interplay of actors, resources, and activities (ARA model) in solution business network
Kleinaltenkamp et al. (2012)	Resource integration as the means through which resource integrators (actors) co-create phenomenologically determined value. Integration requires process(es) and forms of collaboration
Lempinen and Rajala (2014)	The knowledge sharing that happens in the IT service process is part of resource integration, which is a fundamental part of value co-creation in a service process
Löbler (2013)	In a service-dominant network, service (i.e., resource integration) is the activity that keeps the network together, and everything and everyone can become a resource during the integration process
Löbler and Lusch (2014)	Service and, with it, service innovation, which unfold their value throughout the process of resource integration, need an ontology supporting the understanding of process-related dynamics (societal interaction)

(continued)

Authors	Conceptualization
Lusch and Vargo (2012)	SD logic can be better understood by the adoption of a generic, actor-to-actor (A2A) model of resource integrators engaged in mutual service provision
Lusch and Nambisan (2015)	Human actors integrate resources for two primary reasons. First, any resource an actor obtains can never be used in isolation but needs to be combined or bundled with other resource for usefulness or value. Second, all innovation is the result of recombining existing resources Service innovation can be considered the rebundling of diverse resources that create novel resources that are beneficial to some actors in a given context
Lusch and Webster (2011)	Value is created when a customer interacts with the resources and capabilities provided by a relationship with their firm/supplier and other providers of resources
McColl-Kennedy et al. (2012)	Customer value co-creation is a benefit realized from integration of resources through activities and interactions in the customer's service network
Paredes et al. (2014)	To understand value co-creation through resource integration, it is necessary to know what type of resources is available to each party involved in the process
Peters et al. (2009)	Organizational learning (both a structure and process) relates to the creation of value propositions that the firm offer to the market. Organizational learning emerges through the links between the value proposition offered by the firm, value as determined by the customer, and feedback from the customer that results in the co-creation of the value process
Peters et al. (2014)	Resource integration can happen both as a process of emergence and as a process of interaction
Skålén et al. (2015)	Value propositions are created through practices that integrate operand/operant resources into a value creation promise, which is a value proposition. Successful service innovation is not only contingent on having the right resources; established methods and practices for integrating these resources into attractive value propositions are also needed
Smith (2013)	Value co-destruction D process is triggered by a failure of the resource integration process to co-create expected value (resources)
Truong et al. (2012)	Reciprocal value propositions are not the sole prerogative of any one organization, but are the responsibility of a network of suppliers and consumers taking on interchangeable initiating and participating roles
Vargo (2008)	Resource integration is multidirectional (all parties uniquely integrating multiple resources for their own benefit and for the benefit of others) but service beneficiary centered (i.e., both parties in service-for-service exchange)
Vargo and Lusch (2008)	All social and economic actors are resources integrators. It is this unique application of uniquely integrated resources that motivates and constitutes exchange, both economic and otherwise
White and Badinelli (2012)	Both the levels of client skill and client intensity affect the efficiency and quality of the service process

References

Arthur, B. W. (2009). *The nature of technology: What it is and how it evolves*. New York: Free Press.

Ballantyne, D., Frow, P., Varey, R. J., & Payne, A. (2011). Value propositions as communication practice: Taking a wider view. *Industrial Marketing Management, 40*(2), 202–210.

Benavides-Velasco, C. A., Quintana-García, C., & Guzmán-Parra, V. F. (2013). Trends in family business research. *Small Business Economics, 40*(1), 41–57.

Bhattacharya, S., & Basu, P. K. (1998). Mapping a research area at the micro level using co-word analysis. *Scientometrics, 43*(3), 359–372.

Börner, K., Chen, C., & Boyack, K. W. (2003). Visualizing knowledge domains. *Annual Review of Information Science and Technology, 37*(1), 179–255.

Brandenburger, A. M., & Nalebuff, B. J. (2011). *Co-opetition*. New York: Random House LLC.

Callon, M., Courtial, J. P., & Laville, F. (1991). Co-word analysis as a tool for describing the network of interactions between basic and technological research-the case of polymer chemistry. *Scientometrics, 22*(1), 155–205.

Chesbrough, H. W. (2003). *Open innovation: The new imperative for creating and profiting from technology*. Brighton, MA: Harvard Business Press.

Chesbrough, H. W. (2006). The era of open innovation. *Managing Innovation and Change, 127*(3), 34–41.

Cobo, M. J., López-Herrera, A. G., Herrera-Viedma, E., & Herrera, F. (2011). Science mapping software tools: Review, analysis and cooperative study among tools. *Journal of the American Society for Information Science & Technology, 62*(7), 1382–1402.

Cooper, H. M. (1982). Scientific guidelines for conducting integrative research reviews. *Review of Educational Research, 52*(2), 291–302.

Edvardsson, B., Kleinaltenkamp, M., Tronvoll, B., McHugh, P., & Windahl, C. (2014). Institutional logics matter when coordinating resource integration. *Marketing Theory, 14*(3), 291–309.

Edvardsson, B., Skålén, P., & Tronvoll, B. (2012). Service systems as a foundation for resource integration and value co-creation. *Review of Marketing Research, 9*(2), 79–126.

Edvardsson, B., & Tronvoll, B. (2013). A new conceptualization of service innovation grounded in SD logic and service systems. *International Journal of Quality and Service Sciences, 5*(1), 19–31.

Giddens, A. (1984). *The constitution of society: Outline of the theory of structuration*. Berkeley, CA: California Press.

Gidhagen, M., Persson Ridell, O., & Sörhammar, D. (2011). The orchestrating firm: Value creation in the video game industry. *Managing Service Quality: An International Journal, 21*(4), 392–409.

Haase, M., & Kleinaltenkamp, M. (2011). Property rights design and market process: Implications for market theory, marketing theory, and SD Logic. *Journal of Macromarketing, 31*(2), 148–159.

Hakanen, T., & Jaakkola, E. (2012). Cocreating customer focused solutions within business networks: A service perspective. *Journal of Service Management, 23*(4), 593–611.

Hamel, G., Doz, Y., & Prahalad, C. (1989). Collaborate with your competitors. *Harvard Business Review, 67*(1), 133–139.

He, Q. (1999). Knowledge discovery through co-word analysis. *Library Trends, 48*(1), 133.

Hibbert, S., Winklhofer, H., & Temerak, M. S. (2012). Customers as resource integrators: Toward a model of customer learning. *Journal of Service Research, 15*(3), 247–261.

Högström, C., & Tronvoll, B. (2012). The enactment of socially embedded service systems: Fear and resourcing in the London Borough of Sutton. *European Management Journal, 30*(5), 427–437.

Jaakkola, E., & Alexander, M. (2014). The role of customer engagement behavior in value co-creation: A service system perspective. *Journal of Service Research, 17*(3), 247–261.

Jaakkola, E., & Hakanen, T. (2013). Value co-creation in solution networks. *Industrial Marketing Management, 42*(1), 47–58.

Kleinaltenkamp, M., Brodie, R., Frow, P., & Hughes, T. (2012). Resource integration. *Marketing Theory, 12*(2), 201–205.

Korkman, O., Storbacka, K., & Harald, B. (2010). Practices as markets: Value co-creation in e-invoicing. *Australasian Marketing Journal, 18*(4), 236–247.

Lempinen, H., & Rajala, R. (2014). Exploring multi-actor value creation in IT service processes. *Journal of Information Technology, 29*(2), 170–185.

Löbler, H. (2013). Service-dominant networks: An evolution from the service-dominant logic perspective. *Journal of Service Management, 24*(4), 420–434.

Löbler, H., & Lusch, R. F. (2014). Signs and practices as resources in IT-related service innovation. *Service Science, 6*(3), 190–205.

Lusch, R. F., & Nambisan, S. (2015). Service innovation: A service-dominant logic perspective. *MIS Quarterly, 39*(1), 155–175.

Lusch, R. F., & Vargo, S. L. (2012). The forum on markets and marketing (FMM): Advancing service-dominant logic. *Marketing Theory, 12*(2), 193–199.

Lusch, R. F., & Webster, F. E. (2011). A stakeholder-unifying, cocreation philosophy for marketing. *Journal of Macromarketing, 31*(2), 129–134.

McColl-Kennedy, J. R., Vargo, S. L., Dagger, T. S., Sweeney, J. C., & van Kasteren, Y. (2012). Health care customer value cocreation practice styles. *Journal of Service Research, 15*(4), 370–389.

Mele, C., Russo-Spena, T., & Colurcio, M. (2010). Co-creating value innovation through resource integration. *International Journal of Quality and Service Sciences, 2*(1), 60–78.

Moeller, S. (2008). Customer integration - a key to an implementation perspective of service provision. *Journal of Service Research, 11*(2), 197–210.

Muñoz-Leiva, F., Viedma-del-Jesús, M. I., Sánchez-Fernández, J., & López-Herrera, A. G. (2012). An application of co-word analysis and bibliometric maps for detecting the most highlighting themes in the consumer behaviour research from a longitudinal perspective. *Quality & Quantity, 46*(4), 1077–1095.

Ngo, L. V., & O'Cass, A. (2009). Creating value offerings via operant resource-based capabilities. *Industrial Marketing Management, 38*(1), 45–59.

Paredes, M. R., Barrutia, J. M., & Echebarria, C. (2014). Resources for value co-creation in e-commerce: A review. *Electronic Commerce Research, 14*(2), 111–136.

Pels, J., Möller, K., & Saren, M. (2009). Do we really understand business marketing? Getting beyond the RM and BM matrimony. *Journal of Business & Industrial Marketing, 24*(5/6), 322–336.

Persson, O. D., Danell, R., & Wiborg Schneider, J. (2009). How to use bibexcel for various types of bibliometric analysis. In F. Åström, R. Danell, B. Larsen, & J. Schneider (Eds.), *Celebrating scholarly communication studies: A Festschrift for Olle Persson at his 60th Birthday* (pp. 9–24). Leuven: International Society for Scientometrics and Informetrics.

Peters, L. (2012). The role of the knowledgeable customer in business network learning, value creation, and innovation. *Review of Marketing Research, 9*, 127–169.

Peters, L. D., Gassenheimer, J. B., & Johnston, W. J. (2009). Marketing and the structuration of organizational learning. *Marketing Theory, 9*(3), 341–368.

Peters, L. D., Löbler, H., Brodie, R. J., Breidbach, C. F., Hollebeek, L. D., Smith, S. D., et al. (2014). Theorizing about resource integration through service-dominant logic. *Marketing Theory, 14*(3), 249–268.

Pinto, M., Pulgarín, A., & Escalona, M. I. (2014). Viewing information literacy concepts: A comparison of two branches of knowledge. *Scientometrics, 98*, 2311–2329.

Plé, L., & Chumpitaz Cáceres, R. (2010). Not always co-creation: Introducing interactional co-destruction of value in service-dominant logic. *Journal of Services Marketing, 24*(6), 430–437.

Prahalad, C. K., & Ramaswamy, V. (2004). Co-creation experiences: The next practice in value creation. *Journal of Interactive Marketing, 18*(3), 5–14.

Sakata, I., Sasaki, H., Akiyama, M., Sawatani, Y., Shibata, N., & Kajikawa, Y. (2013). Bibliometric analysis of service innovation research: Identifying knowledge domain and global network of knowledge. *Technological Forecasting and Social Change, 80*(6), 1085–1093.

Schumpeter, J. A. (1934). *The theory of economic development: An inquiry into profits, capital, credit, interest, and the business cycle* (Vol. 55). Piscataway, NJ: Transaction Publishers.

Skålén, P., Gummerus, J., von Koskull, C., & Magnusson, P. R. (2015). Exploring value propositions and service innovation: A service-dominant logic study. *Journal of the Academy of Marketing Science, 43*(2), 137–158.

Smith, A. M. (2013). The value co-destruction process: A customer resource perspective. *European Journal of Marketing, 47*(11/12), 1889–1909.

Trott, P., & Hartmann, D. (2009). Why open innovation is old wine in new bottles. *International Journal of Innovation Management, 13*(4), 715–736.

Truong, Y., Simmons, G., & Palmer, M. (2012). Reciprocal value propositions in practice: Constraints in digital markets. *Industrial Marketing Management, 41*(1), 197–206.

Turiera, T., & Cross, S. (2013). *Co-business: 50 examples of business collaboration*. Infonomia.

van Eck, N. J., & Waltman, L. (2007). Bibliometric mapping of the computational intelligence field. *International Journal of Uncertainty, Fuzziness and Knowledge-Based Systems, 15*(5), 625–645.

van Eck, N. J., & Waltman, L. (2010). Software survey: VOSviewer, a computer program for bibliometric mapping. *Scientometrics, 84*(2), 523–538.

van Eck, N. J., Waltman, L., Dekker, R., & van den Berg, J. (2010). A comparison of two techniques for bibliometric mapping: Multidimensional scaling and VOS. *Journal of the American Society for Information Science and Technology, 61*(12), 2405–2416.

Vargo, S. L. (2008). Customer integration and value creation: Paradigmatic traps and perspectives. *Journal of Service Research, 11*(2), 211–215.

Vargo, S. L., & Lusch, R. F. (2004). Evolving to a new dominant logic for marketing. *Journal of Marketing, 68*(1), 1–17.

Vargo, S. L., & Lusch, R. F. (2008). Service-dominant logic: Continuing the evolution. *Journal of the Academy of Marketing Science, 36*(1), 1–10.

Vargo, S. L., & Lusch, R. F. (2011). It's all B2B... beyond: Toward a systems perspective of the market. *Industrial Marketing Management, 40*(2), 181–187.

Waltman, L., van Eck, N. J., & Noyons, E. C. (2010). A unified approach to mapping and clustering of bibliometric networks. *Journal of Informetrics, 4*(4), 629–635.

White, S. W., & Badinelli, R. D. (2012). A model for efficiency-based resource integration in services. *European Journal of Operational Research, 217*(2), 439–447.

Zhu, W., & Guan, J. (2013). A bibliometric study of service innovation research: Based on complex network analysis. *Scientometrics, 94*(3), 1195–1216.

Zimmermann, E. W. (1951). *World resources and industries*. New York: Harper and Row.

Chapter 12
Service Innovation and Smart Cities: Linking the Perspectives

Francesco Bifulco and Marco Tregua

Abstract Cities are innovating thanks to the transformations in smart cities taking place in recent years. Nowadays, cities are emerging as contexts in which different stakeholders merge their resources in activities, by defining and improving the way they focus on the support given to reach common aims. This phenomenon is becoming increasingly prevalent, but an investigation on how to frame it in the service innovation literature is still missing. We move from this gap to link smart cities and service innovation from a theoretical point of view; subsequently, an analysis is performed on documents provided by organisations carrying on smart projects all over the world. A content analysis is used to highlight how the interventions by smart service providers can be considered in service innovation literature and to show how different firms operate in connection with the key elements of service innovation provided by Lusch and Nambisan (MIS Quarterly, 39(1), 155–175, 2015). The results of the analysis are useful to describe, from a practical point of view, how service innovation is taking place in cities and how relevant the roles played by actors involved in smart services are, as these actors favour a better knowledge of the needs to be satisfied, carry resources and skills, evaluate services and provide data. Finally, data play a crucial role when referring to platforms, as the core of smart service provision, leading technology to act as a significant support for service innovation in the city context.

12.1 Introduction

Smart cities are recently being paid increasing attention, and relevant innovations are projected and exploited by both private actors and public institutions. Smart cities emerge as the outcome of interventions aimed at the improvement of services provided through citizens and other stakeholders thanks to the usage of technology

---

F. Bifulco (✉) • M. Tregua
Department of Economics, Management, Institutions, University of Naples Federico II, Naples, Italy
e-mail: francesco.bifulco@unina.it; marco.tregua@unina.it

T. Russo-Spena et al. (eds.), *Innovating in Practice*,
DOI 10.1007/978-3-319-43380-6_12

and big data, which are intended to favour active participation by all actors (Cosgrave et al. 2013). Nowadays, cities are seen as contexts in which different stakeholders apply their resources to activities, and as time goes by, they define and improve ways to support the achievement of common aims. The set of stakeholders is even wider when taking into account the recent research trend of considering metropolitan areas rather than just cities (Yigitcanlar and Lönnqvist 2013).

The research presented in this chapter aims to define smart cities in the framework of innovation literature, as a clear statement on how to read the interventions in smart cities in an innovation context is still missing. This gap is due to the major focus on technological and social aspects of smart projects for cities, rather than on analysing the way innovation is carried out from a managerial-based perspective. Activities that lead cities to become smart are investigated in this chapter through the service innovation approach grounded in service-dominant logic proposed by Lusch and Nambisan (2015). In addition, conceptualisations on smart cities' management are introduced thanks to the four core themes emerging from Lusch and Nambisan's contribution, namely, innovation as an actor-to-actor process, "service as the application of specialised competences" (Lusch and Nambisan 2015: 155), resource liquefaction and density and resource integration.

In order to empower the framing of smart cities' literature in service innovation studies, this investigation focuses on evidence directly related to smart cities and to the process through which urban contexts become smart, so-called smartisation. Making a city smarter means transforming the way in which services are provided, as well as data collection and usage, the business model sustaining services' provision and the expected quality of life, both for citizens and other actors. Useful evidence of the way service innovation can be linked to smart cities will be provided in this research by describing living labs, as contexts that actors are asked to join to stimulate a common cooperation towards innovation-based goals.

The information we used to investigate our empirical contexts was obtained through documents known as official reports, namely, highly detailed dossiers on how innovative activities are carried out in cities that aim to become smart. These documents are provided by both organisations and local agencies working as project leaders in making cities smarter and will be analysed through a content analysis performed through NVivo. The results show how smart cities can be a context in which service innovation takes place through the involvement of a wide range of actors, which are connected by technologies in order to share and combine resources for mutual benefits. This contribution is part of a research stream to depict the main features of the scientific field shaped around both the topic of "smart city management" and the innovation path leading to a new means of public service provision in metropolitan and urban contexts (project ORCHESTRA (ORganization of Cultural Heritage for Smart Tourism and Real-Time Accessibility) in the Italian National Operative Programme 2007–2013).

The chapter is structured as follows: first, the theoretical background is introduced by focusing on smart cities, service innovation and the linkages between them from a theoretical point of view. The research method is then described, followed by the findings and conclusions, both for scholars and practitioners, together with limitations and avenues for future research.

12.2 Theoretical Background

12.2.1 Overview of Smart Cities

Smart cities are complex urban contexts that must be managed in different ways compared to ordinary cities in order to favour smart development. This is quite a recent notion, which refers to a sustainable approach to different aims to be achieved through the usage of ICT (information and communication technology) in providing services (Bélissent 2010), preferably in a sustainable way (Tregua et al. 2015). The idea of smart cities can be considered as the final step of an evolutionary process that started from the so-called digital city, referring to the provision of services through ICT (Ishida and Isbister 2000). As time went by and interests to be taken into account dramatically increased, the perspective was enlarged to embed hardware and software elements (Batty et al. 2012), with the former consisting of infrastructures and ICT technologies and the latter composed of social capital, human resources and participation. The focus on software features led to the definition of a socially inclusive city (Deakin 2009), and this represents a step towards the smart city, where relationships among citizens, technology, service providers and context are several and multidirectional, as actors and objects interact to provide services and to continuously improve how all of the related operations are planned and exploited (Lombardi et al. 2011).

The expected activities leading to smart cities are usually performed through projects from a wide range of actors (Komninos 2006) that play a *decisive role* (Alam 2006); different contributions lead to a network-based innovation perspective in a merger favoured by new technologies and participation of beneficiaries, as they have the best knowledge on their needs (Magnusson 2009) and they hold value perceptions due to their own role (Marsal-Llacuna et al. 2014). Smart cities' projects are context-based and hence differ from one another because of managerial choices (general and public management), aims to be accomplished, citizens' participation, resources to be shared and activities to be carried out to improve life in cities. These differences have an impact on cities' management, but a central idea is common to all urban and metropolitan areas, in particular the focus on quality of life (Anttiroiko et al. 2013), as an element to define and describe goals.

Hence, the wide range of actors needs to be managed and led towards innovation processes and public service improvement; more specifically, one of the possible ways in which to drive urban contexts towards smartisation is given by living labs, as they act as "user-driven innovation ecosystems" (Schaffers et al. 2011: 431), representing the central part of the public–private–people partnership, namely, as a fertile ground favouring opportunities for co-innovation. Living labs are open research and innovation ecosystems (Schaffers and Turkama 2012), which group several actors to shape new ideas, to support decision-making by local agencies and to favour the analysis of local contexts and needs. Moreover, socio-emotional intelligence (Goleman 2006) can emerge thanks to the ecosystem, which drives innovation (Cosgrave et al. 2013). The perspective applied to living labs reflects the conceptualisation of open and social innovation (Chesbrough 2011) as well, because different actors can join or be embedded as time goes by and activities go on, due to the necessity of involving new sources of knowledge (Gallego et al. 2013) to accomplish the smartisation process.

12.2.2 Service Innovation

Service innovation has attracted scholars' attention in recent years, and the debate has focused on elements such as the approaches, actors, models and resources used as leverage. Initially, innovation in services was thought of in comparison to product innovation by Berry et al. (2006), but other scholars have moved the focus to a service-dominant logic framework (Lusch and Nambisan 2015) leading to a deeper focus on resources to be considered when investigating service innovation. In addition, several scholars agree on the idea that service innovation creates new markets, in line with Berry et al. (2006) who suggested that innovations can influence both customers' and providers' behaviour.

The focus then switched again when scholars began to pay more attention to the impact of IT on service innovation as it reshapes the interactions between customers and providers and leads to a better understanding of people's needs, in line with the notion of intimacy (Lukas and Ferrell 2000). Moreover, the core of service innovation studies is the improvement of customer value co-creation (Michel et al. 2008). Rubalcaba et al. (2012) framed service innovation around three main elements, namely, the agent dimension, the sectorial dimension and the activity dimension. Thanks to these three dimensions, it is possible to focus on co-production, interactions among actors and blurring boundaries of specific activities to better describe how service innovation takes place.

The elements considered above are taken into account by Lusch and Nambisan (2015) when debating service innovation, describing it as based on four key elements and broadening the conceptual framework when linking service innovation to the conceptualisations of ecosystems, platforms and value co-creation (Lusch and Nambisan 2015). Moreover, Lusch and Nambisan (2015) took into account the role of technology and thus considered previous studies on service innovation and adopted Akaka and Vargo's (2014) perspective on technology as an operant resource, that is, a factor favouring the deployment of innovation activity in contexts composed of multiple actors, so that innovation boundaries fade away as service innovation does not take place inside the firms, but in the wider ecosystem they are part of. This perspective supports the previous proposition of Windrum and

Koch (2008), who framed service innovation, even in public and social services, in line with the aim of improving the value co-created for—and with—customers.

12.2.3 Linking Smart Cities and Service Innovation

The perspective we adopted aims to frame smart cities within the service innovation literature, as conceptualisations of smart cities are missing from innovation and business perspectives. Moreover, the multidisciplinary approach of service innovation (Rubalcaba et al. 2012) appears to be compliant with studies on cities and cities' management, as several fields of science converge when discussing new ways of providing services in cities. The call for research "toward a more comprehensive, systemic, and social approach to service innovation, such that research covers more social areas, different actors, and public sectors" (Rubalcaba et al. 2012: 705) will be addressed in this investigation.

There are some additional common features between service innovation and smart cities that support our adoption of service innovation as a framework in which to investigate the smartisation process and, more generally, smart cities; first of all, service innovation involves different agents instead of considering only firms or public bodies (Sundbo 2010), as in smart cities' projects due to the contemporary involvement of public agencies, citizens, businesses and other stakeholders to reach the expected innovation goals. In the same framework of service innovation, the role of technology is relevant but not crucial as it is considered "only the medium for a new service" (Sundbo 1997: 436). This statement can be compared to the theoretical evolution in city management, as technology was considered essential in *digital cities*, while it is partnered with all other resources and human contributions in *smart cities*. The role of technology changed in cities' management literature as new labels emerged; when defining digital cities, scholars depict them as a combination of technological tools that support service provision for citizens (Ishida and Isbister 2000), while smart cities involve human and social capital and traditional and modern communication infrastructure (Lombardi et al. 2011). Furthermore, the intervention of external actors involving "both top-down and bottom-up processes" (Rubalcaba et al. 2012: 701) is pivotal in smart cities, as local agencies or businesses in smart city projects commonly act by managing the complex systems they are leading. This participation is fundamental and has led to the huge spread of *e-government* (Anttiroiko et al. 2013) and community-based decision-making processes (Bifulco et al. 2014a; Coe et al. 2001).

Innovation has to be considered in the context of the *open perspective* when approaching smart cities, as people are pivotal in these processes. As a matter of fact, actors are part of these urban areas and can suggest, improve and evaluate smart interventions, while also being expected to benefit from them. The notion of cities linked to innovation bloomed even before the smart cities' concept itself, thanks to Komninos (2006: 13),who defined cities as "territories that bring innovation systems and ICTs within the same locality, combining the creativity" of

Table 12.1 Main elements from Lusch and Nambisan (2015) and "Smart City management" scholars

Key concepts from Lusch and Nambisan (2015)	Approaches emerging from smart city scholars
Innovation as a collaborative process involving a diverse network of actors	"The way for intelligent cities and regions is paved by clusters of innovative organizations forming networks" (Komninos 2006: 14)
Service as the application of specialised competences for the benefit of another actor or the self and as the basis of all exchange	"The interplay between University and Government produces what is here labelled as 'Learning'. Public institutions learn from educational bodies' ways to improve their performance [...]. At the same time, the university system benefits from an efficient management of public goods. The mutual reinforcement of this mechanism generates society-level learning" (Lombardi et al. 2011: 9)
The generativity unleashed by increasing resource liquefaction and resource density	"Intelligent cities and regions constitute advanced territorial systems of innovation, in which the institutional mechanisms for knowledge creation and application are facilitated by digital spaces and online tools" (Komninos 2006: 14) "The existence of an ecosystem opens up new innovative opportunities for large and small regional enterprises to offer new services" (Lee et al. 2013: 6)
Resource integration as the fundamental way to innovate	"Evidently smart city technologies may evolve as they continuously seek to integrate services, systems, infrastructures, and technologies and capabilities into a complex whole" (Lee et al. 2013: 7)

different actors, such as citizens and institutions. Creativity and innovation had already been considered as crucial in public management in the previous decade (Nalbandian 1999). The contributions arise from a network of actors shaping a city, and this can be considered as a set of innovators (Komninos et al. 2007). This perspective is aligned with the notion of service innovation in the framework of service-dominant logic (Lusch and Nambisan 2015). In order to isolate the linkages among service innovation and the field of smart cities, we chose to focus our attention on Lusch and Nambisan's study (2015), as—apart from being the most recent one—it provides a clear description of the elements to be taken into account to depict service innovation activities. We propose a summary of the key concepts from Lusch and Nambisan's model in order to show the way in which the literature on service innovation and smart cities led to a merger between the two topics, which shape our theoretical background (see Table 12.1).

The first concept introduced by Lusch and Nambisan (2015) is the focus on innovation as a collaborative process that is shaped and performed by a network of actors. Starting from the notion proposed by the authors, we drew parallels with the way in which innovation takes place in cities and, in particular, in smart cities' projects. Lots of subjects are involved, and apart from the unavoidable participation of the local agencies, the network of actors is composed of citizens, local firms and sometimes a central institution or a big organisation as coordinator of the project; this happens when European Union or industry players support the development and implementation of this kind of intervention. The network logic for innovation in cities (Komninos 2006) was spread since *intelligent cities* were introduced as an ancestor of the more comprehensive concept of smart cities (Bifulco et al. 2014b). The second notion considered in Lusch and Nambisan's contribution is connected to service and to its meaning as emerging from the usage of different competences that can be beneficial to actors who are directly involved in some activities, as well as to other actors that are not directly involved in these activities. This approach can be investigated in cities' management literature because the interactions that originated from the different actors composing the network—as defined above—are aimed to enable mutual benefits (Lombardi et al. 2011) for the subjects (both directly and indirectly involved) in the urban context. The different competences needed to carry out the activities usually emerge from several actors, as is conceptualised in certain innovation models that are commonly connected to social challenges. The contributions on this topic emerged with an incremental logic, and the first was the "Triple Helix" (Etzkowitz and Leydesdorff 2002), which focused on innovation within the triadic relation among university, industry and government. The model was upgraded when the "Quadruple Helix" came out (Carayannis and Campbell 2009) to take into account the "media-based and culture-based public", moving towards the knowledge integration perspective. Finally, the "Quintuple Helix" (Carayannis et al. 2012) was introduced to incorporate the natural environment of society, leading to an approach that is closer to service innovation.

The third concept emerging from Lusch and Nambisan's study consists of two approaches that are connected to resources: firstly, liquefaction as the opportunity to extract knowledge from its primary source and to recombine it with other resources in order to create innovation opportunities, especially thanks to ICT as a tool to allow a cognitive approach and new social connections, and, secondly, density as the best combination of resources (Normann 2001), which have to be mobilised in a given context with effectiveness and efficiency. Similar considerations have been made in smart city literature when knowledge is created through the usage of tools and in online spaces (Komninos 2006) and even when new services for citizens and other actors in cities can be conceived and provided for (Lee et al. 2013), thanks to cooperation in urban ecosystems.

The last concept among the four proposed by Lusch and Nambisan is resource integration as a requisite to innovate, wherein actors are expected to combine the resources owned to let value emerge, and in order to support the creation of new ones. This phenomenon can be easily described in relation to cities, first of all because of the wide range of actors shaping a metropolitan or urban context. The integration in contexts such as cities takes place among services, infrastructure, systems, technologies and capabilities (Lee et al. 2013), and this mix creates a set of opportunities to enhance cities' development.

Several authors have already linked innovation literature and smart cities, as this kind of projects is totally built around innovation and in particular in innovating service provision in urban contexts. These authors have addressed innovation from a general perspective, without focusing on a specific model or set of actions. Moreover, innovation is considered key to solving complex issues in cities' management (Lu et al. 2015) and is also used in defining a smart city, such as when Nam and Pardo (2011) had underlined the necessity of creating ties among different kinds of innovation. Similarly, Schaffers et al. (2011) provided a focus on open innovation to offer services based on ICT and to involve the whole range of stakeholders in the city. Similarly, new technologies play a crucial role in supporting the generation of several contributions from stakeholders, starting from enabling dialogue among the several actors (Roy 2005).

When considering the above-described proposals, and taking into account the contributions to innovation in cities' management presented in the previous literature, it is possible to describe innovative approaches in urban contexts from a broader and more theoretical point of view; in particular, the usage of key concepts contained in Lusch and Nambisan's (2015) model is useful to depict actors' roles, the set of resources and the way they can be used and the expected results in terms of new resources.

12.3 Objectives and Method

From a theoretical point of view, links between smart cities and service innovation literature have been introduced by leveraging smart cities as the outcome of an innovation process, based on services and aiming at improving the quality of life for citizens and all other actors interacting with a city context. In order to go further in the analysis, we investigate the emerging key elements by looking at what organisations are actually doing in cities, as there are currently no studies to describe the empirical evidence through a theoretical model. We parallel the elements described above from the theoretical background with evidences arising from the official reports issued by firms that play a relevant role in the set of activities leading to smart contexts. The focus is on organisations as pivotal actors in complex systems such as cities; due to this, reports of the activities they have carried out are a suitable source of information to depict both perspectives and ways in which they perform the planned activities.

Empirical evidences are useful to depict these aspects in all phases composing the innovation process and to highlight its key elements through a content analysis (Berelson, 1952) performed using NVivo (version 10.1) on official reports issued by the whole range of organisations providing smart services. It is important to depict scholars' ideas about cities as suitable contexts to investigate innovation in service studies, because in smart cities' projects, there is the simultaneous presence of a leading firm setting the model for innovation and of a wide set of actors supporting it, thereby creating "two main patterns of sequence" (Perks et al. 2012). The role of the firm leading the project is pivotal (Yigitcanlar et al. 2008) in smart city projects, and firms' approach to innovation is considered as the core of the investigation presented in this research.

The empirical context of the analysis was shaped thanks to two criteria related to firms acting in this continuously evolving industry. First of all, a list of firms currently active in smartisation processes was needed; next, a number of documents were selected to create a dataset to be used for the content analysis. The list of firms is frequently updated, and at the time of the analysis presented here, 30 were considered players in smart city industries within a study recently published by Navigant Research (2014). The list of firms represents the cornerstone from which to begin the data collection through official sources available online and to shape a dataset based on smart projects as units of analysis.

Different approaches suggested and implemented by the several corporations acting in smartisation processes around the world can be investigated through official reports. We derived a dataset of 90 documents, three for each of the 30 industry players. During the data collection, the top three reports for each firm were taken into account to build the dataset; the number of documents chosen for each firm depended on the maximum number of reports available for some of the "followers and newcomers" (i.e. latecomers to the market and organisations with low competitive advantage). Moreover, the choice of the same quantity of documents for each firm represents a common approach to carrying out content analysis (Ngwenyama and Nielsen 2003; Campbell and Cornelia Beck 2004).

The data collection was performed through a simple query using the Google search engine, by combining the name of the firm and the words *smart* and *city*. The worldwide scenario connected to smart cities' projects is composed of firms with different backgrounds, and the source we used (Navigant Research 2014, 2Q) provides a clear distinction among these firms on the basis of their position in this emerging business and their core strategy, maturity and number of projects carried out. In particular, it was possible to identify two categories: "leaders" and "followers and newcomers". The "leaders" consist of nine top firms. These organisations are considered leaders in the business context of smart projects as their models are strongly accepted and implemented; moreover, they have proposed guidelines, thus inspiring other firms as well—namely, the "followers and newcomers".

Table 12.2 provides a short brief on these companies in order to describe their backgrounds and some of their projects carried out around the world. It is necessary to underline how different companies can even act in the same city context, due to the high level of specialisation needed for some specific goals.

By developing the description of smart service providers, it is useful to bear in mind that a number of firms have started to compete in this emerging business and improve their conditions; these organisations, which are classifiable as followers, have to fill gaps due to their lower competitive advantage, but are often highly specialised in particular features that are useful in the smartisation process. Finally, the newcomers—viz. the firms that have appeared in this context only recently,

Table 12.2 Key information about the first group of firms ("leaders") among smart service providers

Firm	Core business	Main smart city projects
Accenture	Management consulting	Baerum, Dornbirn, Stuttgart
Cisco Systems	Networking	Kansas City, Las Vegas, Songdo
Hitachi	Information and telecommunication systems	Guangzhou, Malaga, Yokohama
IBM	Technology and consulting	Christchurch, Dubuque, Honolulu
Microsoft	Computer software and electronics	Auckland, Hamburg, Philadelphia
SAP	Enterprises software and services	Masdar, Melbourne, Rio de Janeiro
Schneider Electric	Energy management	Beijing, Dallas, Lyon
Siemens	Engineering	Dubai, Vienna, York
Toshiba	Electronics	Bristol, Kawasaki, Tianjin

sometimes due to their limited specialisation—have to complete using their current offers in order to propose more complete solutions and start competing with the other firms. The firms shaping this second group (see Table 12.3) operate as smart specialised service providers—such as Cityzenith and Urbiotica—while others were followers as they already operate in other industries and recently decided to broaden their activities to smart contexts.

The second group confirmed the possibility of the contemporary presence of more than one firm supporting smart interventions in a single city, and it is even more frequent for these firms to partner with some of the leaders due to the formers' high level of specialisation. Regarding followers and newcomers, even though they started acting as smart service providers recently, it did not affect the data collection as reports were already available.

The documents collected are all from official sources, such as the companies' official websites, the websites of local agencies supported by the company or supranational agencies' websites that have partnered in initiatives through funding and then published the results. The length of reports ranges from 22 to 166 pages, and the issue periods are all from 2011 to 2014 and refer to projects started after 2010. Moreover, all reports investigated are written in the same language, since during the data collection only documents prepared in English were chosen.

The software used for the analysis of the datasets—viz. the two sub-datasets created—is NVivo, which is widely used (Bazeley and Jackson 2013; Krippendorff 2012) due to the range of options it offers for document investigation. A preliminary analysis was conducted to verify that the documents had been correctly uploaded into the dataset, after which a two-level analysis took place. The "word frequency" query started the analysis and provided the chance to define the most important topics considered in the two sub-datasets, namely, the 50 most cited words; the software used in the investigation provided two elements when reporting

Table 12.3 Key information about the second group of firms in smart service industry

Firm	Core business	Main smart city projects
ABB Tropos	Communication systems	Corpus Christi, Oklahoma, Ponca City
AGT International	Information management	Hamburg, Henan, Singapore
Atos	Digital services	Birmingham, Grenoble, Santander
Autodesk	Engineering software	Chicago, Mumbai, San Rafael
Capgemini	Consulting	Abu Dhabi, Oslo, Paris
Cityzenith	Software for smart cities	Amsterdam, Barcelona, San Francisco
Esri	Geographic information system	Cape Town, La Coruña, Phoenix
Firetide	Wireless network	Busan, Drogheda, Scottsdale
Honeywell International	Engineering	Malta, Shanghai, Zapopan Jalisco
Huawei	ICT solutions	Anshun, Langfang, Shenzhen
Itron	Communication networks	Charlotte, Detroit, San Diego
KT Corp	Telecommunications	Dongtan, Gangnam, Jeju
Libelium	Smart cities and M2M (machine-to-machine) platform	Belgrade, Salamanca, San Mateo
Living Plan IT	Software for sustainable cities	Greenwich, Oporto, Paredes
Oracle	Hardware and software	Denver, Lisbon, New York
Orange Group	Telecommunications	Beirut, Doha, Riyadh
OSI Soft	Data management	Donald, Pittsboro, San Leandro
Silver Spring Networks	Smart grid	Copenhagen, Sacramento, San Antonio
Urbiotica	Smart city solutions	Figueres, Hospitalet, Nice
Verizon Communications	Telecommunications	Atlanta, Holyoke, Seoul
Worldsensing	Wireless sensors	Grudziadz, Moscow, Zaragoza

results on word frequency, viz. the absolute frequency of each topic and the weighted percentage related to each. Generally, these two ways of representing results are the same if the topics are looked for one by one, but if the query is performed on "stemmed words", namely, semantic areas grouped automatically by the software to consider simultaneously verbs, singular nouns, plural nouns and adjectives linked to the same root, the "weighted percentage" can be different from the general frequency.

This difference is due to the presence of some words in more than one group of stemmed words. The second level of investigation was "cluster analysis", which led to a focus on the linkages among the most frequent topics that emerged. NVivo enabled us to measure the suitability of using two topics together through statistics-based indices; indeed, the software offers three different ways to determine this probability, namely, Pearson correlation coefficient, Jaccard's coefficient and Sørensen's coefficient. In this research, Jaccard's coefficient is used as it is the most widely applied and because it is the most suitable in comparison to the other two indices, as suggested by several scholars (Iodice et al. 2012). In order to give the highest reliability to the results presented here, we chose to take into consideration only the ties among topics with a Jaccard's coefficient of between 0.9 and 1 (with 1 as the maximum achievable level of likelihood on a range from -1 to $+1$).

12.4 Findings

The analysis of the topics and of the relationships among them enabled us to depict different aspects of the whole industry and of the two different sub-datasets. In particular, the two parts of the entire dataset enabled us to understand differences among the firms, and the elements highlighted are useful to provide several insights into innovation in cities' management, leading to considerations about the model suggested by Lusch and Nambisan (2015) and chosen as our starting point. In order to be aligned with the chosen model, the findings will be presented in four parts, as each represents one of the key elements of service innovation: innovation, service, liquefaction and density and resource integration.

12.4.1 Innovation

The most frequent topics that emerged from the content analysis are related to representation of the innovation networks and are quite numerous, but they emerge in different ways by comparing the nine leader firms and the followers and newcomers. When analysing top firms, "systems" and "systemic approach" are the most common ways to depict how relationships proceed in smart city projects, while the other 21 organisations use "systems" as a "second best" due to their strong preference for the use of "network". Other relevant items emerging from the results are "business" and "innovative", and the usage of these words is common to the two datasets, while the top nine firms prefer to use "grids" when referring to the group of actors they created together with local agencies, local community and other businesses. To sum up, the words cited above are the most used in the whole list, and this result is useful to underline how firms approach the innovation they carry out when transforming cities into smart cities. The necessity of creating relationships in the tangled web of actors and to depict them as a system, a network or a grid is a key element in defining the path towards innovation in cities' management. Finally, in order to ensure that only words with meanings pertinent to this investigation were considered (i.e. to avoid proper nouns), cases in which "systems" and "networks" were used as part of the names of the organisations were purged from the analysis by adding them to the so-called stop list, viz. the list of occurrences not to be taken into account when determining the "word frequency" and "cluster analysis".

The above considerations are confirmed by "cluster analysis", representing the linkages among the words cited and the others shaping the set of the 50 most cited topics. In the schema below, the evidence is represented for both sub-datasets, and the linkages refer only to the words cited above.

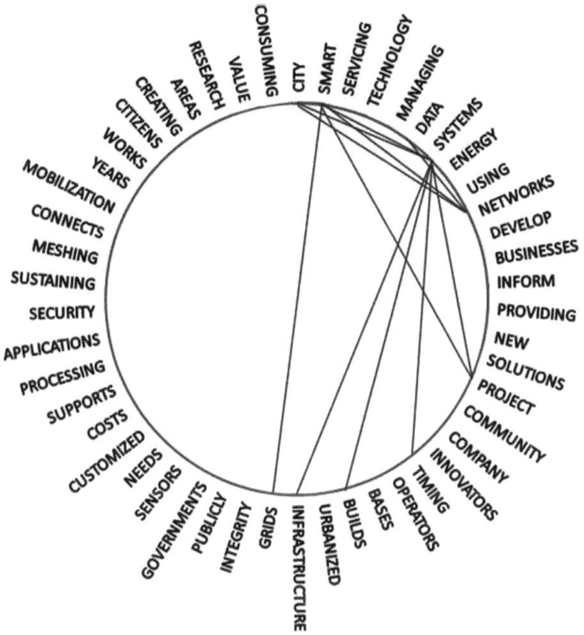

The first series of ties emerges for the group with the word "system" as a stem and is mostly linked to "infrastructure", "time", "city" and "project". The top firms prefer to introduce the notion of "system" together with "project" and "building", and these two findings are useful to underline the approach to smart cities' projects through a systemic view and to look at the "systems" shaped around—and inside—cities. In the second sub-dataset, the organisations reported a linkage between "system" and "infrastructures" to frame the set of actors as an intangible infrastructure to leverage in order to reach the expected aims; moreover, the "systems" are defined as "smart" to better adjust them to the approach to be used when managing actors towards the planned goals and activities.

As emerged from the "word frequency" step, the top nine firms revealed a stronger focus on a systemic approach; this is also revealed via the notion of "networks". This term is less used here than in the second sub-dataset, but the linkages are relevant for top firms and concern "data", "city" and the "smart approach"; these ties portray data as an element that is useful to groups of actors in networks and even as an aim to be reached by the network due to the data integration to be performed among actors, as will be revealed by the results in relation to the words used on the role of technology in resource and service integration.

12.4.2 Service

The results on the most frequent topics highlighted the same words for both sub-datasets, and this provides unique evidence of a similar approach between the two groups of firms. "Service" is the most used word in the whole dataset and leads the cluster of stemmed words such as "servicing", "serviced", "service-based approach", "service vision" and so on. The action of "providing" is another common word within the analysis, as it represents the key action to connect all actors within "smart services". Finally, another important stem is "solution", as a goal and as something to be provided to improve the quality of life for citizens.

The cluster analysis (see diagram below) shows similar results for both sub-datasets and is useful for depicting the linkages among the three elements and with the other relevant issues; the notion of "service" is linked to "providing" for the same reason stated above and has a direct and strong connection to "smart" to better depict the need for smarter services. Even the tie with the notion of "solution" has a connection with the stem "smart", but some differences emerge due to the linkage between "solution" and "management". Indeed the "management of solutions" is particularly stressed in the official reports, unlike in the case of "management of services", and this is relevant in describing the complexity of the approach to solutions compared to services, as the former are related to an approach based on contributions by different actors shaping the innovation systems, while the latter are considered from the perspective of a single organisation or local agency providing something to citizens or to another direct stakeholder in the territorial context considered. Finally, "providing"—apart from being strictly linked to "service"—is connected to "management", which highlights the necessity of monitoring the whole process built to convey services and solutions to citizens.

12.4.3 Liquefaction and Density

The findings about the most frequent topics related to resources and knowledge, as key elements in the innovation process, are numerous, and several differences emerge when comparing the results of the two sub-datasets. The role of information is crucial in this part of the analysis and in the terms that emerge, as confirmed by "data", the most relevant expression in relation to liquefaction and density. "Data" plays a relevant role in city management as this represents both a starting point of strategic analysis performed to define the interventions needed and something to leverage together with the other actors in the decision-making process, as already stated in the first part of the analysis. Another frequently cited notion is "information", which is connected with "data" and is at the core of the relationships among actors partnering in the participative decisions. This is confirmed by the following relevant stem, "community", as the set of actors under this term are recipients of information in order to activate dialogue to favour participation with public agencies and businesses in planning and deploying better services for the local context. Moreover, the fourth relevant element is represented by the word "infrastructure', which is used in the sense of interactive platforms that favour all stakeholders' involvement and participation. It is interesting to note the role of this last finding, as has already been discussed when investigating innovation in the previous lines, but in a different sense. In this case—with a direct impact on the weighted percentage—the usage of the word is slightly dissimilar in the two sub-datasets, even though the general perspective of linking all the actors of the context is confirmed again. Finally, it is necessary to highlight the different usage of "infrastructure" in the two sub-datasets, as the firms shaping the second group use it with a lower frequency. This choice depends on their preference for the notion of "applications", as they used it in the same way, but they aim to underline their specialisation due to a gap in the competition-based perspective.

The words described above have been subjected—as in the previous cases—to cluster analysis. In the following diagram, the linkages are shown for both sub-datasets.

The relationships that emerged enable the description of the context in which the most relevant notions are used.

First of all, the stem "data" has just one connection, with high likelihood only in the sub-dataset of the nine leader organisations; this linkage is with the word "network", and the findings arising from the documents can be easily interpreted as the creation of a "network of data", namely, as the necessity to spread data inside the network of actors within the local context to favour all the connected processes. The issue of the so-called big data has huge relevance in smart city projects in different scenarios, as disseminating data is useful to involve people, to acquire new information from them and to have a benchmark when measuring impacts. "Information" is relevant as well, and more linkages emerge in the sub-dataset of the nine leader firms; "management" of the information is the strongest tie as coordinated activity is needed for acquiring, spreading and leveraging information. The linkage between "information" and "project" confirms this interpretation, as the former is the key element in planning and deploying the latter. Another relevant way to detail how the process of data dissemination proceeds depends on another connection that emerged, namely, that between "community" and "development", as the official documents reported on the relevance of improving and fostering "community" as physical and virtual spaces in which information can be shared to favour participation.

The relationships concerning "infrastructure" are the common point in the two sub-datasets, where the strongest tie is between "infrastructure" and "management" as all firms pay attention to the necessity of performing activities that aim to manage infrastructure in order to take care of the actors involved in the usage of data, in the proposal of solutions and in the decision-making process. Another common approach exists when analysing the ties related to "infrastructure", as this latter is linked to "system" due to the role performed by infrastructure in combining actors from the context of intervention.

12.4.4 Resource Integration

The most common issues included in the official reports investigated in relating to "resource integration" are "integration" and "technology". "Integration" is the most common stem in the set of words analysed in this part, and the idea of integrating resources and actors is considered in the same way in the two sub-datasets; that is, it is relevant for all firms, including "leaders" and others. The latter organisations are even more oriented to the necessity of combining resources, as they used the stem "meshing" many times, in order to underline the necessity of performing integration through online platforms and favouring the combination of resources to create new ones. In addition, the approach to technology is confirmed in both sub-datasets as the stem "technology" is very common; its usage is relevant due to the complexity of the integration and because of the huge number of actors simultaneously involved in the integration process. Moreover, technology is embedded in the notion of smart interventions and through the transformations cities will benefit from getting smarter. One more word among the top cited is "project", as the result of resource integration in terms of elements combined, activities planned and goals to be reached through the usage of technology and, at the same time, the interventions of several actors.

The linkages related to the words introduced above support the in-depth description of the resource integration process in city management towards becoming smart. "Integration" is more relevant in the sub-dataset of the nine leader firms and is linked to the stem "operations", as in the reports on the different phases of the resource integration process that are often described in detail, in order to highlight their relevance and their impacts on the whole process that leads city services towards a smarter condition. Another relevant element, and a difference between the two sub-datasets, is represented by the use of "project"; this is a very common term in the official reports of the "leaders", and lots of relationship with a Jaccard's index of higher than 0.9 emerged. One of these is between "project" and "data", as the role of data is relevant in all different phases of the project, from planning to deployment, as well as during monitoring. Moreover, the stem "project" is closely linked to "infrastructure", as the (virtual) place in which actors interact to perform the different tasks under the "management" of the main actor(s); this last statement is confirmed by the strength of the linkage with the term "management".

In the diagram on the side, we sum up the findings by comparing the results achieved in relation to the two sub-datasets, namely, the one consisting of the nine leader firms and that composed of the followers and newcomers.

A partial common point emerging from the two sub-datasets is related to the word "technology", as it has relevant ties for all firms; however, these ties slightly differ; from the top firms' perspective, "technology" is linked to the stem "business" as there is a continuous call for integration to favour the development of projects. The firms from the second sub-dataset focus on the role of technology as the linkages that emerged are with the stems "smart" and "city". In the first case, the technology is also expected to be smart—that is, smart interventions are favoured that provide all actors with the tools needed to enable interventions oriented towards smart development, as stated regarding the findings related to liquefaction and density of resources. In the second case, the city has to consider technology in the same meaning in which the term "smart" is coupled with "city"—viz. by leveraging technology to transform a city in a smarter context.

12.5 Conclusions and Implications

Starting from the relationships we found between the service innovation conceptualisation by Lusch and Nambisan (2015) and smart cities' scholars' investigations as depicted in the theoretical background, we moved towards a deeper investigation of the key elements by looking at the way in which they emerge from the reports issued by smart service providers. The analysis took place by investigating both guidelines and reports that are useful to support the smartisation processes in the empirical evidence taken into account.

Innovation in smart city projects is approached through an ecosystem perspective, as stated in the literature review (Cosgrave et al. 2013; Lusch and Nambisan 2015) and mainly through an innovation ecosystem path, as the reports present the creation of relationships in the wide range of actors as a necessary step to achieve a set of resources and knowledge that are useful to plan and deploy interventions towards the aim of transforming a city into a smart city.

The logic of projects based on sharing resources and skills as they emerge during the projects themselves is stressed thanks to the role of technology, as it enables the achievement of methods to favour collaboration among partners (Akaka and Vargo 2014); the infrastructures to be conceived to support the shared activities are virtual spaces in which firms act together with other stakeholders, as happens in living labs (Schaffers et al. 2011; Cosgrave et al. 2013), leading to the creation of platforms (Lusch and Nambisan 2015), which is a notion that is taken into account both in literature and in the projects shaping the dataset used in this research. The platforms for smart cities have to be intended as virtual structures that favour the collection, sharing and elaboration of data. More specifically, platforms are ICT systems that combine different elements, such as software, e-services, data and meta-data, to favour continuous improvement in city management, and directly connect stakeholders, towards the continuous improvement of the quality of services provided in an urban context. The multi-stakeholder perspective of innovation takes place through systems of actors and is a necessary step, as recalled in relation to resource integration (Lusch and Nambisan 2015), no matter what the context or which organisations are carrying out the project; the convergence of multiple contributions from different stakeholders leads to the proposal of solutions as a way to provide smart services to the actors interacting with a city, with the aim of improving the quality of life (Anttiroiko et al. 2013).

The role of the actor leading the project is strictly connected to the management of smart services in a complex context and to convey innovative solutions towards the planned aims; involving actors that carry fundamental resources and knowledge from different backgrounds (Rubalcaba et al. 2012) is one of the main tasks. This statement depends on the role of resources and knowledge, as they have to be shared in the innovation ecosystem that emerges in the city context. The technology favours the strategic necessity of creating awareness about the role resources can play in projects and on how all stakeholders can be allowed to leverage resources to contribute to the achievement of the expected goals. Thus, sharing resources and knowledge in cities that aim to become smart is a way to create better solutions for stakeholders, as they are the beneficiaries of smart activities, even if at the same time they contribute to creating the set of elements to propose services and support a bottom-up logic (Coe et al. 2001) within the decision-making process, due to the knowledge they hold (Gallego et al. 2013).

The role of pivotal actors in smart city projects begins by involving actors and collecting resources and knowledge, but in order to propose smart solutions it is necessary to integrate the different contributions, and technology is once again relevant to reach this aim; indeed, technological tools can act as a springboard to enable stakeholders' active participation in this process. As a consequence, the roles played by all actors are several, as they started as instigators of needs to be satisfied (Magnusson 2009), and then became carriers of resources to propose smarter solutions (Bifulco et al. 2014b; Lee et al. 2013), in addition to supporting the decision-making process (Coe et al. 2001) and resource sharing and integration, and ultimately as the ones perceiving services and sustaining the evaluation process (Marsal-Llacuna et al. 2014) towards further improvements. In this complex scenario, another element is relevant, namely, once again, the platform: when looking at projects, organisations commonly refer to the platform as infrastructure or applications, but its centrality is underlined because of the support given to data collection and sharing. Moreover, as new data are collected thanks to the smart services provided, the dataset is continuously enlarged, leading to greater potential results achievable through the technology embedded in the platform. In this way, the notion of platform acquires even more relevance than highlighted by Lusch and Nambisan (2015) (Table 12.4).

When looking at the collaborative approach to innovation, the different backgrounds of each actor can be considered as a strength and can favour the development of e-services for cities, by confirming the opportunity to apply service innovation even in connection with public services (Windrum and Koch 2008). Managers from both public and private organisations providing direct and indirect support to the development of smart services have to involve actors and favour the shared deployment of smart activities, from planning to evaluation. From a similar perspective, some authors have highlighted the relevance of a mix of actors in the so-called helix-based models for innovation from a more general perspective (Carayannis and Campbell 2009; Carayannis et al. 2012; Lombardi et al. 2011; Etzkowitz and Leydesdorff 2002).

Table 12.4 Main elements from the two sub-datasets

Key concept	Leaders	Followers and newcomers	Findings emerging from smart cities' projects
Networking-based innovation	Relationships are mainly represented on the basis of a systemic approach	Relationships are mainly represented on the basis of a network approach	Firms are suggested not to "start developing end-to-end solutions until there is a clear understanding of who will pay for and operate them. Rather, [they should] develop processes to enable open and collaborative innovation with potential clients, NGOs [non-governmental organizations], and academics to ensure solutions that are both functional and economically feasible" (source: Falconer and Mitchell 2012, p. 10) "Geodesign will be the key framework for conceptualising and planning for smart cities; it will assist at every stage from project conceptualising to site-analysis, design specifications, stakeholder participation and collaboration, design creation, simulation and evaluation" (source: ESRI 2014: 4)
	The focus is on the project as centre of gravity, data	Organisations basically focus their interventions on infrastructures	
Services as application of specialised competence	Service provision and service characterisation as "smart" are the most frequent issues from the analysis and they are common to both sub-datasets Moreover, the creation of a "solution" and the "service management" are some more interesting topics from the analysis		"A new professional community—the Urban Systems Collaborative—has formed to foster mutual learning among members of the architecture, planning, engineering, transportation, utilities, information technology, operations research, social sciences, geography and environmental science, public finance and policy, and

(continued)

Table 12.4 (continued)

Key concept	Leaders	Followers and newcomers	Findings emerging from smart cities' projects
			communications profession" (source: Harrison and Donnelly 2012, p. 1) "Our role has always been to help our customers communicate, interact and team up more. We focus on the benefits for people, rather than the technology; we deliver reliable services and create simple experiences" (source: Orange 2014: 3)
Generativity unleashed	Data are considered relevant for all projects and are the core element of the relationships among actors	Data are pivotal in supporting the planning and deployment of projects and are considered from a technological point of view, namely, with reference to big data	"Sharing innovation both at the policy and program level and allowing early use of new approaches will accelerate a new path. This cooperation can be supported through the creation of knowledge- and data- sharing platforms among cities; and it should also allow businesses and citizens to play a greater role in urban planning" (source: Accenture 2013: 8) "The two sides will work together to improve the integration capabilities of each other's security software platforms, actively explore ways to enhance product value and to improve the service network through resource sharing and win–win cooperation" (source: Honeywell 2012: 1)
	The infrastructures, intended both in a physical and in a virtual meaning	Infrastructures are quite often cited and are taken into account as physical means, while the virtual ones are presented as applications	

(continued)

Table 12.4 (continued)

Key concept	Leaders	Followers and newcomers	Findings emerging from smart cities' projects
Resource integration	Resource integration takes place to combine operations to achieve overall development	Resource integration is a key issue in implementing a smart city	"Innovative technology, tools and solutions from Microsoft and our partners can help city leaders analyse, evaluate and tie together the data they need to see the big picture. This integration can help authorities make faster, more informed policy, budget and management decisions" (source: Touch 2013) The company provided to the local government "a full series of products, including server, storage device, switch, security device, firewall, and cloud platform. Professional services, such as planning, service transfer using cloud technology, and O&M [operations and management] enable resource integration at multiple levels, unified service egress, and centralized O&M" (source: Huawei 2013)
	Projects are frequently cited and are based on data to be obtained and shared, in order to set up a proper infrastructure	Projects are mainly considered as a general issue to describe the whole set of actions to be performed, without specific reference to particular goals	
	Technology is considered as the glue to favour resource integration among the actors	Technological features are taken into account since they are smart and have to support the deployment of projects	

The perspective on service innovation is considered as a means to provide benefits in a new way to the wide range of actors that are directly and indirectly taken into account; this idea is also shared by industry players, when they highlight distinct contributions and outcomes for different stakeholders involved in shaping a process of mutual learning. Such a process is also useful in relation to the creation of strategies with a higher potential to improve cities from a perspective similar to platforms, as they are deputed to integrate different services in a sustainable approach. Firms are also aligned regarding the definition of the roles of resources; resources usage leads to relevant outcomes, especially when recalling the advantages offered by advanced ICTs to enhance activities for cities' development and to allow cooperation and empower actors' roles in smart city projects. This focus supports our contribution to answering to the call for research made by Rubalcaba et al. (2012), who underlined the necessity to better depict how actors act when supporting activities framed in a service innovation perspective. Our research highlights different approaches by firms, depending on both internal and external factors (namely, the cities where the projects took—or are taking—place).

Regarding specific characteristics, leaders in smart services business have a focus based on relationships with other actors and on technologies to be used in achieving the expected results. The followers and newcomers emerged as organisations that pay greater attention to technology in a more general sense and using a network approach to the set of actors supporting the process towards achievement of a smart city. Moreover, capabilities are considered as the outcome of resource integration in the notion proposed by the firms investigated, as the process of integration takes place when physical and digital resources with citizens' and other stakeholders' interventions can support the creation of comprehensive solutions. Once more, the notion of platform has to be further brought into focus, because it plays a central role in favouring development of a suitable set of data and improving the way services can be replanned, thanks to this data. A recursive path begins from the platform as service innovation interventions performed in smart cities will benefit from the continuous flow of new data from service provision, leading to better knowledge of how services are provided and how they are linked to one another. This last issue represents an avenue for further research, as detailed data are needed to understand how the services are linked to one another and which elements influence service provision and related experience. Moreover, new insights into the actors themselves can emerge from a deeper analysis of the relationships among firms, as they are often involved together in the deployment of projects in the same city, due to their high specialisation in different areas. New insights are expected to emerge as soon as a wider range of reports is made available, and over a larger time span a longitudinal approach will be suitable to provide a comparison of the approaches adopted by smart service providers to innovation.

References

Accenture. (2013). *Creating prosperous and livable Chinese cities: The New Resource Economy City Index Report*. Retrieved October 31, 2014, from www.accenture.com

Akaka, M. A., & Vargo, S. L. (2014). Technology as an operant resource in service (eco) systems. *Information Systems and e-Business Management, 12*(3), 367–384.

Alam, I. (2006). Removing the fuzziness from the fuzzy front-end of service innovations through customer interactions. *Industrial Marketing Management, 35*(4), 468–480.

Anttiroiko, A. V., Valkama, P., & Bailey, S. J. (2013). Smart cities in the new service economy: Building platforms for smart services. *AI & Society, 29*(3), 323–334.

Batty, M., Axhausen, K. W., Giannotti, F., Pozdnoukhov, A., Bazzani, A., Wachowicz, M., et al. (2012). Smart cities of the future. *European Physical Journal - Special Topics, 214*(1), 481–518.

Bazeley, P., & Jackson, L. (2013). *Qualitative data analysis with NVivo*. London: Sage Publications.

Bélissent, J. (2010). *Getting clever about smart cities: New opportunities require new business models*. Forrester. Retrieved October 31, 2014, from www.forrester.com

Berelson, B. (1952). *Content analysis in communications research*. New York: Free Press.
Berry, L. L., Shankar, V., Parish, J. T., Cadwallader, S., & Dotzel, T. (2006). Creating new markets through service innovation. *MIT Sloan Management Review, 47*(2), 56–63.
Bifulco, F., Tregua, M., & Amitrano, C. C. (2014a). Driving smartization through intelligent transport. *Chinese Business Review, 13*(4), 243–259.
Bifulco, F., Tregua, M., & Amitrano, C. C. (2014b). Smart cities and innovation: A multi-stakeholder perspective. *Journal of Management and Marketing, 2*(1), 27–33.
Campbell, D., & Cornelia Beck, A. (2004). Answering allegations: The use of the corporate website for restorative ethical and social disclosure. *Business Ethics: A European Review, 13*(2-3), 100–116.
Carayannis, E. G., Barth, T. R., & Campbell, D. F. J. (2012). The quintuple helix innovation model: Global warming as a challenge and driver for innovation. *Journal of Innovation and Entrepreneurship, 1*(2), 1–12.
Carayannis, E. G., & Campbell, D. F. J. (2009). Mode 3 and Quadruple Helix: Toward a 21st century fractal innovation ecosystem. *International Journal of Technology Management, 46*(3/4), 201–234.
Chesbrough, H. (2011). *Open services innovation: Rethinking your business to grow and compete in a new era*. New York: Wiley.
Coe, A., Paquet, G., & Roy, J. (2001). E-governance and smart communities: A social learning challenge. *Social Science Computer Review, 19*(1), 80–93.
Cosgrave, E., Arbuthnot, K., & Tryfonas, T. (2013). Living labs, innovation districts and information marketplaces: A system approach for smart cities. *Procedia Computer Science, 16*, 668–677.
Deakin, M. (2009). The Intel Cities community of practice: The eGov services model for socially-inclusive and participatory urban regeneration programmes. In C. Riddeck (Ed.), *Research strategies for eGovernment service adoption* (pp. 83–104). Hershey: Idea Group Publishing.
ESRI. (2014). Smart cities will open up multiple growth opportunities. Infrastructure Today. *Special Supplement Esri India*, August. Retrieved on July 28, 2015, from www.esriindia.com
Etzkowitz, H., & Leydesdorff, L. (2002). *Universities and the global knowledge economy NIP: A triple helix of university-industry-government relations*. New York: Continuum International.
Falconer, G., & Mitchell, S. [CISCO] (2012). *Smart city framework*. Retrieved July 28, 2015, from www.cisco.com
Gallego, L., Hipp, C., & Rubalcaba, L. (2013). Organizational innovation in small European firms: A multidimensional approach. *International Small Business Journal, 31*(5), 563–579.
Goleman, D. (2006). *Social intelligence: The new science of human relationships*. New York: Bantam Books.
Harrison, C., & Donnelly, I. A. (2012). A theory of smart cities. In *Proceedings of the 55th Annual Meeting of the International Society for the Systems Sciences*. The University of Hull, Hull, UK, 17–22 July 2011.
Honeywell. (2012). *Honeywell, Shanghai Yanhua form strategic alliance, enhancing integrated security solutions to build smart cities in India*. Retrieved on October 31, 2014, from www.honeywell.com.cn
Huawei (2013). *Huawei smart city solution*. Retrieved October 31, 2014, from www.enterprise.huawei.com
Iodice, D'Enza, A., & Palumbo, F. (2012). Iterative factor clustering of binary data. *Computational Statistics, 28*(2), 789–807.
Ishida, T., & Isbister, K. (2000). *Digital cities: Technologies, experiences, and future perspectives*. New York: Springer.
Komninos, N. (2006). The architecture of intelligent cities. Intelligent Environments 06. *Institution of Engineering and Technology, 1*, 13–20.
Komninos, N., Passas, I., Tarani, P., & Tsarchopoulos, P. (2007). Four platforms for intelligent cities: Linking virtual spaces and collaborative innovation. In *Proceedings of the 3rd IET International Conference on Intelligent Environments*. Ulm, Germany, 24–25 September 2007, pp. 486–493.

Krippendorff, K. (2012). *Content analysis: An introduction to its methodology* (3rd ed.). Thousand Oaks: Sage Publications.
Lee, J. H., Hancock, M. G., & Hu, M. C. (2013). Towards an effective framework for building smart cities: Lessons from Seoul and San Francisco. *Technology Forecasting and Social Change, 89*, 80–99.
Lombardi, P., Giordano, S., Caragliu, A., Del Bo, C., Deakin, M., Nijkamp, P., et al. (2011). *An advanced triple-helix network model for smart cities performance*. Research Memorandum, University of Amsterdam. Retrieved October 31, 2014, from www.igi-global.com
Lu, D., Tian, Y., Liu, V. Y., & Zhang, Y. (2015). The performance of the smart cities in China—A comparative study by means of self-organizing maps and social networks analysis. *Sustainability, 7*(6), 7604–7621.
Lukas, B. A., & Ferrell, O. C. (2000). The effect of market orientation on product innovation. *Journal of the Academy of Marketing Science, 28*(2), 239–247.
Lusch, R. F., & Nambisan, S. (2015). Service innovation: A service-dominant logic perspective. *MIS Quarterly, 39*(1), 155–175.
Magnusson, P. R. (2009). Exploring the contributions of involving ordinary users in ideation of technology-based services. *Journal of Product Innovation Management, 26*(5), 578–593.
Marsal-Llacuna, M. L., Colomoer-Llinàs, J., & Meléndez-Frigola, J. (2014). Lessons in urban monitoring taken from sustainable and livable cities to better address the smart cities initiative. *Technological Forecasting and Social Change, 90*, 611–622.
Michel, S., Brown, S. W., & Gallan, A. S. (2008). An expanded and strategic view of discontinuous innovations: Deploying a service-dominant logic. *Journal of the Academy of Marketing Science, 36*(1), 54–66.
Nalbandian, J. (1999). Facilitating community, enabling democracy: New roles for local government managers. *Public Administration Review, 59*(3), 187–197.
Nam, T., & Pardo, T. A. (2011). Conceptualizing smart city with dimensions of technology, people, and institutions. In *Proceedings of the 12th Annual International Digital Government Research Conference*. College Park, MD, USA, 12–15 June 2011, pp. 282–291.
Navigant Research. (2014). *Smart cities - smart technologies and infrastructure for energy, water, transportation, buildings, and government: Business drivers, city and supplier profiles, market analysis, and forecasts*. Release date 2Q 2014. Retrieved October 31, 2014, from http://www.navigantresearch.com/research/smart-cities
Ngwenyama, O., & Nielsen, P. A. (2003). Competing values in software process improvement: An assumption analysis of CMM from an organizational culture perspective. *IEEE Transactions on Engineering Management, 50*(1), 100–112.
Normann, R. (2001). *Reframing business: When the map changes the landscape*. Chichester: Wiley.
Orange. (2014). *Smart cities, Orange testimony – Leveraging ICTs for smart sustainable cities*. Retrieved on October 31, 2014, from www.itu.int
Perks, H., Gruber, T., & Edvardsson, B. (2012). Co-creation in radical service innovation: A systematic analysis of microlevel processes. *Journal of Product Innovation Management, 29*(6), 935–951.
Roy, J. (2005). E-governance and international relations: A consideration of newly emerging capacities in a multi-level world. *Journal of Electronic Commerce Research, 6*(1), 44–55.
Rubalcaba, L., Michel, S., Sundbo, J., Brown, S. W., & Reynoso, J. (2012). Shaping, organizing, and rethinking service innovation: A multidimensional framework. *Journal of Service Management, 23*(5), 696–715.
Schaffers, H., Komninos, N., Pallot, M., Trousse, B., Nilsson, M., & Oliveira, A. (2011). Smart cities and the Future Internet: Towards cooperation frameworks for open innovation. In J. Domingue et al. (Eds.), *The Future Internet - Future Internet Assembly 2011: Achievements and technological promise* (pp. 431–446). Berlin: Springer.
Schaffers, H., & Turkama, P. (2012). Living labs for cross-border systemic innovation. *Technology Innovation Management Review, 2*(9), 25–30.

Sundbo, J. (1997). Management of innovation in services. *Service Industries Journal, 17*(3), 432–455.
Sundbo, J. (2010). The toilsome path of service innovation. The effects of the law on the low multi-task capability. In L. Gallouj & F. Djellal (Eds.), *The handbook of innovation and services* (pp. 279–300). Cheltenham: Edward Elgar.
Touch. (2013). *Smart and sustainable*. Spring issue. Retrieved October 31, 2014, from www.onwindows.com
Tregua, M., D'Auria, A., & Bifulco, F. (2015). Comparing research streams on smart city and sustainable city. *China-USA Business Review, 14*(4), 203–215.
Windrum, P., & Koch, P. (2008). *Innovation in public sector services–entrepreneurship, creativity and management*. Cheltenham: Edward Elgar.
Yigitcanlar, T., & Lönnqvist, A. (2013). Benchmarking knowledge-based urban development performance: Results from the international comparison of Helsinki. *Cities, 31*, 357–369.
Yigitcanlar, T., Velibeyoglu, K., & Martinez-Fernandez, C. (2008). Rising knowledge cities: The role of urban knowledge precincts. *Journal of Knowledge Management, 12*(5), 8–20.

Part IV
Experiences: Customer and Value

Chapter 13
Value Proposal Co-Creation in Online Community-Based Idea Contests

Carol Kelleher, Aonghus Ó. Céilleachair, Anu Helkkula, and Joe Peppard

Abstract The purpose of this chapter is to examine why and how participants co-create value proposals in online community-based idea contests (OCBICs), following an open call by service organisations for participation. More specifically, we explore participant motivations, roles, and behaviours as they co-create service ideas that emerge as value proposals. We present a multiple case study of three OCBICs in the global automotive sector. All three OCBICs studied were developed by the same platform host but were otherwise distinct in design and implementation. The findings reveal three propositions in relation to value proposal co-creation in OCBICs: (1) participants are intrinsically and extrinsically motivated to compete and collaborate to co-create value proposals in OCBICs, (2) participants iteratively adopt a number of diverse and overlapping roles when co-creating value proposals in OCBICs, and (3) participant behaviour in OCBICs involves complex negotiations of the contest rules enforced by the host organisation and the norms and values of the community. In order to optimise open service innovation, we conclude that service organisations need to provide participants in OCBICs with an appropriate combination of monetary and nonmonetary incentives and rewards, based on their motivation and expertise. This is one of the few studies to examine the co-creation of value proposals in the context of OCBICs in the global automotive sector.

C. Kelleher (✉) • A.Ó. Céilleachair
Department of Management & Marketing, University College Cork, Cork, Ireland
e-mail: carol.kelleher@ucc.ie; akella365@gmail.com

A. Helkkula
CERS Research Centre, Hanken School of Economics, Helsinki, Finland
e-mail: anu.helkkula@hanken.fi

J. Peppard
EMST European School of Management and Technology, Berlin, Germany
e-mail: j.peppard@cranfield.ac.uk

T. Russo-Spena et al. (eds.), *Innovating in Practice*,
DOI 10.1007/978-3-319-43380-6_13

13.1 Introduction

The rapid growth of collaborative technologies has greatly enhanced cooperation between service organisations and external collaborators, sparking the appearance of dynamic firm-hosted innovation communities, such as Dell's IdeaStorm, InnoCentive, NineSigma, and Threadless (Brabham 2010; Dahlander and Magnusson 2008; Di Gangi et al. 2010; Chesbrough 2011). Online community-based idea contests (OCBICs), namely, "IT-based and time-limited competitions arranged by an organization or individual calling on the general public or target group" (Adamczyk et al. 2012: 335), that facilitate open service innovation by inviting participants to use their "skills, experience and creativity to provide a solution for a particular contest challenge led by an organizer" (Bullinger et al. 2010: 29).

While service scholars have previously studied user-driven service innovation and customer co-creation in relation to new service development (Edvardsson et al. 2005; Hoyer et al. 2010), there is a paucity of studies that specifically examine open service innovation and collaborative value co-creation in the context of OCBICs, and this chapter addresses this gap. The purpose of this chapter is to examine why and how participants co-create value proposals in OCBICs. More specifically, we address participant motivations, roles, and behaviours as they co-create ideas that emerge as value proposals.

Value proposals represent potential ways for integrating resources in the form of potential market offerings. OCBICs enable service organisations to directly assemble and engage with external participants with different technological and cultural backgrounds, levels of expertise, and consumption preferences in order to co-create value proposals that customers or service organisations could not create independently. Value proposals differ from value propositions in that value propositions are market or service system offerings comprising resources that are already integrated from the service provider's perspective and offered to beneficiaries in a prescribed or commercialised form by the service organisations (Frow and Payne 2011, 2014; Kowalkowski 2011). Thus, value proposals present *potential* service applications or resource combinations, while value propositions are *existing* service offerings. In this chapter, we posit that OCBICs represent unique forms of collaborative value co-creation, where participants, who are motivated by a combination of monetary, nonmonetary, and social factors, are inspired to both compete and collaborate to co-create service ideas or potential service applications, which we term value proposals (Hutter et al. 2011).

In this chapter, we first describe the phenomenon of collaborative value co-creation in the context of OCBICs. We then discuss the motivations for and barriers to the co-creation of value proposals in such contexts. Following this, we outline and discuss the findings of three case studies comprising a detailed netnographic analysis of over 3000 participants in three OCBICs in the global automotive sector. We outline participants' motivations, roles, and behaviours in co-creating value proposals in the OCBICs studied. We conclude with three propositions for future research and discuss the implications for researchers and practitioners.

13.2 Literature Review

13.2.1 Co-creating Value Proposals in Online Community-Based Idea Contests

Service organisations can be defined as complex "value-co-creation configurations of people, technology and value propositions connecting internal and external service systems through shared information" (Maglio and Spohrer 2008: 18). Various labels have been applied to collaboration between service organisations and other actors within service systems, including "co-creation", "co-production", and "prosumption" (Kotler 1986; Prahalad and Ramaswamy 2000; Ramirez 1999; Vargo and Lusch 2008).

In an effort to develop more customer-centric value propositions, many organisations are increasingly migrating from "closed" to "open" service innovation approaches, including OCBICs (Chesbrough 2011; Di Gangi et al. 2010). In the context of this chapter, we define OCBICs as "IT-based and time-limited competitions arranged by an organization or individual calling on the general public or target group" (Adamczyk et al. 2012: 335). OCBICs are interesting examples of collaborative value co-creation through "crowd creation" (Howe 2009; Surowiecki 2004) where participants are required to use their "skills, experience and creativity to propose a solution for a particular contest challenge led by an organizer" (Bullinger et al. 2010: 29). OCBICs thereby enable service organisations to implement an open service innovation approach by stimulating and focusing the creativity and skills of participants in ways desired by the firm (Bullinger et al. 2010; Leimeister et al. 2009; Zwick et al. 2008). However, while OCBICs have become increasingly popular, there is limited research to guide theory development and practice in this area (Bullinger et al. 2010).

OCBICs typically comprise four critical elements: purpose, people, policies, and computer systems. The service organisation that hosts the OCBIC typically issues the initial open call for participation and frames the *purpose* of the competition (Sawhney et al. 2005). *People*, or what we refer to as participants, include ordinary users, lead users, and non-users of the service, enthusiasts, and employee representatives. While the host organisation develops and monitors the *policies*, including rules, protocols, and laws that guide participant behaviour, participants also perform a "self" or "community" governing roles in such communities (Libai et al. 2010). Finally, *computer systems* or collaborative technologies are required to promote a sense of togetherness and to facilitate, support, and moderate the social interaction between community members (Preece 2000). Other common characteristics of OCBICs include: a contest proposition, task/topic specificity, specification of the degree of elaboration of the service innovation idea that is required, organisational appearance, specified timeline, recruitment, incentivisation and motivation of participants, provision of toolkits, community functionality, and a rating/evaluation system for participant evaluation of the service innovation ideas contributed by others (Bullinger et al. 2010; Leimeister et al. 2009; Piller and Walcher 2006).

Facilitating value proposal co-creation is both extremely beneficial and challenging for service organisations (Füller et al. 2006; Nambisan 2002) as "customer-to-customer (c2c) co-creation is a dynamic, multi-layered process that is embedded in customers' social contexts" (Rihova et al. 2013). Service organisations need to facilitate and support community members who wish to present, discuss, provide feedback on, and rank and rate participants' service ideas and designs (Füller et al. 2010; Sawhney and Prandelli 2000; Sawhney et al. 2005). Similar to the brainstorming model (Osborn 1963), participants continually interact, engage, and refine value proposals within the OCBIC. This type of co-creative idea generation encourages participants to iteratively and collaboratively rethink and refine value proposals with other participants and to build on the ideas of others (Paulus and Yang 2000). Participants therefore learn from and contribute to the collective knowledge of the OCBIC and encourage other participants to improve, elaborate, test, and sometimes prototype new service ideas (Dolan and Matthews 1993; Hemetsberger and Reinhardt 2006; Nambisan 2002; Sawhney and Prandelli 2000).

On the more negative side, OCBICs may, from the service organisation perspective, increase uncertainty in relation to the service innovation process. New organisational processes may be required to observe and maintain quality and efficiency (Lengnick-Hall 1996). Equally, community members may resign or abdicate their role as value proposal co-creators, which has the potential to disrupt the service innovation process or to discourage other participants (Prahalad and Ramaswamy 2000). In addition, in certain contexts, participants may need to possess higher levels of service and technical knowledge in order to successfully partake in certain value co-creation roles, depending on the complexity of the tasks involved (Von Hippel 1986).

13.2.2 Motivations for Participating in Community-Based Idea Contests

By responding to an open call to participate in an OCBIC, participants actively choose to co-create value proposals with others. Participants in OCBICs emerge as ephemeral, often unstable, temporary groupings that create, share, and appropriate meanings through various rituals for the duration of the OCBIC (Cova 1997). Participants undertake social, economic, and information exchanges with other community members and the service organisation hosting the OCBIC based on a complex set of motivations that include financial, social, technological, and psychological factors (Hoyer et al. 2010).

Numerous researchers have debated the intrinsic and extrinsic motivations that drive participants to contribute to online user innovation communities, including OCBICs (Amabile 1997; Füller et al. 2010; Ryan and Deci 2000). Intrinsic motivation can be described as the performance of an activity for its inherent satisfaction or in other words "the motivation to work on something because it is interesting, involving, exciting, satisfying, or personally challenging" (Amabile 1997: 39). Intrinsic motivations, such as reciprocity, altruism, and reputation, have been found to motivate participants in OCBICs (Lindenberg 2001; Schulz 2010). Franke and Shah (2003) also observed that the desire for personal benefit can be related, not only to material reward, such as the ability to put the service innovations to personal use, but also to intrinsic motivations, such as increased self-esteem as a result of helping other members or increased reputation or status within a community. In other instances, hedonic and altruistic motivations evidenced in OCBICs include fun, intellectual stimulation; the desire to discover, innovate, and create; and the unselfish concerns of community members for the well-being of other members of the community (Schulz 2010). Extrinsic motivation, on the other hand, can be described as the independent outcomes or direct rewards that are obtained when an activity is performed and that can be influenced by service organisations (Ryan and Deci 2000). OCBIC participants may be extrinsically motivated through monetary rewards, bonuses, or promotion possibilities to participate in OCBICs (Amabile et al. 1994; Kim et al. 2008). For example, the desire to reciprocate and to receive direct compensation (such as prizes) is a possible motivator for participating in OCBICs (Leimeister et al. 2009; Ostrom and Walker 2003).

Community norms, namely, the "shared beliefs about what actions and attributes bring respect and approval (or disrespect and disapproval) from oneself and others" (Cancian 1975: 6) also influence participant co-creation of value proposals. For example, Franke and Shah (2003) observed that community norms encouraged community members to assist each other in OCBICs. In addition, community values, namely, "widely held beliefs or sentiment that some activities, relationships, feelings, or goals are to the community's identity and well-being" (Gould 1978: 133), may also impact participant behaviour in such contexts.

Kozinets et al. (2008) noted that some OCBICs seem to collectively adopt a *telo-ludic* orientation (i.e. the online community is oriented towards achieving specific collective value co-creation outcomes), such as the co-creation of value proposals in response to an open call issued by the service organisation that hosts the OCBIC. Alternatively, other OCBICs seem to display a more *communo-ludic* orientation, where value co-creation is a by-product of collaborative value co-creation between OCBIC participants (Kelleher et al. 2011; Kozinets et al. 2008). These two potentially conflicting orientations, i.e. between the commercial objectives of the service organisation hosting the OCBIC on the one hand and the social objectives of some participants on the other, may result in tensions that need to be proactively managed by the host service organisation. In addition, the seemingly contradictory competitive and collaborative nature of OCBICs can act either as a motivator or disincentive for different participants (Bullinger et al. 2010). As a result, service organisations need to design OCBICs that encourage and foster both competitive and collaborative behaviours for the duration of the OCBIC.

13.2.3 User Roles for Participating in Community-Based Idea Contests

Up until recently, service scholars have not considered the roles customers perform in relation to value co-creation, with some notable exceptions. For example, Moeller et al. (2013) identified five roles performed by customers in relation to the collaborative co-creation of value propositions with service organisations, namely, bargain hunting independent, comprehensive help seeker, engaged problem-solver, technology-savvy networker, and self-reliant customiser. In relation to the roles played by OCBIC participants in the collaborative co-creation of value proposals, participants may be regarded as *resources*, *co-creators,* and/or *end users* (Nambisan 2002), depending on their levels of experience and expertise. The role of users as resource integrators or resources for innovation has been extensively documented in the service marketing, management, and innovation literature (Hibbert et al. 2012; Von Hippel 1986).

Participants in OCBICs may focus on specific and sometimes diverse goals depending on whether they are lead users or ordinary users. Lead users, for example, might play a pivotal role in forecasting future customer requirements, as they possess the technical knowledge to create a new solution for their observed needs (Lüthje and Herstatt 2004; Von Hippel 1986). As early adopters of products and services, they are opinion leaders with the potential to promote the mass adoption of a service offering. Such participants are motivated by the promotion of monetary rewards and the competitive nature of the competition, as well as by the challenges involved in problem-solving and experimenting (Lüthje 2004; Magnusson 2009). Ordinary users, which also include creative consumers (Berthon et al. 2006), contribute creative ideas, albeit they may not always be as technically knowledgeable as those proposed by lead users (Kristensson et al. 2004). We will now present our empirical study, where we explore and discuss participants' roles, motivations, and behaviours in the collaborative co-creation of value proposals in three OCBICs in the global automotive sector.

13.3 Methodology

Following Lobler's (2011) assertion that constructionism is an appropriate perspective for studying value co-creation in the context of SD logic, an idealist ontology and constructionist perspective within the interpretivist paradigm frame the current study. Idealist ontology understands reality to be socially constructed by social actors as they try to make sense of their world and other people. It is assumed that no independent "external" objective reality exists independent of our thoughts; therefore, there is "no permanent, unvarying criteria for establishing whether knowledge can be regarded as true" (Blaikie 2010: 95). Social actors are seen to continuously create, re-create, and share interpretations and meanings within their social groups (Blaikie 2010; Berger and Luckmann 1966; Crotty 1998). Our idealist ontological position underpins our epistemological position, namely, the acknowledgement of multiple intersubjective truths. As constructionists, we seek to describe and explain how social actors co-create meaning through their ongoing relations with others in collaborative consumption contexts. As a result, in this research, we prioritise how participants communicate and share their experiences with each other in such contexts (Blaikie 2010).

For our empirical study, we adopted a multiple case study approach. Multiple case studies provide a stronger base for building theory than a single case because the resulting theoretical propositions and elaborations are induced from a variety of empirical evidence (Eisenhardt and Graebner 2007; Yin 1994). Specifically, we studied three OCBICs in the global automotive sector, which provided rich, empirical descriptions of value co-creation and open service innovation. The data provided by the multiple case studies were a critical part of the inductive theory building research approach adopted.

13.3.1 Overview of the Online Community-Based Idea Contests

All three of the OCBIC platforms studied were developed by HYVE Innovation Communities (http://www.innovation-community.de) based in Munich, Germany. While there were slight differences in the toolkits provided to the participants in each of the three OCBICs, the functionality and objectives of the three were quite similar overall. Upon registering with an OCBIC, participants were able to create a profile page where they could post messages and view other participant submissions. Participants could also upload an avatar so that they were easily identifiable when they interacted with other community members. As there were no central forums within the OCBICs, interactions could only take place through participants' profile pages or by providing comments on the submitted ideas and concepts of the other community members. Participation was primarily incentivised by monetary or material rewards. In addition, a points allocation system for community participation acknowledged and listed the most active members in the community and their contributions.

The three OCBICs studied were selected using theoretical sampling as they were considered illuminating in relation to the contemporary utilisation of OCBICs in the global automotive sector. The exploration of three different OCBICs from the same sector provided a greater ability to identify consistent participant engagement patterns across a wider range of contexts. The specific OCBICs studied were Volkswagen's *App My Ride*, BMW's *Tomorrow's Urban Mobility Services* contest, and *Smart's Style Your Smart* design contest (for summary, see Table 13.1 and Table 13.2).

Table 13.1 Characteristics of the three selected online community-based idea contests

Criteria	Volkswagen's App My Ride Contest	BMW Tomorrow's Urban Mobility Services Idea Contest	Smart's Style Your Smart Design Contest
Task specificity	High (seeks apps and ideas for a future Volkswagen infotainment system)	High (new ideas for mobility services in tomorrow's urban areas)	High (design the exterior of a Smart fortwo)
Degree of idea elaboration	Text-based description of idea, supplemented by graphical representation; implementation of application prototype	Text-based description of the idea's functionality, target group, benefits, and uniqueness; supplemented by graphical representations	Presentation of design from four exterior perspectives of the vehicle: front, rear, left, and right side
Toolkit provided	SDK (software development kit) for application prototypes Online form for submitting ideas	Online form for submitting ideas	Online application for mapping designs onto various perspectives of the Smart car
Organisational appearance	Online, community-based environment User profiles, comments, and evaluation sections	Online, community-based environment User profiles, comments, and evaluation sections	Online, community-based environment User profiles, comments, and evaluation sections
Timeline	8-week submission phase	6-week submission phase	7-week submission phase
Incentives	3 monetary prizes ranging from 3000 € to 1000 €; 1 x VIP Trip to Car Presentation; 1 x 6-month internship at VW R&D; 3 x VW car rental weekend vouchers	4 x nonmonetary prizes, including BMW Satnav system, iPod Touch, Mini-Cooper branded bag, Mini-Cooper branded clock	6 x monetary prizes ranging from 1500 € to 300 €
Target group	Application developers; students; drivers and passengers who use infotainment systems; VW owners; non-VW owners; open to all ages	Commuters, city drivers, business people, electric car users; BMW owners, non-BMW owners; open to all ages	Designers, Smart car users, non-Smart car users; open to all ages

Table 13.2 Online community-based idea contest research data summary

Platform	Volkswagen's App My Ride Contest	BMW Tomorrow's Urban Mobility Services Idea Contest	Smart's Style Your Smart Design Contest
Ideas	386	312	52,624 (designs)
Applications	96	N/A	N/A
Members	553	640	9600
Evaluations	1996	8671	608,878
Messages	1291	2741	15,245
Comments	1450	2168	27,329

13.3.2 Volkswagen's App My Ride

The objective of Volkswagen's *App My Ride* OCBIC was to seek "apps" (applications) and ideas for a prototype of a future Volkswagen infotainment system (a portmanteau of information entertainment). According to Volkswagen, the OCBIC's goal was not only to co-create innovative applications but also to involve interested users, coders, and developers in developing apps (http://www.app-my-ride.com/jury-prizes-contest/contest). Volkswagen's *App My Ride* community (http://www.app-my-ride.com) was centred on two separate yet interconnected activities: idea generation and App (application) development. Participants were invited to submit ideas to the community idea "pool", where ideas were subsequently adopted and iteratively developed by other participants. This sharing dynamic emerged as a core value within the community and involved not just the sharing of feedback but also the sharing of detailed technical knowledge and ideas for further development. Volkswagen representatives who actively contributed to the community provided leadership and guidance to the participants. This leadership dynamic proved to be an additional driving force within the community, with members striving to satisfy the requests of other users or the Volkswagen representatives.

13.3.3 BMW's Tomorrow's Urban Mobility Services Online Community-Based Idea Contest

The BMW OCBIC sought new ideas for future mobility services in tomorrow's urban areas. In this sense, BMW was attempting to predict future mobilingy trends in the automotive sector using the experiences and creative potential of participants. According to BMW, this community idea contest (http://www.bmwgroup-ideacontest.com) was the organisation's "first open innovation project, which was launched within BMW Group's latest innovation approach—the Co-Creation Lab". In contrast to Volkswagen, BMW's Tomorrow's Urban Mobility Services OCBIC focused more on idea generation than on concept development. BMW encouraged participants to come up with ideas that could be applicable in the distant future by facilitating a playful and creative dynamic to stimulate community members' imaginations.

While Volkswagen appointed company representatives as leadership figures within the community in order to support participants in the practical development of their ideas, BMW provided community members with more of a free rein to develop fun ideas. As stated by BMW,

> the BMW Group Idea Contest 'Tomorrow's Urban Mobility Services' seeks new ideas for mobility services in tomorrow's urban areas. You can submit your creative ideas regarding innovative services in the area of automotive mobility in cities and metropolitan areas of the future. It is important that you focus your ideas and thoughts on services rather than product solutions and that you keep your creative mind directed to the future decade—and further!

13.3.4 Smart's Style Your Smart Design Online Community-Based Idea Contest

Smart's *Style Your Smart* OCBIC community (http://www.smart-design-contest.com) encouraged contributors' creative and artistic ideas in order to co-create exterior car vinyl designs. While Volkswagen and BMW's OCBICs focused on automotive technology services, Smart's approach emphasised aesthetics, strongly encouraged individuality and provided participants with opportunities to express their creativity by producing exterior designs for Smart's *fortwo* car, in addition to providing participants with a platform to discuss, evaluate, and comment upon each other's designs. The OCBIC provided members with a fun environment to share ideas, including contest-related games that encouraged community members to play and interact with each other. Culture and lifestyle played an important role in interactions between participants and many of the designs provided deep insight into the values held by their creators. The Smart OCBIC was specifically focused on design. In the words of Smart: "Do you love to design things? Or do you just want to have fun with an online platform and evaluate, comment or discuss designs with others? You got it! In this contest, you can create YOUR cool smart for two exterior designs! We invite you to take part in this contest and inspire you to open your mind for your creative designs".

Data for the three case studies was collected using netnographies, interviews, and naturalistic observations between March and June 2010. The data was collected from Volkswagen's App My Ride community in real time, as the OCBIC took place between May until June 2010. Data collection was performed retrospectively on BMW's OCBIC as the OCBIC concluded prior to the commencement of the study. Similarly, Smart's idea competition was conducted between January and February 2010, and so the data collection was also performed in a retroactive manner.

The data collection process involved screening more than 100,000 posts in more than 10,000 profiles and filing all innovation-related content electronically. This method is similar to the non-probability "purposive sampling" approach in ethnography (Belk et al. 1989).

The netnographic data was collected and analysed in accordance with the ethical guidelines for conducting netnographic research (Kozinets 2010). This phase consisted of browsing through the OCBICs' websites, exploring the various community and user profiles, reading messages, and trying to understand the structure, codes, and distinctive ways of communication, in order to obtain a familiarity with each OCBIC. In adopting a netnographic approach (Kozinets 2002, 2010), supplemented with in-depth online personal interviews, a detailed exploration of the language, history, practices, rituals, experiences, motivations, and behaviours of the participants in the three OCBICs was possible.

The data collected from the three OCBICs was transcribed and two authors read all the transcripts independently. The data analysis software tool NVivo was used to collate and prepare the data transcripts and to extract quotes and relevant information. Each transcript was individually and collectively analysed using NVivo, in order to identify the various participant roles, motivations, and behaviours.

In order to supplement the netnographic data and analysis, in-depth interviews were conducted using online instant messaging applications (i.e. cyber interviews) with five highly knowledgeable participants, namely, company representatives and lead users in the three OCBICs studied. The purpose of these cyber interviews was to cross-reference and triangulate the observations recorded during the course of the data collection and analysis stages (Yin 1994). As not all three OCBICs studied had private messaging systems, there was reluctance on behalf of the authors to contact members through their profile pages, despite the fact that they were publicly available, as this would have meant that the exchange of email details or other personal information would have also been accessible to anyone. In order to overcome this obstacle, interviewees were recruited through HYVE's Facebook fan page, which had over 1000 members. Open invitations were sent to fans using Facebook's private mail function in order to ensure that personal contact details were not divulged in an online public space. Fans with diverse occupations were identified, as it was felt that this diversity might reflect the open call that characterises participation in OCBICs and the diverse types of individuals they attract. This was also consistent with the non-probability purposive sampling approach previously mentioned (Belk et al. 1989). The average duration of the interviews was approximately 30 minutes, varying from 20 to 45 minutes. As argued by Giesler and Pohlmann (2003: 274), "cyber interviews not only comment on how things are remembered by informants in terms of a perspective of action, but how things are actively constructed". The interviews thus illuminated diverse perspectives on the nature of participant engagement with the OCBICs, how participants interacted with the OCBIC, and what the cultural values governing each OCBIC were (Eisenhardt and Graebner 2007).

13.4 Findings

13.4.1 Participant Motivations

The findings revealed that participants were motivated by a complex set of competitive, collaborative, extrinsic, and intrinsic factors to co-create value proposals in each of the three OCBICs. Participants who engaged primarily in technical tasks, such as application development, appeared to be motivated by a quest for new knowledge or learning. For instance, one student member of the Volkswagen community commented:

> The feedback you get from others can be helpful in making ideas better or maybe make stuff clearer in an idea, stuff I had taken for granted that people would understand.

However, not all participants were motivated by learning, perhaps due to the time commitments and effort involved in contributing and evaluating ideas. While direct compensation, such as monetary prizes, was an important incentive for some participants, this was not unilaterally the case, and even when this was the case, the perceived value of the monetary prizes differed. For example, a graphic designer participating in the Smart community commented:

> The prize was a lot of money, so I thought I would join and give it my best shot.

Whereas in the case of the BMW community, one participant commented:

> That navigator prize is not thrilling enough for me to present any ideas as yet.

While the majority of participants considered prizes to be an incidental bonus, a minority appeared to be solely motivated by monetary gain, which tended to impact the majority adversely and, in some instances, undermined their contributions and efforts. One striking example of this was in the Smart OCBIC. An unforeseen consequence of providing monetary prizes to participants who achieved the highest activity rankings was that it incentivised certain participants to abuse the system in order to increase their activity points. This resulted in undesirable actions such as rampant "spamming" in the guise of replicated comments and the submission of thousands of designs, with little to no changes in the appearance of each submission. One incensed member of the Smart community commented:

> Nearly 8500 designs from ONE person. 8500 designs with no trace of actual creativity!

In some cases, participants pursued those community members who unequivocally pursued direct compensation, particularly when they felt that their behaviour violated cherished community values such as "originality" and "creativity". As one Smart community member remonstrated:

> It's clear you have no talent and no ideas and you are just using pictures that you find on the Internet and stealing other people's ideas, such a shame for a lousy 800 Euros.

In contrast, intrinsically motivated contributors demonstrated a true passion for user driven service innovation through their contributions, which often incorporated intricate details and elaborate designs. Indeed, many participants related the personal enjoyment that they experienced while exploring, creating, and innovating within the OCBICs. One participant in the Volkswagen community noted:

> I could spend 80 hours coding for this contest, but unless the prize is worth at least 80 hours of my time, the prizes won't be as much of an incentive. Luckily I enjoy coding, so even though the prizes would be a nice thing to get, they wouldn't be the only reason I'd take part.

This quotation also illustrates that direct compensation motives sometimes offset enjoyment-based motives. This is important for firms hosting OCBICs as it is not always feasible to compensate all participant for their contributions.

Although the three online community-based idea contexts studied were in relative infancy due to the limited duration of each competition, it was apparent that friendships formed between participants, and that social interactions, such as appreciation, compliments, and reciprocal pleasantries from peers or from the host service organisations, contributed to the pleasure of the activity. For example, one participant in the Smart OCBIC commented:

> Thanks for leaving your comments—very much appreciated (and encouraging)—very good stuff here on your 'area' too! Good luck!

It is interesting to note that in the BMW OCBIC, no cash prizes were offered, yet the platform had a higher number of posts than Volkswagen's platform, which offered substantial cash prizes. This suggests that many of the participants in BMW's OCBIC were intrinsically motivated to participate.

Finally, other participants, such as students and the actual service organisations hosting the contests, viewed the competition as a platform for self-marketing. Members' profile pages served as a pivotal communication device, and in many instances, the addition of images or avatars to a member's profile facilitated collaboration and was appreciated by other community members, as revealed in the following comment from a BMW community member:

> Thank you for uploading a profile picture—it's nice to know whom we are working with in this community.

Members used their profile pages to highlight their skills and to introduce themselves to the community. For example, profile photos often featured photos of participants' cars and listed their interests and activities. A number of participants even went as far as advertising their firm's services on their profile pages. In this context, self-marketing motives transformed from an intrinsic motivation (personal and social identity) to an extrinsic motivation (monetary driven). All of the OCBICs studied featured a rating system where participants gained points for contributing ideas, posting comments, and so on. The design of this feature also encouraged social engagement through ranking participant activity.

13.4.2 Participant Roles

The findings revealed that participants were motivated to adopt various roles requiring different levels of engagement in relation to developing value proposals. The various participant roles induced were *problem identifiers, idea generators, designers, developers, motivators, leaders, end users, evaluators,* and *lurkers.*

Problem identifiers identified the particular use contexts or everyday issues that they wished to resolve. *Idea generators* comprised ordinary, creative, and lead users, who proposed solutions to problems that they and the *problem identifiers* faced in their everyday driving experiences. The *idea generators* contributed novel insights that typically went beyond the scope of what people with technical knowledge may deem practical. They actively sought to collaborate with other participants who had skills that they did not possess themselves, in order to progress their ideas to the next level.

Designers relied primarily on visual communication devices to contribute content to the community, using, for example, sketches, 3D illustrations, or models. *Designers* concerned themselves with the aesthetic application of the value proposal and were less concerned with the underlying technologies related to the proposed concept. *Designers* expressed their individuality and social identity through the creation of their designs, with some designs focused on cultural enclaves rather than on the OCBIC's host brand.

Developers demonstrated significant technical knowledge, not only within the realm of software programming but also in relation to vehicle electronics. These participants provided technical advice and feedback on how the contributions of *idea generators* and *designers* could be implemented. Some *developers* even demonstrated an interest in incorporating some of the proposed ideas in the modification of their own cars. In other instances, *developers* acknowledged their shortfalls in other disciplines, such as design.

Motivators often played the dual role of *idea generator* and *designer* and actively encouraged other participants to contribute new ideas, concepts, designs, and applications for the betterment of the entire community. Motivators were some of the most socially active members, frequently posting comments relating to encourage other participants. Motivators were particularly important in the OCBICs studied as they pushed the boundaries of what was deemed acceptable in terms of quality standards.

Leaders, typically OCBIC administrators or representatives of the service organisation hosting the OCBIC, also emerged depending on the format and context of the OCBIC. For example, in some instances, representatives from the host service organisation actively provided feedback and leadership to participants. However, in the OCBIC where this was not the case, leadership from within the community was less visible, possibly due to the short duration of the OCBIC and the participants' inability to administer the website of their own accord.

Many participants adopted the role of *end users* when they contributed feedback on the ideas and concepts submitted by others based on their personal experiences of the existing service offerings and contexts. *Evaluators* also provided feedback on the contributions of the other participants, based on their market and cultural knowledge, for example, knowledge of particular national markets, rules, customs, laws, lifestyles, and other cultural aspects that might affect the performance of a proposed idea in a specific market or cultural context. Finally, *lurkers* were participants who appeared to have little interest in direct participation in the community's activities. *Lurkers* may be considered noncommitted but curious participantsor, alternatively, participants or parties who seek to infiltrate a community in order to observe from afar.

13.4.3 Participant Behaviours

Finally, the participants in co-creating the value proposals in the three OCBICs seemed governed by community norms, relationships, and values. In this sense, many participant behaviours appeared to be influenced by community members themselves (participant led) as opposed to being controlled by the host organisation (firm led) (Kelleher et al. 2011). The community members seemed to share a number of values that promoted creativity, diversity, sharing, fun, and humour, amongst others, which positively encouraged opportunity identification and analysis, idea generation, and concept development. The presence of a focal brand and associated brand personality also facilitated this. For example, members who participated in BMW's OCBIC used words relating to "prestige", "status", and "power" in the description of ideas and how they related to the overall brand. As one participant commented in one of the in-depth personal interviews:

> BMW wants people to have fun and express their creativity in the contest, and so those values are demonstrated by members of the community.

In another example from the BMW community, a member contributed an idea relating to park and ride services; however, some participants felt that the idea did not represent the values of the BMW brand or the community members themselves:

> BMW is associated with a lot of prestige and high value because of its comparatively prohibitive cost. People who buy a BMW don't need to queue up for a bus.

Community values, such as respect and courtesy for other members, were evident in all three OCBICs and overcame the competitive rivalries between some community members that emerged from time to time. In a similar ethos to open source development, ideas and opinions were often shared with no compensation expected, except for courteous acknowledgement of assistance. Diversity also emerged as a strong community value in the interactions between community members who were representative of a wide range of cultures, economic circumstances, and lifestyles. One mechanical engineer from the BMW OCBIC commented during a personal interview that:

> Basically, the freedom to create and express yourself and the freedom to express your opinions and views about opportunities and ideas. There are people from all over the world in the contest community, with a wide range of cultural backgrounds, so the freedom to participate is a big part of the contest.

Freedom of expression, originality, and nonconformity were also core values in the three OCBICs studied. Observations suggested that OCBIC members value the freedom to express their creativity, learn new things, and share knowledge openly, often times in a fun and unfettered fashion, within the community. Integrity, where members acted honestly and in good spirit, was central to community activities. Members who did not appear to uphold the value of integrity were often called out and confronted. For example, one user commented:

> Would you mind not stealing other people's ideas?!!! I upload my design in front of my daughter's eyes and 15 minutes later you also upload a design after you have spotted it. Pretty lame if you ask me!

Different relationships were observed between participants in the three OCBICs studied. Some participants exhibited a rather self-centred view towards community interaction, which led to very little trust building between members, leaving participants feeling unfilled. On the other hand, the wide range of skills and knowledge available within the OCBICs provided many opportunities for collaborative relationships to grow and mature, resulting in more meaningful user engagements. Groups of members collaborated on ideas and designs or offered each other assistance with technical issues. The social and emotional engagements between community members resulted in close bonds forming between some members. Unfortunately, the bonds that formed between participants tended to come to a definite end once the OCBIC ended, and most participants were aware of this in advance. As a result, users offered sentimental parting messages to fellow community members. For example, one user expressed his sadness that the OCBIC was ending and that the community might disperse:

> Hi my friend, [username]. Today is the final day of this contest. I think we will all miss each other from tomorrow onwards. The last 2 months we stayed here like a community. Anyway, it's time to leave. Thanks for your support and valuable comments. Wish you all success and good luck!

13.5 Discussion

In this section, we summarise the findings by presenting and discussing three propositions induced from the three OCBICs studied relating to participants' motivations, roles, and behaviours in co-creating value proposals in OCBICs.

Proposition 1 Participants are intrinsically and extrinsically motivated to compete and collaborate to co-create value proposals in online community-based idea contests.

The findings suggest that participants' behaviour and propensity to co-create value proposals in OCBICs were influenced by a complex set of intrinsic and extrinsic motivations, including learning, direct compensation, self-marketing, and social motives (Ardichvili 2008; Dholakia et al. 2009; Hibbert et al. 2012; Leimeister et al. 2009). Some participated in the OCBICs because they wished to enhance their reputation or status within and beyond the community (Ardichvili 2008). For other participants, the possibility of personal gain was a key motivator, in addition to other factors, such as identification with the community and its goals (Ardichvili 2008). The findings also indicated that prizes acted, in some cases, as negative reinforcement. In these instances, extrinsically motivated participants' desire for monetary rewards led to greed and cheating behaviour, which superseded or negatively impacted the quality contributions of others.

We posit that the range of diverse motivations and behaviours observed is not surprising—rather, it reflects the value of diversity and freedom encouraged by such OCBICs. Certain participants appeared to be primarily interested in sharing their technical expertise to co-create the functional and technological aspects of the value proposition. Both specialist and non-specialist participants freely exchanged knowledge and assisted other participants in a manner similar to open source communities; however, some specialist participants had disconnected relationships with other community members and were only interested in the specified tasks.

It would seem, therefore, that the true value of OCBICs lies in diversity as well as in the similarities of the participants' motivations, expertise, cultural backgrounds, and geographical provenance (Simmie 2003). Despite the overtly competitive nature of such OCBICs, all participants contribute to the collective knowledge and co-creation of value proposals within the community (Hemetsberger and Reinhardt 2006; Wikström 1996). While some participants were often willing to help others with their creations, the social relationships and interactions between the participants became the cornerstone of the community friendships and trust that sometimes extended beyond the timescale of the original OCBIC itself. In addition, participants shared personal experiences and cultural preferences relating to their consumption of related products and services, which provided the basis for members' feedback on other members' ideas and concepts. The ranking of participants' levels of engagement recognised high levels of commitment and contribution, as well as increased the potential for winning material rewards. In this way, even the most basic social interactions within OCBICs can take the form of competitive practice as the boundaries between competitive action and co-operative interaction become increasingly blurred. In such OCBICs, many participants, who were potential rivals, choose to cooperate with other participants, reflecting the paradox of collaboration and competition inherent in such OCBICs (Hutter et al. 2011).

Our findings lead us to conclude that participants engage with OCBICs as part of a finely balanced social system. Community or obligation-based motivations, such as reciprocity, altruism, and reputation, occur when participants act on principle or when they become socialised in the norms of a group or community (Dholakia et al. 2004; Lindenberg 2001). Kozinets (2007: 208) suggests that community members "under the caring- and sharing ethos of community prefer to distribute their works through the gift economy", and in OCBICs, this appears to be the case, at least for those motivated by the desire to collaborate rather than to compete. Alternatively, it might be beneficial for individual community members motivated by self-interest to reveal innovation-related information in such OCBICs, because their actions may encourage improvements by others or expectations of reciprocity and improve one's reputation (Franke and Shah 2003).

Proposition 2 Participants iteratively adopt a number of diverse and overlapping roles when co-creating value proposals in online community-based idea contests.

The findings demonstrate that different participants acted either as "resources", "co-creators", and "end users" in the OCBICs (Nambisan 2002). In addition, participants adopted a number of additional and sometimes overlapping roles in the three OCBICs studied, namely, problem identifier, idea generator, designer, developer, motivator, leader/governor, end user, evaluator, and lurker. When adopting these roles, participants adopted some of the internal organisational roles typically involved in traditional new service development. Many problem identifiers, designers, and developers engaged with the challenges and tasks posed in the OCBICs in a mechanistic and work-like manner. Other participants, often end users and evaluators, displayed some of the characteristics of neo-tribes and were driven by their adoration of cars in general or by the automotive brand that was hosting the OCBIC (Cova 1997).

We posit that the question of when and how to involve participants in the co-creation of value proposals is directly linked to task objectives, for example, generating ideas, evaluating and refining concepts, specifying product features, or creating prototypes. In OCBICs, participants provide feedback iteratively and reveal tacit knowledge during the service innovation process (Chesbrough 2011) as they adapt, provide multiple diverse user perspectives, and interact with other users adopting the various roles. In this way, a community of motivated users, who co-create value proposals in OCBICs, replaces the traditional stage gate process of new service development.

Proposition 3 Participant behaviour in online community-based idea contests involves complex negotiations between the contest rules enforced by the host organisation and the community norms and values.

It is clear that the automotive manufacturers in the three OCBICs studies sought to adopt a service logic and to support participants' value co-creation processes to co-create value proposals (Gronroos and Helle 2010; Gronroos and Ravald 2011). While creation of reciprocal value is the basis for all business (Gronroos and Ravald 2011), contested value perspectives emerged in the OCBICs studied which led to tensions between the host organisation and participants and between participants themselves. While service organisations hosting OCBICs have a telo-ludic and primarily commercial orientation to co-creating value proposals (Kozinets et al. 2008), they need to acknowledge the social and communal motivations of the assembled crowd of OCBIC participants. The findings reveal that community norms and values play an important role in the nature of participant interactions and relationships in the OCBICs studied because they frame "what actions and attributes bring respect and approval (or disrespect and disapproval)" (Cancian 1975: 6; see also Füller et al. 2006; Nguyen et al. 2006). Furthermore, in most cases, the participants successfully negotiate the fine line between the competition rules imposed by the host organisation and the OCBICs' norms and values (Schau et al. 2009), and as in all communities, the fringes will consist of members who do not play by the rules, regardless of the possible implications. For example, a number of problematic issues arise in some of the OCBICs, which impinge on the values of the community, including copyright issues, intellectual property rights, and the stealing of other participants' contributions, which then needs to be addressed by the service organisation hosting the OCBIC or by the community itself. While OCBIC administrators had the ability to discipline community members, self-regulation within the community occurred when participants perceived that others breached the OCBIC regulations. For example, in one OCBIC, members rallied against deviant participants who included copyrighted images in their submissions. Consistent breaches of community and OCBIC laws resulted in unfavourable experiences for other "law-abiding" participants. Considering how community norms and values were established over a 6–9-week period, it is possible that participant engagements would be even more intimate and worthwhile over a longer time. We therefore suggest that firms integrate OCBIC events into existing company-sponsored communities, with established hierarchies, norms, and values, rather than setting up new communities around each separate OCBIC event.

We posit that the participants are not oblivious to the fact that by co-creating value proposals in OCBICs, they are contributing lucrative ideas and concepts for relatively little reward. As one member commented to another:

> I guess this competition is an example of "crowdsourcing" in action, but I knew that at the start and it doesn't bug me. If it makes you angry, do something else. Life is too short to fill it with bitterness. I hope you find another fulfilling outlet for your energy. Good luck!

This quotation encompasses all the good that participants see in OCBICs; it is this "fulfilling outlet for energy" that enables members to express their creativity to a global audience, enables them to engage in activities related to their interests, and gives them an opportunity to meet like-minded individuals from a diverse range of cultural backgrounds. Despite this, service organisations cannot ignore the desires of the diverse community of "customer-developers" who participate in OCBICs and need to engage with them respectfully and facilitate their social needs in addition to the service organisation's own commercial objectives (Di Gangi et al. 2010). Figure 13.1 presents participant motivations, roles, and behaviours as value proposals are co-created in OCBICs.

13.6 Managerial Implications

This chapter has examined why and how participants engaged in three OCBICs in the global automotive sector, which type of roles they adopted, and how they behaved in co-creating value proposals based on their and other participants' ideas. As a managerial implication, there is little doubt that OCBICs can contribute a valuable array of ideas and concepts for service innovations. Even if OCBICs may serve as a platform for co-creating value proposals, service organisations need to consider the limitations inherent in such open innovation approaches, including the fact that individual participants may not have the requisite expertise or may be incorrect about facts or knowledge. Pitta et al. (1996) noted that OCBIC participants are not employees; rather, they have their own personal interests and concerns that may not necessarily be shared by the service organisation hosting the OCBIC, and they can expect little direct reward for their contributions.

The challenge for service organisations in using OCBICs also lies within service organisations themselves and the internal management of such OCBICs (Adamczyk et al. 2012). OCBICs may not be prioritised internally by certain service organisations, particularly in those organisations that previously employed a closed approach to service innovation. Even where service organisations prioritise such initiatives, Magnusson (2003) warns that participants should not be used as a convenient replacement for professional designers and developers. He argues that it is a mistake to view participant contributions as advanced and ready-to-implement

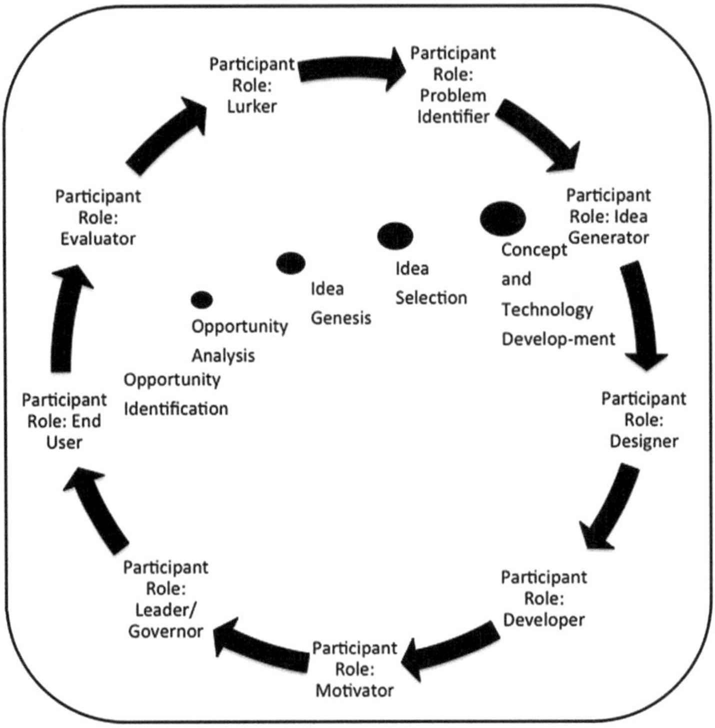

Fig. 13.1 Co-creation of the value proposals in online community-based idea contests

concepts, rather they need to be integrated with the organisation's existing resources and processes (Björk et al. 2010). Instead, service organisations should consider participant contributions as a catalyst for the service innovation and development process. In this manner, participants can be regarded as sources of inspiration and user insight rather than as implementers of new products and services. While OCBICs provide service organisations with a platform for inspiring service innovations, the identification and selection of ideas are, in itself, not sufficient to realise service innovations.

Prior to the value proposals that emerge from OCBICs becoming adopted by the service organisation and eventually being commercialised as value propositions, idea screening and evaluation present significant challenges to the service innovation process (Reinertsen 1999). In particular, the large volume of ideas that are generated by participants exacerbates such tasks. Therefore, service organisation's hosting OCBICs need to develop standardised processes for identifying and describing the ideas submitted (Di Gangi et al. 2010). Through the provision and use of toolkits, participants can undertake "needs-based" tasks, which can then be transferred to the firms for "solutions-based" processes of implementation via the OCBIC (Jeppesen 2005; Prugl and Schreier 2006; Von Hippel and Katz 2002).

Jeppesen (2005) argued that the benefits obtained in the interchange of need-related information via the toolkit method were likely to be offset by community support costs.

While the service organisations that hosted the three OCBICs examined in this study provided participants with mechanisms to provide feedback on contributions, the process involved a simple voting system for community evaluation which ranked ideas by popularity. In many instances, the votes had no bearing on the jury decision, which meant that internal experts had the significant task of screening ideas on a case-by-case basis. However, participant-to-participant help functions, technical knowledge sharing, and collaboration features were all features of the OCBICs studied, which can be viewed as a promising means for service organisations to reduce the responsibility of support and to create conditions for better toolkit use within such innovation-related communities.

In terms of managerial implications, we posit that service organisations should proactively recruit willing participants for their OCBICs and provide participants with an appropriate combination of monetary or nonmonetary rewards. Service organisations should not just consider the insights of "expert" participants but should also recognise the valuable contributions that "ordinary" participants can make to the service innovation process. While non-experts may have difficulties engaging with the technical issues that may arise in a particular OCBIC, they can act as inspiration for creativity and innovation. Finally, it is clear that OCBIC participants are not oblivious to the fact that they are contributing possibly lucrative service ideas and concepts to the service organisation for relatively little reward; however, they may be willing to do so if they obtain relevant social, monetary, and other rewards.

13.7 Limitations and Future Research

This chapter has examined participant motivations, roles, and behaviours in co-creating value proposals in three OCBICs in the global automotive sector. The data analysed was primarily qualitative in nature and therefore open to a greater deal of interpretation. More research is required to develop methods to screen initial ideas more efficiently, so that the internal experts may be able to spend their time and resources on ideas and concepts that are worth serious consideration.

In addition, while all three OCBICs were different in design and implementation, they were all hosted by the same marketing company and all focused on the global automotive sector. A comparison between other OCBICs from different sectors and designed by different companies might yield additional and different insights in relation to the co-creation of value proposals in OCBICs.

Future research might also examine the ethical issues relating to the potential exploitation and property rights of "working consumers" (Cova and Dalli 2009; Cova et al. 2011; Ritzer and Jurgenson 2010; Zwick et al. 2008). Future studies might explore the nature of participants' attitudes towards service organisations who conduct service innovation initiatives such as OCBICs or other "crowdsourcing" activities in order to determine whether users believe they are adequately compensated for their contributions. This topic may also have significant implications for service organisations in terms of their public relations and customer relationship management strategies.

Additional research focusing on the design and development of an incentive assessment framework for OCBICs would also be a worthwhile avenue for investigation, as more enticing incentives might encourage more participation (Leimeister et al. 2009). The data suggests that incentives must not be restricted to prizes. OCBICs may provide a useful starting point for understanding social reward systems, for example, VIP access to special community content or special administrative powers for the most active community members. These social reward systems also have implications for consumer behaviour research. For example, future research might explore whether consumers of high-end, luxury brands are more likely to be enticed by extra privileges, such as VIP access, in comparison with users who are not as driven by "status"-oriented products.

13.8 Conclusion

The findings of this multiple case study involving three different OCBICs in the global automotive sector suggest that the co-creation of value proposals in OCBICs is determined by participants' intrinsic and extrinsic motivations and the multiple roles that are adopted by participants when co-creating value proposals. Participant roles were performed as part of a finely balanced set of community norms and values, which sometimes created tensions with the commercial motivations of the service organisation hosting the OCBIC. Factors, such as reciprocity, altruism, reputation, and community identity, seemed to impact the nature of participant engagement in the OCBICs in addition to other factors, such as the presence or absence of rewards or other monetary incentives.

This research highlights the complexities of incentivising and rewarding service consumers and aligning their objectives with those of the service firm. While service organisations hosting OCBICs issue an open call to both expert and non-expert participants to co-create value proposals, the biggest challenge is the provision of meaningful incentives for participants to compete or collaborate throughout the competition. Participant input is a valuable resource for service organisations as they identify and analyse opportunities, provide ideas and feedback, and collectively evaluate the finalised service innovation from an end user perspective. These resources need however to be respected, appropriately facilitated, and rewarded using a combination of monetary and nonmonetary incentives. Finally the research has applicability for service organisations who wish to build and facilitate brand communities or tribes to sustain and encourage ongoing service innovation and user-driven service development with lead and other consumers over time.

References

Adamczyk, S., Bullinger, A., & Moslein, K. M. (2012). Innovation contests: A review, classification and outlook. *Creativity and Innovation Management, 21*(4), 335–359.
Amabile, T. M. (1997). Motivating creativity in organizations: On doing what you love and loving what you do. *California Management Review, 40*(1), 39–58.
Amabile, T. M., Hill, K. G., Hennessey, B. A., & Tighe, E. M. (1994). The work preference inventory: Assessing intrinsic and extrinsic motivational orientations. *Journal of Personality and Social Psychology, 66*(5), 950–967.
Ardichvili, A. (2008). Learning and knowledge sharing in virtual communities of practice: Motivators, barriers and enablers. *Advances in Developing Human Resources, 10*(4), 541–554.
Belk, R. W., Wallendorf, M., & Sherry, J. F. (1989). The sacred and the profane in consumer behaviour: Theodicy on the odyssey. *Journal of Consumer Research, 16*(1), 1–38.
Berger, P. L., & Luckmann, R. (1966). *The social construction of reality*. London: Penguin Press.
Berthon, P. F., Leyland, F. P., McCarthy, I., & Kates, S. M. (2006). When customers get clever: Managerial approaches to dealing with creative consumers. *Business Horizons, 50*(1), 39–47.
Björk, J., Boccardelli, P., & Magnusson, M. (2010). Ideation capabilities for continuous innovation. *Creativity and Innovation Management, 19*(4), 385–396.
Blaikie, N. (2010). *Approaches to social enquiry - Advancing knowledge*. Cambridge, UK: Polity Press.
Brabham, D. C. (2010). Moving the crowd at Threadless. *Information, Communication and Society, 13*(8), 1122–1145.
Bullinger, A. C., Neyer, A. K., Rass, M., & Moeslein, K. M. (2010). Community-based innovation contests: Where competition meets cooperation. *Creativity and Innovation Management, 19* (3), 290–303.
Cancian, F. M. (1975). *What are norms? A study of beliefs and action in a Maya community*. New York: Cambridge University Press.
Chesbrough, H. W. (2011). Bringing open innovation to services. *MIT Sloan Management Review, 52*(2), 85–90.
Cova, B. (1997). Community and consumption: Towards a definition of the 'linking value' of a product or services. *European Journal of Marketing, 31*(3/4), 297–316.
Cova, B., & Dalli, D. (2009). Working consumers: The next step in marketing theory? *Marketing Theory, 9*(3), 315–339.
Cova, B., Dalli, D., & Zwick, D. (2011). Critical perspectives on consumers' role as 'producers': Broadening the debate on value co-creation in marketing processes. *Marketing Theory, 11*(3), 231–241.
Crotty, M. (1998). *The foundations of social research*. London: Sage.
Dahlander, L., & Magnusson, M. (2008). How do firms make use of open source communities? *Long Range Planning, 41*(6), 629–649.
Dholakia, U. U., Bagozzi, R. P., & Pearo, L. K. (2004). A social influence model on consumer participation in network and small-group based virtual communities. *International Journal of Research in Marketing, 21*(3), 241–263.
Dholakia, U. M., Blazevic, V., Wiertz, C., & Algesheimer, R. (2009). Communal service delivery: How customers benefit from participation in firm-hosted virtual P3 communities. *Journal of Service Research, 12*(2), 208–226.
Di Gangi, P. M., Wasko, M. M., & Hooker, R. E. (2010). Getting ideas to work for you: Learning from Dell how to succeed with online user innovation communities. *MIS Quarterly Executive, 9*(4), 213–228.
Dolan, R., & Matthews, J. (1993). Maximizing the utility of customer product testing: Beta test design and management. *Journal of Product Innovation Management, 10*(4), 318–330.
Edvardsson, B., Enquist, B., & Johnston, R. (2005). Cocreating customer value through hyperreality in the prepurchase service experience. *Journal of Service Research, 8*(2), 149–161.

Eisenhardt, K. M., & Graebner, M. E. (2007). Theory building from cases: Opportunities and challenges. *Academy of Management Journal, 50*(1), 25–32.
Franke, N., & Shah, S. (2003). How communities support innovative activities: An exploration of assistance and sharing among end-users. *Research Policy, 32*(1), 157–178.
Frow, P., & Payne, A. (2011). A stakeholder perspective of the value proposition concept. *European Journal of Marketing, 45*(1/2), 223–240.
Frow, P., & Payne, A. (2014). Developing superior value propositions: A strategic marketing imperative. *Journal of Service Management, 25*(2), 1–21.
Füller, J., Bartl, M., Ernst, H., & Muhlbacher, H. (2006). Community-based innovation: How to integrate members of virtual communities in new product development. *Electronic Commerce Research, 6*(1), 57–73.
Füller, J., Faullant, R., & Matzler, K. (2010). Triggers for virtual integration in the development of medical equipment from a manufacturer and a user's perspective. *Industrial Marketing Management, 39*(8), 1376–1383.
Giesler, M., & Pohlmann, M. (2003). The anthropology of file sharing: Consuming napster as a gift. *Advances in Consumer Research, 30*, 273–279.
Gould, T. F. (1978). Value conflict and development: The struggle of the professional Zairian woman. *The Journal of Modern African Studies, 16*(1), 133–139.
Gronroos, C., & Helle, P. (2010). Adopting a service logic in manufacturing – conceptual foundation and metrics for mutual value creation. *Journal of Service Management, 21*(5), 563–590.
Gronroos, C., & Ravald, A. (2011). Service as business logic: Implications for value creation. *Journal of Service Management, 22*(1), 5–22.
Hemetsberger, A., & Reinhardt, C. (2006). Learning and knowledge-building in open-source communities: A social-experiential approach. *Management Learning, 37*(2), 187–214.
Hibbert, S., Winklhofer, H., & Temerak, M. S. (2012). Customers as resource integrators: Toward a model of customer learning. *Journal of Service Research, 15*(3), 247–261.
Howe, J. (2009). *Crowdsourcing: How the power of the crowd is driving the future of business.* New York: Crown Business.
Hoyer, W. D., Chandy, R., Dorotics, M., Krafft, M., & Siddarth, S. S. (2010). Consumer cocreation in new product development. *Journal of Service Research, 13*(3), 283–296.
Hutter, K., Hautz, J., Fuller, J., Mueller, J., & Matzler, K. (2011). Communitition: The tension between competition and collaboration in community-based design contests. *Creativity and Innovation Management, 20*(11), 3–21.
Jeppesen, L. B. (2005). User toolkits for innovation: Consumers support each other. *Journal of Product Innovation Management, 22*(4), 347–362.
Kelleher, C., Whalley, A., & Helkkula, A. (2011). Collaborative value co-creation in crowd sourced online communities – acknowledging and resolving competing commercial and communal tensions. In R. W. Belk, K. Grayson, A. M. Muñiz, & H. Jensen Schau (Eds.), *Research in consumer behavior* (pp. 1–18). London: Emerald Group.
Kim, J. W., Choi, J., Qualls, W., & Han, K. (2008). It takes a marketplace community to raise brand commitment: The role of online communities. *Journal of Marketing Management, 24* (3-4), 409–431.
Kotler, P. (1986). The prosumer movement: A new challenge for marketers. *Advances in Consumer Research, 13*(1), 510–513.
Kowalkowski, C. (2011). Dynamics of value propositions: Insights from service-dominant logic. *European Journal of Marketing, 45*(1-2), 277–294.
Kozinets, R. V. (2002). The field behind the screen: Using netnography for marketing research in online communities. *Journal of Marketing Research, 39*(1), 61–72.
Kozinets, R. V. (2007). Inno-tribes star trek as wikimedia. In B. Cova, R. V. Kozinets, & A. Shankar (Eds.), *Consumer tribes* (pp. 194–209). Oxford: Butterworth-Heinemann.
Kozinets, R. V. (2010). *Netnography: Doing ethnographic research online.* London: Sage Publications.

Kozinets, R. V., Hemetsberger, A., & Schau, H. J. (2008). The wisdom of consumer crowds: Collective innovation in the age of networked marketing. *Journal of Macromarketing, 28*(4), 339–354.

Kristensson, P., Gustafsson, A., & Archer, T. (2004). Harnessing the creative potential among users. *Journal of Product Innovation Management, 21*(1), 4–14.

Leimeister, J. M., Huber, M., Bretschneider, U., & Krcmar, H. (2009). Leveraging crowdsourcing: Activation-supporting components for ideas-based competition. *Journal of Management Information Systems, 26*(1), 197–224.

Lengnick-Hall, C. A. (1996). Customer contributions to quality: A different view of the customer-oriented firm. *Academy of Management Review, 21*(3), 791–824.

Libai, B., Bolton, R., Bügel, M. S., de Ruyter, K., Götz, O., Risselada, H., & Stephen, A. T. (2010). Customer-to-customer interactions: Broadening the scope of word of mouth research. *Journal of Service Research, 13*(3), 267–282.

Lindenberg, S. (2001). Intrinsic motivation in a new light. *Kyklos, 54*(2–3), 317–342.

Lobler, H. (2011). Position and potential of service-dominant logic - evaluated in an 'ism' frame for further development. *Marketing Theory, 11*(1), 51–73.

Lüthje, C. (2004). Characteristics of innovating users in a consumer goods field: An empirical study of sport-related product consumers. *Technovation, 24*(9), 683–695.

Lüthje, C., & Herstatt, C. (2004). The lead user method: An outline of empirical findings and issues for future research. *R&D Management, 34*(5), 553–568.

Maglio, P., & Spohrer, J. (2008). Fundamentals of service science. *Journal of the Academy of Marketing Science, 368*(1), 18–20.

Magnusson, P. R. (2003). Benefits of involving users in service innovation. *European Journal of Innovation Management, 6*(4), 228–238.

Magnusson, P. R. (2009). Exploring the contributions of involving ordinary users in ideation of technology-based services. *Journal of Product Innovation Management, 26*(5), 578–593.

Moeller, S., Ciuchita, R., Mahr, D., Oderkerken-Schroder, G., & Fassnacht, M. (2013). Uncovering collaborative value creation patterns and establishing corresponding customer roles. *Journal of Service Research, 16*(4), 471–487.

Nambisan, S. (2002). Designing virtual customer environments for new product development: Toward a theory. *Academy of Management Review, 27*(3), 392–413.

Nguyen, L., Torlina, L., Peszynski, K., & Corbitt, B. (2006). Power relations in virtual communities: An ethnographic study. *Electronic Commerce Research, 6*(1), 21–37.

Osborn, A. F. (1963). *Applied imagination*. New York: Scribner.

Ostrom, E., & Walker, J. (2003). *Trust and reciprocity: Interdisciplinary lessons for experimental research*. New York: Russell Sage Foundation.

Paulus, P. B., & Yang, H. C. (2000). Idea generation in groups: A basis for creativity in organizations. *Organizational Behavior and Human Decision Processes, 82*(1), 76–87.

Piller, F., & Walcher, D. (2006). Toolkits for idea competitions: A novel method to integrate users in new product development. *R&D Management, 36*(3), 307–318.

Pitta, D. A., Franzak, F., & Katsanis, L. P. (1996). Redefining new product development teams: Learning to actualize consumer contributions. *Journal of Product and Brand Management, 5*(6), 48–60.

Prahalad, C. K., & Ramaswamy, V. (2000). Co-opting customer competence. *Harvard Business Review, 78*(1), 79–87.

Preece, J. (2000). *Online communities: Supporting sociability, designing usability*. Chichester: Wiley.

Prugl, R., & Schreier, M. (2006). Learning from leading-edge customers at the Sims: Opening up the innovation process using toolkits. *R&D Management, 36*(3), 237–250.

Ramirez, R. (1999). Value co-production: Intellectual origins and implications for practice and research. *Strategic Management Journal, 20*(1), 49–66.

Reinertsen, D. G. (1999). Taking the fuzziness out of the fuzzy front end. *IEEE Engineering Management Review, 28*(1), 51–57.

Rihova, I., Buhalis, D., Moital, M., & Gouthro, M. B. (2013). Social layers of customer-to-customer value co-creation. *Journal of Service Management, 24*(5), 553–566.

Ritzer, G., & Jurgenson, N. (2010). Production, consumption, prosumption: The nature of capitalism in the age of the digital prosumer. *Journal of Consumer Culture, 10*(1), 13–36.

Ryan, R. M., & Deci, E. L. (2000). Self-determination theory and the facilitation of intrinsic motivation, social development, and well-being. *American Psychologist, 55*(1), 68–78.

Sawhney, M., & Prandelli, E. (2000). Communities of creation: Managing distributed innovation in turbulent markets. *California Management Review, 42*(4), 24–54.

Sawhney, M., Verona, G., & Prandelli, E. (2005). Collaborating to create: The internet as a platform for customer engagement in product innovation. *Journal of Interactive Marketing, 19*(4), 4–17.

Schau, H. J., Muniz, A. M., & Arnould, E. J. (2009). How brand community practices create value. *Journal of Marketing, 73*(5), 30–51.

Schulz, C. C. (2010). *Organizing user communities for innovation management.* Wiesbaden: Gabler.

Simmie, J. (2003). Innovation and urban regions as national and international nodes for the transfer and sharing of knowledge. *Regional Studies, 37*(6–7), 607–620.

Surowiecki, J. (2004). *The wisdom of crowds.* New York: Anchor Books.

Vargo, S. L., & Lusch, R. F. (2008). Service-dominant logic: Continuing the evolution. *Journal of the Academy of Marketing Science, 36*(1), 1–10.

Von Hippel, E. (1986). Lead users: A source of novel product concepts. *Management Science, 32*, 791–805.

Von Hippel, E., & Katz, R. (2002). Shifting innovation to users via toolkits. *Management Science, 8*, 821–833.

Wikström, S. (1996). The customer as co-producer. *European Journal of Marketing, 30*(4), 6–19.

Yin, R. K. (1994). *Case study research: Design and methods* (2nd ed.). Newbury Park, CA: Sage.

Zwick, D., Bonsu, S. K., & Darmody, A. (2008). Putting consumers to work: 'Co-creation' and new marketing govern-mentality. *Journal of Consumer Culture, 8*(2), 163–196.

Chapter 14
Customer-Centric Service Innovation in Public–Private Value Networks

Inka Lappalainen

Abstract This chapter aims to deepen the understanding of customer-centric service innovation in the value networks of public and private actors. In practice, this topical issue has attracted growing interest, though it has not been studied to any great extent within service innovation research. To narrow the gap, a conceptual framework is built, driven by recent studies on public service innovation and the service-dominant logic (S-D logic) perspective. The emphasis is on innovation as complex networked value co-creation by resource re-bundling, embedded in an everyday service exchange among different actors. Customers as beneficiaries are an active and necessary group of actors in co-innovation, for which reason a specific focus is placed on deepening the understanding of customer value in use. This focus enables the development of more extended service concepts and platforms, transparent value-creation processes, and the strategic management of novel resource integration at different levels of service ecosystems. The contribution of an external facilitator, as an intermediary of innovation activity in value chains and networks, is also examined. The framework is illustrated with two empirical case studies. The main findings contribute to the topical debate on the linkages between S-D logic and innovation and provide conceptual starting points for its further development. They also provide a tool for collaborative reflection among practitioners in the public–private interface.

14.1 Introduction

The purpose of this chapter is to study customer-centric service innovation in public–private interfaces. As in the private sector, the public sector has also been faced with ongoing transformation in the innovation landscape, towards *co-creation and information- and network-centricity* (Langergaard and Scheuer 2012; Lusch and Nambisan 2015; Lusch et al. 2010; Sundbo 2011). Dyadic

---

I. Lappalainen (✉)
VTT Technical Research Centre of Finland Ltd, Espoo, Finland
e-mail: inka.lappalainen@vtt.fi

T. Russo-Spena et al. (eds.), *Innovating in Practice*,
DOI 10.1007/978-3-319-43380-6_14

partnerships have been extended to service value chains and networks across different public stakeholder groups, and they also extend to collaboration between public and private actors. However, within service innovation research, neither public services nor the public–private interface has been studied to the same extent as private service (Hartley 2005; Kallio et al. 2013; Langergaard and Scheuer 2012).

In addition, while previous research has focused more on the economic features of value co-creation, recent studies have emphasized the integration of social and economic elements (Möller and Rajala 2007; Sundbo 2011). Sundbo (2011), for example, with his theoretical concept of value chain innovation, focused on the innovation process, consisting of dynamic interaction and participatory practices in goal-oriented multi-actor settings. Similarly, increasing interest in the social construction of value co-creation processes has emerged in the service-dominant logic (S-D logic) framework. It stresses social interaction as the basis of service exchange and innovation, shaped by culture-historically modified values, roles, and related expectations of all the actors involved (Chandler and Vargo 2011; Edvardsson et al. 2011; Lusch et al. 2010).

Nevertheless, the growing interest in S-D logic lacks empirical evidence, particularly from innovation perspectives. Following Lusch and Nambisan (2015), service innovation is here defined as complex network- and information-centric value co-creation by resource re-bundling in novel ways and embedded in the everyday service exchange among different actors, including beneficiaries such as customers. A deeper understanding of the mechanisms for enhancing service innovation is still needed. To narrow these research gaps, the present chapter examines customer-centric service innovation in public–private value networks both theoretically and empirically. As a starting point, the conceptual frame of Lusch and Nambisan (2015) is applied, illustrating the systemic nature of service innovation. Their tripartite service innovation framework consists of *service platforms, the value co-creation processes, and the service ecosystems* (ibid.).

In the public context, the conceptual frame needs to be supplemented with considerations linked to the broader context of the political and regulatory environments of different levels. In the chapter, this topic is discussed critically by combining recent studies on S-D logic with previous service innovation research in the public context (Hartley 2005; Langergaard and Scheuer 2012). Considering societal issues, the concept of the *service ecosystem*, which is gaining ground, seems relevant. By this, Lusch and Nambisan (2015: 7) refer to "A self-adjusting system of loosely coupled social and economic (resource-integrating) actors connected by shared institutional logics and mutual value creation through service exchange". When focusing on public–private interfaces, institutional logics with strong regulatory aspects play an essential role in the underlying construction of service concepts or platforms and in the value-creation processes in service ecosystems. Thus, they are also manifested in innovation activities in terms of the pursued relationships, competencies, and knowledge.

The roles of actors, and particularly those of the customers (or users), in service exchange and innovations, have attracted increasing attention among researchers and practitioners (Carlborg et al. 2014; Edvardsson et al. 2012; Sundbo and Toivonen 2011). However, until recent years a direct participation of citizens as customers in service development has been rather limited in the public sector. Usually, it has taken place and been legitimated by means of representative democracy, formal procedures of hearing, and service-specific feedback mechanisms (Elg et al. 2012; Hennala et al. 2012). The development of public services is characterized by long-term, multistage planning prescribed by laws and predefined by decision-making procedures (Kallio et al. 2013). Nonetheless, citizens as customers have shown an increasing interest in societal issues and the willingness to take an active part in decisions related to their day-to-day lives.

According to Edvardsson et al. (2012), it is essential to capture information on value in use in the terms of context and situ in order to understand customers and the aspects of value co-creation that are critical to them. This chapter applies their novel framework of data collection methods, classified in two dimensions. Firstly, data is classified either as in situ (data captured in a customer's use situation) or ex situ (data captured outside the use situation). Secondly, data is categorized as either in context or ex context, which concerns a resource constellation available for customers to enable value co-creation (ibid.). In addition, increasingly complex value-creation and innovation activity open up possibilities for new actors: innovation intermediaries (Lappalainen et al. 2012; Lusch and Nambisan 2015). In-depth studies of this external facilitation in multi-actor settings are still rare (Kallio et al. 2013). In the present study, the role of knowledge-intensive business service (KIBS) companies as intermediaries (den Hertog 2002) is examined in particular.

This chapter focuses on customer-centric service innovation in public–private value chains and networks. I will examine how customer-centricity is interpreted, manifested, and facilitated in the ongoing transformation towards collaborative service innovation in public–private multi-actor settings. My empirical study is based on a case-study approach and includes two Finnish cases, which illustrate the value network perspective. The chapter provides application of service innovation frameworks with empirical illustrations, which also open up possibilities for further conceptual development regarding the complex and dynamic service innovation in public–private interfaces.

14.2 Theoretical Background

With increasingly complex customer needs and economic pressures, there is a need to examine and develop more systemic service platforms, multiform value co-creation processes, and effective and dialogical resource integration among multiple actors in service ecosystems. Within these dynamic ecosystems, the customers, as part of a specific network of public and private actors, constitute their own networks. They can be interpreted as subsystems in the whole ecosystem (Elg et al. 2012). Given this, there is a need to point out customer-centricity by examining the expanding collaborative service development from the customers' perspectives, in terms of service concepts/platforms, value co-creation processes, and service ecosystems.

14.2.1 Towards Developing Systemic Customer-Centric Service Concepts and Platforms

In order to define a concept of value in the public service context from a customer perspective, this study follows the definition of Lehtonen and Tuominen (2011), which differentiates the type of value for the customer as *public or private*. In comparison with private companies, public organizations aim not to produce profit, but public value, which means ensuring an equal basis for the citizens' well-being and balancing the competing interests of different stakeholders (Hartley 2005; Langergaard and Scheuer 2012). Therefore, public service innovation includes strong social and value aspects (Ballantyne and Varey 2006). The private value of customers is co-created when they directly participate in service exchange. In service research, it is broadly agreed that service organizations as providers can only make value propositions (what and how) for customer needs. The (private) value of the service is always co-created and thus realized in use and uniquely experienced by the customers (Vargo and Lusch 2008). However, public and private values might be *contradictory*; when seeking the collective good, public actors may be unable to contribute to specific individual needs (Rosenthal and Peccei 2007). Due to these two existing value-creation logics and the related balance of power relations, the customers are seen and legitimated in various ways in the public context (Elg et al. 2012; Langergaard and Scheuer 2012; Lehtonen and Tuominen 2011; Rosenthal and Peccei 2007). Furthermore, both public and private values are very challenging to evaluate, being typically realized in the long term and co-created among several actors.

A characteristic of public service is sector-specific concepts for specific customer needs, social and health-care services providing an example. For customers, value propositions of different providers may appear to be confused and even controversial and overlapping. Lehtonen and Tuominen (2011: 237) refer to the "dilemma of distributed customer (user)" (Elg et al. 2012). Despite the increasingly complex customer needs acknowledged by researchers and practitioners, it seems difficult to combine the competing interests for mutual benefit in a novel way and to reconstruct *extended or even systemic service concepts* (Hennala et al. 2012; Sundbo and Toivonen 2011). Lusch and Nambisan (2015) suggest that organizations design their offerings as *service platforms* that, according to their definition, enable not only daily service exchange and value co-creation but novel scalable solutions. Furthermore, the platform should be structured in a layered, modular manner, consisting of tangible and intangible resources and facilitating multiple interactions by actors and resources (ibid.). Such extended and explicit service concepts or platforms would also enable customers to evaluate and choose between different value propositions and resources of service providers in relation to their own needs and resources.

However, the *freedom or obligation* of the customer participation in such a novel value co-creation might set constraints, which characterizes public service exchange and co-innovation (Lehtonen and Tuominen 2011). Customers do not (necessarily) have actually choices or exit options (cf. primary schooling) or, as mentioned earlier, the public good is prioritized over private interests (Rosenthal and Peccei 2007). It has also been claimed that sometimes customers cannot choose the correct service for them, or clearly articulate or verbalize their needs and wishes, particularly in the long term. Institutional logics, with laws and regulations, guide what kinds of protocols and rules are set for (re)defining networked value promises with beneficiaries (Kallio et al. 2013; Langergaard and Scheuer 2012; Lusch and Nambisan 2015). They may play a critical role as hindrances or catalysts when aiming at extended service concepts and modular service platforms.

On the other hand, public actors, particularly focusing on well-being, have been increasingly acknowledged customers (with knowledge, skills, relationships) playing a crucial role in the value co-creation. An example is the emphasis on behavioural changes towards a healthier lifestyle (Elg et al. 2012; Hennala et al. 2012). Preventive, holistic, and life-cycle perspectives have emerged as topical issues in order to support citizens in taking care of themselves more proactively. These trends can be interpreted as a gradual change towards customer-centric service logic in the public context.

Thus, understanding customer value in use becomes increasingly important. Different customer roles with relevant data collection methods have been examined by emphasizing both dynamic value-creation situations and static contextual resource constellations (Edvardsson et al. 2012). Customers contribute in several ways to service development. So as to enable the co-construction of extended service concepts, the customer role as *the correspondent* seems relevant, as it provides information that often takes the form of ideas and solutions captured by methods such as the lead-user method (Edvardsson et al. 2012). Here, the information is based on actual customer needs and is anchored in real-life situations by indicating when and what kind of value is experienced (ibid.). This kind of approach has also been referred to as direct participation (Lehtonen and Tuominen 2011). Lusch and Nambisan (2015) go further by identifying the role of *ideator*, which "reflects the capability of beneficiaries of service offerings (i.e., customers) to bring knowledge about their needs and unique context to the firm (or organization) and then to integrate that with knowledge about how they use existing market offerings to envision new services".

When developing modular service concepts for several beneficiaries, customers as *testers* might be relevant, in order to collect information that is difficult to acquire other than by simulating a situation and the resource context (Edvardsson et al. 2012). Methods such as prototyping, role playing, or creating a 3-D environ ment provide various alternatives for "safe" testing, even if they cannot replace direct links to real service situation and use context (ibid.). In order to capture radical new ideas, Edvardsson et al. (2012) suggest inviting customers as the *dreamers* with little or no experience, either in a use context or in a service situation. They refer to various creativity techniques so as to enable brainstorming and out-of-the-box thinking. Notably, the value that an actor creates or co-creates does not necessarily directly contribute to the usage of a specific service or offering. Instead, it may benefit the broader context in terms of public value, beyond experienced private value (e.g. the social setting with other actors in the service ecosystem) (Langergaard and Scheuer 2012; Lusch and Nambisan 2015).

14.2.2 Towards Developing Customer-Centric and Networked Service Processes

A service process can be defined as the complex chains of interactive activities that enable value co-creation between customers and workers of service organizations (Lusch and Nambisan 2015). This definition is also relevant in the public sector, especially when the interactive, dynamic, and intentional nature of service is emphasized. A service process can be considered to be a prototype or model for unique value co-creation processes, with various encounters to fulfil customer needs in a purposeful manner (Edvardsson 1997).

There are two main reasons that, in the public context, drive the traditional, more or less dyadic service processes between customers and providers towards expanded and co-developed processes between various beneficiaries. First, the institutional and innovation landscape is changing and requires more systemic service concepts and platforms. Second, the increasingly complex customer needs have to be answered with diminishing public resources (Langergaard and Scheuer 2012; Lusch et al. 2010; Sundbo 2011). These developments make it crucial to create supportive environments for resource integration and value co-creation in novel ways. Focusing on mechanisms that facilitate interactions between diverse actors and enhancing the transparency of activities in an increasingly complex service ecosystem are essential (Lusch and Nambisan 2015). Critical activities emerge at the interfaces of different actors when the service processes cross sectors or organizational borders of public, private, and third sectors. The activities of different providers, with various roles and responsibilities, have to be co-constructed, and complex knowledge flows between beneficiaries have to be coordinated.

Thomson's multidisciplinary studies on cross-sectorial collaboration supplement the S-D logic view of service innovation. Social constructivism included in these studies is consistent with the basic premises of S-D logic. Thomson defines cross-sectorial collaboration as "a process in which autonomous actors interact through formal and informal negotiation, jointly creating rules and structures governing their relationships and ways to act or decide on the issues that brought them together; it is a process involving shared norms and mutually beneficial interactions" (Thomson 2001: 83; see also Thomson and Perry 2006). In complex cross-sectorial collaborations between public, private, and third sector actors, the following critical dimensions can be identified: (1) governing, (2) administration, (3) autonomy, (4) mutuality, and (5) norms of reciprocity and trust. Awareness of the dynamic and systemic nature of cross-sectorial collaboration is essential; equilibrium among all the dimensions of collaboration is needed (Thomson and Perry 2006; Thomson 2001). Instead of dominant hierarchal exchange, inter-functional and networked exchange is needed within and between service actors. Furthermore, dialogical interaction is a necessity when building trust, commitment, reciprocity, and empathy between actors (Ballantyne and Varey 2006). These are central building blocks in terms of reducing complexity and transaction costs. They are particularly important when uncertainties and risks grow—a situation that is typical when moving from provider–customer dyads to value networks (Thomson and Perry 2006).

Service encounters between front-line workers and customers have attracted growing interest in service innovation research (Sørensen et al. 2013; Sundbo and Toivonen 2011). Public service innovations are characterized by relatively minor changes and adaptations to existing service concepts and processes, but they are nonetheless vitally important (Hartley 2005; Kallio et al. 2013). Innovation activity among front-line workers and customers is typically a more or less informal practice embedded in dynamic and interactive value co-creation processes. However, without legitimation, enabling mechanisms and dialogue between top-down and bottom-up activities, there is a risk that innovation potential and the increasingly important system perspective are neglected (Lusch and Nambisan 2015; Sørensen et al. 2013).

Customers can enact different roles to contribute to the contextual and situational understanding of value co-creation processes in multi-actor networks. It is reasonable to suppose that the roles of "correspondent, tester, and dreamer" introduced by Edvardsson et al. (2012) are also applicable in the development of prerequisites for service activities and processes. However, even more relevant could be the fourth role (ibid.): *reflective practitioner*, whose experiential information is based on the actual resource context. Furthermore, associated methods, such as participatory design, living labs, and the conversational approach, are primarily beneficial when the actual service performance is highly dependent on the integration of other resources in service networks. In these, information is collected after the actual service situation, in order to enable reflection and further elaboration of ideas. However, the time gap can cause difficulties in benefiting from detailed information, thus disturbing the reconstruction (ibid.).

Therefore, it seems that co-innovation via critical reflection, co-construction, and experimentation as part of the daily value-creating activities needs further deepening. In their public sector innovation review, Langergaard and Scheuer (2012) crystallized public sector service innovation as *co-evolutionary*, referring to iterative, interactive, and heterogeneous processes initiated both bottom-up and top-down. Goals are also often relatively unclear, resulting in innovation that is typically possible to identify afterwards rather than as intended outcomes of the implementation of the specific plan (ibid.). The so-called rapid experimenting models have emerged as alternatives to linear stage gate models; they aim at testing the emerged idea immediately with customers. If the idea receives approval, it is developed further (Toivonen 2010). Recent studies indicate that citizen-customers are increasingly seen as collaborative innovators or designers (Brand 2005; Hennala et al. 2012; Kallio et al. 2013; Langergaard and Scheuer 2012; Lusch and Nambisan 2015). The service encounters thus provide a dynamic interactive arena for innovations in terms of challenging the current way of carrying out and modifying service activities iteratively by experimenting, reflecting, and developing further in a relevant manner.

14.2.3 Towards Developing Customer-Centric and Networked Service Ecosystems

The service ecosystem concept emphasizes the importance of common organizational structures and sets of principles in facilitating resource integration and re-bundling between the actors involved (Lusch and Nambisan 2015). In terms of innovation potential, there is a need to enhance both the *structural flexibility* (in organizational forms) and *structural integrity* (in the nature of relationships) of the service ecosystem (ibid.). At the same time, it is crucial to manage potential conflicts between the two—an issue that is characteristic of public–private interfaces.

Public organizations have often been considered problematic from the innovation standpoint: legislation, local politics, a communal decision-making culture, risk aversion, and organizational inertia are typically seen as the main inhibitors (Hartley 2005; Lappalainen et al. 2012). Innovation activity is rendered complex by differing, even conflicting, drivers related to policy, professional, and managerial issues (Langergaard and Scheuer 2012). According to Hartley (2005), there is an ongoing shift towards a *networked governance paradigm* in the public sector. It emphasizes the enabler role of policymakers in both transformational and incremental innovations in service exchange in multi-actor networks. Manifestations of previous paradigms—New Public Management and the traditional public administration—still coexist, causing challenges particularly when the goal is radical innovations (Hartley 2005; Langergaard and Scheuer 2012).

Even though S-D logic has not discussed the problematics of the public sector separately, its general emphasis on the importance of institutions and institutionalization is in line with the above-described views. The underlying logic of institutions may function as an enabler or a hindrance in shaping value creation and resource integration among actors (Chandler and Vargo 2011; Edvardsson et al. 2011). As regards the phenomenon of *institutionalization*, Chandler and Vargo (2011: 10) state that "the notion of a service ecosystem is a fundamental aspect of value co-creation because it acknowledges how large-scale social structures and institutions evolve relative to the individual service efforts of actors, dyads, triads, and complex networks".

Furthermore, in the public sector, diffusion of innovations is often weak and local in nature. It is suggested that, in a public context, different networked and customer-centric models would be the most effective way of creating innovations that are cross-sectorial and cross-regional (Hartley 2005; Brand 2005). For private companies, multi-actor networks may serve as important arenas for renewal: they provide short- and long-term business potential, networking and benchmarking possibilities, and a possibility of exercising influence on local politics. Consequently, increasingly complex networked resource integration in novel ways calls for systemic and strategic management, which is often claimed to be left unconsidered, particularly in a changing public context (Thomson and Perry 2006). For example, strategic repositioning in value networks may enable novel resource re-bundling in service concepts, platforms, and processes, or even at the systemic level.

Systemic dialogical coordination is essential for diffusing novelties in terms of efficiency and effectiveness, as well as for boosting radical cross-sectorial innovations. As Möller and Rajala (2007) illustrate in their integrative framework, effective network management is strongly related to value-creation logics (efficiency/ effectiveness), based on primary ontological knowledge needs (utilization/ creation). In their recent study, Lusch and Nambisan (2015) contribute to the managerial issues, studied mainly at the business network level, by extending the examination to the ecosystem level in the framework of S-D logic and innovation. They suggest co-constructing a *common* (service) *mindset* and *architecture of participation* aligned to *maintain* shared institutional logics as the main coordination mechanisms. In the light of the argumentation above, these suggestions sound somewhat static, though they might also imply a more transformative view of the general prerequisites, given various institutional settings. As a consequence of the growing complexity of interaction and knowledge flows, Lusch and Nambisan (2015) also discuss how digital infrastructures can facilitate collaboration and value co-creation in the ecosystem.

Increasing network- and information-centricity opens possibilities for new actors to become involved. Lusch and Nambisan (2015) refer to *new types of innovation intermediaries* in coordinating and facilitating innovation activities across value chains, networks, and ecosystems. In a recent study, we have been able to identify both internal (organizational change agents) and external (independent actors) innovation intermediaries (Kallio et al. 2013). Our findings are aligned with previous studies. First, they show that internal intermediaries play a crucial role, particularly in complex, multilevel public sector innovations (Kallio et al. 2013; see also Langergaard and Scheuer 2012). Second, they confirm the importance of external agents, particularly knowledge-intensive business services (KIBS). Our studies, as well as earlier studies, indicate that an innovating organization can gain fundamental benefit from using an external facilitator. The contribution of KIBS in innovation can be classified in three main roles: (1) facilitators (supporting the innovation process), (2) carriers (transferring the innovations emerging), and (3) sources (creating innovations themselves) (den Hertog 2002; see also Lappalainen et al. 2012). However, the contribution of KIBS as external innovation intermediaries in multi-actor networks still requires further research.

14.3 Objectives and Methods

In this chapter, the main objective is to build a *holistic framework* of how customer-centricity is interpreted, manifested, and facilitated in the ongoing transformation towards collaborative service innovation in public–private interfaces.

The research questions are as follows:

1. How is customer-centric service innovation described and manifested in a value network of public, private, and third sector actors in terms of service concepts and platforms, processes, and the ecosystem?
2. How does the external facilitation contribute to customer-centric service innovation in a value network of public, private, and third sector actors?

The methodological choices are based on enabling a critical reflection and the co-construction of a holistic framework on service innovation as a systemic change, with an emphasis on customer-centricity and cross-sectorial collaboration. Thus, a case-study approach (Eisenhardt 1989), including two cases from two Finnish cities, was adopted. Three main characteristics of this approach were followed, as these are also in line with the so-called dialogic paradigm in service research (Tronvoll et al. 2011). First, the research questions were (only) tentatively driven according to the main theoretical perspectives specified during the research process. Second, two cases were selected from the broader interview study focusing on public organizations that have been involved in a service development programme led by an external facilitator. The programme is described later in this section. Third, the dynamic and iterative process of qualitative data collection and analysis was carried out, with a new framework gradually emerging from the empirical data in a dialogue with the relevant literature (Eisenhardt 1989).

Long-term and cross-sectorial involvement was the core of service development in the selected cases at both strategic and practical levels. Furthermore, the two cases were chosen because the ongoing collaborative customer-centric service development could be approached from a value network perspective. Case A focused on *supporting growth and the learning path of children*, and Case B focused on *the service integration supporting the customer's mental health rehabilitation*. Even though the two case contexts as cities have different historical backgrounds and future views, they share the same challenges, which are related to increasing and complex service needs with efficiency pressures.

The main empirical data were collected by thematic interviews in two separate phases, in May and June 2012 and in April and May 2013, as a part of broader research project. Altogether, 14 people were interviewed individually or in pairs. They represented service directors, service managers, and staff involved in the local service development. The thematic interviews followed the themes discussed in the theory section. On the one hand, insights and manifestations of customer-centricity in daily practices were gathered. On the other hand, customer-centricity in the ongoing change was mapped from the viewpoints of service concept/platform, process/activities, and the system. Interviews took approximately one and half hours. They all were tape-recorded and later transcribed for analysis. In the following, the cases are presented in greater detail.

Case A is located in a typical medium-sized and vitally growing Finnish city (with a population of 30,000), near to the Helsinki metropolitan area. There are strong quantity and quality increases in service needs with scarce public resources, and therefore, service innovation in terms of concepts, processes, and ecosystem is called for. The study focused on supporting growth and the learning path of children. It was topical in the city, due to an ongoing integration process of day-care, preschool education, and primary school services into one organizational and administrative unit within the educational services. Previously, day-care and preschool services had been linked to social services. The change reflects a national trend in Finland, with the emphasis on resource integration and customer orientation in public services. In this case, the main goals of the integration were set as follows:

1. Shared mindset considering the growth and the learning path of children
2. Shared customer-centric working culture and administration
3. Support and rehabilitation service through childhood and adolescence
4. Collaboration practices between educational and social service providers
5. Resource efficiency and effectiveness

The change targeted a total of 320 workers in the city. In addition, indirectly it influenced private and third-sector actors collaborating with the city and a wide range of citizens as customers. After the political decision at the end of 2010, a representative steering group and working groups were established in order to plan a new administrative model for the service organization. The primary focus was to improve the coherency of preschool services, either as a part of day-care or primary school, based on collective bargaining, organization models, and management practices. The planning process was organized in an open and collaborative manner, though with a rather tight time schedule considering local decision-making procedures with conflicting interests between professions. The change was implemented administratively at the beginning of 2012 and in practice in August 2012. The change organization continued to support the multilevel cultural and managerial change.

Figure 14.1 illustrates the main actors in the value network of Case A from customers' perspective. In addition, the interviewees as informants are indicated in the figure.

Case B is located in a medium-sized Finnish city (population 54,700) that has faced substantial industrial restructuring within the last 15 years. Consequently, people have suffered from long-term unemployment, with spillovers that have led to increased and more complex needs for special services, including mental health.

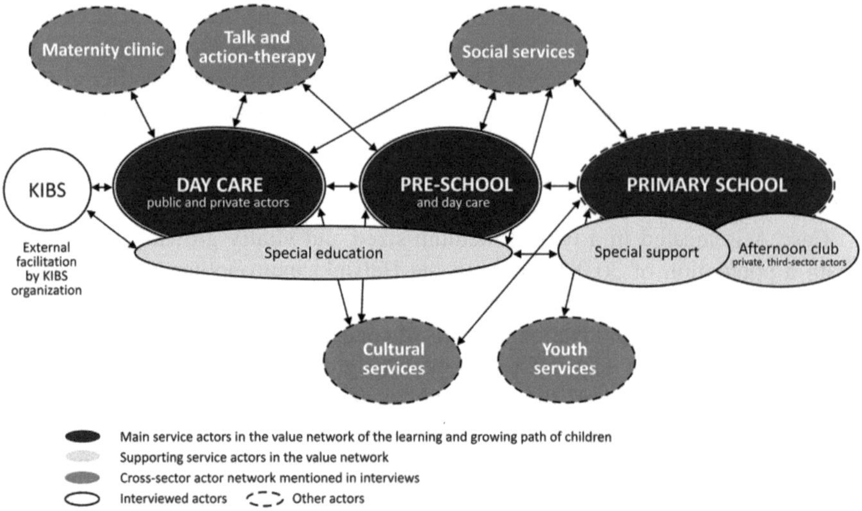

Fig. 14.1 Value network actors in Case A: supporting growth and the learning path of children

Novelties are called for in terms of extended customer promises, service networks, and systemic resource integration. They would be important not only for economic reasons but also from the customer-centric perspective, to answer the changing needs and to support the empowerment of citizens.

The study focused on the service integration supporting the customer's mental health rehabilitation in a value network, illustrated in Fig. 14.2. The service provider has a long and rich history in the local public service ecosystem in supporting customers' comprehensive mental rehabilitation. The provider has a wide customer base, which has been grouped into long-term customers with housing service on the one hand and into interval, reception, and acute customers on the other. The strength of the service has been low threshold, flexible, and 24/7 opening hours; committed, multi-skilled, and experienced workers; and agile customer-centric practices. In addition to a medical staff of 18 people, complemented by a part-time doctor and rehabilitation specialists, the provider has a broad collaboration network of various local service providers to support the comprehensive rehabilitation of customers.

However, in the past, the role of the service provider in the network has been somewhat blurred. As a consequence, the provider has faced challenges in answering more complex customer needs, in a situation characterized by changing regulations, scarce resources, and poorly managed network collaboration. In fact, all the outpatient mental health services of the city had lacked long-term strategic management that would have been a crucial prerequisite for the service provider to create value with beneficiaries. The new director, with strong local experience in different positions of the service ecosystem, acknowledged not only those multilevel challenges but also the opportunities for service innovation. The move

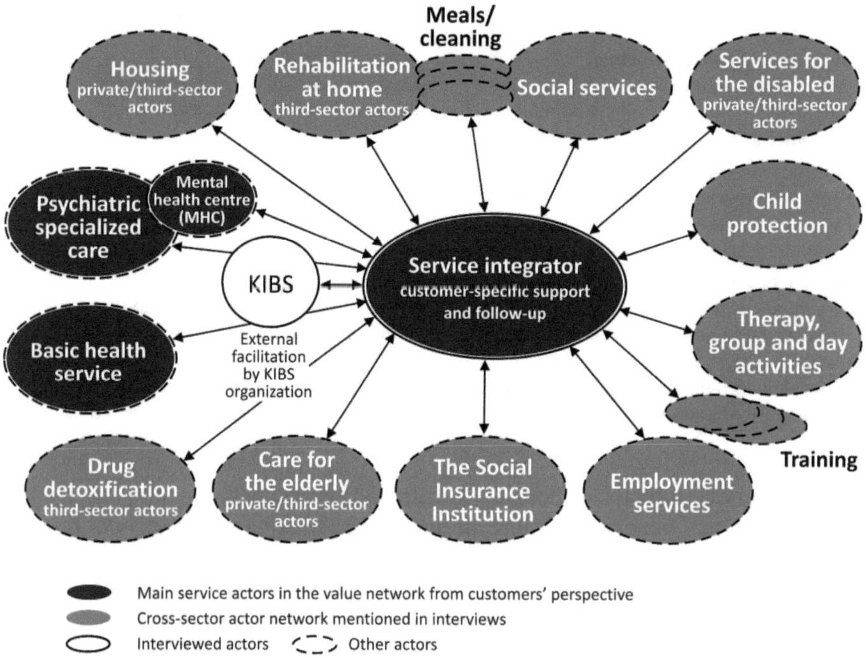

Fig. 14.2 Value network and service integrator in Case B: supporting the customers' mental health rehabilitation

to modern premises provided an ideal concrete boost for co-innovation and enhanced the role of the provider as a *service integrator* in the multi-actor value network.

Figure 14.2 illustrates the main actors in the value network of Case B from customers' perspective. The interviewees as informants represented the manager and the staff of the service integrator and the director of the outpatient mental health service of the city.

In both cases, the service directors have adopted the role of *internal innovation intermediary* (Kallio et al. 2013; Lusch and Nambisan 2015). Simultaneously with their local collaborative innovation activities, the service directors were involved in the one-year service development programme for Finnish cities, coordinated by a KIBS organization. This *external innovation intermediary* coordinated a programme for the service directors from several cities. This programme included service-oriented knowledge and tools, multilevel practical experiences presented by specialists, collegial networking, and individual and subgroup support for chosen local experiments. In the present study, those local experiments are examined from the customer-centric co-innovation perspective. In Case A, only two of four directors, one responsible for day care and preschool and the other responsible for special education, were able to participate in the entire programme. This has influenced the focus of my research in terms of interviewees and results. In Case B,

Table 14.1 The analysis frame for the case studies modified from Lusch and Nambisan (2015)

Service concepts/platforms: Private and public value Extended value propositions and modularity Customer as co-ideator vs. constraints	Aims and challenges Development actions taken with support of external KIBS organization
Service processes: Roles and responsibilities in actor networks Process and knowledge transparency Customer as co-innovator	Aims and challenges Development actions taken with support of external KIBS organization
Service ecosystem: Structural flexibility and integrity Institutional logics: regulations, policies, professions Strategic coordination: common mindset and architecture of participation	Aims and challenges Development actions taken with the support of external KIBS organization

the new service director responsible for the outpatient mental health service was involved in the same programme during 2012. Based on the second research question, my interviews also included specific questions concerning external facilitation. This facilitation was provided by two representatives of the KIBS organization. So they as external intermediaries and those three service directors involved in the one-year programme were asked about the aims and implementation of the programme as well as the contributions to the local service development.

The empirical data was analysed by applying qualitative content analysis with the help of the holistic framework of customer-centric service innovation. A specific focus was customer-centricity in terms of the service concepts/platforms, processes, and ecosystem discussed in the theoretical part. The contribution of the external facilitator to the local innovation activity studied in the value networks is also summarized. The analysis was firstly conducted case by case. Secondly, I compared the cases by defining similarities and differences between cases in order to specify and ensure internal coherency of my interpretations. The analysis was conducted iteratively through classifications, and testing and redefining empirical data through dialogue with the summarized key theoretical constructions presented in the Table 14.1.

14.4 Results

The main findings are presented case by case in Tables 14.2 and 14.3, in the framework of customer-centric service innovation. The externally facilitated transformation towards extended service concepts/platforms and transparent value co-creation processes in public–private value networks are illustrated, and their linkages to the wider service ecosystem are pointed out.

As the main findings in Table 14.2 show, Case A considers a complex and long-term transformation to enhance support for the growth and learning path of children

Table 14.2 Main findings of Case A: supporting growth and the learning path of children in the value actor network

Service concepts/platforms – Private and public value – Extended value propositions and modularity – Customer as co-ideator vs. constraints	**Strategic aim to support growth and learning path of children** – Increased and more complex customer needs for individual special support: difficulties in ensuring continuous support along the growth and learning path of children – Needs to enhance the adopted life-cycle approach and customer-centricity, particularly across sectors, among service providers: good service-specific tools for specifying value propositions from individual to group and organizational levels, but a lack of coherence between providers in the value networks (due to different laws and regulations) – Freedom of choice between providers, but not guaranteed in practice. Freedom of usage until age 7, when school starts **Clarifying extended value propositions and concepts with support of external KIBS** – Building mutual understanding and redefining value propositions in critical interfaces between value network actors – Some embryos of extended service concepts within the value network enabled by organizational change – Redefining collaboratively service-specific concepts step by step (by utilizing observation, regular customer surveys, customer-specific plans/discussions/evaluations, daily interaction with families and children)
Service processes – Roles and responsibilities in actor networks – Process and knowledge transparency – Customer as co-innovator	**Improving transparency within and between activities/actors in the service value networks** – More coherent and explicit value co-creation processes between actors from education, preschool, school, and special support were needed in order to support customer-specific life-cycle orientation – Lack of arenas for sharing ideas and diffusing lessons learned across value networks to enable radical innovations – Daily customer encounters are acknowledged as an arena for continuous innovations: ongoing experiments to co-innovate even more in a reflective, experiential, and systematic manner **Redefining critical interfaces between actors with support of external KIBS** – Process modelling (interfaces, responsibilities, knowledge transfer) and emerged renewed practices in value networks

(continued)

Table 14.2 (continued)

	– Service-specific process modelling and identifying development needs: transparency, coherency, equal service – Customer-centricity, process thinking, and evaluative development practices have been adopted
Service ecosystem – Structural flexibility and integrity – Institutional logics: regulations, policies, professions – Strategic coordination: common mindset and architecture of participation	**Change process from centralized towards decentralized, regional organization and management** – Challenges in effective resource integration in the value networks due to complex operative and strategic management – Staff focused on developing at the level of daily practice: challenge to diffuse and contribute to systemic level – Lack of systematic analysis of alternative scenarios and follow-ups: value in use/effectiveness of complex service chains **Resource integration and new management models with support of external KIBS** – Participatory change process: core/representative working groups, open regional groups enabling structural flexibility and integrity – Politically biased administrative and cultural change: resistance and confusion when making sense of new responsibilities, collaboration models and resource management. Sector-specific laws and regulations are experienced as a hindrance – Multilevel co-development experiments supporting change on operative and strategic levels: from suboptimization towards common systemic service culture and management

in the multi-actor networks. The need for systemic, extended service concepts is acknowledged in a strategic manner. Though, in this early phase of the transformation, the emphasis was still on redefining service-specific concepts collaboratively. However, there were already some concrete examples of pursuing co-creation of shared value propositions among several providers cross-sectorially.

The three-step model for special education needs provides one concrete manifestation from co-construction of common evaluation criteria and a plan for customer-specific support between various actors from special education, day-care, and primary schools to enhance seamless support in growing paths of children and their families. Similarly, networked collaboration practices with more transparent knowledge flows across sectors were co-constructed in critical interfaces, such as transition phases in schooling.

Difficulties were experienced in renegotiating mutual roles and responsibilities due to conflicting interests among different professionals and sector-specific regulations. Due to organizational change, some of those boundaries could have been

Table 14.3 Main findings of Case B: supporting the customer's mental health rehabilitation in a value network with a service integrator

Service concepts/platforms – Private and public value – Extended value propositions and modularity – Customer as co-ideator vs. constraints	**Strategic aim to renew support for customer's mental health rehabilitation** – Towards low-threshold and preventive service, daily performance activation, and from individual to family scope – Increased multiple deprivations result in complex service needs: cross-sector systemic service concepts are needed – Voluntary in law and practice: service is about gradually co-constructing customer commitment and mutual dialogue. Self-determination of customers confused in actor networks: legal rights of customers are increased, but not obligations **Redefining customers and clarifying value propositions with support of external KIBS** – The director as an internal innovation intermediary started the development work by spending time with staff and customers (observing, questioning, chatting), studying reports/statistics, and interviewing service providers and partners – Redefining with staff the main customer segments with prerequisites as a basis for customer-specific value propositions – Making value proposition criteria more explicit enables better mutual dialogues, evaluation, and refocus with customers and partners
Service processes – Roles and responsibilities in actor networks – Process and knowledge transparency – Customer as co-innovator	**Improving transparency within and between activities/actors in the service value network** – Increased need for cross-sector collaboration models; networked practices were developed. Challenges were faced with ownership and adoption (role confusion and optimization) due to project-based development work – Complex/unclear service chains with scarce knowledge-sharing practices: decreased public and private value, misuse – Need for boost to support customers to take ownership of their well-being, not only as a value co-creator but as a co-innovator of prerequisites for service activities and networked processes **Redefining critical interfaces between actors with support of external KIBS** – Specifying collaboration and communication practices between partners: repositioning as an integrator in better supporting the customer's comprehensive and effective rehabilitation (in terms of private and public value)

(continued)

Table 14.3 (continued)

	– Collaborative process modelling based on customer segments within the service integrator – Continuous practical, collaborative development by reflecting and experimenting with customers
Service ecosystem – Structural flexibility and integrity – Institutional logics: regulations, policies, professions – Strategic coordination: common mindset and architecture of participation	**Change process towards a service integrator in the complex service ecosystem** – Lack of coordination over complex value networks: scarce negotiation power and partnerships with private providers – Complex resource management and prioritizing policy: strong economic and prioritizing pressures – Need for strategic management and systemic mindset in outpatient mental health service in the service ecosystem **Prerequisites for dialogical and networked value co-creation with support of external KIBS** – Strategically managed participatory change process: empowerment, mutual appreciation, all competence in use, visibility – Developing (strategic) coordination and collaboration with service providers – Gradually systematized knowledge architecture serving all beneficiaries, from customers to politicians

resolved, and new practices have already been ideated and tested with customers, such as customer-specific daily arrangements for children who need constant individual special support. Although from an ecosystem perspective the administratively driven change process was challenged by conflicting interests between professions and policies, open planning process and supportive networked forums were enabled better cross-sectorial co-construction and diffusion of new practices. For that, the external facilitation of KIBS provided collaborative tools for focused targets and service-oriented change management approaches. So far, the service directors seem to adopt a systemic service mindset, while for most of the actors, the mutual benefits of the complex transformation, particularly from customer and value network perspectives, appear to be somewhat vague. However, renewals indicate the gradual emergence of a shared customer-centric educational culture within the value networks.

As the main findings in Table 14.3 indicate, the service provider is taking important steps in adopting a new role as a service integrator in supporting the customer's mental health rehabilitation in dynamic value networks. The specified need-based service concept as a low-threshold service has led to increases in customer volume and clarified collaboration with partners. Resource needs and activities are co-defined with customers and other providers in a more dialogical manner. However, challenges were faced to scale up and build alternative long-term service chains with network of providers (cf. module-based platform).

The service director applied the tools and lessons learned provided by the external KIBS organization to boost service and customer orientation both at the service integrator and network levels. She enhanced coordination and collaboration with service providers by visiting all the providers to set the ground rules, initiate continuous open dialogue and regular meetings, and establish the coordination team. Strategically guided and practically facilitated collaborative development resulted in more transparent service processes internally and gradually clarified collaboration practices with knowledge flows within multi-actor networks. Consequently, continuous practice-based co-innovating with customers by reflecting and experimenting was explicated as essential and legitimated practice. Group therapies provide one ideal manifestation of that kind of multi-actor activity. Thus renewals regarding future-oriented prerequisites for service exchange, networked management, and novel resource integration seem to enhance both customer-specific rehabilitation and systemic effectiveness (private and public value).

14.5 Conclusions and Implications

This study has aimed to provide a deeper understanding of the ongoing transformation towards customer-centric service innovation in public–private interfaces. The conceptual framework is driven from recent studies on the S-D logic, with the emphasis on systemic innovation from the points of view of both economic and social construction. When focusing on public–private interfaces, institutional logics with strong regulatory aspects play an essential role and are manifested in innovation activities in terms of the pursued relationships, competencies, and knowledge. Thus, the applied tripartite service innovation framework of Lusch and Nambisan (2015) is critically discussed and supplemented, together with previous service innovation research in the public context (Hartley 2005; Langergaard and Scheuer 2012). The holistic conceptual frame has been empirically illustrated in two case studies. Case A concerned supporting growth and the learning path of children in the value actor network, and Case B supporting the customer's mental health rehabilitation in a value network with a service integrator.

To conclude, even though Case A and Case B differ in their scope and the context of service innovation activities, we can also identify similarities between them in terms of evolving value propositions and service concepts, value co-creation processes, and systemic changes *towards a deeper service-, customer-, and network-centricity*. In the light of the tripartite framework of service innovation suggested by Lusch and Nambisan (2015), the results illustrate three issues in particular. First, the co-construction of extended value propositions and service concepts among customers and several service providers appears to be in an active experimentation phase. Challenges are faced considering modularity and scalability in particular. Second, the findings show that practices that enables more customer-centric and cross-sectorial service chains have emerged gradually. Third, the prerequisites towards not only effective but also novel resource integration and re-bundling have been developed at both strategic and operative levels. Structural flexibility and integrity have been enhanced by means of both new organizational arrangements and dynamic networked forums. Therefore, my cases reveal empirical indications of the *implementation of S-D logic in customer-centric service innovation in the public–private interface*.

The empirical findings of this study are also aligned with the recent studies on the significance of the dialogical and legitimated interplay between *bottom-up and top-down innovation activities* (Hartley 2005; Lappalainen et al. 2012; Saari et al. 2015; Sørensen et al. 2013). It seems that increasingly, also in the public–private interface, innovations are understood to be created in everyday social and value-creating activities among different actors, including beneficiaries such as customers. In my study, even though the daily customer encounters are comprehended as an arena for continuous collaborative innovations, more sophisticated ways of capturing information of value in use experienced by customers are gradually tested and sought after (Edvardsson et al. 2012). However, as claimed by Rosenthal and Peccei (2007), there were also indications in my empirical findings that *public and private value* are perceived as contradictory, thus causing confusion in networked service practice. Therefore, it has become evident, also in the public–private interface, that, in order to develop the prerequisites for successful value co-creation, it is necessary to utilize the customers in alternative roles in the service innovation. The inclusion of other stakeholders in the wider value-creation network is equally important (Kallio et al. 2013; Edvardsson et al. 2012; Elg et al. 2012; Hennala et al. 2012; Langergaard and Scheuer 2012; Lusch et al. 2010).

Cross-sectorial and networked coordination via re-negotiated rules and open platforms and architectures of participation seems to be a necessity for resource integration and re-bundling as well as scaling up novelties at ecosystem level, as suggested by Lusch and Nambisan (2015). As has been claimed to be a characteristic of innovation in the public sector, my study also show that conflicting interests related to policy, professional, and managerial issues have been experienced as a significant hindrance to developing novel ways for value co-creation in a cross-sectorial and networked manner (Langergaard and Scheuer 2012). These *institutional idiosyncrasies* cannot be ignored when studying service innovation in the public–private interface.

In addition to bottom-up innovation activities, a more systemic service mindset has been adopted among interviewed directors, who enacted the role of important *internal innovation intermediaries*. They have adopted a coordinative role between strategic and practice levels by enhancing diffusion of emerged extended service concepts and collaborative practices in value networks and supporting continuity across project-based experiments. My empirical findings are supported by the recent study by Saari et al. (2015), who point out the central role of middle managers by examining their “bridging” activities in public service innovation. Regarding the contribution of the KIBS organization as an external innovation intermediary, the empirical results indicate that external facilitation might be multiform in nature. It can be targeted at a multilevel change in the service ecosystem and then step by step at specific interfaces. The external facilitation can also be targeted at the service integration and gradually be expanded to value networks of the service ecosystem. To conclude, the *external innovation intermediary* can serve as a kind of *facilitative value co-creation platform* for the service directors, by means of service-oriented knowledge and tools, multilevel practical experiences, collegial networking, and individual and subgroup support for chosen local experiments (Lusch and Nambisan 2015). Our study confirms the findings of previous studies on KIBS organizations playing several important roles as external innovation intermediaries (den Hertog 2002; Kallio et al. 2013; Lappalainen et al. 2012).

In terms of conceptual validity, the service innovation framework (Lusch and Nambisan 2015) that I have supplemented in this study seems promising in the rather little studied context of public and private actors. The empirical results are discussed with interviewees to improve their internal validity (Kvale 1996). Further examination could be targeted at gaining understanding from multi-actor perspectives on co-innovation in complex value networks. In terms of transferability, my supplemented service innovation framework needs further empirical verification in different service contexts. For practitioners, the study provides a holistic frame for reflection in terms of the transformation towards a service mindset and collaborative innovation when tackling topical societal challenges.

Acknowledgements The author wishes to thank the case organizations for fruitful cooperation. The author is also grateful to Research Professor Marja Toivonen from VTT, the reviewers of the article, and the editors of the book for their valuable comments. The research has been funded by TEKES (Finnish Funding Agency for Technology and Innovation) and VTT.

References

Ballantyne, D., & Varey, R. J. (2006). Creating value-in-use through marketing interaction: The exchange logic of relating, communicating and knowing. *Marketing Theory, 6*(3), 335–348.

Brand, R. (2005). The Citizen-innovator. *The Innovation Journal: The Public Sector Innovation Journal, 10*(1), 1–10.

Carlborg, P., Kindstrom, D., & Kowalkowski, C. (2014). The evolution of service innovation research: A critical review and synthesis. *The Service Industries Journal, 34*(5), 373–398.

Chandler, J. D., & Vargo, S. L. (2011). Contextualization and value-in-context: How context frames exchange. *Marketing Theory, 11*(1), 35–49.

den Hertog, P. (2002). Co-producers of innovation: On the role of knowledge-intensive business services in innovation. In J. Gadrey & F. Gallouj (Eds.), *Productivity, innovation and knowledge in services: New economic and socio-economic approaches*. Cheltenham: Edward Elgar.

Edvardsson, B. (1997). Quality in new service development: Key concepts and a frame of reference. *International Journal of Production Economics, 52*(1–2), 31–46.

Edvardsson, B., Kristensson, P., Magnusson, P., & Sundström, E. (2012). Customer integration within service development—A review of methods and an analysis of in situ and ex situ contributions. *Technovation, 32*(7), 419–429.

Edvardsson, B., Tronvoll, B., & Gruber, T. (2011). Expanding understanding of service exchange and value co-creation: A social construction approach. *Journal of the Academy of Marketing Science, 39*, 327–339.

Eisenhardt, K. M. (1989). Building theories from case study research. *Academy of Management Review, 14*(4), 532–551.
Elg, M., Engström, J., Witell, L., & Poksinska, B. (2012). Co-creation and learning in health care service development. *Journal of Service Management, 23*(3), 328–343.
Hartley, J. (2005). Innovation in governance and public services: Past and present. *Public Money & Management, 25*(1), 27–34.
Hennala, L., Konsti-Laakso, S., & Harmaakorpi, V. (2012). Challenges of bringing citizen knowledge into public sector service innovation. In H. Melkas & V. Harmaakorpi (Eds.), *Practice-based innovation: Insights, applications and policy implications* (pp. 255–275). Berlin: Springer.
Kallio, K., Lappalainen, I., & Tammela, K. (2013). Co-innovating in public services: planning or experimenting with users? *The Innovation Journal: The Public Sector Innovation Journal, 18* (3), article 7.
Kvale, S. (1996). *InterViews. An introduction to qualitative research interviewing*. Thousand Oaks, CA: Sage Publications.
Langergaard, L. L., & Scheuer, J. D. (2012). Towards a deeper understanding of public sector innovation. In L. A. Macaulay, I. Miles, J. Wilby, Y. L. Tan, L. Zhao, & B. Theodoulidis (Eds.), *Case studies in service innovation* (pp. 167–193). New York: Springer.
Lappalainen, I., Kallio, K., & Toivonen, M. (2012). Facilitating service development in a multi-agent environment: Different roles of consultancy. In J. C. Spohrer & L. E. Freund (Eds.), *Advances in the human side of service engineering* (pp. 276–285). Boca Raton, FL: CRC Press.
Lehtonen, M., & Tuominen, T. (2011). Multiple voices of the user in public sector services. In J. Sundbo & M. Toivonen (Eds.), *User-based innovation in services* (pp. 227–250). Cheltenham: Edward Elgar.
Lusch, R. F., & Nambisan, S. (2015). Service innovation: A service-dominant logic perspective. *MIS Quarterly, 39*(1), 155–175.
Lusch, R. F., Vargo, S. L., & Tanniru, M. (2010). Service, value networks and learning. *Journal of the Academy of Marketing Science, 38*(1), 19–31.
Möller, K., & Rajala, A. (2007). Rise of strategic nets – New modes of value creation. *Industrial Marketing Management, 36*(7), 895–908.
Rosenthal, P., & Peccei, R. (2007). The work you want, the help you need: Constructing the customer in Jobcentre Plus. *Organization, 14*(2), 201–223.
Saari, E., Lehtonen, M., & Toivonen, M. (2015). Making bottom-up and top-down processes meet in public innovation. *The Service Industries Journal, 35*(6), 325–344.
Sørensen, F., Sundbo, J., & Mattsson, J. (2013). Organisational conditions for service encounter-based innovation. *Research Policy, 42*(8), 1446–1456.
Sundbo, J. (2011). Value chain innovation: An actor-network-theory based model for innovation at the interface between service and other economic sectors. In J. Sundbo & M. Toivonen (Eds.), *User-based innovation in services* (pp. 71–98). Cheltenham: Edward Elgar.
Sundbo, J., & Toivonen, M. (Eds.). (2011). *User-based innovation in services*. Cheltenham: Edward Elgar.
Thomson, A.M. (2001). *Collaboration: Meaning and measurement*. Ph.D. dissertation, Indiana University, Bloomington, IN.
Thomson, A. M., & Perry, J. L. (2006). Collaboration processes: Inside the black box. *Public Administration Review, 66*(Special Issue 1), 20–32.
Toivonen, M. (2010). Different types of innovation processes in services and their organizational implications. In F. Gallouj & F. Djellal (Eds.), *The handbook of innovation and services* (pp. 221–249). Cheltenham: Edward Elgar.
Tronvoll, B., Brown, S., Gremler, D. D., & Edvardsson, B. (2011). Paradigms in service research. *Journal of Service Management, 22*(5), 560–585.
Vargo, S. L., & Lusch, R. F. (2008). Service-dominant logic: Continue the evolution. *Journal of the Academy of Marketing Science, 36*(1), 1–10.

Chapter 15
Business Model Design and Value Co-creation: Looking for a New Pattern

Angela Caridà, Monia Melia, and Maria Colurcio

Abstract Innovation requires openness and collaboration and firms' abilities to design business models that enable higher resource density for value creation. The framework depicts both the dimensions of a co-creative business model and how they interact to integrate resources for enhancing value creation and capture. It empirically contextualizes the framework, by illustrating the Formabilio case, to better demonstrate how it works in practice. The study provides a detailed and granular approach to co-creation design to sustain the competitiveness of small- and medium-sized enterprises in a complex and increasingly digital-based ecosystem. More specifically, it offers new knowledge about mechanisms and methods for aligning the contributions of different actors to better understand how value is co-created through actors' resource-integrating efforts and captured by the lead company. It addresses relevant gaps in current knowledge on business model design and on resource integration and the value co-creation practice.

15.1 Introduction

The success of new organizational forms is often ascribed to the development of novel business models (George and Bock 2011; Franke et al. 2008): changes in the competitive environment force manufacturing firms to adopt new business models and modus operandi increasingly focused on partnering with different actors within a complex ecosystem. As competitiveness now relates to enterprise networks (Lee et al. 2010), collaboration is a necessary path, especially for small- and medium-sized enterprises (SMEs) (Colurcio et al. 2012; Goss 1991; Rothwell and Dodgson 1992), to access new resources (e.g. new technological and relational opportunities) and develop successful competitive strategies for accessing new markets (Colurcio et al. 2012; Ulaga and Eggert 2005).

A. Caridà (✉) • M. Melia • M. Colurcio
Department of Legal, Historical, Economic and Social Sciences, University Magna Graecia of Catanzaro, Catanzaro, Italy
e-mail: angela.carida@unicz.it; monia.melia@unicz.it; mariacolurcio@unicz.it

T. Russo-Spena et al. (eds.), *Innovating in Practice*,
DOI 10.1007/978-3-319-43380-6_15

Collaboration within and through a network of firms requires a global and open firm perspective to help actors (e.g. customers, suppliers and distributors) to integrate their own resources in any relevant practice (Storbacka et al. 2012), to create value (Colurcio et al. 2014; Vargo 2008; Vargo and Lusch 2008) and to match their standpoints (Rampersad et al. 2010). Afterwards, openness and collaboration have forced firms to design business models that enable higher resource density for value creation (Normann 2001) by shifting from a traditional business model based on a hierarchical system and competition into one based more on collaboration and social networking.

This transformation is now enhanced by the Internet and interactive technologies that allow for techniques and tools (Caridà and Colurcio 2013; Schau et al. 2009) to mobilize actors' resources by allowing, independent of location, their best combination for creating the optimal value/cost result (Normann 2001). Technology allows for interaction between the actors of the firm's network, but does not ensure that interactive value formation (Echeverri and Skålén 2011) and the related positive outcomes—value co-created through resource integration—take place in practice. Value co-creation is a process of multiple interactions and multidirectional resource integration (Vargo and Lusch 2008) that includes the idea of reciprocity (Storbacka et al. 2012) by requiring the dynamic alignment/matching of actors' resources (Caridà and Colurcio 2013; Echeverri and Skålén 2011). In this view companies are called to rethink the design of business models as a strategic process (Teece 2010) to support the practice of resource integration and value co-creation.

The practices of integrating resources and the methods for designing and configuring the resource integration process remain fuzzy (Kleinaltenkamp et al. 2012: 59): 'a focal actor wishing to engage in co-creation needs to design an open business model that permits other actors to influence specific design elements in such a way that other actors have the opportunity to participate in specific activities' (Storbacka et al. 2012). The business model construct defines the availability of resources in the network and the practices by which an actor interacts with other actors in order to increase the density of resources for several actors simultaneously (Storbacka et al. 2012). This definition stresses the role of business model design as a fundamental phase to shape collaborative loci for integrating resources and improving value co-creation practices (Storbacka et al. 2012).

It is a blueprint for how to run a business (Osterwalder et al. 2005), as a bundle of practices that involves new forms of cooperation and collaboration (Zott et al. 2011: 1032) for influencing and shaping collective and individual action (Mason and Spring 2011: 1038). This approach shifts the business model construct towards networked value creation (Zott and Amit 2008) and frames it as a central concept in explaining how value is co-created (Storbacka et al. 2012) and captured (Zott and Amit 2008). Both these functions are critical for the sustainability of the business itself (Chesbrough 2007).

Despite the fact that the literature on business models could enrich the existing discussion on value co-creation (Frow et al. 2015) and that many authors call for contributions on tools, processes and practice encouraging effective value co-creation (Barczak 2012; Edvardsson et al. 2014; Korkman et al. 2010;

McColl-Kennedy et al. 2012) and on how to design and configure the integration process (Kleinaltenkamp et al. 2012), the topic of the business model has received little attention in the marketing literature: 'previous discussions do not sufficiently address business models for purposeful co-creation design' (Frow et al. 2015). Indeed, some recent contributions (Frow et al. 2015; Storbacka et al. 2012) clearly address the strong link between business model design and firms' abilities to identify and manage new opportunities for value creation in a networked business context.

This chapter aims to offer a strategic framework for designing a business model that supports firms' abilities to manage new business opportunities based on networked value creation. Drawing from previous contributions on the topic, this study frames the dimensions of a co-creative business model. The chapter analyses the interconnected elements that constitute it, how they interact to integrate resources and the consequences of their doing so (value created and captured). To better show how firms might use the framework we propose, we illustrate how Formabilio, the company involved in the study, applied it to co-create and capture value. Formabilio is an Italian start-up producing furniture, lighting and home accessories designed by creative minds from all over the world in a participatory and eco-friendly way.

The purpose of this empirical contextualization is to demonstrate how the framework works in practice; it offers a detailed and granular approach to co-creation design (Frow et al. 2015) to sustain the competitiveness of SMEs in a complex and increasingly digital-based ecosystem. More specifically, this chapter provides new knowledge about mechanisms and methods for aligning the contributions of different actors to better understand how value is co-created through actors' resource-integrating efforts and captured by the lead company. It addresses relevant gaps in current knowledge on business model design and on resource integration and value co-creation practice.

It contributes from both a theoretical and a managerial perspective. First, it outlines the field of designing a business model for co-creating value that is still in an emergent stage (Frow et al. 2015) and provides a wide range of opportunities for scholars to undertake future studies. It extends the extant literature on resource integration that has been rather abstract in nature and provides a few implications for managers and practitioners to realize its potential value in practice.

Accordingly, it can support practitioners in better designing and managing business models based on the resource integration mechanism and value co-creation practice, which according to many authors are one fundamental source of competitive advantage. Moreover, it offers some insights on how to exploit interactive technologies to mobilize and enhance the best combination of the actors' resources for creating a collaborative and networked environment in a traditional and not open business context such as the Italian manufacturing one.

The remainder of this chapter is organized as follows. First, we review the literature discussing previous contributions on business model design for co-creating value. Second, we present and contextualize the research construct through a qualitative case to show it in practice. Then we discuss the research contributions and theoretical and managerial implications. Finally, we address the limits of the study and suggestions for further research.

15.2 Theoretical Backbone

The chapter outlines the main issues covered by studies on business models and on the design of a business model for enabling resource integration and value co-creation.

15.2.1 Designing a Business Model for Co-creating and Capturing Value

Since the mid-1990s, interest in business models has grown considerably among scholars and practitioners (Berglund and Sandström 2013; Zott et al. 2011). The advent of the Internet is one of the main reasons for such an interest (Amit and Zott 2001; Zott et al. 2011), as well as global competition, the development of flexible manufacturing technologies and industrial and technological convergence (Berglund and Sandström 2013). However, despite the interest and the literature debate, there is no general consensus regarding the definition (Berglund and Sandström 2013; Zott et al. 2011), nature, structure and evolution of business models (Morris et al. 2005). It is a very fragmented and multidisciplinary concept that includes different elements according to the researchers' field of interest: economic (Afuah and Tucci 2003; Stewart and Zhao 2000), operations (Osterwalder et al. 2005; Timmers 1998) and/or strategy (Hamel 2000; Chesbrough and Rosenbloom 2000). The business model concept has been defined as a statement, a description, an architecture, a conceptual tool or model (Baden-Fuller and Morgan 2010), a framework, a pattern and a set of activities (Morris et al. 2005; Zott et al. 2011). According to Kajanus et al. (2014), it is a vehicle that closes the gap between abstraction and practice within strategic management and business. The business model manifests itself in the practices in which the focal actor engages, which influence how the focal actor relates to other actors (Storbacka et al. 2012).

The lack of a convergent and well-defined theoretical construct on the business model (George and Bock 2011) contributes to increasing the gap between theory and managerial practice. To close this gap, many authors frame the business model as a set of components or building blocks (Afuah and Tucci 2003; Amit and Zott 2001; Kajanus et al. 2014; Osterwalder and Pigneur 2010; Timmers 1998; Zott et al. 2011), as well as linkages between these components and dynamics (Afuah and Tucci 2003). Components can refer to resources, business actors, their role and their potential benefits (Timmers 1998) and activities (Zott et al. 2011), as well as to their interrelationships. These latter are enabled by technology and transcend the firm's boundaries to create (Amit and Zott 2001), deliver and capture value (Chesbrough and Rosenbloom 2000; Teece 2010). In other words, the business model can be viewed as a new unit of analysis (Zott et al. 2011), as a system-level concept that goes beyond the boundaries of the focal firm (Afuah and Tucci 2003) and is enabled by technology, centred on activities and focused on value co-creation (Storbacka et al. 2012) as well as on value capture (Zott et al. 2011). Therefore, its design should illustrate the relationships that the actor has with other actors in its value network, the resources available in the network and the practice of integrating them, as well as the performance of the actor in terms of value created and captured.

This network-centric approach to business models (Storbacka et al. 2012) is new and opens up interesting opportunities for designing business model innovation, thus for framing new business logics and/or for defining different business models totally based on value co-creation in existing and traditional businesses (Schneider and Spieth 2013). Indeed recently, some authors (Frow et al. 2015; Nenonen and Storbacka 2014; Storbacka et al. 2012) have clearly addressed the strong link between business model design and firms' abilities to manage co-creation opportunities: 'co-creation can be one over-riding theme' (Storbacka et al. 2012: 59) in the business model literature. A co-creation design framework assists firms in identifying new opportunities for co-creation, as through the business model it defines the practices by which an actor interacts with other actors in order to increase the density of resources for several actors simultaneously (Frow et al. 2015; Storbacka et al. 2012). To support co-creation, the business model has to ensure change in the focal actor's practices, which is necessary for aligning the other actors' practices and for involving them in specific activities (Frow et al. 2015). For example, Storbacka et al. (2012) stress the importance of the configurational fit of the subsequent design elements (i.e. principles, resources and capabilities) and dimensions (i.e. markets, offerings, operations and organization) to support the firm in catching value co-creation opportunities: 'it is likely to require new types of capabilities such as more modular offering design, modular and flexible manufacturing and delivering, the ability to manage partners and alliances' (Storbacka et al. 2012: 64). A business model for value co-creation creates a new firm configuration totally based on the idea of creating value not only with and for customers but with and for all the actors integrating their own resources in the network (Vargo and Lusch 2011).

15.2.2 Research Methodology

The study adopts an abductive research approach (Dubois and Gadde 2002) characterized by an iterative process of systematic combinations and inferences that match theory with reality. It is particularly appropriate when pursuing theory development, that is, refining existing theories rather than inventing entirely new ones. The research process, undertaken over 6 months, consisted of four phases.

The first phase related to research planning and exploration of co-creative business model dimensions. In this phase, we identified the key dimensions on the basis of an extensive literature review. The second phase concerned the co-creative business model framework development. The third phase related to the empirical contextualization of the co-creative business model that we proposed through the analysis of the Formabilio case. Through this empirical contextualization, we illustrate how the framework works in practice, by supporting firms in co-creating and capturing value. We adopted the single case-study method as it appears the most appropriate in order to deeply investigate and understand complex and dynamic phenomena (Yin 2009), such as the one under investigation. Furthermore, it solves the problem of contemporaneity between the phenomenon's evolution and its analysis (Yin 2009), allowing us to explain how or why a social phenomenon works in its real-world context (Yin 2009).

Data has been collected in two stages. The first stage related to non-participant observation of the mechanisms of the digital platform Formabilio.com. We carried out, moreover, a detailed search and analysis of information, articles, documents and publications available through the web and other sources. This research stage allowed us to better understanding elements constituting Formabilio's co-creative business model, their interaction to integrate resources and the consequences of their doing so (value created and captured).

The second step has been focused on direct interaction with all actors of the Formabilio network. Direct interviews and online questionnaires have been conducted to understand the actors' roles, the activities they develop within the community and the different typologies of value co-created. The interviews were tape-recorded and transcribed verbatim. Data and information were selected, analysed and interpreted on the basis of the research framework that we explained above.

15.3 The Business Model Design Framework for Co-creating and Capturing Value

Based on the relevant supporting literature, Fig. 15.1 provides a theoretical framework for designing business models encouraging effective value co-creation. Our framework is consistent with a network-centric approach to business models (Storbacka et al. 2012) to emphasize the shift from focal firm to networked value creation, as well as with the idea that the key to understanding value co-creation in practice is to focus on the actors involved, their actions and interactions (Edvardsson and Tronvoll 2013). It explains the practice through which a firm integrates the resources of specific actors to create value. Accordingly, it defines the elements and the working mechanism of the business model from a co-creation and resource integration view, to assist firms in the mutual alignment of actors' practices, for enhancing co-creation opportunities, value co-created and captured.

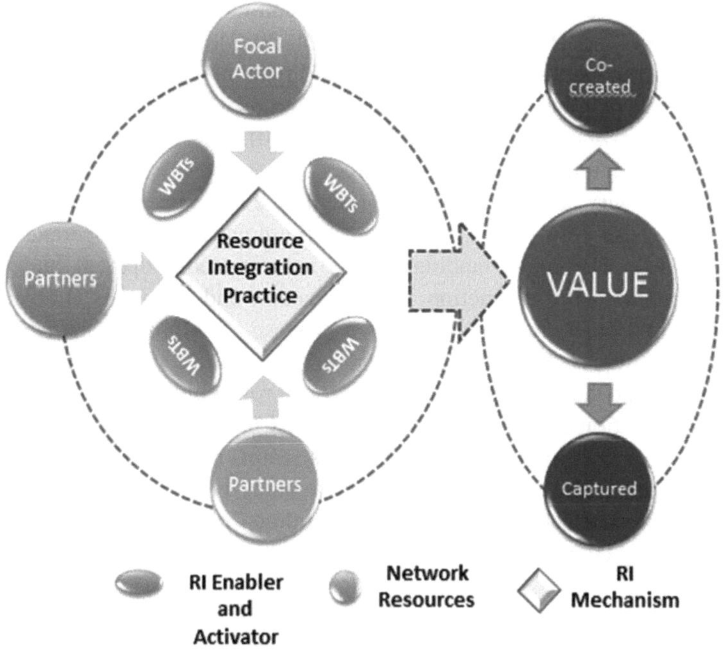

Fig. 15.1 The business model framework: Key elements and working mechanism

We explain the key elements of the framework: (1) resources, including human/ non-human actors and their role within the network; (2) the resource integration mechanism, including rules, principles and activities; and (3) the value created and captured by the lead firm.

15.3.1 Network Resources

Resources are critical for co-creation (Akaka and Vargo 2014; Storbacka et al. 2012; Vargo and Akaka 2012); they are not simply in existence but rather must come into being (Löbler 2013; Edvardsson et al. 2014; Pels et al. 2009; Vargo and Lusch 2011; Zimmermann 1951). Indeed, 'to be a resource is a property of the specific entity, which comes and goes in parallel with the integration process resources' (Löbler 2013: 424). Resources come into being through the involved actors' resource activities in practice (Löbler 2013; Edvardsson et al. 2014; Pels et al. 2009; Vargo and Lusch 2011; Zimmermann 1951).

Things, persons, machines, money, institutions or concepts only become so when they are integrated through interaction to perform an intended activity (Löbler 2013). Resources can be physical and non-physical entities (Vargo et al. 2010) as well as human and non-human (Edvardsson and Tronvoll 2013). Human resources are dynamic and related to the actor, whereas non-human resources are static and related to the object (Edvardsson and Tronvoll 2013). Actors are generic entities (i.e. social and economic) who have the capability to integrate resources (Vargo and Lusch 2008). They are engaged in an exchange relationship and act as resource integrators to carry out activities (Mele et al. 2010).

While some actors act on other resources to foster new ways of creating value, when they are very active and more representative of an operant resource (Löbler 2013), others are very passive and more representative of an operand resource (Löbler 2013). The understanding of the actor as both operant and operand resources replaces the distinction between actors and resources (Löbler 2013). According to Löbler, the framework we propose does not distinguish between actors and resources. However, all actors act as resource integrators (Vargo and Lusch 2008), and they can obtain a specific social role within the network according to their expectations. According to Akaka and Chandler (2011), social roles are resources that actors integrate through their attempts to co-create value. They can be both operant and operand resources, influencing or being influenced by value networks. Our view is consistent with these authors' idea that social roles are a set of practices that connect actors to enact resource integration and value co-creation activities.

Actors can play a diverse set of roles in resource integration and service innovation (Lusch and Nambisan 2015). For example, the focal firm can enact the role of facilitator or orchestrator in the actors' value-creation process. It can influence value creation through orchestrating interaction—and hence resource integration—by taking the role of an inspirator, a facilitator and an attendant within the value emergence process (Gidhagen et al. 2011: 404). Similarly, service dominant logic identifies three broad roles—ideator, designer and intermediary—depending on the nature of service exchange and the type of resource integration achieved (Lusch and Nambisan 2015). Most studies on value co-creation relate more to the customers' roles (Cova and Dalli 2009; Prahalad and Ramaswamy 2004; Sawhney et al. 2005); therefore, the analysis of the other actors' roles has been missed so far.

15.3.2 RI Enabler

Technology is a critical component in value creation (Maglio and Spohrer 2008) and innovation, as it enables resource density and increases use value (Storbacka et al. 2012). Information technology (IT) allows for new combinations of digital and physical components to create digital or digitally enabled innovations (Lusch and Nambisan 2015). These innovations are novel market offerings (Yoo et al. 2010), technically feasible and economically viable.

Accordingly, the understanding of the IT role is fundamental to better design of a co-creative business model (Orlikowski 1992). Technology is a product and an enabler of human action; it can be considered as both an outcome (operand resource) of and input (operant resource) affecting the way in which value is determined (Akaka and Vargo 2014; Orlikowski 1992), as well as the practices and processes that increase the ability to integrate and apply resources to create value (Arthur 2009). Technology is the practical application of knowledge (Lusch and Nambisan 2015). Although it could have been included in the section on network resources, we put technology in a separate section to stress its key role in connecting, integrating and empowering (Ng et al. 2011) actors to be resource integrators and value co-creators. It can be defined as cross-cutting resources because it is interrelated with each of the other resources and has the potential to influence almost every aspect of service. In this view, technology, innovation, and service are interlinked (Lusch and Nambisan 2015). Specifically, we focus on Internet-based technologies, as they have profoundly influenced how value is created and managed. They facilitate collaboration between actors, allowing for the broad application of techniques and tools to stimulate the participation and engagement of the entire firm network in the value-creation process (Caridà et al. 2014; Schau et al. 2009). Web-based technologies (WBTs) allow actors access to additional resources, creating new exchangeable resources through integration (Vargo and Lusch 2011): through their interactions, partners integrate knowledge, skills and other resources to put together a network of firms possessing a set of competencies capable of offering a value innovation that is an innovative value proposition which enables higher value co-creation (Mele et al. 2010).

15.3.3 RI Mechanism/Practice

RI is the means through which resource integrators (actors) co-create phenomenologically determined value through process(es) (Kleinaltenkamp et al. 2012).

Its working is moderated by fitting any particular potential resource with other potential resources, which entails the removal of resistance to resource utilization and the beneficiary's ability to integrate those resources (Vargo and Lusch 2011: 184). In sum, it relates to the incorporation of an actor's resources into the processes of other actors in accordance with their expectations, needs and capabilities and implies a social and cultural process that enables an actor to become a member of a network (Gummesson and Mele 2010).

The RI mechanism relates to the actors' operation on available resources (Caridà et al. 2015). It implies a continuous process including a series of activities (Payne et al. 2008) and requires the definition of procedures, such as explicit rules, principles, precepts and instructions (how actors may interface) to prevent conflicts and guide the collaboration involved in resource integration (Edvardsson et al. 2014). Regulative, normative and cognitive rules (institutions) foster a shared awareness (Nambisan and Sawhney 2007) that is critical to capitalize on the diverse expertise and capabilities of the network actors. In other words, rules affect the actors' activities and interactions during resource integration for value co-creation (Edvardsson et al. 2014). Activities are the active things that actors carry out through a specific set of interactions (Ballantyne and Varey 2006; Peters et al. 2014) by integrating resources from multiple sources (Mccoll-Kennedy et al. 2012). They can be referred to as the Co-s for innovation (Russo-Spena and Mele 2012).

15.3.4 Value

Value is an abstract concept with specific meanings that vary according to context (Sweeney and Soutar 2001). It is the potential outcome that actors can extract from the interaction. Thus, it is co-created and assessed in use (Gummerus 2013; Laamanen and Skålén 2014). As Nenonen and Storbacka (2014) and Storbacka (2011) emphasize, value can be analysed according to the customers' (value creation) and the firm's perspectives (value capture). Value creation is a prerequisite for value capture (Gosselin and Bauwen 2006), as long-term value capture depends on the customer's positive evaluation of the relationship with the firm (Storbacka 2011).

Previous research (Nenonen and Storbacka 2014) focused more on value capture from a dyadic perspective (customer–firm relationship) rather than from a network perspective. We consider economic revenue as an appropriate measure of value capture that emerges from the relationship with the entire actors' network instead of only with customers. In other words, value co-creation is a prerequisite for value capture. Differently from the value captured, the value co-created is not limited to the economic or financial aspects but encompasses different interrelated dimensions. The first is functional, which is related to the ability of the product/service to fulfil its functional purposes (Sheth et al. 1991). It includes service excellence, price and time savings (Lee and Overby 2004) and the ease and speed of achieving a task effectively and conveniently (Pura 2005). The second is social, which is related to the benefits derived through interactions, for example, emotional support, self-esteem and social enhancement (Sicilia and Palazón 2008). The third is emotional, which is related to the utility derived from the affective states—confidence, excitement or fear (Sheth et al. 1991)—generated by a product or a service (Pura 2005; Sweeney and Soutar 2001). Finally, value can be epistemic, which is related to the capacity to arouse curiosity, to provide novelty and/or to satisfy a desire for knowledge (Pura 2005; Sheth et al. 1991). It includes the concept of learning as the gaining of knowledge (Bourdeau et al. 2002).

15.4 The Framework Application: Formabilio

Box in the following illustrates the background to Formabilio, the company participating in the research. It applies the framework outlined above and rebuilds the business logic of an existing and very traditional business, one in Italian furniture and home accessories.

Founded in 2013, Formabilio is a furniture brand headquartered in Treviso, Italy. It sells online furniture, lighting and home accessories, which are designed and produced in a participatory and eco-friendly way through the digital platform formabilio.com. Formabilio is a contest provider that organizes idea contests to allow fresh ideas to be shared in a networked context, including talented designers from all over the world, manufacturing companies with a strong craftsmanship tradition and value ethics and design lovers (customers, brand lovers, etc.). Original and innovative design ideas are submitted online through a digital platform, but only the best ones, which are chosen by the community and evaluated by a jury of experts, are turned into exclusive, eco-friendly products, 100 % made in Italy and sold in the store online. The community's judgement is based on the number of votes, the average of the opinions (e.g. I don't like it, do it better; good, I like it; and great), as well as the reputation of voters.

The Formabilio business model is based on co-design, co-production on demand, online sales and revenue participation. These practices are enabled by the online platform; it allows the entire network—designers, customers and companies—to be involved in the process of value creation. Eighteen months from its launch, the Formabilio community accounted for:

- 4521 designers
- 137,311 brand lovers
- 11,401 ideas submitted
- 50 online contests
- 805,812 community votes
- 100 products

In the following, we explain how Formabilio applied the co-creation design framework to lead to actors' resource integration for enacting its intended activities to co-create value.

15.4.1 Network Resources

Formabilio is the hub of the network. The company enhances the integration of resources among all the players involved in the project, establishing connections and relationships through the digital platform formabilio.com. Formabilio orchestrates the interaction and guides the resource integration through the practice of idea contests to achieve original, valuable and realizable ideas.

> "*Formabilio is a showcase for good design. It pays attention to experimentation, encourages creativity and fosters sharing and comparison among experts*". Chiara, Formabilio designer

All of the actors interacting in the network form a valuable resource and act as resource integrators. They perform the roles of both resource provider and value beneficiary, integrating human (e.g. creativity, knowledge, skills, time, effort, etc.) and non-human (e.g. the platform's technical infrastructure) resources.

The main actors involved within the network are designers and the manufacturing companies that produce the selected ideas (partners). Both of them act as primary resource integrators, as they provide input for the design and the further realization of new products through new materials and new applications. The knowledge exchange within the community allows for the optimization of products and solutions. The following extracts illustrate the nature of these discussions.

> "*G: Very good! I like your idea. Probably you can use a part of it for the lamp backing. Anyway my vote is 5, the maximum*". Bruno, designer

> "*S: I think you're right. I can regain more space. Thank you so much for your suggestion.*" Giancarlo and Silvia, designers

> "*I remember a conversation between designers, during which one of them provided some suggestions to better fit an idea. The proposal has been changed according to the suggestions and has been successfully evaluated in the following contest*". Denis, CEO Euroline—partner

Formabilio plays an intermediary role within the network, by inspiring, facilitating and helping actors to develop activities and play different roles. It starts and reinforces relationships among the actors in the community. It supports the learning process within the community by channelling and allowing the matching of the actors' skills and knowledge to improve their effectiveness as resource integrators and to set up new ways of creating value.

> "*Formabilio goes between us and designers to solve problems or to redefine some details of the product. It bridges the gap between the actors of the community, as it interprets the needs of both supplier and designers. Furthermore, it organizes events for personally interacting with designers and other companies*". Denis, CEO Euroline—partner

> "*Formabilio is a good opportunity to have some feedback about my work from experts and design lovers*". Elena, Formabilio designer

The company provides a digital space for connecting people and making their ideas come true. The actors' interactions are not limited to the online context; Formabilio periodically organizes events (e.g. Be Our Special Guest at Milano Design Week 2015) to strength existing relationships and create new ones. Events are planned to publicize Formabilio products and to increase both the visibility and the recognition of designers as professional and creative actors and of partners as Italian traditional companies. Furthermore, events are designed to share ideas and create new knowledge in a friendly co-working space.

> "The Soul Salad is an event open to all, where we have lunch together, sharing ideas and great food, meeting people, organizations, best practices, innovative and inspirational change". Maria Grazia, Formabilio co-founder

All actors involved in the Formabilio network integrate their own resources through different activities and with a different degree of intensity. They integrate both human and non-human resources to create new and sustainable solutions. The main resources that actors integrate within the Formabilio community are creativity, knowledge, technical and artistic competences, as well as the 'made in Italy' culture. Such resources can be meant as a sort of shared values between partners. Actors are continually encouraged to act on, recombine and improve existing resources, as well as to create new ones. This mechanism is well recognized within the community as a suitable way to shift ideas into new and original products.

"Formabilio allows us to offer to the creativity of designers all our skills, competences and all our 60 years' experience and tradition, to further their opportunities and ours". Katia, Manager Ivo Fontana Mobili—partner

15.4.2 RI Enabler and Initiator

Technology is the key element for developing an osmotic environment, where actors can learn, share their knowledge and integrate their resources to co-create value. The digital platform (e.g. the architecture of participation) implemented by Formabilio has been designed on three main elements: the e-commerce, the contest provider and the community that is continually involved in the value-creation process through an advanced gamification logic.

The digital infrastructure has a user-friendly interface; it has been architected to facilitate the easy incorporation of diverse resources, in other words to assist designers and partners in achieving their objectives and compelling tasks. Indeed, it has been designed as a road map that clearly explains what and how activities can be developed and how the participants will be rewarded for their contributions.

The rules of engagement, relating to the working mechanism of the idea contest (e.g. idea submission, idea evaluation, idea production, rewards), are clearly specified to enhance openness and transparency, whereas specific tools, such as a blog, designers' personal pages and an experiential video telling the company's history, are used to ensure the visibility and recognition of the community's members.

Formabilio.com is a virtual plaza for information and opinion sharing and for collaborating. It is both a relationship facilitator/enabler and an interaction initiator, as it stimulates the actors' creativity and interaction (1-month contest), drives participants' activities and coordinates actors and their service exchanges. It can be viewed as an actor that triggers or initiates innovation through resource integration.

"Formabilio perfectly reflects the mechanisms and potential of the network, as a new tool, fast and simple. To "do" design in all its phases: design, research, production and marketing". Paolo, designer

15.4.3 RI Mechanism

The RI mechanism adopted by Formabilio relates to the practice of the online idea contest. It is a well-known method, which is based on the nature of competition to encourage users' participation, inspire their creativity and increase the quality of the ideas submitted. In other words, it is a useful practice to get valuable and new resources from external actors.

> *"Formabilio? It's a modern and successful idea that promotes the interaction between creatives and industries, preserving Made in Italy and its craftsmanship heritage as assets of our country. Formabilio is an opportunity, inspiration and exchange of views"*. Laura, designer

Community participants can be involved in one or more tasks according to their social role, purpose and capabilities. Activities are designed to enhance the logic of collaboration and value co-creation among all actors involved in the project. They range from simple to complex ones. Activities can refer to the mere collection of information through reading and non-participant observation of discussions, comments, etc., as well as compliance with the community through liking ideas. More complex activities require the actors' cognitive effort; they relate to product purchase and idea co-evaluation, which is based on community votes and on the analysis of an idea's technical feasibility by experts, manufacturing companies and the Formabilio founders. Other activities requiring high actor engagement relate to idea co-design, which is based first on the creative contribution of the designer (new idea submission) and in the second stage on the collective actions of the community members, who provide comments and suggestions to improve and better fit ideas to market needs. Comments and suggestions for idea improvement foster co-learning and thus the creation and sharing of new knowledge and of new product solutions.

> *"I discovered Formabilio quite by accident. I saw the spirit of the competition, the quality of products and especially the opportunity to share, discuss and refine my skills, so I couldn't resist getting involved"*. Livia, Formabilio designer

Differently from designers, potential customers or made in Italy furnishing lovers or more simply brand lovers do not affect decisively the creative process or the phase of co-ideation and co-design. They are more involved in the phases of idea co-evaluation and co-testing.

> *"I like it, it is an interesting project. I'll vote for it"*. Andrea, Formabilio design lover

> *"Good job! Your projects are always original and interesting, but please, could you replace the green colour with another one? It is too common and often used"*. Davide, Formabilio community member

Co-ideation within specific projects is the main mechanism used by Formabilio to gather the co-generation of ideas. It is based on the definition of specific guidelines driving the design and submission of ideas, such as particular themes relating to raw material or events (e.g. wood, Valentine's Day, etc.) and technical instructions about the proposal (e.g. the final product must be easily assembled and installed by the buyer and packaged with packaging of up to 230 cm in length and weighing up to 30 kg each per package).

The clear definition of rules and precepts about the idea-evaluation process and the clarity of the voting system and of the community's purposes (e.g. to produce and share value with all players involved in the project) have key roles in the system's sustainability. They ensure the working mechanism of the community, as they clarify what activities actors can do and how, as well as why they should integrate their resources and co-create value (rewards policy).

15.4.4 Value

Consistent with the idea that value is different for each referent and must be assessed separately (Lusch and Nambisan 2015), the digital platform allows actors to create and achieve many types of value: economic and financial, social, emotional, cognitive and epistemic.

The company's goal is to produce and share value with all players involved in the project; indeed, the sustainability of its business model is based on the fair distribution of revenues to all the network actors. From 2013 to 2014, the company's revenues increased from 9141 € to 89,327 €.

For designers, value relates first of all to the opportunity to increase their professional competences (epistemic value) as well as to obtain trust and reputation (social and emotional value) as a 'professional designer' from peers, firms and potential customers (e.g. each Formabilio product is marked with the designer's and manufacturer's names). Moreover, they also extract economic value from the collaboration, as they gain a monetary reward for each sold product that they designed (a fee of 7 % on all sold products).

> *"I believe Formabilio is a good chance to keep in touch with people who want to share ideas and knowledge. It allows young professional designers to stand out and cooperate with potential customers and firms"*. Maddalena, Formabilio designer

> *"Democracy is what I like the most in Formabilio, in other words the chance to participate in a contest with enthusiasm and see your own idea produced and especially on sale online"*. Carlo, Formabilio designer

> *"Formabilio, first individually and then as a group, gave us the chance to express ourselves as designers and to realize our ideas"*. Andrea, Formabilio designer

The partners are small, highly specialized Italian craft firms who are very sensitive to environmental, sustainability and made in Italy issues. For them the value depends first on business and network opportunities: the Formabilio community allows them to access a wider network and to get in touch with other companies and partners. This type of social value leads to—or is a precondition for—economic and financial value. Moreover, value is epistemic relating to knowledge, as it is concerned with tackling new technological and market frontiers and coping with the challenge of competitiveness.

"Formabilio is the future. It gives us the opportunity to work with designers who have new ideas and, above all, the opportunity to enter into a global market". Paolo, Live In—partner

"We have no doubt that the furniture industry is suffering from the recession and that there is a need for change and revolution. A part of this change is shown in the very innovative development and sale system applied by Formabilio. We grew up in terms of relationships with other suppliers and designers. We have information about materials and manpower costs that before were not easily accessible; furthermore, the reference system enacted by Formabilio allowed us to be engaged in other projects outside the community". Denis, CEO Euroline—partner

Actors, such as manufacturing companies, are networked with each other and can discuss issues that are also beyond their actual business context (new working processes, new materials, market trends, etc.). They have the opportunity to learn about working methods and to have direct contact with the market.

"Through Formabilio we have direct feedback from customers about products. My company is a components supplier. Until now, we had mainly B2B relationships instead of direct relationships with customers. Our engagement in the Formabilio family made us more aware and interested in customers' opinions. It marks our route and pushes us to interact more with designers and other suppliers—within the supply chain—to refine products according to the customers' preferences". Denis, CEO Euroline—partner

"The traditional mistrust within the supply chain has been replaced by the culture of collaboration. I have the opportunity to improve my work by learning working techniques that belong to interrelated businesses (supplier companies) and that till a year ago I perceived as very far from my work". Denis, CEO Euroline—partner

Value for brand lovers can be cognitive, epistemic, relational, social, ludic and emotional too. It depends on their own engagement, system of values, role and expectations. Brand lovers are enthusiastic people—customers, design lovers and the innovation sensitive—who co-create value by providing comments, votes and suggestions. These activities increase their own knowledge (e.g. about products, processes, designers, firms, materials, events) and fun. More specifically, value for them is linked to their involvement in the process of idea co-evaluation and selection; indeed, as their level of participation increases, they perceive higher service quality, feel greater satisfaction and are more motivated and committed to co-creation. The value for these customers is complex and relates to the opportunity to co-create a product solution that matches their needs in terms of design, price, quality and sustainability (e.g. work processes, materials, respect for employees, innovative culture).

15.5 Discussion

This chapter provides a detailed and granular approach to designing a business model for co-creation. It frames a strategic framework to support SMEs, operating in existing and traditional businesses, in managing new business opportunities based on networked value creation.

The study provides more depth on how through resource integration mechanisms—based on specific rules and activities—network resources (e.g. human/non-human actors and their role within the network) can be mobilized and unfrozen to enable higher resource density for creating and capturing value. We empirically contextualize the framework to demonstrate how it works in practice. The Formabilio network is a useful case to illustrate how a new business model designed for value co-creation can work in a mature, traditional and not open industry, as in the Italian manufacturing business context. This network-centric approach to business models (Storbacka et al. 2012) opens up interesting opportunities for creating new companies. Indeed, Formabilio is completely different from other firms (e.g. Mulino Bianco, Starbucks, Lego, etc.), which use crowdsourcing just to integrate their existing business strategy. It is a start-up company that is totally based on and inspired by value co-creation and the resource integration logic. The technical infrastructure—that is, the digital platform—provides the context for the actors to experience value (Chandler and Vargo 2011; Vargo et al. 2008). It allows this co-creation business model to work by taking a dual role: facilitator or enabler, as it connects actors with other actors and other resources, and initiator or actor, as it triggers the development of new types of capabilities, which relate to organizational structural flexibility e.g. different ways of organizing actors to suit varied innovation opportunities in a network (Lusch and Nambisan 2015), such as more modular design offerings, modular and flexible manufacturing and delivery as well as the ability to manage partners and alliances (Storbacka et al. 2012).

The practice of the online idea contest, which we consider the main RI mechanism, has been adopted by Formabilio to coordinate, integrate and synchronize the actors' contributions. It is the core element of the architecture of participation that is largely determined by shared institutional logics, as well as by means and rules. The online contest mechanism affects the working activities and processes of all the network actors. It allows the actors to operate on the available resources more in unison, without a strong command-and-control structure (Lusch and Nambisan 2015), as it extends the actors' business boundaries and prioritizes interaction and sharing of information, knowledge, competences, etc., with new actors in a changeable relationship system. Indeed, as the quotes above highlighted, suppliers—who traditionally focus on a narrow business approach—cooperate with other actors (e.g. Formabilio, designers, other suppliers, furniture experts and customers) from the design to the product prototyping phase. The network actors' heterogeneity makes this case unique and more complex in comparison to other examples (e.g. Nel Mulino che Vorrei, Nutella, Dell, etc.) in which companies focus on dyadic relationships (e.g. company and consumers, company and experts, company and fans). Indeed, Formabilio encompasses the point of view of many participants who are actively engaged in different activities (co-design, co-evaluation, co-production, etc.) within the entire business ecosystem.

Formabilio offers a connection between competences and relationships and orchestrates interaction and guides resource integration by taking the role of an inspirator (e.g. defining new themes for the contest), a facilitator (defining specific rules) and an attendant within the value emergence process. The definition of activities, of the working procedures, rules and precepts, enhances a shared awareness (Nambisan and Sawhney 2007) and/or a shared worldview (Lusch and Nambisan 2015) that ensures that actors can interpret resource integration opportunities to better combine and mobilize resources for a particular situation or goal. This is fundamental to improving the community members' effectiveness as resource integrators (e.g. the recombination of existing resources and creation of new ones) and their opportunity to extract value from the community itself. Indeed, the greater density of resources corresponds to enhanced value (Lusch et al. 2010; Normann 2001).

The sustainability of the Formabilio business model is ensured by the fair distribution of value to all the network actors. The architecture of participation, which defines the means by which the participants will realize value from the exchange, is inspired by a win–win logic. It enables a diverse set of value types—economic and financial, social, emotional, cognitive and epistemic value—to fit the different actors' expectations. These depend on the actors' roles and thus on the nature of service exchange and on the type of resource integration achieved.

15.6 Main Implications

This chapter provides new knowledge about mechanisms and methods for aligning the contributions of different actors to better understand how value is co-created through the actors' resource-integrating efforts and captured by the lead company. It addresses relevant gaps in current knowledge on business model design and on resource integration and the value co-creation practice.

It contributes from both theoretical and managerial perspectives. First, it relates to the field of designing business models for co-creating value that is still in an emergent stage (Frow et al. 2015). It addresses the need of academia to design through the business model a supportive environment for resource integration and value co-creation (Frow et al. 2015; Storbacka et al. 2012) by focusing on mechanisms that enhance resource density (Lusch and Nambisan 2015). Therefore, it defines the actors' key roles in the network and the rules of exchange through the digital platform (rules and activities to interact with and through the platform) and describes the nature of value co-created and/or captured by each actor's role.

It extends the extant literature on resource integration that has been rather abstract in nature and provides a few implications for managers and practitioners to realize its potential value in practice.

This chapter calls for a new and more strategic role of the business model to replace episodic activities for innovating with a structured, codified and recursive mechanism to enhance continuously the idea co-generation process and to create value for all the actors engaged. It emphasizes that the cooperation, interaction and integration of resources are not simply techniques or tools to put on; they concern the definition of a practice and of its elements, as people, value systems, culture and openness. Accordingly, it can support practitioners in better designing and managing a business model based on the resource integration mechanism and value co-creation practice, which according to many authors are fundamental sources of competitive advantage. Moreover, it offers some insights on how to exploit interactive technologies to mobilize and enhance the best combination of actors' resources for creating a collaborative and networked environment in a traditional and not open business context such as Italian manufacturing.

References

Afuah, A., & Tucci, C. (2003). *Internet business models and strategies*. Boston: McGraw Hill.

Akaka, M. A., & Chandler, J. D. (2011). Roles as resources: A social roles perspective of change in value networks. *Marketing Theory, 11*(3), 243–260.

Akaka, M. A., & Vargo, S. L. (2014). Technology as an operant resource in service (eco) systems. *Information Systems and e-Business Management, 12*(3), 367–384.

Amit, R., & Zott, C. (2001). Value creation in e-business. *Strategic Management Journal, 22*(6-7), 493–520.

Arthur, W. B. (2009). *The nature of technology: What it is and how it evolves*. New York: Free Press.

Baden-Fuller, C., & Morgan, M. S. (2010). Business models as models. *Long Range Planning, 43* (2), 156–171.

Ballantyne, D., & Varey, R. J. (2006). Creating value-in-use through marketing interaction: The exchange logic of relating, communicating and knowing. *Marketing theory, 6*(3), 335–348.

Barczak, G. (2012). The future of NPD/innovation research. *Journal of Product Innovation Management, 29*(3), 355–357.

Berglund, H., & Sandström, C. (2013). Business model innovation from an open systems perspective: Structural challenges and managerial solutions. *International Journal of Product Development, 18*(3), 274–285.

Bourdeau, L., Chebat, J. C., & Couturier, C. (2002). Internet consumer value of university students: E mail-vs.-Web users. *Journal of Retailing and Consumer Services, 9*, 61–69.

Caridà, A., & Colurcio, M. (2013). Viral marketing communication: Just sales or more? *Business Systems Review, 2*(1), 99–110.

Caridà, A., Colurcio, M., & Melia, M. (2014). Rethinking and improving the healthcare service through interactive web technologies. In E. Baglieri & U. Karmarkar (Eds.), *Managing consumer services: Factory or theater?* (pp. 191–210). Cham: Springer.

Caridà, A., Edvardsson, B., & Colurcio, M. (2015). Modes of resource integration driving value co-creation: An empirical investigation in virtual brand communities (VBCs). In

E. Gummesson, C. Mele, & F. Polese (Eds.), *Proceedings of Naples Forum on Service, June 9–12*. Italy: University of Naples.
Chandler, J., & Vargo, S. L. (2011). Contextualization: Network intersections, value-in-context, and the co-creation of markets. *Marketing Theory, 11*(1), 35–49.
Chesbrough, H. (2007). Business model innovation: It's not just about technology anymore. *Strategy and Leadership, 35*(6), 12–17.
Chesbrough, H., & Rosenbloom, R. S. (2000). *The role of the business model in capturing value from innovation: Evidence from XEROX Corporation's Technology spinoff companies*. Boston: Harvard Business School Press.
Colurcio, M., Edvardsson, B., & Caridà, A. (2014). Resource integration in innovation processes: A literature review. In W. Ahram, Karwowski, & T. Marek (Eds.), *Proceedings of the 5th International Conference on Applied Human Factors and Ergonomics AHFE 2014*, Kraków, Poland, 19–23 July.
Colurcio, M., Wolf, P., Kocher, P. Y., & Russo Spena, T. (2012). Asymmetric relationships in networked food innovation processes. *British Food Journal, 114*(5), 702–727.
Cova, B., & Dalli, D. (2009). Working consumers: The next step in marketing theory? *Marketing Theory, 9*(3), 315–339.
Dubois, A., & Gadde, L. E. (2002). Systematic combining: An abductive approach to case research. *Journal of Business Research, 55*(7), 553–560.
Echeverri, P., & Skålén, P. (2011). Co-creation and co-destruction: A practice-theory based study of interactive value formation. *Marketing Theory, 11*(3), 351–373.
Edvardsson, B., Kleinaltenkamp, M., Tronvoll, B., McHugh, P., & Windahl, C. (2014). Institutional logics matter when coordinating resource integration. *Marketing Theory, 14*(3), 291–309.
Edvardsson, B., & Tronvoll, B. (2013). A new conceptualization of service innovation grounded in SD logic and service systems. *International Journal of Quality and Service Sciences, 5*(1), 19–31.
Franke, N., Gruber, M., Harhoff, D., & Henkel, J. (2008). Venture capitalists' evaluations of start-up teams: Trade-offs, knock-out criteria, and the impact of VC experience. *Entrepreneurship Theory and Practice, 32*(3), 459–483.
Frow, P., Nenonen, S., Payne, A., & Storbacka, K. (2015). Managing co-creation design: A strategic approach to innovation. *British Journal of Management, 26*(3), 463–483.
George, G., & Bock, A. J. (2011). The business model in practice and its implications for entrepreneurship research. *Entrepreneurship Theory and Practice, 35*(1), 83–111.
Gidhagen, M., Persson Ridell, O., & Sörhammar, D. (2011). The orchestrating firm: Value creation in the video game industry. *Managing Service Quality: An International Journal, 21*(4), 392–409.
Goss, D. (1991). *Small business and society*. London: Routledge.
Gosselin, P. D., & Bauwen, A. G. (2006). Strategic account management: Customer value creation through customer alignment. *Journal of Business and Industrial Marketing, 21*(6), 376–385.
Gummerus, J. (2013). Value creation processes and value outcomes in marketing theory: Strangers or siblings? *Marketing Theory, 13*(1), 19–46.
Gummesson, E., & Mele, C. (2010). Marketing as value co-creation through network interaction and resource integration. *Journal of Business Market Management, 4*(4), 181–198.
Hamel, G. (2000). *Leading the revolution*. Boston: Harvard Business School Press.
Kajanus, M., Iire, A., Eskelinen, T., Heinonen, M., & Hansen, E. (2014). Business model design: New tools for business systems innovation. *Scandinavian Journal of Forest Research, 29*(6), 603–614.
Kleinaltenkamp, M., Brodie, R., Frow, P., & Hughes, T. (2012). Resource integration. *Marketing Theory, 12*(2), 201–205.
Korkman, O., Storbacka, K., & Harald, B. (2010). Practices as markets: Value co-creation in e-invoicing. *Australasian Marketing Journal, 18*(4), 236–247.

Laamanen, M., & Skålén, P. (2014). Collective–conflictual value co-creation: A strategic action field approach. *Marketing Theory, 15*, 381–400.
Lee, E. J., & Overby, J. W. (2004). Creating value for online shoppers: Implications for satisfactions and loyalty. *Journal of Consumer Satisfaction, Dissatisfaction and Complaining Behavior, 17*, 54–67.
Lee, S., Park, G., Yoon, B., & Park, J. (2010). Open innovation in SMEs-An intermediated network model. *Research Policy, 39*(2), 290–300.
Löbler, H. (2013). Service-dominant networks: An evolution from the service-dominant logic perspective. *Journal of Service Management, 24*(4), 420–434.
Lusch, R. F., & Nambisan, S. (2015). Service innovation: A service-dominant logic perspective. *MIS Quarterly, 39*(1), 155 175.
Lusch, R. F., Vargo, S. L., & Tanniru, M. (2010). Service, value networks and learning. *Journal of the Academy of Marketing Science, 38*(1), 19–31.
Maglio, P. P., & Spohrer, J. (2008). Fundamentals of service science. *Journal of the Academy of Marketing Science, 36*(1), 18–20.
Mason, K., & Spring, M. (2011). The sites and practices of business models. *Industrial Marketing Management, 40*(6), 1032–1041.
Mccoll-Kennedy, J. R., Vargo, S. L., Dagger, T. S., Sweeney, J. C., & Van Kasteren, Y. (2012). Health care customer value co-creation practice styles. *Journal of Service Research, 15*(4), 370–389.
Mele, C., Russo Spena, T., & Colurcio, M. (2010). Co-creating value innovation through resource integration. *International Journal of Quality and Service Sciences, 2*(1), 60–78.
Morris, M., Schindehutte, M., & Allen, J. (2005). The entrepreneur's business model: Toward a unified perspective. *Journal of Business Research, 58*(6), 726–735.
Nambisan, S., & Sawhney, M. (2007). A buyer's guide to the innovation bazaar. *Harvard Business Review, 85*(6), 109–116.
Nenonen, S., & Storbacka, K. (2014). Management of customer assets for increased value capture in business markets. *Management Decision, 52*(1), 101–121.
Ng, I., Guo, L., & Ding, Y. (2011). *The use of information technology as value cocreation: The role of contextual variety and means drivenness* (The University of Exeter Business School Discussion Papers in Management). Exeter: University of Exeter Business School.
Normann, R. (2001). *Reframing business: When the map changes the landscape*. Chichester: Wiley.
Orlikowski, W. (1992). The duality of technology: Rethinking the concept of technology in organizations. *Organization Science, 3*(3), 398–427.
Osterwalder, A., & Pigneur, Y. (2010). *Business model generation: A handbook for visionaries, game changers, and challengers*. New York: Wiley.
Osterwalder, A., Pigneur, Y., & Tucci, C. L. (2005). Clarifying business models: Origins, present, and future of the concept. *Communications of the Association for Information Systems, 15*, 1–40.
Payne, A., Storbacka, K., & Frow, P. (2008). Managing the co-creation of value. *Journal of the Academy of Marketing Science, 36*(1), 83–96.
Pels, J., Möller, K., & Saren, M. (2009). Do we really understand business marketing? Getting beyond the RM and BM Matrimony. *Journal of Business and Industrial Marketing, 24*(5/6), 322–336.
Peters, L. D., Löbler, H., Brodie, R. J., Breidbach, C. F., Hollebeek, L. D., Smith, S. D., et al. (2014). Theorizing about resource integration through service-dominant logic. *Marketing Theory, 14*(3), 1–20.
Prahalad, C. K., & Ramaswamy, V. (2004). Co-creating unique value with customers. *Strategy and Leadership, 32*(3), 4–9.
Pura, M. (2005). Linking perceived value and loyalty in location-based mobile services. *Managing Service Quality, 15*(6), 509–538.

Rampersad, G., Quester, P., & Troshani, I. (2010). Managing innovation networks: Exploratory evidence from ICT, biotechnology and nanotechnology networks. *Industrial Marketing Management, 39*(5), 793–805.
Rothwell, R., & Dodgson, M. (1992). European technology policy evolution: Convergence towards SMEs and regional technology transfer. *Technovation, 12*(4), 223–238.
Russo-Spena, T., & Mele, C. (2012). Five Co-s in innovating: A practice-based view. *Journal of Service Management, 23*(4), 527–553.
Sawhney, M., Verona, G., & Prandelli, E. (2005). Collaborating to create: The Internet as a platform for customer engagement in product innovation. *Journal of Interactive Marketing, 19*(4), 4–17.
Schau, H. J., Muñiz, A. M., Jr., & Arnould, E. J. (2009). How brand community practices create value. *Journal of Marketing, 73*(5), 30–51.
Schneider, S., & Spieth, P. (2013). Business model innovation: Towards an integrated future research agenda. *International Journal of Innovation Management, 17*(01), 237–247.
Sheth, J. N., Newman, B. L., & Gross, B. L. (1991). Why we buy what we buy a theory of consumer behavior. *Journal of Business Research, 22*, 159–170.
Sicilia, M., & Palazón, M. (2008). Brand communities on the internet: A case study of Coca-Cola's Spanish virtual community. *Corporate Communications: An International Journal, 13*(3), 255–270.
Stewart, D. W., & Zhao, Q. (2000). Internet marketing, business models, and public policy. *Journal of Public Policy and Marketing, 19*(2), 287–296.
Storbacka, K. (2011). A solution business model: Capabilities and management practices for integrated solutions. *Industrial Marketing Management, 40*(5), 699–711.
Storbacka, K., Frow, P., Nenonen, S., & Payne, A. (2012). Designing business models for value co-creation. *Review of Marketing Research, 9*, 51–78.
Sweeney, J. C., & Soutar, G. N. (2001). Consumer perceived value: The development of a multiple item scale. *Journal of Retailing, 77*(2), 203–220.
Teece, D. J. (2010). Business models, business strategy, and innovation. *Long Range Planning, 43*, 172–194.
Timmers, P. (1998). Business models for electronic markets. *Journal on Electronic Markets, 8*(2), 3–8.
Ulaga, W., & Eggert, A. (2005). Relationship value in business markets: The construct and its dimensions. *Journal of Business-to-Business Marketing, 12*(1), 73–99.
Vargo, S. L. (2008). Customer integration and value creation: Paradigmatic traps and perspectives. *Journal of Service Research, 11*(2), 211–215.
Vargo, S. L., & Akaka, M. A. (2012). Value co-creation and service systems (re)formation: A service ecosystems view. *INFORMS Service Science, 4*(3), 207–217.
Vargo, S. L., & Lusch, R. F. (2008). Service-dominant logic: Continuing the evolution. *Journal of the Academy of Marketing Science, 36*(1), 1–10.
Vargo, S. L., & Lusch, R. F. (2011). It's all B2B... beyond: Toward a systems perspective of the market. *Industrial Marketing Management, 40*(2), 181–187.
Vargo, S. L., Lusch, R. F., & Akaka, M. A. (2010). *Advancing service science with service-dominant logic: Clarifications and conceptual development* (Research and innovations in the service economy, Handbook of service science). Berlin: Springer.
Vargo, S. L., Maglio, P. P., & Akaka, M. A. (2008). On value and value co-creation: A service systems and service logic perspective. *European Management Journal, 26*(3), 145–152.
Yin, R. (2009). *Case study research: Design and methods. Applied social research methods.* London: Sage.
Yoo, Y., Henfridsson, O., & Lyytinen, K. (2010). Research commentary-The new organizing logic of digital innovation: An agenda for information systems research. *Information Systems Research, 21*(4), 724–735.

Zimmermann, E. W. (1951). *World resources and industries: A functional appraisal of the availability of agricultural and industrial materials*. New York, NY: Harper & Brothers.
Zott, C., & Amit, R. (2008). The fit between product market strategy and business model: Implications for firm performance. *Strategic Management Journal, 29*(1), 1–26.
Zott, C., Amit, R., & Massa, L. (2011). The business model: Recent developments and future research. *Journal of Management, 37*(4), 1019–1042.

Part V
Experiences: Multi Actors and Context

Chapter 16
Seeking New Ways of Innovating in Industry-Research Collaboration Practice

Maaria Nuutinen, Marko Seppänen, Anssi Smedlund, and Eija Kaasinen

Abstract This chapter sheds light on the dynamics of innovation activity and its prerequisites for industry-research collaboration practice. It focuses on a particular intentional attempt to change the prerequisites of innovations. This new kind of research programme aimed to improve the competitiveness of the Finnish metals and engineering industry through enhancing innovations and industrial renewal. Three theoretical lenses were used in reflecting collaboration practice and learning in the programme. Two examples of changes in collaboration activity were presented and analysed within the broader context of the programme. The findings present how the characteristics of the programme and its management practices seemed to construct favourable conditions for co-innovating. The interpretations were founded on the reflections made by the four persons in charge of leading the programme. The chapter is concluded by summing up the results in terms of insight gained, as well as a general framework that aims to support practical efforts to create better prerequisites and conditions for industry-research collaboration in innovation activity.

16.1 Introduction

Focusing on it (experience) opens a whole new world for design ... but where do we find the courage to differentiate? It requires very strong self-confidence in a company.

We have to understand that this is not going to change with just bringing a consultant from outside our company and having another workshop—again...

User experience is a strategic issue ... it's a hard thing, but should be managed softly.

M. Nuutinen (✉) • E. Kaasinen
VTT Technical Research Centre of Finland Ltd., Helsinki, Finland
e-mail: maaria.nuutinen@vtt.fi; eija.kaasinen@vtt.fi

M. Seppänen
Tampere University of Technology, Tampere, Finland
e-mail: marko.seppanen@tut.fi

A. Smedlund
Aalto University, Helsinki, Finland
e-mail: anssi.smedlund@aalto.fi

T. Russo-Spena et al. (eds.), *Innovating in Practice*,
DOI 10.1007/978-3-319-43380-6_16

These quotations are from company representatives, expressed during a two-day meeting of a research programme in August 2014, slightly after the three-year point during a five-year collaboration effort involving five research institutes and eight companies. There have been hundreds of such discussions aimed at comprehending how to create competitive advantage-based user experience and changing the way of thinking and doing—first in those seven companies, and then for accelerating renewal in the whole Finnish metals and engineering industry (MEI). This process took the form of discussions on what this requires from our understanding with regard to competitive advantages, customer value, innovation activity, user and customer experiences, marketing and selling, and so on. This shared journey included dozens of case studies and tests, thousands of hours spent on interviews, and innumerable experiments and workshops—all potentially contributing to learning a new way of innovating.

The research programme builds a frame for collaborative innovation activity that reflects the current theoretical and managerial discussion beyond user experience, such as open innovation and service logic. The programme also assumes a format within the broader industry-research collaboration culture of a prevailing innovation policy atmosphere. In this programme, the aim of challenging the present methods of collaborating and thinking on R&D was explicitly stated from the very beginning. Furthermore, there was a bigger change taking place in the Finnish innovation system—the introduction of the new public/private partnership funding model described later in this chapter.

This was reflecting a balance shift between demand-based innovation policy and science/technology-based policy in favour of the former, and thus aimed to promote growth and renewal of the economy and to generate employment (Nikulainen and Tahvanainen 2009). The protracted critical economic situation in Finland (as well as in many other countries) emphasises the need to ensure the impact of innovation activity. This progress calls for recognition of the boundaries of innovation activity constructed through the interactions and interpretations of the people participating. Here we focus on how research programme players themselves could pursue the creation of better prerequisites for innovation by developing a frame of reflection.

Participation in any research programme can be understood as a normal collaboration practice between research and industry, with already learned and entrenched ways of working, communicating, and collaborating, with a long history behind it: it is equivalent to an instrument we have gained mastery over and of which we are not so aware anymore. Referring to the well-known classic example of *hammering* (Gherardi 2000; Polanyi 1962), the focal awareness is on one's own duties and responsibilities in one's work, and the learning and innovating combined in the programme elicit only subsidiary attention. When everything goes smoothly, the instrument is not given attention. But how can we thereby enhance learning and co-innovating? Can we increase awareness of the instrument and its properties? According to Polanyi, knowledge arises when a breakdown occurs and reflexive activity interferes. The creation of reflexive, investigable, and theoretical knowledge requires that something previously usable must now be unusable.

Thus, we formulate a further question: *What are the conditions in the programme that were enhanced to reflect 'the rule-based operations' that produce a normal situation in industry-research practice?*

In a more general sense, the development of the new funding instrument can already be regarded as the result of the reflective learning within the innovation system, pursuing the enhanced industrial impact of research. However, has there been a permanent change—did learning take place? If we look at this question within this programme context, the question can be formulated as follows: *Has the logic of innovation changed? Can we recognise hints of a new kind of industry-research-innovation practice worth supporting?*

This chapter focuses on this particular intentional attempt to change the prerequisites of innovation using a new kind of research programme. The analysis aims to understand the characteristics of the programme and its management that seemed to construct favourable conditions for *co-innovating*. The analysis is based on the use of three theoretical lenses: a research programme as (1) an innovation platform, (2) a way towards new business and innovation logic, and (3) a learning community. These lenses provide a frame of reference for reflecting the programme. We will interpret the results using insight and questions aiming to support practical efforts to create better prerequisites and conditions for industry-research collaboration in innovation activity. Further, this chapter aims to contribute by providing an example of how to narrow the gap between theory and practice. We aim to settle under 'the interpretivism umbrella' as our ontological and epistemological standpoint, but due to our pragmatic viewpoint, this is not totally coherently achieved.

In the following, we will start by describing briefly the conditions and aims of the research programme. After a short section devoted to methods, we will turn to the analysis by introducing the three theoretical 'lenses' and defining the criteria/questions for reflection on the programme. Two examples from the research programme are then analysed in accordance with the criteria. In the final part of the paper, we will discuss the practical implications in line with the aims of the tool section and summarise briefly the theoretical contribution of the chapter.

16.2 The Strategic Centres for Science, Technology, and Innovation

The Strategic Centres for Science, Technology, and Innovation established in Finland are public/private partnerships for speeding up innovation processes. Their main goal is to thoroughly renew industry clusters and create radical innovations (http://www.tekes.fi/en/programmes-and-services/strategic-centres/) (see also Nikulainen and Tahvanainen 2009). The Strategic Centres (SHOKs) can be characterised as open innovation environments. The SHOKs were designed to develop and apply new methods of cooperation, co-creation, and interaction. Each SHOK has companies and research organisations as shareholders. The shareholders of each SHOK jointly define the research strategy of the SHOK. The research strategy is then operationalised into research programmes with research organisations. In these programmes, companies and research units work in close cooperation, with the volume divided equally between the companies and the research organisations. The programmes are, on average, 40 % funded by the participating companies. The rest of the funding comes from public funding organisations and the research organisations. At the time of writing this article there were five SHOKs in progress, which were started during the years 2007–2009. The work described in this chapter has been carried out in a research programme by the SHOK Finnish Metals and Engineering Competence Cluster (FIMECC).

During the operation of the SHOKs, they have been regularly externally evaluated. In 2011, the Federation of Finnish Technology Industries (Teknologiateollisuus) carried out an evaluation of four SHOKs that were most relevant for the industry. This evaluation included FIMECC, as well as SHOKs, focused on energy, the forestry sector, and ICT (Annala and Ylä-Jääski 2011). Managing directors and technology directors of the five most heavily investing and the five most strategically relevant companies of each SHOK were interviewed. In addition, a web survey was conducted for other companies. The companies agreed that SHOKs have succeeded in the goal of creating industry-driven research and innovation cooperation. SHOKs, as such, have become communication and cooperation forums where new knowledge is created in unison. New partnerships have been evolved between companies in various branches. Some companies have extended their customer network and some their research network. Even large companies have found new, stunning partners. Firm bilateral partnerships have also grown. The companies were of the view that the SHOK programmes were an easy way to network with relevant partners.

In 2012, the Ministry of Employment and Economy in Finland commissioned the evaluation of the SHOKs from a consortium consisting of consulting companies and individual business research experts (Lähteenmäki-Smith et al. 2013). The evaluation covered the policy, strategy, and activities of all six SHOKs independently, as well as an analysis of SHOKs as a policy instrument. In the evaluation by Lähteenmäki-Smith et al. (2013), the conclusion was that the SHOKs had managed to define strategic research agendas, and by promoting those, the SHOKs had produced new instruments for innovation and research policy. Tensions were identified between the short-term interests of industry and the longer-term perspective needed for the promotion of cutting-edge scientific research. The centres seemed to have multiple and often internally contradictory objectives, often leading to inadequate steering and performance guidance. The ownership of the SHOK model was also regarded as unclear in this evaluation.

The user experience theme was considered to be highly relevant for industry renewal, but at the same time, it was relatively difficult for the industry (Lähteenmäki-Smith et al. 2013). The ability to collect resources and critical mass seemed to be the drivers for FIMECC. FIMECC has managed to increase industry-research collaboration. FIMECC has strengthened already existing relationships, but there has also been more interdisciplinary research and collaboration. In some sectors, including FIMECC, the catalysing role of SHOK activity was assessed as essential. In these sectors, the SHOKs had brought together and gradually built a research ecosystem in a new research and development area. The SHOKs had helped to create more systematic, R&D-intensive activity and networking.

In addition to these formal evaluations, the public discussion around the SHOKs has grown for and against. At the time of finalising this article, the future of the SHOKs seems unsure, since the new Finnish government made a definition of policy to discontinue the funding instrument that facilitated SHOK programmes. At the same time also other instruments of public research funding are being reformulated. SHOKs such as FIMECC that have grown influential actors in their field may still continue as innovation platforms in new forms like FIMECC as DIMECC (Digital, Internet, Materials & Engineering Co-Creation). This kind of change makes it even more relevant to achieve understanding of the conditions of innovating at 'grass-roots level'.

16.3 Research Programme on User Experience and Usability in Complex Systems

The research programme being studied was the User Experience and Usability in Complex Systems (UXUS) programme, respective to FIMECC. The planning of the programme was conducted at the same time as FIMECC was taking shape. UXUS received funding at the end of 2010 and the kick-off was held at the beginning of 2011. The programme ended as planned at the end of 2015. This paper focuses on the first three and half years of its five-year schedule. Five research institutes, as well as five metal and engineering industry (MEI) companies operating in a business-to-business (B2B) environment, participated from the start, and three more companies joined in from the second year onwards. The total volume of the programme was raised to its final amount of 17.6 million euros. The programme plan stated its aim as follows: 'The overall target of the programme is to increase the competitiveness of the Finnish metals and engineering industry (MEI) by radically renewing practices.... The central research challenges are to create extensive understanding of user and customer experience in complex systems and of how to realise user and customer experience orientation in the companies' design and collaboration practices in innovation networks'. The programme was multidisciplinary and reflected the discussions of the time on service science, innovation management, human-centred design, and design management. UXUS was structured as four work packages, three of them offering a specific viewpoint to UX (or the object of research), and one focusing on the programme and its reflection.

The company representatives were from research and development functions, and many of them have an industrial design or engineering background. The companies' products varied quite a lot: steel, lifts, forklifts, cranes, and factory and process automation. Many researchers had earned their PhD (e.g. in the fields of engineering, business, or psychology), and in addition there were plenty of doctoral candidates and research assistants.

16.4 Methodology

Our methodological choice was based on our interest in what is really happening in innovation activity. The prerequisites for innovation are not something that are static or 'out there', but something that is socially constructed when various parties meet or come into contact (Chesbrough 2011; Edvardsson et al. 2010). Therefore, constructivist or practice-oriented studies (Peñaloza and Venkatesh 2006; Russo-Spena and Mele 2012) of these kinds of situations are suggested and focus on how they shape cultural beliefs.

The object of the study was the research programme as a particular case of socially constructing industry-research practice. Focusing on practice allowed us to investigate empirically how contextual elements shape knowledge, as well as how competence is built around a contingent logic of action (Corradi et al. 2010). The authors of this chapter were persons who all contributed to writing the programme plan and 'were assigned to' manage and lead the programme with the company representatives.

The overall research strategy can be characterised as an ethnographically spiced case study conducted by theoretically supported self-reflection on the part of practitioner researchers. The research strategy was inspired, firstly, by the idea of the 'reflective practitioner' by Schön (1983). According to him, reflective practice is 'the capacity to reflect on action so as to engage in a process of continuous learning'. Secondly, it has ethnographical features. According to Maréchal (2010: 43), 'autoethnography is a form or method of research that involves self-observation and reflexive investigation in the context of ethnographic field work and writing'. Thirdly, we propose that it is a form of practitioner research (e.g. Jarvis 1999) of industry-research practice, since the researchers and the company people were resolving the problems of the companies using action research (Reason and Bradbury 2001) and managing the programme closely interlinked with the academic research work with multiple methods on the side.

The main sources of data were self-reflection by four persons who were responsible leaders of the programme. In addition, the programme plan, the way in which it was updated during the first three years, and the research reports, as well as several more-or-less formal evaluations during the programme, were all utilised. The notes made by the authors themselves during the programme were also applied.

The analysis of the programme had two main phases. During the first phase, we formulated theoretical lenses and they were applied to defining overall criteria and more detailed questions reflecting the instrument, the research programme. In the second phase, two examples of activities and change regarding the overall criteria were described, and then the theoretical lenses were utilised in interpreting the characteristics of the programme, which enabled common activities and change—meaning learning. This phase also included tentative interpretation of the cultural beliefs of the broader collaboration practice challenged.

The adoption of the various theoretical viewpoints for the research programme is regarded as essential in order to better understand the conditions that the whole programme is creating for innovation and the renewal of innovation practices. The selection of the theoretical lenses was influenced by the fact that they formulated already-applied frames of reference in the programme plan and the SHOK context, the theoretical discussion among them is still lively, and the main focus is on various levels of activity. They therefore demonstrate contextually shaped and at least partly shared knowledge for the authors and the participants, offering a shared conceptual instrument for reflective learning. The theoretical lenses, with the first reflections of the programme, are described in the following.

16.5 Theoretical Lenses for Shedding Light on Industry-Research Practices

16.5.1 The Research Programme as an Innovation Platform

The research programme at hand realises 'open innovation ideology' at least in its external form: several organisations involved, a multidisciplinary approach, industry-research collaboration, a wide publicity model that required openness in research and development efforts, and so on. On this basis, we can conclude that basic prerequisites for innovation exist. Within this discussion, the research programme can be further mirrored as a form of innovation platform.

FIMECC, which is a limited company owned by major firms and research institutes in industry, boldly stated that it provides a world-class innovation platform 'to create new international research networks, new top science, new application-driven research contents and new business benefits'. This kind of platform thinking has gained a strong foothold in management practice and among scholars, simultaneously improving speed to market, cost, differentiation, and quality (Mäkinen et al. 2014; Gawer and Cusumano 2014). Platform thinking is the process of identifying and exploiting shared logic and structure in activities and offerings to achieve leveraged growth and variety (Sawhney 1998). During the past few years, this approach has been applied in various dimensions, such as brand-, customer-, collaboration-, technology-, and, finally, process- and product-based platforms, in order to enable the power of sharing a similar underlying logic. Platforms enable joint value creation and the integration of complementarities (Cusumano and Gawer 2002; Mäkinen et al. 2014).

Building and continuously improving the performance of the platform requires proper practices to support how its operations are managed and led. Gawer (2010) suggested four means of platform leadership, which basically are in four areas of managerial decisions: (1) firm scope, (2) technology design and intellectual property, (3) external relations with complementers, and (4) internal organisation. Further, she considers a hypothesis that the internal organisation of firms (such as platform leaders) and the external organisation of firms (i.e. the organisation of the sector or ecosystem of firms) are interrelated and mediated by the platform technology organisation. Parker and van Alstyne (2014) have shown how platform management can engage in open innovation by offering default agreements that grant platform partners a right to build on the results without negotiating. The default agreement between partners reduces risk and increases profits as network effects grow. As an innovation platform, the UXUS programme internally implemented several of these means. Its scope was rather manageable (number of participants and thus people), it had a joint consortium agreement for the whole programme that clarifies IPR issues, and there were fairly clear mechanisms as to how the programme should operate with external parties. FIMECC also executed its strategic research agenda by launching and hosting ambitious research programmes in its strategic research areas. These programmes were coordinated using a portfolio approach, in order to maintain continuous learning in addition to ensuring the transfer of research results to the partner companies.

Another theme of the innovation platform is the degree of openness. Due to the fact that 50–75 % of research funding came from the government budget, FIMECC was merely canalising public funding. The research results of the programmes are open to all industries, but the results of the companies are only public to the extent that a company decides to describe them, in addition to the required general description of the object of research and the general results achieved. There are some special conditions on patentable findings and their utilisation added to possible compensation between partners. The UXUS programme was still a rare case, since it acted according to a wide publicity model, which means that the results from the companies are also public, in spite of their nature or content right after the end of the programme. Eisenmann et al. (2011) reviewed factors that motivate decisions to open or close mature platforms. For FIMECC (or DIMECC) at the platform provider level, these decisions deal with (1) interoperating with established rival platforms, (2) licensing additional platform providers, or (3) broadening sponsorship. It seems that currently, there are no rival platforms in Finland, but conversely this innovation platform model has received interest and is also taking root in other European countries. Secondly, SHOK companies are collaborating to ensure the transferability of results between programmes. Thirdly, in addition to the national funding agency, SHOK firms have been broadening their funding base to EU and foreign companies to enable top-level research with the best possible talents. Open innovation also means that the managers in networked businesses must open their platforms and must be willing to sacrifice profits from the current platform to subsidise other, previously unknown participants.

In the UXUS programme, the openness for new company partners to join in was greatest at the beginning, when the planning was struggling to find enough company interest to enable a sufficient amount of research organisation-based contributions in relation to the ambitious vision of the SRA. The interest on the researcher side was moderate due to the topical theme, but flamed up with the strong lead of a forerunner company. The interest in joining in was increasing from the beginning—and three companies did so—but the limitations in increasing the programme volume significantly reined in this development. Moreover, due to the funding rules, joining in was possible only during a period change, of which there were only two during the programme—in the second year and at the end of the third year.

We can conclude that a research programme should constitute a platform that creates supportive frames enhancing collaborative open innovation activity as the first criterion for a new way of innovating. Such questions remain open as:

- To what extent do current consortium agreements foster larger participation in programmes?
- Is the current level of openness enough to ensure wide-scale learning and to benefit society?
- Is the decision-making mechanism designed so that it enables the making of breakthrough results within the programme? Or do the firms canalise their really strategic research somewhere else?
- Is the participant selection done properly to support innovativeness? How are the results measured and do the evaluation results affect programme preparations?
- What characteristics of an innovation platform does the programme have?

16.5.2 The Research Programme as a Way Towards a New Business and Innovation Logic

The transformation of society at large towards being an information society has been ongoing since the introduction of commercial connections to the Internet in the mid-1990s. Value networks that are formed to produce modular products as efficiently as possible emerged. There are also other types of networks for incremental and radical innovation among companies (Smedlund 2009). The transformation of innovation activities towards a networked format has taken longer compared to networked production, but the paradigm shift is evident. Popular examples of consumer products, such as mobile devices, have shown that new products can be introduced to the market faster when most of the complementary technology is brought in with licensing, and core key technology is developed in collaboration with many companies. It is nowadays impossible to develop a product without networks: all products have outsourced technology incorporated in them. According to the idea of open innovation (Chesbrough 2003), not only production but also innovation is taking place in the networks *between* companies instead of *inside* companies. In the new, networked innovation paradigm, collaboration on new knowledge creation with other firms is as important as new knowledge creation in-house.

The notion of value has also changed in industry and society. The perception of value is contingent and does not reside in a given product or service, but rather in the use context of the product (Ramirez 1999). The mass of total value is no longer produced at the supplier site with functions of design, development, manufacturing, and delivery, but in collaboration with the customer and user of the product during its usage. Internal processes of companies and usage of the product are no longer visibly detached, but parallel. This means that products and services are constantly re-innovated and redeveloped in the customer–supplier relationship. The technology as such does not sell itself in the market anymore without complementary services and products connected to it: the value of a stand-alone product is not more than the sum of its parts.

The paradigmatic change in understanding how value is created highlights the role of business models, as well as customers' and users' active participation in every step of the process when a new product or service is brought to market. As presented in the service-dominant logic (SDL) theories (Vargo and Lusch 2004), products merely convey a service, and it is the service that actually creates the added value that will then turn into the profits of a company. Within the context of the manufacturing industry, this means that it is more important to design products and services that integrate with the customer's system of other products and services, rather than design products that perform best as stand-alone features. The effect and role of the product in the customer's use context is more important when compared to the suboptimised technical performance of the product.

Value co-creation does not only happen between suppliers and customers, but also among customers, facilitated by a product or service from a supplier. In fact, building platforms that enable multisided markets (e.g. the Uber taxi service or AirBnB hotel service) to emerge where participants connected to the platform co-create value together is becoming the most successful business model in the information economy, because of the potentially exponential scalability (Smedlund and Faghankhani 2015). The role of the supplier has become to design digital or physical venues for experiences that consist of an inseparable bundle of product and service, and the customer is part of this context (Payne et al. 2008).

Despite the changes in society, the manufacturing-based innovation paradigm is still rather strong (Barcet 2010; Mele et al. 2014). This paradigm is technology-oriented and is focusing on coming up with the best possible solution for a highly focused problem. The output of a manufacturing-based innovation is a new product or a new service, and the attention is drawn to developing companies' internal innovation processes (Mele et al. 2014; Nuutinen and Ojasalo 2014). The Finnish metals and engineering industry is not an exception, but the interest in more open innovation exists, as, for example, our research programme shows. Mainstream business thinking is claimed to be gradually shifting from goods-dominant logic to service-dominant logic (Grönroos and Ravald 2011; Kingman-Brundage et al. 1995). The actual realisation of this new way of thinking in companies' everyday practice—how they make sense of their business world—is still rare to meet in many contexts (Nuutinen and Ojasalo 2014).

The idea of networked innovation has already had a significant impact on how innovation emergence is understood, and its effects are visible particularly in any public-funded support for precompetitive innovation activity. UXUS was not an exception. At first sight, the effects of these other current discussions on service science and value co-creation were not so evident here. One reason is that there was a parallel programme planned and in progress, focusing particularly on service business. The UXUS programme was built round a challenge to create competitive advantage based on user experience (UX), a concept that originates from the human-centred design tradition and refers within that discussion to users' emotionally laden experience and value-related judgments concerning the technologies used and the technological environment. But the value creation discussion is very much related to the question about how the user experience can be turned into a competitive advantage for the Finnish MEI. The present understanding of the main term of the programme, 'user experience', directs the focus deep into the customers' processes and towards value-in-use over time (Edvardsson et al. 2010) and a rather broad conception of 'value'. Professionals in different research and development fields have understood that the user has to be taken into account in design as a comprehensive person, not just as a user of a product or service (Arhippainen 2009). UX emphasises that products or services need to support users' hedonic needs, such as stimulation and self-expression, in addition to their pragmatic ones (Hassenzahl 2004). Designing for UX aims at broader views of users' emotional, contextual, and dynamically evolving needs and the impact of users' previous experiences on new experiences. Thus, the targets of design broaden from value of efficiency, safety, productivity, and so on, to those of experience. In this context, that means designing for experience when working in a customer company's processes.

In this case, what kind of support could the research programme have created? In the light of this discussion, the favourable conditions for innovation should have supported service-logic thinking in practice. Summarising the above, this means, for example, customer-centricity and results in value-in-use over time—or in other words, how value and its creation is understood and how it shows in beliefs in business and innovation.

On the basis of these discussions, we can conclude a further criterion: namely, the ability of the programme to challenge the present business and innovation logic in a way that enhances change towards service-dominant logic, value co-creation, and, of course, open innovation.

Further questions from these points of view to evaluate a programme are:

- SD logic:
  - How does the programme enable the participant companies to see the role of their products and/or services in the market as complementing customers' sociotechnical systems?
  - Do the participating companies understand the difference between providing technological solutions and providing experience-conveying solutions?

- Value co-creation:
  - Does the programme facilitate innovation activities:

    Among the participating companies?
    Between research institutes and companies?
    Inside companies between various divisions and units?
  - Does the programme improve the (UX-sensitive) capabilities of participating companies to:

    Sell existing products and/or services to customers?
    Incrementally develop products or services with customers?
    Radically renew products or services with customers?
- Open Innovation:
  - Does the programme help companies to find new systemic problems that are shared across the whole industry (new ecosystem creation)?
  - Does the programme help companies to combine the commercialisation process with the R&D process?
  - Does the programme help companies to develop competencies to make use of freely available industrial knowledge?

16.5.3 The Research Programme as a Learning Community

The third perspective on the research programme is learning community. Has the programme created favourable conditions for learning a new way of innovating? The act of innovating and learning in the programme was only part of the participants' everyday life in their work practice, but also a part of the broader industry-research practice. We are looking at the programme as a 'system of practice, existing in the world of tacit knowledge' (Gherardi 2000: 215). In other words, tacit knowledge is simply usable but becomes the object of reflection when a breakdown occurs. The new funding instrument provided 'a breakdown' and opened potential for learning. But does the learning continue?

We approach learning through participation in practice as something that takes place in a flow of experience, with or without our awareness of it (Gherardi 2000). Participating in a practice is to learn the logic of that practice—*sens pratique* (Bourdieu 1990; Gherardi 2000). Participating in a practice is a way of acquiring knowledge in action—but also of changing or perpetuating such knowledge.

In order to take a closer look at our programme practice, we can utilise perhaps the best known theory on situated learning and practices, namely, Lave and Wenger's theory on legitimate peripheral participation (LPP) and its key concept, communities of practice (Lave 1991; Lave and Wenger 1991; Wenger 1998/1999). The primary focus of the LPP theory is on learning as social participation. LPP gives a good frame for analysing learning in collaboration practice instead of as something that takes place in an individual mind. Participation refers to a process of being active participants in the practices of social communities and constructing identities in relation to these communities (Wenger 1998/1999). According to Lave and Wenger, the development of identity, knowledge, and skills is a part of one and the same process, where the level of participation defines access to knowledge. The learning is not one-sized. In practical action, both the individual and the community shape themselves and each other. Motivation for learning comes from the effort of developing the 'identity of a master', which gives full membership of the community (Lave 1991; Lave and Wenger 1991).

Lave and Wenger developed the concept of the community of practice (CoP) as a 'set of relations among persons, activity, and [the] world, over time and in relation with other tangential and overlapping communities of practice' (Lave and Wenger 1991: 98). According to a later definition, 'Communities of practice are groups of people who share a concern, a set of problems, or a passion about a topic, and who deepen their knowledge and expertise in this area by interacting on an ongoing basis' (Wenger et al. 2002: 4). The CoP can be seen as a form of self-organisation that matches neither organisational boundaries nor friendship groups (Corradi et al. 2010). It is based on sociality among practitioners and on the sharing of practical activities. So, sociality is the dimension within which interdependencies arise among people engaged in the same practices (Corradi et al. 2010). Compared with teams, CoPs are social systems that are much less tightly connected and defined in terms of interaction, shared tasks, and pre-established deliverables (Euerby and Burns 2014). CoPs may share some common high-level goals: an effective community can generate several benefits for an organisation—over time, effective communities build practices and expertise among their members and promote the flow of information and practice between members (Euerby and Burns 2014).

There were several characteristics of CoP in the programme. The programme group clearly developed an identity: the identity of the UXUS programme and the differences between it and other ongoing programmes were repeatedly pointed out. The core group developed certain rules of collaborating, which the newcomers only gradually became aware of. In addition, the persons responsible for leading the programme had superior access to sources of knowledge because of their duties, such as coordinating work between tasks and with companies.

To conclude, a research programme should create a community of practice that enhances learning a new way of innovating. This is the third general criterion for reflecting the conditions of innovating in industry-research collaboration created in the UXUS programme. For further guidance on planning a research programme, we can formulate the following questions:

- Do people participating in a programme share a concern, set of problems, or passion about a topic, and do they deepen their knowledge and expertise in the area by interacting on an ongoing basis?
- Can we find characteristics of a community, identity development, or learning?
- Are a community and its individuals shaped by each other?

What makes a community of practice succeed depends on the purpose and objective of the community, as well as the interests and resources of its members. Wenger identified seven actions that could be taken in order to cultivate communities of practice. On the basis of these, we can formulate further questions:

- Does the programme design allow the community to evolve naturally?
- Does it create opportunities for open dialogue within and with outside perspectives?
- Does it welcome and allow different levels of participation?
- Does it develop both public and private community spaces?
- Can we focus on the value of the community?
- Is there a combination of familiarity and excitement?
- Did we succeed in finding and nurturing a regular rhythm for the community (Wenger et al. 2002)?

Next, we will turn to demonstrating, with two examples, how the theoretical lenses can further be utilised in reflecting the programme. In the second phase, two examples of activities and change towards the overall criteria were described, and then the theoretical lenses were utilised in interpreting the characteristics of the programme that enabled common activities and change—meaning learning.

16.6 Results: Learning a New Way of Innovating

16.6.1 *Example 1: Change in Object of Research*—From Interaction Tools to UX Mindset and Voice of Customer

Our project in one of the work packages started with quite a traditional setup: we had three companies that all had specific user interaction concepts in mind to be developed in the programme. User interaction concepts represented a specific work package in which researchers and company representatives together worked on the company-specific case studies. Even if the focus was on new interaction concepts, the novelty value was taking targeted user experience (UX) as the starting point of the design. Common challenges were identified in developing methods to define the targeted user experience and in interpreting the user experience target-to-design implications. Defining a common approach to user experience-driven design became a common task in all the cases. The cases during this phase were interaction oriented, as can be seen from the case names: 'mobile interaction with an elevator', 'gesture-based interaction in a metals workshop', and 'remote control of cranes'.

During the next phase of the programme, the understanding of user experience had grown beyond user interaction solutions. Two additional companies joined the team at this point. We set up new company case studies in which the focus extended from interaction concepts to concepts of operation, thus including in the focus the overall work environment and work processes. The new company cases were also more future-oriented, and they aimed to renew concepts of operation with user experience-driven innovation. User experience was seen as a renewing factor in design, and a new design approach for the user experience-driven innovative leap was developed. The names of the new cases reflect how the focus had widened and had become more future-oriented: 'A day at future factory' and 'Future ship bridge concepts'.

At the time of conducting the analysis, we were just starting the third, and the final, phase of our programme. The user experience understanding gained enabled proceeding with UX work at the next level: to widen the scope from concepts of operations to organisational practices and contacts with customers. This wider scope showed in the planned new case studies. The case studies still included interaction concepts and the development of operational concepts, but there were new cases that focus on developing the user experience mindset in the organisation. One of the cases focused on a whole development project in the company, and the aim was to support the adoption of the user experience mindset within the project organisation. Another case developed a user experience style guide that also aimed to develop the user experience mindset in the organisation. The new cases also focused on understanding customers and engaging them in codesign activities. One case was developing *UX Playroom*—a codesign space where it is possible to step into the user's shoes and experience a day at future factory. The *voice of customer* theme gathered together several companies to develop methods to monitor the user experience in the field.

16.6.2 Example 2: Booklets—Learning How to Communicate Results

At the beginning of the programme, progress was documented in the internal summary report of the company cases and in the form of conference papers typical for this kind of programme. A significant step was taken when the evolving understanding of UX and its role in MEI within the programme was published in the form of booklets, the first being published in 2012 (Roto et al.) and the second in spring 2014 (Roto et al.) (and the third in 2015, Roto et al.). There were several reasons behind the choice of publication mode. Firstly, focusing on experiences in the programme should also apply to the programme's own publications—they should create positive experiences for readers. Secondly, the quite condensed, richly illustrated, and rather simple but tempting way of presenting the key issues of UX were regarded as essential for reaching the most important target group—those managers and personnel in companies who do not yet understand the meaning of UX. Thirdly, the funding condition that required wide publicity for the results of the programme encourages publishing something faster than 'true' research results can publish. The limited space in this kind of booklet publication benefited learning within the programme group: it necessitated summarising the key messages and findings into short sentences and a simplified form, which definitely contributed to developing a shared understanding.

The first of the booklets focused on why UX is important in the business-to-business context and what it means. The content included definitions of UX, both from literature and the one agreed within the programme group. The attempt was to cover issues from all the work packages. The graphic layout drew attention as a 'glimpse through a hole'—illustrating the fact that the content only introduces the world of UX and does not give direct answers to everything. There were more dramatic suggestions for the booklet cover and name: for example, 'The shot of UX', with a picture of a drink.

The second volume focused more on the question of 'how'. The booklet started with a game-like content sheet, which illustrates that UX thinking should affect everything going on in a company—but that the way one proceeds can be unique. The change towards experience-driven thinking can be achieved through baby steps and following various paths. The game metaphor continues throughout the second booklet.

All of the booklets are based on the ideation made by the programme group. The participants contributed both to the content and to the way of presenting it, and the editor group, together with a graphic designer, developed the ideas further, created something new, and asked for additional contributions from the others, and so on. The process was iterative and supplementary. However, there was a clear difference between the processes of the volumes: in volume one, the responsibility for the content remained more in the hands of the editor-in-chief, but in volume two, the programme group produced more, and the responsibility for the process was more equally shared within the editor group; in addition, the work with content was clearly enhanced by the graphic designs. Enhanced visualisation attracted the intended audience and the booklet received very much positive feedback.

16.6.3 Interpreting the Supporting Characteristics of the Programme

In this chapter, we adopt three different viewpoints of the research programme, in order to better understand the conditions that the whole programme is creating for innovation and the renewal of innovation practices. We recognised three general criteria for reflecting the conditions of innovation in industry-research collaboration and for new innovation practice to emerge. They are:

- As realising an open innovation platform: a research programme should constitute a platform that creates supportive frames enhancing collaborative and open innovation activity.
- As challenging present business thinking and innovation practice—and leading them towards a new logic: the ability of a programme to challenge current

business and innovating logic in a way that enhances change towards service-dominant logic, value co-creation, and open innovation.
- As building a learning community: a programme should constitute a community of practice that enhances learning a new way of innovating.

The introduction of the SHOK pursuing realisation of the open innovation platform model at the beginning of the programme planning created favourable conditions for a new practice to emerge, since it broke the old practice. At that time, the model was rather unformulated and left room for a new kind of programme planning. But was there also something in the programme that enhances breaking 'the rule-based operations' that produce a normal situation in industry-research collaboration practice? Can they support a change of logic in innovating? Did learning take place? Are there signs of a new innovation practice that realises the above criteria, and do we understand how to support them?

The two examples presented demonstrate learning towards a broader understanding of utilising UX-driven thinking in company operations and co-creating value. The progress towards openness is evident in both examples—in the first example, inside companies' cross-functional boundaries and, in the second example, openness outside. In the UXUS programme, we have experienced a common learning process in which we have learned how the user experience is reflected in various corporate operations. In this respect, the facilitating conditions for learning have been a working approach with concrete focused cases, close cooperation between industry practitioners and researchers, and long-term cooperation that has made it possible to proceed all the way from interaction concepts to the user experience mindset in organisation and real-time customer understanding.

In addition, the content of the programme was not independent from the learning results, but contributed to a change in value conceptions. Although the programme mainly concentrated on product development, the focus on the user experience challenged the present conceptions of value in the companies. Basically, if value is embedded in a product, the object of innovation is also directed to the product. Defining the targeted experiences and designing for them represent moving the focus further towards deeper value-in-use in customers' processes. The object of work (or innovation activity) broadened from products towards experiences that everyone has in touchpoints with the company. This can be interpreted as a way of realising service logic in practice. Furthermore, the focus shift from usability to experience has challenged the pure 'hard' (efficiency, safety, etc.) conceptions of value. There was also evidence of changes in correspondence in innovation activity—meaning who participates, when, how, and so on. In one company, the design and marketing teams were integrated into one, resulting in close collaboration.

When further reflecting on the conditions for new innovation practice to emerge, we recognised six of seven actions that could be taken in order to cultivate communities of practice as defined by Wenger et al. (2002). There was room for the community to evolve naturally, that is, support shifts in focus when the interests, goals, and members changed, as is natural for CoP dynamic (1) (Wenger 1998/1999). The vision- and problem-driven programme plan, periodically updated according to the lessons learned, together with financial flexibility between the work packages, allowed for this. The regular programme events and invited outside speakers or participants created opportunities for open dialogue with inside as well as outside perspectives (2). The various levels of participation (3) were clearly evident in the programme group: (a) the core group, which has also taken on leadership roles in guiding the group; (b) the active group, which attended and participated regularly, but not to the level of the leaders; and (c) the peripheral group which, while they were rather passive participants in the community, still learned from their level of involvement, such as in development projects in the companies. There were efforts to develop both public and private community spaces using wiki and social media (4), but these attempts were not highly successful, whereas the regular meeting served well as a space where all members shared, discussed, and explored ideas, in addition to offering time for private exchanges. They also offered a regular rhythm for the community (5). In the beginning, the meeting was held monthly, but later every second month, allowing the members to regularly meet, reflect, and evolve. The rhythm maintained the anticipated level of engagement to sustain the vibrancy of the community, yet not so fast-paced that it became unwieldy and overwhelming in its intensity. The structure of the meetings, which included both presentations and different workshops on, for example, the booklet, was a way of combining familiarity and excitement (6). In this way, they offered the expected learning opportunities and opportunities for members to shape their learning experience together by brainstorming and examining the conventional and radical wisdom related to their topic. The least evident of the actions taken was focus on the value of the community (7). The programme collaboration did not explicitly offer opportunities for participants to explicitly discuss the value and productivity of their participation in the group. This kind of discussion existed in work package-specific meetings and organisation-oriented meetings. However, the lack of discussion at programme level created some tensions between the various groups. Table 16.1 gives further examples of the results and the use of the theoretical lenses.

To summarise, there were several potential issues at the beginning, in the structure and the management practices of the programme, that joined breaking 'the rule-based operations' that produce the normal situation in industry-research practice and that also seemed to support a new practice emerging roughly as categorised in Table 16.2. When we recognised the positive characteristics of the programme, we shed light on implicit cultural beliefs in conceptions held by the other party, and roles and rules of a jointly funded research programme with undertones in the innovation logic of broader industry-research collaboration practice (see Table 16.2). The beliefs might counteract the aims of any new kind of research programme, but awareness of them can open further possibilities for supporting innovation activity. We claim that they reflect the illusory reality that limits and constrains our thinking and collaborating, even though they would never exist as more than prejudices. However, recognising these allowed us to interpret further the emerging new innovation practice worth supporting (Table 16.2).

Table 16.1 Examples of the evaluation

Criteria	Evidence (on the transformation towards)	Characteristics of the programme	
		Enhancing	Counteracting
Realising an open innovation platform	Companies shared openly their results with others in the programme meetings. They utilised the results and experiences of others	The focus also included company practices, not only concept development The companies were not direct competitors The programme group was multidisciplinary Use of concrete cases	The differences between companies' products and business Change in 'mindset' is not easy to recognise and explain to others
Challenging current business thinking and innovation practice towards new logic	The focus of designing for experiences shifted from HTI concepts to all touchpoints the customer contacts; the customer can therefore also be seen as a user (e.g. user of a contract) New collaboration between marketing and R&D	The term user experience was rather new in the industry at the outset of the programme, and ready-made definitions for it did not exist in the industrial B2B context Focusing on experience is in line with the ideas of broader value Shared cases The need for change in 'mindset' was a shared target The new partners joining during the second period brought stronger future orientation with them	Industrial service business development was part of another ongoing programme, and overlapping was not encouraged, thus limiting some rather obvious actions and research topics Participants from the companies mainly represent product design
Building a learning community	UXUS was referred to as a unique and positive example for the other programmes	Identity creation was supported by a logo competition The programme plan provided room for changing the focus, based on what has been learned The way of managing and collaborating in the programme was developed gradually	The changes in people taking part in the programme Most of the participants also have other projects going on Company representatives were also gatekeepers and the only contact points inside the firm: larger participation could have distributed the results quicker and wider

Table 16.2 Summary of the characteristics of the programme and tentative interpretation of elements of the broader collaboration practice challenged

Criteria	Characteristics of the programme and its management facilitating learning a new way of innovating	Elements of 'the logic of innovating'	
		Examples of the implicit beliefs of 'old logic'	Hints of new logic worth of supporting
Realising an open innovation platform	Introducing a SHOK concept aiming to increase the industry's impact on research, collaboration in the spirit of 'open innovation', and more long-term strategic research in order to enhance the sustainable competitiveness of industry A new kind of cost and funding structure enabling flexibility to revise and adjust research activities during the programme Wide publicity of the results made dissemination of the results to industry possible. However, it is rather difficult to show the impact of wide publicity	*Structure, management, and plans of research programme—beliefs on how to ensure high-quality and efficient research*	
		The research should be planned beforehand in detail; a clear plan and distribution of work according to the participants' responsibilities (e.g. in organisation-specific tasks) ensure efficiency and high quality of the research and management; the main task is to ensure that the plan is realised and the promised deliverables produced The main determinant of the collaboration is exchange typically between two parties: when companies invest in research, they expect specific work from the researchers. The researchers, in turn, expect to get access to collect 'real' data for their own 'scientific' purposes, including outside the particular interest of the company	The research plan should create a rather open frame for collaborative sensemaking of the topic, particularly in the beginning. Then the programme plan (and inspiring publications, such as booklets) can be used as a tool to enhance and document the learning by redefining it during the programme This strategy provides clear benefits in rather long-term research programmes, allowing participants to follow their interests and cling to emerging opportunities This also supports a broader impact: the participants can become 'the ambassadors' of the whole message—not just their own topic. On the downside, joining the programme in the middle could be difficult if the rules are not clear for those outside the inner group

(continued)

Table 16.2 (continued)

Criteria	**Characteristics of the programme and its management facilitating learning a new way of innovating**	**Elements of 'the logic of innovating'**	
		Examples of the implicit beliefs of 'old logic'	**Hints of new logic worth of supporting**
Challenging current business thinking and innovating practice towards new logic	A long planning period with several cross-organisational workshops organised in collaboration between industry and research, but with a strong industry leader Industry-research pairs in leading work packages, which were considered very useful for both parties due to mutual learning Room for new issues to emerge; mutual confidence between participants emerged and enabled open discussions Interdisciplinary collaboration both in industry and academia	*Roles and expertise of different parties: beliefs on the object of collaborating and useful tools*	
		The objectives of research and industry are different and counteract when the former pursues bigger benefit in the form of generalised knowledge and the latter practical benefits of their own Theories and research are owned by researchers and they are superior in them. The use of particular research methods produces knowledge that is only useful for research	The research task should be seen as a shared effort targeted at collaboratively making sense of the object of research without too much emphasis on the organisational background of the participants. For example, scientific papers can be co-authored; doctoral students can come both from companies and universities In this process, multiple theoretical approaches, as well as methodical solutions, even varying in their ontological and epistemological background, can be utilised 'as tools' in enhancing a common theory–practice dialogue benefiting both industrial innovations and knowledge creation
Building a learning community	Community building, for instance, by having frequent programme-wide meetings Practical relevance/ practical problem-driven research The UX theme was rather new and nobody 'owned' it, and making sense of UX in	*Motives and values of participants: beliefs on the main interest of the parties in participating the collaboration*	
		The motives are permanent and are not affected by the nature and progress of the particular collaboration The targets of the companies are totally	In overcoming boundaries between the disciplines and organisations and enhancing real interdisciplinary work, the programme should construct an arena for

(continued)

Table 16.2 (continued)

	enhancing competitiveness of the industry was a target requiring different perspectives The development of the programme's 'own language', logo and publication mode (booklet), which created cohesion and a shared identity among the participants	different from those of the researchers, and all parties are seeking ways to shape their participation accordingly, although a certain investment in collaboration is regarded 'as part of the game' There are many lasting implicit generalisations on the beliefs of the other party, such as 'the work done for companies is something less important and steals time from 'real research'', but these are something that should just be accepted	a learning community to emerge Awareness of hidden tensions between the new and the old culture of innovating can support this task, but the real development is energised by committing to the challenging targets enforced by shared experiences within the community. This gradually builds the basis for a new way of innovating by demonstrating its benefit, such as in radical innovations, how practical problems inspire theory-building, and how seeing the practical impact is really valued by researchers

16.7 Discussion and Conclusions

The paper aimed to contribute by recognising new ways to ensure the impact of innovation activity by identifying conditions that seem to construct favourable conditions for innovation in this particular collaborative context, by utilising three theoretical perspectives. The main questions were:

1. What are the issues in the programme that combined to break 'the rule-based operations' that produce the normal situation in industry-research practice?
2. Has the logic of innovating changed?
3. Can we recognise hints of a new kind of industry-research innovation practice that is worth supporting?

Furthermore, this article also aimed to contribute to the overall motive of the book, by providing an example of how to narrow the gap between theory and practice, which was demonstrated in the use of theories reflecting the programme. Two examples of learning were described, and the theoretical lenses were utilised in interpreting the characteristics of the programme, to enable learning following the deeper interpretations of the construction of a new kind of practice worth

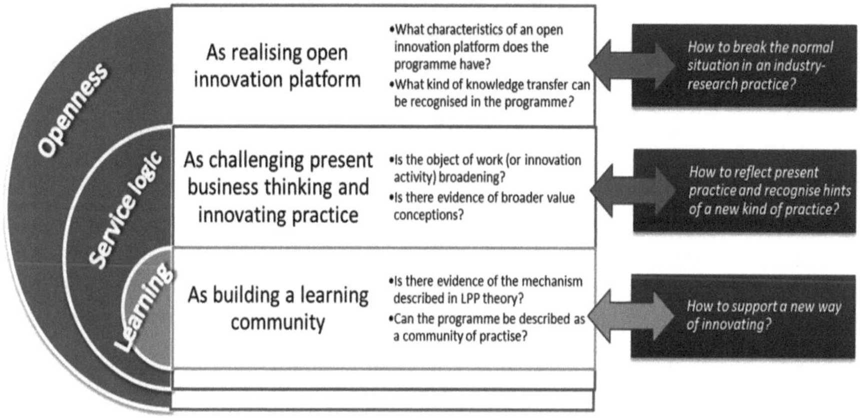

Fig. 16.1 The framework for planning and evaluating a research programme

supporting. Figure 16.1 summarises the guidance for planning a research programme on the basis of the analysis. The questions formulated in the theoretical sections can also be utilised for evaluating conditions for innovation in industry-research collaboration. The selection of the theoretical lenses might vary according to the aims and content of the programme, but the questions described on the right in Fig. 16.1 can be applied even then by research programme players. Following the steps we described in this chapter, they can create better prerequisites for innovation by developing a frame of reflection of their own. The new emerging way of innovating based on this kind of reflective practice is a path towards profound impact when it supports continuous learning (Norros 2004).

As noted by Gherardi (2000: 216–217), 'the logic of practice is necessary for the order and continuity of an organization. Practical knowledge is kept within the habitus, which as the historical product of previous individual and collective practices, produces historical 'anchors' and ensures the correctness of practices and their constancy over time more reliably than formal and explicit rules. At the same time, the replication of the logic of practice contributes to its transformation simply making it explicit'. We increased the knowledge on co-innovating practice in jointly funded research programmes by reflecting on our own.

The theories we utilised are widely discussed, and the first two were also clearly affected behind planning the new funding instrument, as well as the UXUS programme. LPP theory and the CoP concept have interested many since their introduction, and interest in them from innovation research has increased lately. The theory is criticised, for example, by Engeström (1995) for offering little for qualitative change in work, which typically represents 'a leap into the unknown' (p. 87), and by Kivinen and Ristelä (2001) for being too generous to offer a basis for viable and practical solutions for promoting learning. Moreover, as Gherardi (2009:121) noted, referring to several other authors (Brown and Duguid 1991, 2001; Gherardi et al. 1998; Swan et al. 2002; Roberts 2006) reversing the emphasis between the two terms, from community of practice to practices of a community is proposed. According to her, what is emphasised in the former case is that the community constitutes the container of knowledge and the community pre-exists its activities. In the latter, it is the activities themselves that generate a community, in that they hold together a configuration of people, artefacts, and social relations. 'Knowledge, therefore, is not an 'asset' of the community, but rather an activity (a 'knowing'), and an activity that itself constitutes practice ('knowing in practice')' (Gherardi 2009:121).

In our research, the role of these theories was to provide various perspectives for evaluating the programme, thus providing *a frame for reflecting and becoming more aware of industry-research collaboration practice—as well as for learning more about the conditions this has created for innovation and for new innovation practices to emerge.*

Our reflection of the UXUS programme showed that many of the 'objective' characteristics counteract each other in terms of the objectives and guidance that can be formulated on the basis of the current theories, such as the size of the UXUS programme supporter communication and collaboration, in other words, 'openness' inside the programme, enhancing community development. The three-period model that supported focusing on the subject instead of on writing funding applications (and arguing about the shares of the budget) made access to the programme more difficult. However, these together limited openness for new members. Similar examples from the programme are numerous.

The recommendations by Lähteenmäki-Smith et al. (2013) included building stronger ecosystems based on co-location, such as 'SHOK campuses' or Living Labs. Cross-disciplinary and sector-transgressing themes and research substance could be a way to focus on future success areas, in line with 'grand challenges' thinking. The evaluation revealed indications that the SHOKs had enabled the integration of new partners and broader consortia and partnerships. Sometimes, however, the breadth of the partnerships had been won at the expense of the depth and intensity of the collaboration. Often the SHOKs had become too large to allow efficient collaboration. On the basis of our study, we would like to emphasise the role of the learning community. For example, the location was not critical in our case, but rather regular meetings and actions that supported identity creation were. In the evaluation of the FIMECC SHOK by Lähteenmäki-Smith et al. (2013), several stakeholders emphasised that the set-up for collaboration is totally different from previous instruments, as the initiative for various research topics comes much more often from the industry. This perhaps has led, in FIMECC, to the tendency towards incremental innovation. New research areas able to transform and renew the industry had not yet shown their impact at the time of the evaluation. Our programme showed that in this kind of rather long collaboration, there is room for learning to think about broader strategic issues and new ways of co-innovating more probable leading to radical innovations and lasting impact.

Good company–company cooperation was seen as specific to FIMECC, but good industry–academy cooperation was also praised. FIMECC was seen as a tool to get closer to open innovation. The internationalisation of FIMECC was seen as a clear improvement need, similar to most other SHOKs. Our own reflection supported these notions.

The emergence of the UXUS community as a new kind of industry-research practice, or practice of community, which has evidently boosted many favourable phenomena during the programme also could have its downside. It may limit the broader effects of the programme even we strove for finding ways to communicate the constructed, partly tacit knowledge of the role of UX-driven thinking in business renewal (see e.g. Nuutinen and Koskinen 2015). However, an even more promising future avenue for spreading the impact rests on the members of this community as ambassadors of the new industry-research practice in innovation. There lies real potential for redeeming the promise of renewing the Finnish MEI, as envisaged in the programme plan—and even more. The seeds are already growing, and now we have to ensure that they flourish even the programme has ended.

When planning similar platform models for supporting innovation and renewal, we suggest carefully considering the questions presented at the end of the description of each of the theoretical lenses, and familiarising oneself with the experience in SHOKs.

To conclude our guidance for planning and managing research programmes and new innovation systems supported by public funding, we would like to emphasise three points. Firstly, to seek multilevel theoretical guidance and utilise frames of references that take into account issues from measurable objectives and structures to the construction of the conceptions and dynamics of the practice. Secondly, pursuing openness in collaboration, networked or ecosystemic value co-creation in business thinking, and a learning community are good tentative suggestions at this time, as general guidelines for managing an innovation programme, but the ways of realising them in innovation practice should always be discovered in that particular context. Thirdly, dare to challenge current practices and leave room and time for learning.

References

Annala, K., & Ylä-Jääski, J. (2011). *Strategiaa, huippuosaamista, keskittymiä. Strategisen huippuosaamisen keskittymät CLEEN, FIMECC, Metsäklusteri ja TIVIT yritysnäkökulmasta* [Strategy, top knowledge, centres. Strategic centres CLEEN, FIMECC, Forest Cluster and TIVIT from the industry viewpoint]. Helsinki: Teknologiateollisuus (The Federation of Finnish Technology Industries).

Arhippainen, L. (2009). *Studying user experience: Issues and problems of mobile services – Case ADAMOS: User experience (im)possible to catch?* Oulu: University of Oulu.

Barcet, A. (2010). Innovation in services: A new paradigm and innovation model. In F. Gallouj & F. Djellal (Eds.), *The handbook of innovation and services* (pp. 49–67). Cheltenham: Edward Elgar.

Bourdieu, P. (1990). *The logic of practice*. Stanford, CA: Stanford University Press.

Brown, J. S., & Duguid, P. (1991). Organizational learning and communities-of-practice: Toward a unified view of working, learning, and innovation. *Organization Science, 2*(1), 40–57.

Brown, J. S., & Duguid, P. (2001). Knowledge and organization: A social-practice perspective. *Organization Science, 12*(2), 198–213.

Chesbrough, H. (2003). *Open innovation: The new imperative for creating and profiting from technology*. Boston: Harvard Business School Press.

Chesbrough, H. (2011). *Open services innovation*. San Francisco: Jossey-Bass, Wiley Imprint.
Corradi, G., Gherardi, S., & Verzelloni, L. (2010). Through the practice lens: Where is the bandwagon of practice-based studies heading? *Management Learning, 41*(3), 265–283.
Cusumano, M., & Gawer, A. (2002). The elements of platform leadership. *MIT Sloan Management Review, 43*(3), 51–58.
Edvardsson, B., Gustafsson, A., Kristensson, P., & Witell, L. (2010). Customer integration in service innovation. In F. Gallouj & F. Djellal (Eds.), *The handbook of innovation and services* (pp. 301–317). Cheltenham: Edward Elgar.
Eisenmann, T. R., Parker, G., & van Alstyne, M. W. (2011). Platform envelopment. *Strategic Management Journal, 32*(12), 1270–85.
Engeström, Y. (1995). *Kehittävä työntutkimus. Perusteita, tuloksia ja haasteita* [Developmental work research. Arguments, results and challenges]. Helsinki: Painatuskeskus.
Euerby, A., & Burns, C. M. (2014). Improving social connection through a communities-of-practice-inspired cognitive work analysis approach. *Human Factors, 56*(2), 361–383.
Gawer, A. (2010). The organization of technological platforms. In N. Phillips, G. Sewell, & D. Griffiths (Eds.), *Technology and organization: Essays in honour of Joan Woodward* (Research in the Sociology of Organizations, Vol. 29, pp. 287–296). Bingley: Emerald Group Publishing.
Gawer, A., & Cusumano, M. A. (2014). Industry platforms and ecosystem innovation. *Journal of Product Innovation Management, 31*(3), 417–433.
Gherardi, S. (2000). Practice-based theorizing on learning and knowing in organizations. *Organization, 7*(2), 211–223.
Gherardi, S. (2009). Introduction: The critical power of the 'Practice Lens'. *Management Learning, 40*(2), 115–128.
Gherardi, S., Nicolini, D., & Odella, F. (1998). Towards a social understanding of how people learn in organizations. The notation of situated curriculum. *Management Learning, 29*(3), 273–297.
Grönroos, C., & Ravald, A. (2011). Service as business logic: Implications for value creation and marketing. *Journal of Service Management, 22*(1), 5–22.
Hassenzahl, M. (2004). The thing and I: Understanding the relationship between users and product. In M. A. Blythe, K. Overbeeke, A. F. Monk, & P. C. Wright (Eds.), *Funology. From usability to enjoyment* (Vol. 3, pp. 31–42). Netherlands: Springer.
Jarvis, P. (1999). *The practitioner-researcher. Developing theory from practice*. San Francisco: Jossey-Bass, Wiley Imprint.
Kingman-Brundage, J., George, W. R., & Bowen, D. E. (1995). "Service logic": Achieving service system integration. *International Journal of Service Industry Management, 6*(4), 20–39.
Kivinen, O., & Ristelä, P. (2001). *Totuus, kieli ja käytäntö. Pragmatistisia näkökulmia toimintaan ja osaamiseen* [Truth, language and practise. Pragmatist perspectives to action and expertise]. Helsinki: WSOY.
Lähteenmäki-Smith, K., Halme, K., Lemola, T., Piirainen, K., Viljamaa, K., Haila, K., et al. (2013). *"Licence to SHOK?" – External evaluation of the Strategic Centres for Science, Technology and Innovation*. Publications of the Ministry of Employment and the Economy. 1/2013, Ministry of Employment and the Economy. Edita Publishing. http://www.tekes.fi/globalassets/julkaisut/licence_to_shok.pdf
Lave, J. (1991). Situated learning in communities of practice. In L. Resnick, J. Levine, & S. Teasley (Eds.), *Perspectives on socially shared cognition* (pp. 63–82). Washington, DC: American Psychological Association.
Lave, J., & Wenger, E. (1991). *Situated learning: Legitimate peripheral participation*. Cambridge: Cambridge University Press.
Mäkinen, S. J., Seppänen, M., & Ortt, J. R. (2014). Introduction to the Special Issue: Platforms, contingencies and new product development. *Journal of Product Innovation Management, 31* (3), 412–416.
Maréchal, G. (2010). Autoethnography. In A. J. Mills, G. Durepos, & E. Wiebe (Eds.), *Encyclopedia of case study research* (pp. 43–45). London: Sage.

Mele, C., Colurcio, M., & Russo-Spena, T. (2014). Research traditions of innovation: Goods-dominant logic, the resource-based approach, and service-dominant logic. *Managing Service Quality, 26*(6), 612–642.
Nikulainen, T., & Tahvanainen, A-J. (2009). *Towards demand based innovation policy? The introduction of shocks as innovation policy instrument* (ETLA discussion paper, No. 1182). http://hdl.handle.net/10419/44488
Norros, L. (2004). *Acting under uncertainty. The core-task analysis in ecological study of work* (VTT Publications 546). Espoo: VTT.
Nuutinen, M., & Ojasalo, K. (2014). Enhancing service innovation in a business-to-business context – Four questions for SMEs transforming to service logic. *International Journal of Quality and Service Sciences, 6*(4), 290–308.
Nuutinen, M., & Koskinen H. (Eds.), (2015). *User experience and usability in complex systems - UXUS 2010–2015* (FIMECC Publication Series; 8, 225 p). Tampere: FIMECC Oy. http://issuu.com/vttfinland/docs/fimecc_115_uxus_?e=5313536/31352366
Parker, G., & Van Alstyne, M. W. (2014). *Innovation, openness, and platform control* (October 3, 2014). Available at SSRN: 10.2139/ssrn.1079712
Payne, A. F., Storbacka, K., & Frow, P. (2008). Managing the co-creation of value. *Journal of the Academy of Marketing Science, 36*(1), 83–96.
Peñaloza, L., & Venkatesh, A. (2006). Further evolving the new dominant logic of marketing: From services to the social construction of markets. *Market Theory, 6*(3), 299–316.
Polanyi, M. (1962/1958). *Personal knowledge*. London: Routledge & Kegan Paul.
Ramirez, R. (1999). Value co-production: Intellectual origins and implications for practice and research. *Strategic Management Journal, 20*, 49–65.
Reason, P., & Bradbury, H. (2001). Introduction: Inquiry and participation in search of a world worthy of human aspiration. In P. Reason & H. Bradbury (Eds.), *The handbook of action research* (pp. 1–14). London: Sage.
Roberts, J. (2006). Limits to communities of practice. *Journal of Management Studies, 43*(3), 623–39.
Roto, V., Nuutinen, M., & Smedlund, A. (2014). *UX – How? A glimpse of user experience for B2B industry*. Issue 2. http://uxus.fimecc.com/sites/uxus.fimecc.com/files/ux_booklet2_web-2.pdf
Roto, V., Smedlund, A., Passera, S., & Nuutinen, M. (2012). *UX – Why? What? A glimpse of user experience for B2B industry*. Issue 1. http://uxus.fimecc.com/sites/uxus.fimecc.com/files/uxbooklet_issue1_online_2_0.pdf
Roto, V., Nuutinen, M., Kaasinen, E., Smedlund, A., & Seppänen, M. (2015). *Industry expedition to the UX world*. Issue 3. https://issuu.com/vttfinland/docs/uxus_booklet3_e-version_eng
Russo-Spena, T., & Mele, C. (2012). Five Co-s in innovating: a practice-based view. *Journal of Service Management, 23*(4), 527–553.
Sawhney, M. S. (1998). Leveraged high-variety strategies: From portfolio thinking to platform thinking. *Journal of the Academy of Marketing Science, 26*(1), 54–61.
Schön, D. A. (1983). *The reflective practitioner: How professionals think in action*. New York: Basic Books.
Smedlund, A. (2009). Network approach to fundamental tasks in knowledge-based organizations. D.Sc. dissertation, Helsinki University of Technology. Faculty of Information and Natural Sciences. Department of Industrial Engineering and Management. http://lib.tkk.fi/Diss/2009/
Smedlund, A., & Faghankhani, H. (2015). *Platform orchestration for efficiency, development and innovation*. In Paper accepted for conference: 2015 48th Hawaii International Conference on System Sciences (HICSS 2015) (pp. 1380–1388)
Swan, J., Scarbrough, H., & Robertson, M. (2002). The construction of 'communities of practice' in the management of innovation practice. *Management Learning, 33*(4), 477–496.
Vargo, S. L., & Lusch, R. F. (2004). Evolving to a new dominant logic for marketing. *Journal of Marketing, 68*(1), 1–17.
Wenger, E. (1998/1999). *Communities of practice: Learning, meaning, and identity*. New York: Cambridge University Press.
Wenger, E., McDermott, R., & Snyder, W. M. (2002). *Cultivating communities of practice: A guide to managing knowledge*. Boston, MA: Harvard Business School Press.

Chapter 17
Practice-Based 'Inside-Out' Innovation in Public Service: A Regional Child Welfare Agency

Monica Pedrazza and Sabrina Berlanda

Abstract The aim of this chapter is to describe an inside-out, practice-based innovation process in the public sector. The innovation is concerned with the improvement of the quality of the internal vs. external and cross-disciplinary information/data transfer on foster youth in a regional child welfare agency in Italy. The process is an open one; it occurs at the third sector level of the public service agency. Different professional groupings are involved in the inter-institutional decision-making process on children; thus, local and role-specific knowledge is needed to resolve problems that arise at a more complex inter-institutional level, such as the joint evaluation of children's behaviours. The practitioners' need to reduce intra-team conflict rates turned out to be the activating influential factor of this innovation. The employee-driven innovation process developed through three steps: (1) in-service training led by psychologist, (2) staff assessment, and (3) development of a new tool for the systematic observation by residential youth workers of children's behaviour. They can be referred to as 'educators', 'front-line community educators', 'practitioners' or youth workers. The new tool adopted reduced interpersonal and team conflicts at intra-service level, according equal dignity to both beginners and to experienced workers when reporting on children's behaviour. Unintended and positive outcome of the development of the new tool emerged at the adoption/diffusion stage: the large number of observations of each child, available in real time, allowed youth workers to be effective both in intra-service and interservice information transfer. Comparative studies should be carried out in order to identify commonalities between social, health and educational services' innovation influential factors and outcomes. According to De Vries et al. (2014), innovation's outcomes are seldom analysed; we've contributed to fill this research gap.

M. Pedrazza (✉) • S. Berlanda
University of Verona, Verona, Italy
e-mail: monica.pedrazza@univr.it; sabrina.berlanda@univr.it

T. Russo-Spena et al. (eds.), *Innovating in Practice*,
DOI 10.1007/978-3-319-43380-6_17

17.1 Introduction

Innovation in the public sector has often been neglected in the mainstream of innovation studies, even if the article by Geoff Mulgan (2007) entitled 'Ready or not? Taking innovation in the public sector seriously' has strongly raised the matter and a *Handbook of Innovation and Services* was published in 2010 (Gallouj and Djellal 2010).

Wallin and von Krogh (2010) state that innovation can be seen as a process that covers the creation of relevant knowledge for the development and introduction of something new and useful in organisations and services. The simplest definition for the public sector is that innovation is about new ideas that work at creating public value (Mulgan 2007). According to Cunningham et al. (2005:51), the here presented innovation process, can be seen as a 'deliberate change of behaviour at the level of institutions that includes a new or improved service, process, technology, or administrative tool'. Third sector organisations (NGO), according to the principle of subsidiarity, often supply for residential home services for foster youth in public services (Cunningham et al. 2005).

In this study we focused on residential youth workers who are the front-line community practitioners.

Community staff were concerned about the following issues:

- A perceived polarisation in evaluations of negative children's behaviour (e.g. aggression, worrying introversion)
- Confrontations between individuals and teams over the evaluation of children's behaviour, which often set up a vicious circle of division and mistrust
- Information transfer to other professionals, primarily to social workers who guide and co-ordinate the decision-making process on the young people in the care of the agency

The first aim of the present chapter is to follow the three steps of the innovation process which involved the development of a new tool for the collection and transfer of data on young people in residential care: (a) formative, restorative and normative training sessions run by a psychologist and then (b) the development of the tool and (c) its implementation.

Academic researchers ran the sessions in seven teams, one session a week. Each team was responsible for the residential and/or daytime service for children or young people in residential care.

The second aim is to present a new way of looking at interpersonal conflict in the workplace, as result of direct, first-order consequence of a co-developed innovation process. Academic researchers' epistemic positioning in this analysing work of practitioners' practices, during training sessions, according to Evered and Louis's (1981) distinction, refers to the 'inquiry from inside'. Reading 'from within' practices and following the changes in performing practices, due to the innovation process occurred, allowed practitioners and academic staff to identify the new emergent and negotiated order and meaning of actions (evaluations).

17.2 Theoretical Background

17.2.1 Innovation in Public Sector

According to Mulgan (2006), there are many national strategies to support innovation in business and technology, but no comparable strategies exist at national level to support innovation in social public services. Severe innovation deficits exist in several public services fields, especially in child welfare. Local communities at regional and provincial level, NGOs and foundations are more likely to drive and support innovation in the delivery of services for young people in residential care. We will present an innovation process which started at provincial third sector organisation level and involved all of the front-line residential youth workers in the oldest NGO responsible for residential and daytime homes for young people in the province of Trento in Northern Italy. There is especially a lack of training in multidisciplinary and dispersed teams (Sole and Edmondson 2002) such as those carrying out the programmes of young people in residential care.

In the recent systematic review of the literature on innovation in public sector, De Vries et al. (2014) analysed 158 articles and books (from 1990 to 2014) and showed that most studies do not pay attention to the outcome of innovation processes. Innovation's outcomes are seldom analysed or even recognised, since innovation is considered as a value itself, so further information are usually not required and sometimes even not investigated. The simple activation of any type of innovation seems to be an important ritual for public organisations, and thereby legitimacy is often achieved (De Vries et al. 2014; Meyer and Rowan 1977).

According to De Vries et al. (2014: 5), innovation in public sector (Osborne and Brown 2005; Rogers 2003) has been defined as 'the introduction of new elements into a public service—in the form of new knowledge, a new organization, and/or new management or processual skills, which represents discontinuity with the past'. Data showed that most studies are concerned with administrative, technology-driven innovation processes. As also in our study, De Vries et al. (2014) identified *conflict* (Fig. 17.1) as one of the influential organisational factors of innovation, and it mainly emerged during the generation phase of the process.

17.2.2 Third Sector 'Inside-Out' Open Innovation

The child welfare agency in which the innovation process described here took place can be thought of as a 'community-of-practice' (Brown and Duguid 1991; Lave and Wenger 1991). Two aspects of innovation in child welfare have to be considered:

1. Innovation in child welfare requires co-constructive practice (Hull and Saxon 2009) through purposive inflows and outflows of knowledge (Chesbrough 2003, 2006) at each step because of the large number of professionals involved

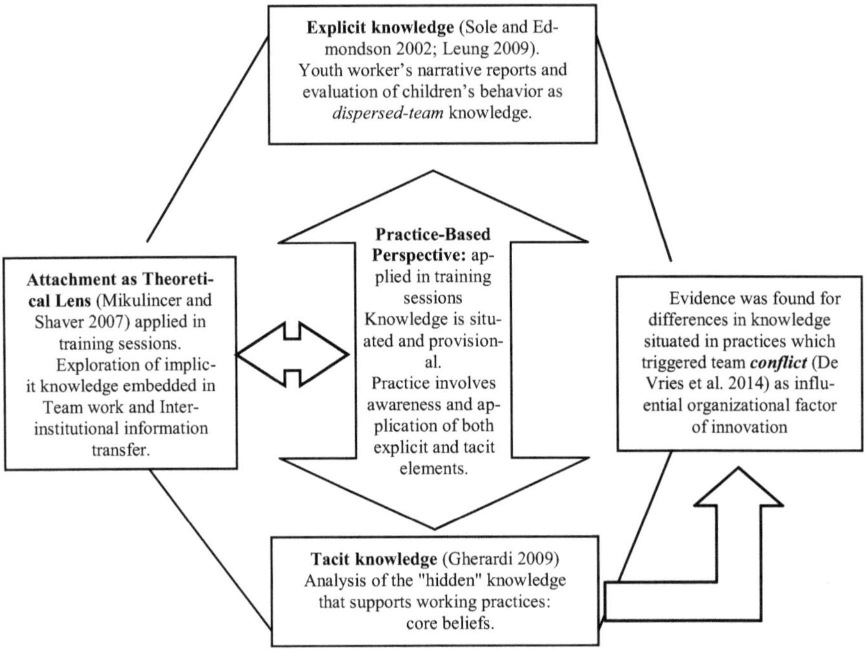

Fig. 17.1 Chart outlining the theoretical framework of the inside-out open innovation process

(Nouwen et al. 2012). These professionals, front-line residential youth workers, physicians, psychologists, paediatricians and juvenile court members belong on the one hand to role-based communities (Lave and Wenger 1991; Wenger 1998) where members have similar jobs, and, on the other hand, they have to cooperate within the child welfare agency where they act as a collection of professionals who are interdependent, despite their different institutional affiliations: they have a common task, share responsibility for outcomes of child welfare programmes, see themselves and are seen by others as a socio-institutional entity embedded in one or more public organisations and manage their relationships across organisational boundaries. They all collaborate by contributing to the development, updating, evaluation and concluding of the programmes designed for young people in residential care. Practitioners' and professionals' knowledge (Davenport and Prusak 1998a, b) differ in relation to theoretical frameworks, tools, daily practices and settings, and these differences (Tsoukas and Vladimirou 2001) often introduce discontinuities in the formal flow of information about service users.

2. According to Chesbrough (2006), there are two relevant aspects of open innovation: the *outside-in* one where organisations integrate external knowledge in their own practice and the *inside-out* aspect where professionals transfer their own knowledge into other sites in order to help others to accomplish goals. The innovation process we will describe can thus also be seen as an open one (Chesbrough 2003, 2006; Fuglsang 2008; Gassmann 2006; Gassmann and

Enkel 2004; Hilgers and Ihl 2010; Lee et al. 2012; Perkmann and Walsh 2007). Knowledge at each intra-service level (third sector/public health service/public social service/public educational services/public) has to be integrated with knowledge available in the external context at the interservices level, within the agency (Wallin and von Krogh 2010). In fact third sector-driven innovation has to be integrated with an open system of knowledge and values which needs to refer to two different sources of data and relevant information: (a) information from the health, social and educational service delivery context and (b) information from public bodies which support the biological parents' new needs. Public sector innovation is motivated to improve service performance (De Vries et al. 2014) and to add value in terms of public benefit, but in the field of *helping professions*, great importance must also be accorded to negotiation and dispute resolution (Zapf 2002) in order to meet the needs of professionals and practitioners for subjective well-being at work, which in turn represents the fundamental prerequisite of public social and health services. According to Lee et al. (2012), public services often give priority to the outside-in type of open innovation. Not surprisingly they are typically concerned with protection and handling of confidential data. In our study researchers and youth workers co-developed an inside-out type of innovation, filling thereby a research and literature gap, crossing residential care service boundaries and making knowledge available to social workers and to other components/services of the *dispersed teams* (Sole and Edmondson 2002) of the agency.

17.2.3 The Practice-Based Knowledge

Scholar's interest in CoP (Gherardi et al. 1998; Gherardi 2009; Roberts 2006; Swan et al. 2002) shifted, during the last decade, from the concept of 'community of practice' to 'practices of the community'. The second label implies that the activities themselves generate and constitute the community. The child welfare agency we dealt with does not really pre-exist as a community. It comprises a number of dispersed teams, in Sole and Edmondson's (2002) terms, such as youth workers, social workers, teachers and medical professionals, and generates itself as a community through interwoven activities and practices intersubjectively created, aligning different professionals and artefacts (Czarniawska 2004).

According to Evered and Louis (1981), we adopted the perspective of the 'inquiry from inside' in order to intercept the temporality and processuality of practices performed in the dispersed team of youth workers. The in-depth analysis of 'dispersed' patterns of activities (Geiger 2009) highlights the practice-based knowledge of the community as a whole. We thus analysed activities such as assessment and evaluation, needed on the part of all practitioners and professionals involved in the common practice of updating youth's programmes. The emerging common dialogue within the agency supposes the common intention towards the benefit and well-being of the users such as children themselves and their family of origin (Harmaakorpi and Mutanen 2008). Routines (Bechky 2003; Carlile 2002; Dougherty 2004), such as formal and informal communication among professionals and practitioners, create the shared knowledge of the agency. According to Ellström (2010), the practice-based innovation emerges from a cyclical process of learning. The implicit knowledge is concerned with the subjective interpretation of norms and prescribed work processes, developing a sort of routinised performance (Hedberg 1981) which can sometimes be interrupted by problems (Fig. 17.1). Facing problems triggers questions and doubts, which in turn, enact the development of new ways of understanding and handling the prescribed tasks. We identified the tacit (Tsoukas and Vladimirou 2001) and the explicit components of the practice-based knowledge of the youth workers' team, and as often is the case, according to Dougherty (2004), innovation occurred at work boundaries: in fact when dispersed teams (Sole and Edmondson 2002) work together, they have to decontextualise and recontextualise relevant knowledge in order to 'translate' their knowledge and switch on a process of sharing and meaning co-creation.

17.2.4 Insights for Care Services from Attachment Theory

Attachment theory is well recognised as helpful for health and social professionals and caseworkers in identifying the best ways to ensure a child's safety, stability and well-being (Bennett 2008; Bennett et al. 2012; Bifulco et al. 2008; Howe 1999; Lawler et al. 2011; McMillen 1992; Schofield and Beek 2006; Wilkins 2012). The widespread references to attachment research findings in the recent social work literature about children in the care of social and health services are often linked with improvements in the ability front-line social workers and practitioners to recognise children's attachment style, with educational treatment and with the subsequent evaluation of both children's behaviour and parental skills.

Both perspectives are thus deeply connected to recursive and institutionalised activities in child welfare.

In our study seven community staff were concerned with frequent confrontations over different evaluation of children's negative behaviour. Researchers and youth workers co-developed an in-depth reflective process on evaluation reporting practices of youth behaviour in order to address important differences in practitioners' threat perception which in turn triggered team conflicts. The framework of attachment theory explains the link between attachment style and caregiving: it distinguishes between secure and insecure (avoidant or anxious) attachment styles. Insecure, *controlling caregiving* consists in the adoption of a domineering, uncooperative stance during the caregiving process, whereas insecure, *compulsive caregiving* is the tendency to merge with the needy partner.

We argue that practitioners' attachment styles and caregiving styles are individual difference variables that need to be recognised as important in orienting practitioners' perceptions and evaluations of threat-related issues in interpersonal relationships with children. Residential youth workers perceive threat each time they have to cope with negative children's behaviour. This triggers different affective reactions, which often bear the features of defensive strategies.

17.2.5 The Child Welfare Agency: A 'Dispersed Team'

The work of the residential youth workers consists of specific work processes which follow formally organised procedures and protocols. These are based on both explicit knowledge such as theories or models and on implicit work processes rooted in subjective and personal interpretations of the work and the way it is performed in practice (Ellström 2010) (Fig. 17.1). The agenda that has been set by the third sector organisation staff relates to the residential youth workers' implicit knowledge. This can be understood as what Klein and Bloom (1995), referring especially to helping professions and social work, call 'practice in wisdom'. According to these authors, this is value-driven knowledge emerging from experience with service users, which connects with scientific knowledge when particular techniques, strategies and methods are needed.

The practice-based perspective underlines the provisional and situational nature of organisational knowledge (Lave and Wenger 1991; Wenger 1998). Situated knowledge is defined as shared by co-workers who are co-located, despite task based or role differences (Sole and Edmondson 2002). It cannot be shared by those working in different sites or crossing institutional boundaries, even when they have the same role. We adopt the construct of situated knowledge (Bourdieu 1977; Wenger 1998) in order to acknowledge how a dispersed team (Sole and Edmondson 2002) learns and innovates. The seven teams we dealt with are part of a larger *dispersed team*, the child welfare agency. Each team uses its local-specific knowledge to resolve problems that arise in the environment. However, when these teams need knowledge situated at a site other than where the problem arose, they must first recognise local-specific practices in which that knowledge is embedded before they can use it (Sole and Edmondson 2002). Residential youth workers, from third sector organisations, have to cooperate with other professionals such as neuropsychiatrists, paediatricians, clinical psychologists, teachers and social workers, in order to accomplish the goal set out by the child welfare agency. The agency itself doesn't have a stable institutional seat; teams are dispersed and meet at institutionalised and conventional places, only when decisions and/or programmes' updates are needed. Sole and Edmondson (2002: 31) found that dispersed team knowledge is a high valuable resource, 'but varied in its ease depending on differences in teams' members *awareness* of relevant knowledge and differences in its *appropriability*'. Sole and Edmondson (2002) suggest that manager of dispersed teams, in our case social workers, should be able to enhance both awareness and appropriability of knowledge. Awareness should be heightened through both technological and social means. New tools and development of network in support of inter-institutional communication could trigger recognition and easy recollection of relevant knowledge offering new knowledge-sharing opportunities.

Professionals' different educational background often hinders knowledge-sharing practices (Strauss and Olivera 2000) and communication in cross-functional and dispersed teams. Data are needed for updating projects and intervention programmes. As component of the dispersed team of the child welfare agency, residential youth workers receive information from the family of origin and from teachers, paediatricians, neuropsychiatric physicians and psychologists if and when needed. They combine this information with observation data on each child and organise an individual educational programme tailored to the needs of the young person concerned and the goals of the social service.

17.3 Objectives and Methods

17.3.1 Research Context

The regional child welfare service we focus on contracts out residential and daytime care services for young people to third sector organisations. Outsourcing has become the main strategy used in this area of public service delivery (Borzaga and Fazzi 2010).

In Italy the actors involved with child welfare work in day centres and residential settings include public stakeholders such as health-care services, social service for youth, services for the family and for family preservation/reunification and private and third sector organisations providing residential homes and daytime centres for young people. Children from 0 to 18 years eligible for residential care often come from multiproblem domestic/family contexts in which parents already receive different forms of economic and social support.

17.3.2 Research Design

The residential youth workers, the central staff of the third sector organisations and academic researchers are the main actors in the practice-based innovation process presented here. According to the triangulation principle, which underpins multi-method research procedures, researchers and practitioners agreed in avoiding being overreliant on a single research method, employing consequently more than one method in order to investigate the research question (Deacon et al. 1998).

We look at state child welfare from a 'user-oriented' constructivist perspective, with the assumption that the main goal of residential homes is to offer the best possible conditions for the socio-psychological development and well-being of the young people in care. In order to achieve this goal, residential youth workers' knowledge production and their ability to share it with other professionals have to

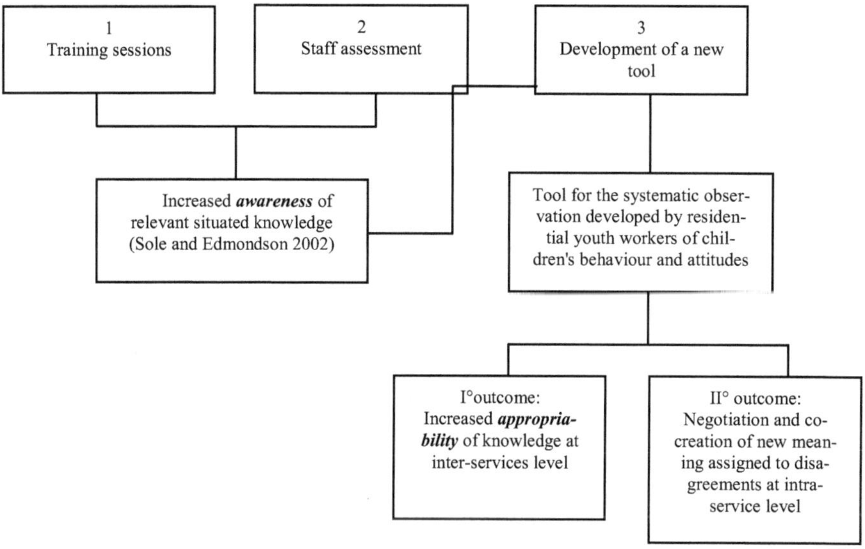

Fig. 17.2 Flow chart and phases of the innovation process

be assessed and implemented. High rates of intra-team conflict hinder the proper achievement of the third sector organisation's targets and objectives.

We therefore developed three research phases (Fig. 17.2), involving academic researchers and 120 residential youth workers of APPM[1] responsible for residential care services for young people.

1. Exploratory phase: content analysis of qualitative transcripts of meetings with supervisors. Psychologist-led training session facilitating reflective activities on:

   - Residential youth workers' evaluations of young people's negative behaviour
   - Perceived critical issues in communications, internal and external

   Through the practice lens, training sessions were devoted to formative, restorative and normative objectives (Davenport and Prusak 1998a, b). It is likely that any one session will cover all of the three functions to some degree. The main method that session leaders used was supportive feedback on practitioners' performance and reflective practices. Training sessions provided a forum for staff to receive support in gaining an overview of their work processes also through the lens of attachment theory.

   In-service training in the caring professional context supports practitioners in carrying out emotionally demanding tasks (Karvinen-Niinikoski 2004). Tacit-shared knowledge is analysed, and reflective activities support practitioners' trust in the organisation and the confidence they feel in their ability to cope with perceived threat. Tacit knowledge has to be *deconstructed* and *reconstructed*,

[1]APPM: Associazione provinciale per i minori, Italian regional NGO

providing insights for all practitioners. The way practitioners perform their routines is analysed and shared.

2. Residential youth workers' *staff assessment phase* (quantitative method, assessing magnitude of constructs): the practitioners rated their attachment styles with regard to the skills which support psychological processes in managing up and down and peer-to-peer interpersonal relationships.

   Attachment theory adopted as a theoretical framework in in-service training (Proctor 1987) in personal social services and health services allows a deeper understanding of the psychological processes and core beliefs related to the perception of threat. The main issue referred to by residential youth workers was interpersonal and team conflicts about different ways of perceiving, assessing and reacting to a given threatening event concerning negative children's behaviour. Different residential youth workers' perceptions and evaluations of children's negative behaviours were in fact based on reports which were mutually contradictory and which triggered internal team conflicts.

   Contradictory conclusions in outcome evaluations of educational interventions placed significant constraints on the whole decision-making process. Reflecting on their tacit knowledge through the lens of attachment theory helped the residential youth workers develop a number of ways of recognising features and issues underlying their interpersonal and team disputes. They chose to treat with respect and dignity those individual differences in perceptions of threatening events, which had previously caused conflict.

3. Practice-based innovation (Brown and Duguid 1991; Buech et al. 2010) and development of the new tool for children's behaviour assessment.

   Finally, but not least importantly, residential youth workers developed a new tool for the systematic observation of children's behaviour and attitudes. The implementation of the tool helped residential youth workers to develop new routines both for gathering data on young people's behaviour and for evaluating the outcomes of their educational interventions. New reporting and recording procedures were given equal value with practitioners' evaluations, convergent or divergent. The new structure and content of organisational knowledge production and the concomitant modified decision-making process enhanced the staff's mutual recognition of the value of the different contributions to children's safety.

   A child's negative behaviours are evaluated in relation to the perceived associated threat. They are relevant because on the one hand they divide opinion among the practitioners about what stage the programme has reached in relation to the overall intervention; on the other hand, there are signs of unease and discomfort in the child. Practitioners need to analyse their often divergent evaluations, in order to reduce the possibility of their triggering conflict within the team.

   The social constructionist view allows the enrichment of knowledge production and the sharing of ideas in organisational contexts by the consideration of relevant social variables such as power, beginners' and experts' roles and implicit norms. According to Nouwen et al. (2012), different opinions and the subjective meanings assigned by co-workers to practices can influence the team's decision-making process. Nouwen et al. (2012) emphasise that an effective decision-making process is related to leadership, team learning and team architecture. The latter is of interest because it relates to the role of committed professionals, to trust and to alignment (Drath et al. 2008)

The influence of a single team member varies according to their relative power, which in turn depends on their trustworthiness and on their degree of conformity to the majority or adherence to minority stances (Leung 2007). What residential youth workers observe, record and pass on about children's social competence, attitudes and behaviours is contained in narrative reports which should ensure a smooth and timely hand over at the end of each working period. Each written and verbal report potentially contributes material to the process of updating the social worker's overall plan for each child in the care of the agency.

17.3.3 Data Analyses

The transcripts of meetings with trainers were submitted to a content analysis (Krippendorff 2012) through NVivo 8 software, which provided 20 key 'at-risk' behaviours, as indicators of different types of discomfort and maladaptive behaviour on the part of young people. The list was evaluated by five independent judges (practitioners with at least 10 years of service), who identified the most frequent ones. A 13-item 'Scale to Assess the perceived Threat presented by children's negative behaviour' (from now on SAT) was developed. The practitioners identified the following critical behavioural and attitudinal issues: shouts, physical and verbal aggression, swearing, depressive mood, introversion, refusal to take part in discussion, homesickness, thefts, self-injury, escapes from care institution, sexual acting out and serious grief.

17.4 Results

17.4.1 Attachment Theory and 'Practice Lens' for Supervision Activities

Training sessions with psychologists allowed residential youth workers to explore the forms and content of the discomfort they felt in handling with the data and information on the young people. This important aspect of their daily practice (Gherardi 2009) was analysed from two different points of view: (1) within the relationship with co-workers at an interpersonal level and in teamwork and (2) from an interservices perspective, looking at issues of cooperation with other professionals (health and/or social services).

Practitioners also reported problems in sharing data on, and evaluations of, specific behaviours within their weekly team meeting (95 %) and in their reports on cases and discussions with social workers (65 %). In addition 73 % underlined their difficulty in using reports as exclusively narrative forms of data collection and transfer. They (86 %) referred to the reports as informal and as not sufficiently structured for providing data to the range of other professionals and for combining their observations on the young people with other type of information. Moreover, they perceived other professionals, such as health professionals and social workers, who are used to using formal and validated tools for the assessment of the young people, as feeling uncomfortable in communicating with residential youth workers (78 %).

Low levels of structuration in facilitating reflective activities and interactions in meetings with supervisors allowed researchers and practitioners to share information on their difficulties in contributing their knowledge appropriately and effectively to the decision-making processes.

Recursive practices ascribable to two main relational contexts were systematically analysed during meetings with supervisors: (1) caregiving styles (social interactions between residential youth worker and child) and (2) relational issues and confrontation strategies concerning different perspectives and perceptions of children's negative behaviour (social interactions between staff). Both issues offered the opportunity for residential youth workers to exhibit two quite-polarised caregiving styles and two correspondingly different perspectives on evaluating and responding to children's negative behaviour (secure subjects vs. insecure subjects). Using a socio-psychological perspective (attachment theory), the in-service training session enables practitioners to reflect on emerging divergences in perception and evaluation of threat. Looking for compromise between differing evaluations, residential youth workers experience conflict with teammates. Asymmetric stances arise and practitioners unduly favour affiliation to teammates supporting their own view. Oversensitiveness to peers who defend different stances makes practitioners willing to engage in team conflict, ascribing negative (Table 17.1) value to statements which differ from the own ones. The psychologist leading reflective activities

Table 17.1 Pattern of attribution of meaning and value to the subjective evaluation of children's negative behaviour

Working practice	Evaluation of threat	Meaning	Attributed value	Consequences
Perception of children's negative behaviour	From less threatening ⇩⇧		Positive when convergent	Unequal treatment of information reported and subsequent unequal contributions of data to the interim or final decision-making process Perceived disparities in treatment of residential youth workers' personal contributions to information flow Resentment and stress
Matching evaluation		Contradictory evaluations		
Report	⇧⇩ To more threatening		Negative when divergent triggering team conflict	

in-service training sessions can support, and often activates (Chesbrough 2006), practice-based innovation as in the case study presented here. The long tradition of in-service training practices in knowledge management as described by Karvinen-Niinikoski (2004) supports practitioners in their effort to gain new perspectives in knowledge production and sharing practice. According to Engeström (1987), the need for change often arises from contradictions in the work system itself (Fig. 17.1). According to Karvinen-Niinikoski (2004) and Mezirow (1991), change may start in the systematic deconstruction of each simplest daily routine, triggering recognition and recollection of the implicit knowledge segments which are rarely discussed or explicitly documented and remain often unrecognised. In Table 17.1 we synthesise results of the collective and systematic deconstruction of work practices: practitioners understand to what extent their perception of threat can modify the subsequent evaluation of negative children's behaviour. They also identify the shared meaning they ascribe to divergences. Table 17.2 shows results of the reconstruction of work practices: practitioners ascribe a new meaning to divergences sensibly reducing team contrapositions and stress.

Different evaluations are interpreted as a function of practitioners' caregiving and attachment styles. Residential youth workers became aware that their subjective ways of handling important interpersonal issues affected their cognitive responses through selective attention, that is, individuals tending to pay attention to particular features of the stimuli they are exposed to and to ignore others. What people select depends on the accessibility of particular patterns contained in their representations both of self and others. The latter are quite stable in time, but may differ in different social domains.

Residential youth workers who interpret a child's failure to listen or to co-operate as indicating a lack of respect are likely to react with anger and distress and to evaluate the child's behaviour as potentially threatening. Those who interpret the same failure to co-operate as being related to some as yet unknown issues or simply as arising from stress and discomfort may feel sorry rather than angry. In this way their evaluation will be much more empathic or compassionate and may also

Table 17.2 Emerging new patterns of attribution of meaning and value to working practice

Working practice	Evaluation of threat	Meaning	Attributed value	Consequences
Perception of children's negative behaviour	From less threatening ⇩⇧		Recognition and acceptance of divergences	Equal treatment of information reported and subsequent equal contribution of data to the interim or final decision-making process Perceived equal opportunities to contribute to organisational knowledge Well-being and comfort
Matching evaluation		Complementary evaluations	Positive (associated to sensible reduction of team conflict)	
Report	⇧⇩ To more threatening			

be supportive. Practitioners are now aware that they can choose whether to further a dispute about their different perceptions of threat or to reinterpret divergences as outcomes of their own different ways of being sensitive to the children's relational needs and responsive to their emotional states.

The new ascription of meaning (Table 17.2) to the output of their recursive practice allowed them to pose new questions, for example, about their attachment styles and about how their internal representation of others (positive vs. negative) and their self-concept (positive vs. negative) may underlie projective mechanisms in the perception of others. These multifaceted interests enriched and supported the residential youth workers' reflective practices.

Practitioners recognise that the different ways they perceive, evaluate and respond to threats largely depend on self-serving and biased defensive strategies. They realise that different opinions on threats are associated with different evaluations of it, and this leads team members to attribute positive 'added value' to conformity and convergence with the majority view in the team.

Divergent evaluations and minority stances were simply overlooked or labelled as irrelevant and thus dismissed (Table 17.1). Divergences in the perception and evaluation of important relational patterns between users and practitioners usually triggered interpersonal conflicts and negatively affected teamwork, sometimes hindering the complex process of making decisions about children. Convergent evaluations of children were often the reason for giving respect to co-workers, while divergent evaluations were sufficient motivation for showing contempt or resentment.

Social psychologist Harré (1993) states that people hand out marks of respect and contempt not only for success and failure in social life but also on the basis of complex systems of activities in daily life in public environments such as workplaces. Attributing respect on the basis of opportunities to agree and contempt on the basis of emerging differences have been acknowledged as risk factors: added value (positive or negative attributions) is attributed to psychological processes which inherently differ in content and structure, even though pertaining to the same recursive practice socially recognised as an institutionalised practice (Gherardi 2009).

Deconstruction and reconstruction practices enabled them to recognise internal working models as cognitive representations of the self and of the generalised other (Holdsworth and Morgan 2007; Mead 1934) which function as prototypical exemplars of any type of social interaction, activating expectations and attitudes and suggesting specific attributions. This recognition occurs in the implicit dimension of their work process as opposite to the explicit one which is formally codified, organised and prescribed. According to Ellström (2010), the implicit dimension of work processes is conceived as the way practitioners perceive their working practices, how they co-ordinate their activities with co-workers and how they perform them in practice. Residential youth workers took into account that what we called 'added values', that are the attributions of positivity or negativity to the perceptions by co-workers of crucial aspects of their work practices, must be separated from data and reports on children's behaviour and that they need to be redefined in order to contribute to a more comfortable teamworking climate.

17.4.2 Staff Assessment

Residential youth workers completed the 13-item SAT, and adult attachment style was assessed using the Italian validated version of the Attachment Style Questionnaire (Feeney et al. 1994; Pedrazza and Boccato 2010, 2012).

The aim of the staff assessment phase is to measure the awareness and mindfulness with which practitioners face the broad range of children's at-risk behaviour. This is the issue on which practitioners often encounter interpersonal conflicts and team confrontations. We asked residential youth workers to rate how threatened they feel facing the 13 identified negative behaviours of minors. Attachment theory is the theoretical framework which provides models of emotion regulation in interpersonal context. Such models enable us to understand to what extent individual differences in the relevant variables determine different evaluations of perceived danger and threat.

Exploratory factor analysis (EFA) was applied to identify the factor structure of the SAT. Three dimensions explained 61.42 % of variance; they were labelled as follows: *introversion*, *physical aggression* and *verbal aggression*. *Introversion* accounted for 24.05 %, *physical aggression* for 18.98 % and *verbal aggression* for 18.41 % of the total variance. *Introversion* refers to such behaviours as isolation or refusal to talk, serious grief and to feelings such as depression and homesickness. *Physical aggression* concerns physical violence, self-injury, theft, escape and sexual acting out. *Verbal aggression* is related to shouting and swearing. Table 17.3 shows that residential youth workers perceive high risks in facing children's *physical aggression*, and *introversion* is considered less dangerous. *Verbal aggression* is evaluated as the least threatening of all the negative behaviour on the part of children. As Table 17.3 shows, the differences in SAT means are all significant. According to previous research findings in attachment theory up to data literature (Mikulincer and Shaver 2007), we also observed that secure practitioners perceive *introversion* and *verbal aggression* as less threatening than do insecure practitioners (Table 17.3). The statistical difference is significant. Differences in rating *physical aggression* are statistically not significant, but insecure practitioners rate *verbal aggression* as much more threatening than *physical aggression*. Explicit

Table 17.3 Descriptive statistics: total sample, secure vs. insecure residential youth workers

	Sample ($n = 120$)			Secure ($n = 41$)		Insecure ($n = 43$)		
Measure	*M*	SD	Sig	*M*	SD	*M*	SD	Sig
1. Introversion	3.02	1.01	0.000	3.87	0.95	4.81	1.15	0.030
2. Physical aggression	4.09	1.09	0.000	2.76	1.46	2.70	1.07	n.s.
3. Verbal aggression	2.59	1.14	0.000	2.66	0.64	3.23	0.76	0.050

well-known rules regulate everyday life in the community so that the responses to and consequences of physical aggression are usually stipulated by the agency, and very detailed, harsh penalties are typically imposed on offenders. Insecure practitioners may therefore feel to some extent relieved of their responsibility because they must simply apply the agency's regulations. The reason insecure practitioners perceive verbal aggression as much more threatening than physical may be their assumption that they are left alone to provide containment, propose resolutions, settle disputes and restore a peaceful climate in the community.

The differences in the opinions and attitudes on which evaluations are based cannot be completely reduced or eliminated, but the introduction of a tool which allows systematic observations became very important for the residential youth workers.

17.4.3 The Development of the Tool as Result of a Practice-Based Innovation

The results of the self-acknowledgement procedures paved the way to an employee-driven innovation process in which teams developed a shared tool, implemented by software for the systematic observation of children's behaviour. The introduction of a systematic observation tool as radical innovation according to Ellström (2010) entails a break with the practitioners' previous, institutionalised, knowledge production (Leung 2007). They believed that developing and implementing the tools allowed them to feel much more effective, because it facilitated a systematic, broad distribution of responsibility for the evaluation of outcomes among practitioners. The narrative mode of producing reports on the young people in residential care can be thus implemented by systematic observations assigning to each practitioner the same value and significance in the decision-making process. By the same token, internal and external conflicts triggered by disagreements on evaluation can also be more easily tackled and settled.

The tool was implemented by the use of software because of the great amount of information it permits team members to gather. At any moment one can turn to the summary diagram which shows the trends in a child's behaviour. When the practitioner files a progress report concerning the child's adjustment or issues related to his/her maladjustment, both positive and negative behaviours and attitudes can be rated. Each variable is rated on a 4-point scale (0 corresponds to its absence, 1 = low perceived intensity up to 3 = high intensity). Data can be summarised in order to show how each caseworker perceived the intensity of the variable, observed twice daily. This guarantees that each caseworker's views, differing but yet of equal value, are represented in reports on children.

Diagram downloads are available for any selected period, from just a few days to 3 months or longer.

The tool allows residential youth workers to insert two different observations per day, and daily observations are always made by two different practitioners in order to assure the widest representation of views. A large amount of information and understanding emerges from the opportunity this tool offers to match each observation with different events the observed child experiences. For example, it is often the case that a child who has adapted to life in the residential home is invited to visit the biological parents. The visit to the family of origin, which confronts the child again with past negative experiences such as affect deprivation, abuse and so on, often causes dysfunctional or maladaptive behaviour to become worse. In such a case, the return to the residential home is typically associated with a more or less severe exacerbation of maladaptive behaviours. Given the frequent and often severe opposition of parents to juvenile court decisions, this form of evidence in the observation of the child's behaviour can be very useful in minimising the child's pain and the consequent maladaptive responses.

The large number of observations of each child is available at any moment, for intra-service teamwork or interservices information transfer, and the diagram can be attached to social service reports for the court, for the family of origin and for physicians or mental health professional and school officers.

The possibility of rapidly synthesising data and presenting them in a schematic form allows practitioners to be effective in inter- and intra-services communication. Each observation diagram can be easily matched with any type of event which happens to the child. The absence of particular events associated with a worsening of the child's behaviour would shift the practitioner's focus from the child's external environmental and social context to internal issues and experiences which could have activated the decreasing effectiveness of his/her adaptive strategies.

17.5 Conclusion and Implications

The present study is focused on an inside-out open innovation in daily evaluation practices, on the part of a third sector organisation providing residential care for foster youth in a local child welfare agency. The activating factor of the process resides in practitioners' shared and deep motivation to reduce intra-team divergences. The complex innovation process for the development of the tool increased practitioners' well-being. Further research should be addressed to identify to what extent supporting well-being can free up resources for innovation. Positive affect seems to increase the scope both of attention and cognition (Fredrickson 1998); it also makes cognitive additional material available for processing and for increasing flexibility (Isen 1999). In this regard economic recognitions or promotions are of limited interest (Mergel and Desouza 2013). According to Lee et al. (2012), public services often give priority to the outside-in type of open innovation; in fact, they are typically concerned with protection and handling of confidential data. In order to compensate this often-neglected issue, we examined an inside-out type of innovation; its adoption and implementation crossed the residential care service boundaries where it was developed, thus making knowledge available to social workers and to other components/services of the *dispersed teams* (Sole and Edmondson 2002) of the agency.

Public organisations promote innovation in a quite ambivalent manner: on the one hand, innovation is seen as positive despite type of organisation and resources, in fact it often seems to legitimate and increase the perceived reputation and value of public organisations itself (De Vries et al. 2014); on the other hand, professional groupings which are typically well established in public sector are often inherently resistant against innovation (Cunningham et al. 2005). Our study contributes to fill a literature gap on outcomes of innovation in public sector: in fact, according to De Vries et al. (2014), outcomes of innovation are seldom reported because of the self-serving, pro-innovation bias affecting researchers (Rogers 2003), institutions and organisations which typically rest on the increased public positive consideration accorded to them.

As also in the local child welfare agency this study is focused on, the up-to-date and systematic review on innovation in the public sector by De Vries et al. (2014) highlights that innovation mainly takes place at a local government level. Cunningham et al.'s (2005) review provides evidence that the complexity of the public sector and the diversity of its organisational arrangements and service processes can represent a huge barrier which has to be overcome in order to adopt and implement innovation.

Despite the still-increasing attention scholars (Somech and Drach-Zahavy 2000) and organisations pay to individual innovators, identifying, i.e. the psychological and social antecedents of individual innovative behaviour (Mack et al. 2008; Wu et al. 2014), there is little interest in studying teams', working groups' or even more complex organisational environments' innovative behaviour. We therefore focused our attention on two different influential factors of innovation: individuals (team members at intra-group level) and groups (youth workers' group vs. social workers' group, at inter-group level). At the generating stage of the innovation process (De Vries et al. 2014; Rogers 2003), interpersonal, intra-team *conflict* emerges as dominant factor.

According to Kuipers et al. (2014), scholars and institutions should refer both to intended vs. unintended and positive vs. negative outcomes of the innovation process. Unintended and positive outcome of the development of the new tool emerged at the adoption/diffusion stage: the large number of observations of each child, available in real time after the adoption of the tool, allowed youth workers to be effective both in intra-service and interservice information transfer. At systemic level the impact of innovation can often be seen as an unpleasant perturbation by professionals not directly involved in the generation stage of innovation.

Outcomes at intra-team level can be listed as follows:

1. Decrease of interpersonal and intra-team contrapositions
2. Dissemination of relevant information within the team and crossing team and institutional boundaries

3. Shifting from a pure qualitative methodology to a more complex perspective including quantitative methods and instruments to monitor youth behaviour
4. Increased knowledge and awareness of processes and interpersonal dynamics in teamwork

Outcome at inter-institutional level emerges as the increased appropriability (Sole and Edmondson 2002) on the part of social workers, physicians, mental health professional, school officers and court members of the knowledge, created and disseminated through the implementation of the new tool within the third sector organisation.

As the outcome of effective reflective practices, the new tool contributes to a reduction in the number of potential conflict issues at both interpersonal and team level. It supports the reinterpretation of differences of perception between co-workers of threats that is of negative children's behaviours, viewing them as complementary ways of applying their youth social work skills. It supports practitioners in regarding their individual differences as differing ways of perceiving adolescents' behaviour, deserving equal dignity rather than uncontrolled, biased rejection.

The in-depth reflections on working practices, such as the evaluation of problematic interpersonal affective issues with both children and co-workers, enabled practitioners to integrate the new knowledge into their own daily practices and to receive more respect, at both the individual and the interpersonal level. Rather than striving for changes in their attachment insecurity, they integrated the new knowledge and understanding of their own working practices so as to allow both secure and insecure attached residential youth workers to contribute in their own way to the knowledge management and decision-making processes. Thus, the interservice information flow and knowledge production on the young people were both improved: data-gathering methods, data exploitation and interservices data dissemination were all simplified. The new way of processing information and knowledge also responds to the need of users for rapid and effective intervention when children need it.

In improvisation, according to Ellström's practice-based innovation cyclical process, practitioners perceive their subjective asymmetry (Edvardsson et al. 2011) in the co-ascription of value to working practices. They then develop a new interpretation of their recursive, institutionalised practices, following the principle of equal dignity in order to restore a more balanced subjective contribution to the work. Comparative studies should be carried out in order to identify commonalities between social, health and educational service innovation's influential factors and outcomes.

References

Bechky, B. A. (2003). Sharing meaning across occupational communities: The transformation of understanding on a production floor. *Organization Science, 14*(3), 312–330.

Bennett, C. S. (2008). Attachment-informed supervision for social work field education. *Clinical Social Work Journal, 36*, 97–107.

Bennett, S., Mohr, J., Deal, K. H., & Hwang, J. (2012). Supervisor attachment, supervisory working alliance, and affect in social work field instruction. *Research on Social Work Practice, 23*(2), 199–209.

Bifulco, A., Jacobs, C., Bunn, A., Thomas, G., & Irving, K. (2008). The Attachment Style Interview (ASI): A support-based adult assessment tool for adoption and fostering practice. *Adoption and Fostering, 32*(3), 33–45.

Borzaga, C., & Fazzi, L. (2010). Processes of institutionalization and differentiation in the Italian third sector. *Voluntas: International Journal of Voluntary and Nonprofit Organizations, 22*(3), 409–427.

Bourdieu, P. (1977). *Outline of a theory of practice*. Cambridge: Cambridge University Press.

Brown, J. S., & Duguid, P. (1991). Organizational learning and communities-of-practice: Toward a unified view of working, learning and innovation. *Organization Science, 2*(1), 40–57.

Buech, V. I. D., Michel, A., & Sonntag, K. (2010). Suggestion systems in organizations: What motivates employees to submit suggestions? *European Journal of Innovation Management, 13* (4), 507–525.

Carlile, P. R. (2002). A pragmatic view of knowledge and boundaries; Boundary objects in new product development. *Organization Science, 13*(4), 442–455.

Chesbrough, H. (2003). The era of open innovation. *Sloan Management Review, 44*(3), 35–41.

Chesbrough, H. (2006). *Open innovation: The new imperative for creating and profiting from technology*. Boston, MA: Harvard Business School Press.

Cunningham, P., Grant-Pearce, C., Green, L., Miles, I., Rigby, J., & Uyarra, E. (2005). In sickness, in health and in innovation: NHS DIRECT—A health sector innovation study. *Administration, 53*(3), 42–65.

Czarniawska, B. (2004). *Narratives in social science research. Introducing qualitative methods*. London: Sage Publications.

Davenport, T. H., & Prusak, L. (1998a). *Working knowledge-how organizations manage what they know*. Boston, MA: Harvard Business School Press.

Davenport, T., & Prusak, L. (1998b). Learn how valuable knowledge is acquired, created, bought and bartered. *The Australian Library Journal, 47*(3), 268–272.

De Vries, H. A., Bekkers, V. J. J. M., & Tummers, L. G. (2014). *Innovation in the public sector: A systematic review and future research agenda*. Speyer: EGPA Conference.

Deacon, D., Bryman, A., & Fenton, N. (1998). Collision or collusion? A discussion of the unplanned triangulation of quantitative and qualitative research methods. *International Journal of Social Research Methodology, 1*(1), 47–63.

Dougherty, D. (2004). Organizing practices in services: Capturing practice-based knowledge for innovation. *Strategic Organization, 2*(1), 35–64.

Drath, W., Mccauley, C., Palus, C., Vanvelsor, E., Oconnor, P., & Mcguire, J. (2008). Direction, alignment, commitment: Toward a more integrative ontology of leadership. *The Leadership Quarterly, 19*, 635–653.

Edvardsson, B., Tronvoll, B., & Gruber, T. (2011). Expanding understanding of service exchange and value co-creation: A social construction approach. *Journal of the Academy of Marketing Science, 39*, 327–339.

Ellström, P.-E. (2010). Practice-based innovation: A learning perspective. *Journal of Workplace Learning, 22*(1/2), 27–40.

Engeström, Y. (1987). *Learning by expanding: An activity-theoretical approach to developmental research*. Helsinki: Orienta-Konsultit Oy.

Evered, R., & Louis, M. R. (1981). Alternative perspectives in the organizational sciences: "Inquiry from the inside" And "inquiry from the outside". *The Academy of Management Review, 6*(3), 385–395.
Feeney, J. A., Noller, P., & Hanrahan, M. (1994). Assessing adult attachment. In M. B. Sperling & W. H. Berman (Eds.), *Attachment in adults: Clinical and developmental perspectives* (pp. 122–158). New York: Guilford Press.
Fredrickson, B. L. (1998). What good are positive emotions? *Review of General Psychology, 2*(3), 300–319.
Fuglsang, L. (2008). Capturing the benefits of open innovation in public innovation: A case study. *International Journal of Services Technology and Management, 9*(3/4), 234–248.
Gallouj, F., & Djellal, F. (2010). *The handbook of innovation and services: A multidisciplinary perspective*. Cheltenham: Edward Elgar Publisher.
Gassmann, O. (2006). Opening up the innovation process: Towards an agenda. *R&D Management, 36*(3), 223–228.
Gassmann, O., & Enkel, E. (2004, July 6–9). Towards a theory of open innovation: Three core process archetypes. In *Proceedings of the R&D Management Conference*, Lisbon, Portugal.
Geiger, D. (2009). Revisiting the concept of practice: Toward an argumentative understanding of practicing. *Management Learning, 40*(2), 129–144.
Gherardi, S. (2009). Introduction: The critical power of the 'practice lens'. *Management Learning, 40*(2), 115–128.
Gherardi, S., Nicolini, D., & Odella, F. (1998). Towards a social understanding of how people learn in organizations. The notation of situated curriculum. *Management Learning, 29*(3), 273–297.
Harmaakorpi, V., & Mutanen, A. (2008). Knowledge production in networked practice-based innovation processes—Interrogative model as a methodological approach. *Interdisciplinary Journal of Information, Knowledge, and Management, 3*, 87–101.
Harré, R. (1993). *Social being* (2nd ed.). Oxford: Basil Blackwell.
Hedberg, B. (1981). How organizations learn and unlearn. In P. Nystrom & W. H. Starbuck (Eds.), *Handbook of organizational design* (Vol. 1, pp. 3–27). London: Cambridge University Press.
Hilgers, D., & Ihl, C. (2010). Citizensourcing: Applying the concept of open innovation to the public sector. *International Journal of Public Participation, 4*(1), 67–88.
Holdsworth, C., & Morgan, D. (2007). Revisiting the generalized other: An exploration. *Sociology, 41*(3), 401–417.
Howe, D. (1999). *Attachment theory for social work practice*. Basingstoke: Palgrave Macmillan.
Hull, D. M., & Saxon, T. F. (2009). Negotiation of meaning and co-construction of knowledge: An experimental analysis of asynchronous online instruction. *Computers and Education, 52*, 624–639.
Isen, A. M. (1999). Positive affect. In T. Dalgleish & M. J. Power (Eds.), *Handbook of cognition and emotion* (pp. 521–539). New York: Wiley.
Karvinen-Niinikoski, S. (2004). Social work supervision: Contributing to innovative knowledge production and open expertise. In N. Gould & M. Baldwin (Eds.), *Social work, critical reflection and learning organisation* (pp. 23–39). Aldershot: Ashgate Publishing.
Klein, W. C., & Bloom, M. (1995). Practice wisdom. *Social Work, 40*(6), 799–807.
Krippendorff, K. (2012). *Content analysis: An introduction to its methodology*. Thousand Oaks, CA: Sage Publications.
Kuipers, B. S., Higgs, M. J., Kickert, W. J. M., Tummers, L. G., Grandia, J., & Van der Voet, J. (2014). The management of change in public organisations: A literature review. *Public Administration, 92*(1), 1–20.
Lave, J., & Wenger, E. (1991). *Situated learning: Legitimate peripheral participation*. Cambridge: Cambridge University Press.
Lawler, M. J., Shaver, P. R., & Goodman, G. S. (2011). Toward relationship-based child welfare services. *Children and Youth Services Review, 33*(3), 473–480.

Lee, S. M., Hwang, T., & Choi, D. (2012). Open innovation in the public sector of leading countries. *Management Decision, 50*(1), 147–162.
Leung, L. C. (2007). Reflective practices: Challenges to social work education in Hong Kong. *Social Work Education, 26*(6), 632–644.
Leung, Z. C. S. (2009). Knowledge management in social work: Types and processes of knowledge sharing in social service organizations. *The British Journal of Social Work, 39*(4), 693–709.
Mack, W. R., Green, D., & Vedlitz, A. (2008). Innovation and implementation in the public sector: An examination of public entrepreneurship. *Review of Policy Research, 25*(3), 233–252.
McMillen, J. C. (1992). Attachment theory and clinical social work. *Clinical Social Work Journal, 20*(2), 205–218.
Mead, G. H. (1934). *Mind, self, and society*. Chicago: University of Chicago Press.
Mergel, I., & Desouza, K. C. (2013). Implementing open innovation in the public sector: The case of Challenge.gov. *Public Administration Review, 73*(6), 882–890.
Meyer, J. W., & Rowan, B. (1977). Institutionalized organizations: Formal structure as myth and ceremony. *The American Journal of Sociology, 83*(2), 340–363.
Mezirow, J. (1991). *Transformative dimensions of adult learning*. San Francisco, CA: Jossey-Bass.
Mikulincer, M., & Shaver, P. R. (2007). *Attachment in adulthood: Structure, dynamics, and change*. New York: Guilford Press.
Mulgan, G. (2006). The process of social innovation. *Innovations, 1*(2), 145–162.
Mulgan, G. (2007). *Ready or not? Taking innovation in the public sector seriously*. London: Nesta. http://www.nesta.org.uk/publications/provocations/assets/features/ready_or_not_taking_innovation_in_the_public_sector_seriously.
Nouwen, E., Decuyper, S., & Put, J. (2012). Team decision making in child welfare. *Children and Youth Services Review, 34*, 2101–2116.
Osborne, S. P., & Brown, K. (2005). *Managing change and innovation in public service organisations*. Abington: Routledge.
Pedrazza, M., & Boccato, G. (2010). Attachment style questionnaire: A contribution to the Italian validation of the scale. *Ricerche di Psicologia, 1*, 1–18.
Pedrazza, M., & Boccato, G. (2012). Training on attachment styles increases secure adult's attributions of control. *TPM, 19*(2), 69–79.
Perkmann, M., & Walsh, K. (2007). University-industry relationships and open innovation: Towards a research agenda. *International Journal of Management Reviews, 9*(4), 259–280.
Proctor, B. (1987). Supervision: A co-operative exercise in accountability. In M. Marken & M. Payne (Eds.), *Enabling and ensuring: Supervision in practice* (pp. 21–34). Leicester: National Youth Bureau and Council for Education and Training in Youth and Community Work.
Roberts, J. (2006). Limits to communities of practice. *Journal of Management Studies, 43*(3), 623–639.
Rogers, E. M. (2003). *Diffusion of innovations* (5th ed.). New York: Free Press.
Schofield, G., & Beek, M. (2006). *Attachment handbook for foster care and adoption*. London: British Association for Adoption and Fostering.
Sole, D., & Edmondson, A. (2002). Situated knowledge and learning in dispersed teams. *British Journal of Management, 13*(S2), S17–S34.
Somech, A., & Drach-Zahavy, A. (2000). Understanding extra-role behavior in schools: The relationships between job satisfaction, sense of efficacy, and teachers' extra-role behavior. *Teaching and Teacher Education, 16*, 649–659.
Strauss, S. G., & Olivera, F. (2000). Knowledge acquisition in virtual teams. In M. A. Neale & E. A. Mannix (Eds.), *Research on managing groups and teams* (Research on managing groups and teams, Vol. 3, pp. 257–282). Bingley: Emerald Group Publishing Limited.
Swan, J., Scarbrough, H., & Robertson, M. (2002). The construction of 'communities of practice' in the management of innovation practice. *Management Learning, 33*(4), 477–496.

Tsoukas, H., & Vladimirou, E. (2001). What is organizational knowledge. *Journal of Management Studies, 38*(7), 973–993.
Wallin, M. W., & von Krogh, G. (2010). Organizing for open innovation: Focus on the integration of knowledge. *Organizational Dynamics, 39*, 145–154.
Wenger, E. (1998). *Communities of practice: Learning, meaning, and identity*. New York: Cambridge University Press.
Wilkins, D. (2012). Disorganised attachment indicates child maltreatment: How is this link useful for child protection social workers? *Journal of Social Work Practice, 26*(1), 15–30.
Wu, C., Parker, S. K., & de Jong, J. P. J. (2014). Need for cognition as an antecedent of individual innovation behavior. *Journal of Management, 40*(6), 1511–1534.
Zapf, D. (2002). Emotion work and psychological well-being. A review of the literature and some conceptual considerations. *Human Resource Management Review, 12*, 237–268.

Chapter 18
Service Innovation in Translational Medicine

Sergio Barile, Francesco Polese, Marialuisa Saviano, and Luca Carrubbo

Abstract Innovation implies collaboration among business actors and is fundamental in value cocreation networks. A significant context is the healthcare, where Translational Medicine (T-MED) aims at decreasing barriers between clinical research and medical treatments. The Translational Medicine is the integration and optimization of inputs in basic research improving the patient care, advancing and speeding up the process of offering new protocols, therapies, and practices. This conceptual contribution analyzes the Service Innovation in Translational Medicine using the paradigms of the Service Science (SS), Complex Service Systems, and Viable Systems Approach (VSA). It aims to increase the comprehension of relationships and complexity reduction processes in Translational Medicine and in Service Innovation in healthcare and improving of understanding the constraints to the innovation supporting Translational Medicine approach, seeking strategic solutions with a holistic view of healthcare networks.

18.1 Introduction

This work interprets innovation and, especially, Service Innovation as the result of systemic interaction of more actors in the environment. In general, innovation represents not simply a goal but an always-active dynamic process that is defined through various theoretical frameworks and that emerges through resource integration among the actors involved. Innovation is not a linear process but rather a set of connections in a network of networks, involving many actors able to share a purpose, which is founded on healthcare service improvement and value cocreation.

S. Barile (✉)
La Sapienza University of Rome, Rome, Italy
e-mail: sergio.barile@uniroma1.it

F. Polese • M. Saviano • L. Carrubbo
University of Salerno, Salerno, Italy
e-mail: fpolese@unisa.it; msaviano@unisa.it; lcarrubbo@unisa.it

T. Russo-Spena et al. (eds.), *Innovating in Practice*,
DOI 10.1007/978-3-319-43380-6_18

The innovation development process in service (in particular) continuously stimulates the actors concerning questions about their roles, role enhancement (fostering their contribution in service provision), information, and resource sharing. Using value cocreation frameworks, the co-innovation process can be interpreted as a mutual exchange of resources not only among actors operating inside service systems but also among service systems in a specified context in any field of interest.

Healthcare represents one of the most important and critical issues for any worldwide community. The complexity in healthcare is generally recognized and due to the number of engaged actors, to its dynamics concerning service exchanges and the multiple possible observation perspectives such as public health, economic constraints, and overall sustainability. Therefore, Service Innovation in healthcare is a very interesting topic that deserves to be investigated.

It is commonly agreed that healthcare must be organized more functionally with respect to interested actors and to the availability of resources offered by actors involved within the provision process of related services. This issue is a challenging goal to reach because it addresses political priorities of a community, funds allocation, technological advances, territorial constraints, and, above all, the health of citizens. In this view, sustainable solutions ought to be pursued models and methods should be sought that are capable of being replicable, scalable, and versatile in different contexts worldwide.

Service Innovation in healthcare is strongly based upon resource integration and cocreation and might be effectively investigated by adopting a systems perspective focused on complex phenomena such as the Viable Systems Approach (VSA) and Service Science (SS) try to do.

By contextualizing healthcare in the SS framework (Maglio et al. 2006), it is possible to understand the relevance of learning as a strategic tool of management that leads to the recognition of continuing innovation as an incremental and dynamic process. However, Service Innovation in complex contexts calls for more holistic interpretation tools, such as the ones offered by systems thinking.

The application of VSA (Barile 2008; Golinelli 2000, 2010) in the context of healthcare aims to identify conditions for achieving the ultimate goal of the system as a whole, in order to improve the overall level of a population's health and the responsiveness of the system to people's needs and expectations (Saviano et al. 2014).

According to VSA and SS, within the healthcare service system, each actor (considered an integrate and active part of the system), working in a functional and effective way with respect to its own purpose, through the development of specific innovative processes, inevitably contributes to the improvement of the quality level of the healthcare service system as a whole.

Based on these considerations, in the healthcare service, innovation is intended as the main source of efficiency and success and the first step for achieving the shared purpose. Many actors (for instance, pharmaceutical companies) invest a large part of their earnings to improve the solutions currently available on the market, with the dual purpose of solving the greatest possible number of problems related to human health and to earn an acceptable profit. This stimulates the creation of research centers, specialized laboratories, international organizations, and new private partnerships for continuing innovation.

Considering the healthcare context as a complex, integrated and interconnected service system helps to understand how important it is to facilitate and implement its innovative processes in service and to understand how multi-actor contributions affect the overall performance of the system as a whole. In this context, one driving concept for wiser innovation management in healthcare is Translational Medicine.

Translational Medicine (T-Med) tries to integrate the inputs of basic research to optimize them and improve in practice the service of patient care, advancing and speeding up the process of offering new protocols, therapies, and practices. T-Med has regulatory and commercial implications because of a large sector of the biopharmaceutical industry that is focused on the development of new therapeutic modalities and is largely concerned with the management, optimization, and reduction of time to market. To be translated is mostly and primarily informational, allowing a quick concretization of an idea or solution in tools or procedures, successively adopted for experimentation and later for medical practice in a hospital.

It is possible to use the VSA and SS as lens to increase the comprehension of the complexity in T-Med (in particular in T-Med relationships) and the concept of value cocreation as the ultimate purpose of T-Med.

Following the introduction, this work is divided in three sections, each illustrating in more detail the purposes and goals of the research. First, the relevance of Service Science and value cocreation in emerging innovation in healthcare is presented. Second, this section highlights healthcare evolution within innovation contexts, examining interactions adopting the VSA framework and analyzing the T-Med within a systems view. Third, management implications and conclusions including Service Innovation insights in T-Med are presented.

18.2 Service Systems, Innovation, and Value Cocreation in Healthcare

18.2.1 Service Systems and Service Innovation

A service-oriented culture implies the recognition of service as a key success factor of all organizations. This culture creates and promotes actions aimed at meeting the expectations of stakeholders and efficiently fitting with their needs; the service orientation might be accomplished, for the organization, within the long-term policies (supporting and recognizing organizations' behaviors and attitudes) and service systems that can create and provide excellence in service (first of all in business).

Service represents a goal, a purpose that is based on a significant change characterizing the inner essence of all organizations and businesses (Spohrer et al. 2007; Vargo and Lusch 2006). Service systems, according to the first definition of the Service Science, represent configurations of people, technology, value propositions, and shared information that can cocreate value through service provision, such as language, laws, measures, and methods (Spohrer et al. 2007). Each service system is, then, simultaneously a provider and user of services, structured as a value chain, a network of value, or a system value (Vargo et al. 2008). This derives from a sort of service mutual exchange between provider-users, forcing new types of relationships that can be seen nowadays (Capunzo et al. 2013).

Service systems are able to foster connections and interactions between the various actors (Vargo and Lusch 2004, 2006) involved in a process of exchange, following several channels of communication between businesses, consumers, and various stakeholders (Mele and Polese 2011; Gummesson and Polese 2009). When a Service System is strongly interconnected and characterized by multi-actor interactions (Barile et al. 2012a, b), it is interpreted as an open system (Golinelli 2010), capable of improving its equilibrium through the acquisition, sharing and supplying of resources.

As the world becomes smarter—*the concept of SMART comes from the English words specific, measurable, agreed, realistic, and timely*—service systems ought to become people-centric, information-driven, and e-oriented, and collective satisfaction should encourage actors to cooperate and innovate (Qiu et al. 2007). The role of technology in service systems guarantees the evolution of interactions from the value proposal toward the user and vice versa; in this sense, innovation increases its relevance in service, and it becomes a cocreated and dynamic concept that starts at different points and moments of the value network.

We can identify two major analyses of the relationship that exists between innovation and service (Gadrey and Gallouj 2002; Rullani 2005; Tether 2003). The first one, defined alternatively as a technologist (Bryson and Monnoyer 2004) or assimilation approach (Coombs and Miles 2000), supposes that innovation and service concepts and dynamics are fundamentally similar to manufacturing firms' innovation dynamics, based on technology. This implies that technological development represents the key factor of innovation and quality improvement processes. Against it there is another approach defined, *service-oriented* that observes how innovation in service differs substantially from the manufacturing archetype (Den Hertog 2000; Sundbo and Gallouj 2000) due to the specific nature of services and to related implications.

Avoiding traditional differences in interpretation (that no longer matters in the above-described emerging service logic) and to manage and succeed in challenging service scenarios, great attention must be focused on innovation activities characterized by high levels of complexity (Schilling 2008). Technological developments, human resources, and organizational innovations require such focus (Nahapiet and Ghoshal 1997). Technological innovation in service is a phenomenon involving different types of firms engaging in both manufacturing and services, both public and private. One of the most striking aspects of this phenomenon is the investment in the acquisition and development of new systems or new software applications, activities that generally involve an R&D function. Hence, attention to the processes of technological innovation has grown due to the emergence of new technologies that are producing changes—*more evident than in the past*—in relationships between supply and demand actors.

Innovation may result in an experimental process that fosters continuous learning obtainable *by doing, by using, by failing, and by interacting* (Gronhaug and Kaufmann 1988) that, in service logic, may address significant innovation dimensions such as service definitions and related value propositions, actor interfaces, service delivery systems, and technologies (Den Hertog 2000). Innovation, if well designed and managed, thus can enable costs reductions, quality improvements, technical performance advancements through a wise management of innovation sources, both internal (e.g., R&D, production, engineering, and marketing proposals) and external (e.g., deriving from relationships with universities and other public/private research centers, scientific journals, and companies).

Service Innovation derives from cooperating technological aspects along with social and organizational relationships, together with business and market interactions. Service Innovation is thus "codetermined" (see Fig. 18.1) by technological aspects, by socio-organizational relationships (referred to social science), by business processes (deriving from Business Administration and Management), and by supply and demand (for global economy and market), all enabling business competitiveness.

With the evolution in technology and interconnected exchange activities, the relevance of emerging innovation can be analyzed using the concept of value cocreation. With this concept, it is possible to go beyond the classical interaction in which one actor is the provider and the other is the receiver. Nobody has the absolute responsibility to "innovate," but all actors certainly can contribute to stimulate dynamic innovation in the system.

Recent advances in Service Research have led to a common point of view from the perspective of service; several researchers have agreed on the logic of mutual benefit, shared value, and mutual satisfaction as key leverage items for harmonious development of a successful proposition, particularly if referring to the provision (but also to the production) of a given service.

So this, value cocreation logic is defined under a win-win logic, taking into account the interaction among different entities represented by various service systems and the desire to gain collective mutual satisfaction in which active contributions are multiple, integration is maximized, and complementarity is essential (Polese et al. 2011). "Win-win" interactions develop only through the promotion and maintenance of relationships with stakeholders or through a common desire to encourage cocreation engagement and exchanges (not-opportunistic behavior, long-term relationships, and shared values). Successful developments convey that "life consists of a network of relationships within which it interacts" (Capra 1997) and confirm that "life is a network of relationships, as is business" (Gummesson 2008).

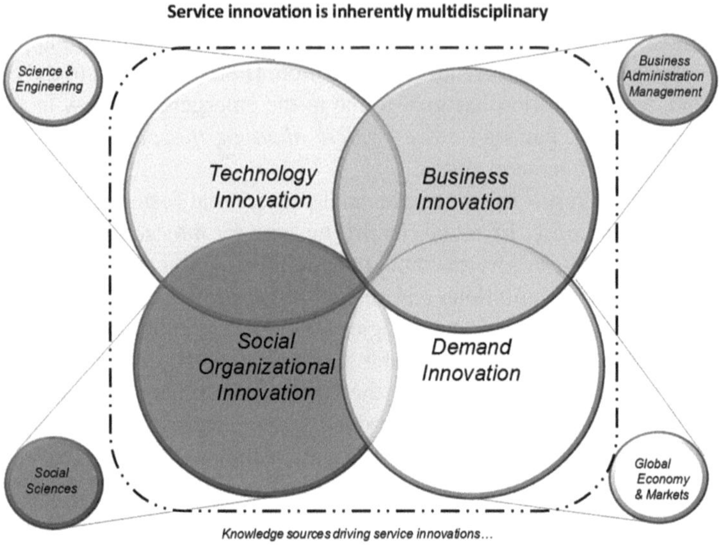

Fig. 18.1 Service Innovation insights (*Source*: Polese et al. 2009b)

From this point of view, qualifying involved operators does not matter; distinguishing between the supplier, customer, and user is irrelevant. What matters instead is highlighting the role of relationships and common interests promoting collaboration and the achievement of a higher level of satisfaction, that is, overcoming the logic of B2B, B2C, or C2C and analyzing more carefully the characteristics and advantages of the connections (Gummesson and Polese 2009). Some recent advances in Service Research have led to a different interpretation of interorganizational relationships, defining all relationships as Actors-to-Actors (A2A) (Wieland et al. 2012).

18.2.2 Innovation in Healthcare Service Ecosystem and Value Cocreation for Patient Empowerment

A healthcare ecosystem (as any National Healthcare System—NHS—appears to be) can provide a single global service toward users (as patients) in terms of mutual exchange, in which each actor can contribute to the proposition and the fruition of healthcare service. In a more ample set of relationships, service systems are embedded in a more general service ecosystem. Such embedding is due to the vanishing of the system's boundaries and engagement with an increased number of actors/entities including citizens, enterprises, doctors, clinicians, and so on, stabilizing relationships and fulfilling needs and expectations (see Fig. 18.2).

Each actor may not only access useful and needed resources but also release resources, creating a prolific service exchange and, consequently, stimulating a

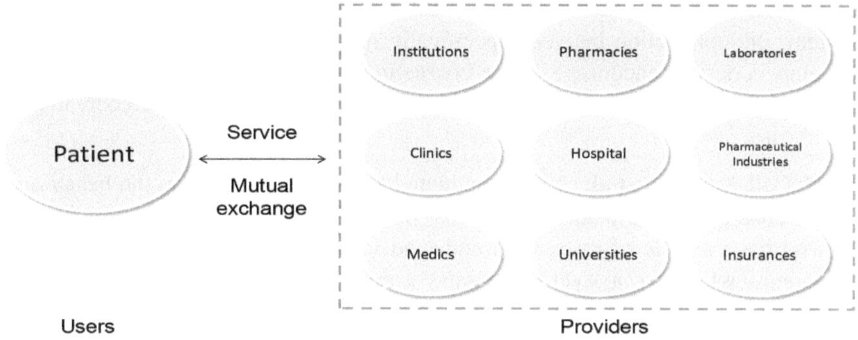

Fig. 18.2 Service mutual exchange in healthcare (*Source*: Capunzo et al. 2013)

value creation and diffusion for the overall benefit of the healthcare system as a whole (Iandolo et al. 2013).

The suggested service and systems view highlights the role of relationships as promoters of innovation, competitiveness, viability, and survival within service systems (Polese et al. 2009a) and, of course, within healthcare service systems.

As mentioned above, this view proposes service culture as the enabler and fundamental base of "smart" healthcare, capable of validating experiences and medical initiatives to benefit all involved actors (IBM 2009; Meyer and Müller 2006). This facilitates connectivity, real-time assessment, standard metrics, and metadating and leading to safer diagnoses, increased clinical knowledge, and more efficient hospitals, helping them to improve collective well-being sustainably (Saviano et al. 2010).

However, technology is complex to transfer, even more in healthcare because it is a turbulent, challenging environment (Castiaux 2012). Solutions ought to be scalable, replicable, and versatile for different contexts and users; key problems can be severe. Consider financial issues (such as rising costs and lack of funds), limited access to information, and coordination problems. With reference to this last aspect, processes, components, and even people in healthcare systems are not always coordinated, and errors of duplication or missing information (or wrong information) often occur (Polese 2013). Because many players are involved in healthcare, each with different priorities and goals, it is necessary to overcome such divergences before beginning operations in terms of language, education, culture, and purposes. Currently, technology transfer and integration is making healthcare systems smarter (IBM 2012), that is, with better, faster, and more detailed information within the involved actors, reducing errors and inefficiencies in the transfer to allow the system to capture, manage, and turn data into relevant information in real time (Carrubbo et al. 2013).

The need to share critical resources to achieve a common final goal led actors operating in healthcare to increase their collaboration on service provision. This collaboration contributes to better performance in general and to mutual benefits in terms of value; the increasing involvement of all actors is intended to be a real cocreation exchange. Starting from a known configuration of a defined service system, the connection between specifically involved actors and opportunities to develop codes to encourage value cocreation logic and dynamic innovation is evident. There are several contributions in the literature on value cocreation in healthcare:

- McColl-Kennedy et al. (2009) conducted an empirical study on the behavior of the customers of a healthcare service divided into two categories: individuals who have a style of cocreation and tend to show a higher quality of life and patients who have a style of "passive compliant" or "isolated controller" that tend to be associated with a lower quality of life.
- In a later work, McColl-Kennedy et al. (2012) identify different practices of value cocreation in healthcare and describe how customers can contribute to value creation through activities in the course of managing their healthcare. The authors propose a typology of healthcare "Customer Value Co-creation Practice Styles" (CVCPS) that is especially useful because it allows one to show the connections of the various practices to quality of life and their potential application to healthcare. Furthermore, their work displays empirically how clients participate effectively in the cocreation of value through activities and interactions with other actors in their care network.
- Rehman et al. (2012), after conducting a study of the healthcare sector in developing countries, show that there is a relationship among the participation of the consumer in the process of providing a health service, the perceived quality of the service and value cocreation.
- Nambisan and Nambisan (2009) provide a theoretical framework that analyzes value cocreation in healthcare consisting of four alternative models of value cocreation for the consumer: partnership, open source, group support, and the spread. The authors also analyzed the implications of the different models for a health service.
- Chakraborty and Dobrzykowski (2013) identify the practices of supply chain management that can affect value cocreation in healthcare and identify some elements that support the cocreation of value, such as the dialog between the various actors of the supply chain, information access within the supply chain, risk-benefit analysis, and transparency.
- In a subsequent work, Chakraborty and Dobrzykowski (2014) analyze supply chain collaboration in healthcare (hospitals) and note that it has a significantly positive effect on the cocreation of value; the cocreation of value in turn has a significantly positive effect on the performance of the hospital (both clinical and financial).
- Nordgren (2009) shows that the productivity of the service in the context of healthcare should be evaluated based on not only cost reduction and the results achieved but also values such as health, quality of life, accessibility, trust, communication, suffering, and avoidable deaths.

From these sentences, we can mean that healthcare agencies and actors do not produce a tangible asset but rather perform a service: collective healthcare and its sustainability. The work they perform is not aimed at producing a simple product; rather, it is targeted at the affirmation in a collective orientation of healthcare as a public value. This approach necessarily implies the involvement, empowerment, and awareness of all engaged actors, such as firms, suppliers, patients, and individuals in value cocreation project.

Delivery of such services necessarily requires the active participation of various stakeholders, including institutional actors (local health authorities, hospitals, districts, nursing homes, municipalities, and volunteer associations) entrusted with service care and delivery; other national and local agencies concerned with resource allocation planning and control (region, state, and local entities); actors who are currently in charge of medical and scientific training (public administrations, professional associations, scientific societies, trade unions of category, and universities); citizens/patients; providers of goods and services of healthcare actors; and others.

In recent decades, there has been a handover of the provision of service to its coproduction, shared between different healthcare providers and between them and the users, changing the paradigm of the "medicine of waiting" to one of "medicine initiative oriented" or "active health promotion." In this context, it is essential to build a "common sense" of medicine, enabling all individuals to take control of an issue that has implications and aspects relevant in their daily lives and, therefore, requires extensive testing and acceptance.

The above passages support the concept and need of a logic of patient empowerment. In particular, the reorganization of healthcare systems is designed to place the citizen at the center of the system and of its programming. The intent is to make patients not simply aware of the care process and the reasons for which business decisions are made but active actors in the process of value cocreation through a willingness to share information with other patients, medical staff, and clinicians and a capability to collect and properly manage data and information relevant to the history and traceability of diagnostic or therapeutic, self-medication, medical checkup, and other preventive behaviors and proactive activities (patient empowerment). More generally, patient empowerment logics, protocols, and models provide the meaning by which citizens can communicate in the care process, proactively participate in the care process, prevent illnesses and behave in healthy and conscious manners, and ultimately directly promote their health achievements. However, the effectiveness of this process is closely linked with effective interaction and cooperation between the various actors and resources involved and is only possible through the implementation of mechanisms for coordination at the management, political, social, economic, and information technology levels. Certainly, the logic of patient empowerment could be a real example of dynamic innovation in healthcare systems.

The work of organizations in healthcare, aimed at affirming a collective orientation as the recognition of health as a public value, necessarily requires the involvement and awareness of a heterogeneous variety of actors as health system stakeholders (Bjerrum and Linde 2011). They thus contribute to the creation of public health through the sharing of goals and pathways, transforming the paradigm of clinicians and doctors as passive recipients of patients' needs to proactive actors engaging with patients for their benefit (Nambisan and Nambisan 2009). There is the necessity of introducing a "democratization" process in medicine, in which partnership is the only alternative to perishing for every stakeholder (Turner 2011). The effectiveness of this process is closely related to the interaction and cooperation between these parties, which is only possible through the implementation of strong coordination mechanisms at operational, political, social, economic, ethical, and legal levels (Payne et al. 2008).

18.3 VSA as Lens for Actors' Engagement in Healthcare

VSA is a theoretical framework for the analysis and interpretation of systems behaviors, derived from systems thinking and based on a constructivist approach. According to VSA, each entity (i.e., individual, community, and organization) can be described as a Viable System (VS) whose ultimate purpose is to survive within its specific own context (Barile 2008; Golinelli 2005). Moreover, the constructivist approach states that reality cannot be perceived as something objective, independent of the person who experiences it, because it is him who creates, builds, and invents what he believes exists (Von Glasersfeld 1984). This implies that, with specific reference to organizations, in analyzing phenomena, a central role is played by the individual that leads these organizations and orients their processes, i.e., the decision-maker.

According to the systems vision, any entity can be interpreted as a structure consisting of a set of subordinate and interconnected components. Such a structure, if aimed, allows the emergence of a system, which in turn interacts with other systems organizing resources, skills, and expertise to achieve a satisfying outcome for all, which is survival (Barile 2008; Golinelli 2005).

VSA is an appropriate interpretative lens within the context of healthcare because the systems vision recognizes the principle of interdependence that characterizes the operation of systems and creates circular cause-effect relationships between events and phenomena (Barile et al. 2012a, b, c). This type of systems organization results really integrated, and no actor operating within can operate in isolation; each node can have a specific connection with others, allowing stronger relationships useful in cooperating for a global competitiveness, considering not the qualifications of the individual entities but highlighting the relevance and the roles of the relationships among them.

The healthcare service delivery process is progressively abandoning the original doctor-patient relational approach centered on human values and needs, with the result that client evaluation of effectiveness is undermined (Saviano et al. 2010). In this sense, there has been an evolution in structures, services, and needs, as demonstrated by the transition to the current provider-client relationship, centered on the principle of patient autonomy and a more qualitative, customized service (Saviano et al. 2010). To realize such interactions, rules sharing and compliance are critical. This requirement is especially true in healthcare, and several systems theories address it.

A number of theoretical statements made by VSA scholars might be applied to healthcare system functions and operations. Thus, the development of system *consonance* (potential harmonic relationship) and *resonance* (effective harmonic interaction) between elements of the same system and among systems themselves represents critical elements in the government of multi-subjective entities (Polese 2013). In particular, in multi-stakeholder environments, progressive cognitive reorientation and convergence toward a same aim (overall health) implemented by all participating actors represents the key to success. A conceptual framework for analyzing and developing conditions of consonance in healthcare provider-client interaction processes highlights the need for recovering a harmonious relationship between the ethical/emotional and economic rationales in healthcare system governance (Saviano et al. 2010). In particular, we use the relational key of VSA as a general conceptual framework for applying the information variety model to analyze the nature and the dynamics of the healthcare provider-client relationship, pointing out evident links with SS theoretical constructs (Saviano et al. 2010).

In the study of the VSA, factors that are relevant to the present work include considerations concerning the complexity of the systems and their management, of the relationship between consonance (structural) and resonance (systemic), and concerning the consequences of interpreting the phenomena in terms of organizations' behavior (Barile et al. 2012a). Given that two identical systems do not exist, each is distinguished by the characteristics of its "structural" elements and (in the case of a same structure) by the presence of improvements in knowledge (technicalities), operational experience (practicalities), skills (abilities), and iterations (even unknowingly) as these mature. It is important to understand what changes can be considered to facilitate the development and implementation of synergy. Thus, decision-makers should govern and manage the system by accomplishing an ongoing learning process that allows it to adapt itself constantly to the changing requirements of a growing context complexity. The intent is to allow the system to contribute to the improvement of the physical and psychological conditions not only of the patients directly involved in the service process but also of the entire social context, at the same time increasing public awareness of healthcare that benefits community wellness (Polese 2013; Saviano et al. 2013).

In recent years, healthcare was the ground for a series of changes that are mutated principles, goals, objectives, and operating modes. The way forward is a healthcare system in which the state, regions, universities, businesses, and individuals work together to strengthen the conditions necessary for the viability, sustainability, effectiveness, and efficiency of health value in health systems (Barile et al. 2012b).

In a system-based logic, the healthcare system can be interpreted as a complex service ecosystem built around the following considerations:

- It is a formalized set of connected and interacting parties.

- It is a set of elements, complex, dynamic, integrated, and finalized.
- The final and shared goal is to ensure an adequate level of health for a served community/population.
- Its functionality is dependent on the influence and pressure from numerous supra-systems (stakeholders).
- It is influenced by external changes and stimuli to maintain equilibrated (survival) conditions.
- It facilitates interaction, thus allowing actors to improve.
- The level of its performance improvement in quality of service offered is for the benefit of all of the participants (win-win logic).

All of the above considerations affect the analysis of complex phenomena, the design of (complex) service systems that operate in complex environments and is declinable in different fields of application (Barile et al. 2012a), among them healthcare, which is considered a real example of a complex service ecosystem. A healthcare service ecosystem is an organization of people, institutions, and resources aimed at providing healthcare services for a community.

In conclusion, the healthcare organizations framed with the systems framework of relationships can also be defined as a Viable System (Golinelli 2000, 2005, 2010), part of a wider context that opens its borders to the external environment (Barile et al. 2012a, c) by tightening relationships with various actors who work there (Barile 2008; Golinelli et al. 2002). The real and effective interaction between various operators, companies, and communities necessarily involves the participation of the mentioned different actors within the process of cocreation of sustainable value for the benefit of all.

18.4 Emerging Innovation from Translational Medicine (T-Med)

Translational Medicine (T-Med) is a discipline within biomedical and public health research that aims to improve the health of individuals and the community by "translating" findings into diagnostic tools, medicines, procedures, policies, and education; translational research is the bridge between academic science and clinical practice. Such a bidirectional path "from bench to bedside" and back (Abraham et al. 2012; Marincola 2003) advocates robust, bidirectional information flow (Zerhouni 2005) and more effective collaboration involving academia, industries, and patients (Anderson 2012; Littman et al. 2007).

One prevalent description of T-Med, first introduced by the Institute of Medicine's Clinical Research Roundtable, highlights two roadblocks: the first translational block (T1) prevents basic research findings from being tested in a clinical setting and the second translational block (T2) prevents proven interventions from becoming standard practice (see Fig. 18.3).

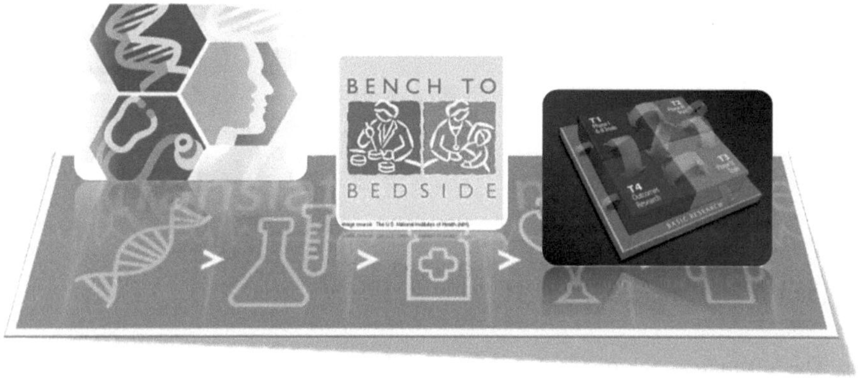

Fig. 18.3 T-MED, a long evolution

T-Med consists of transferring the results of basic research into useful clinical applications, focusing on patient-oriented research to understand the mechanisms of the disease and its specific treatment. It realizes synergies between basic research and clinical research because not only can knowledge obtained through basic research be conveyed to the application stage but also clinical applications may apply an important stimulus to fundamental research to enable real progress in the medical field. The result of this process is the creation of a bidirectional flow between the patient and the laboratory, accelerating the transfer of information and knowledge gained through scientific research to clinical practice, thus improving the conditions of patients.

The mutual exchange of information between basic research and clinical research allows, therefore, the evaluation of different pathophysiological aspects under experimental conditions and the application of this knowledge to human beings through clinical research (see Fig 18.4).

Hence, T-Med's principal goal is to speed the development of new compounds or medical protocols and/or treatments to improve the patient's quality of life. To achieve this purpose, T-Med calls for a synergy between epidemiology, basic research, and clinical trials and is strongly based upon innovation management and research development in medicine. For this reason, a managerial view of T-Med is particularly useful in terms of insights for researchers and clinicians involved in improving healthcare service.

Actually, the problem is that if "in times of abundance, efficiency may not be the highest priority, and scientists might have the chance to indulge the luxury of speculative adventures in the world of the unknown in these times of restricted funding opportunity, it behooves us to select our scientific challenges parsimoniously by constantly confronting our intuitions with the reality of human pathology" (Marincola 2003). In other words, application criteria must ensure positive results in a framework of appropriateness, financial sustainability, interventions equity, and integration.

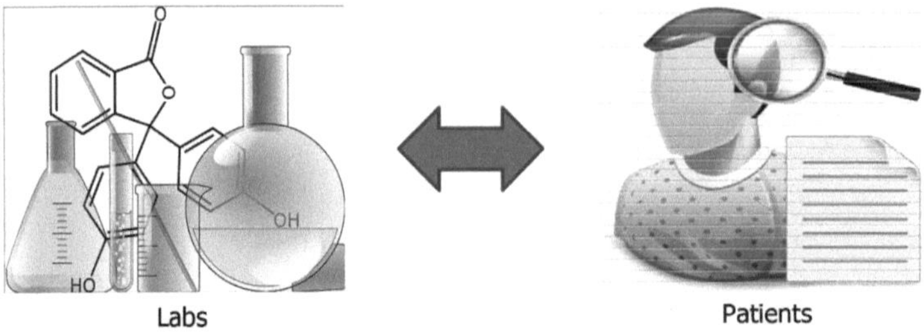

Fig. 18.4 T-Med effects

The ultimate goal of T-Med may be identified in the development of new treatments and insights for the improvement of health across populations (Sarkar 2010; Woolf 2008). This implies that T-Med (also called translational research) does not only aim to produce values and bring them to the patient. Its essence lies in validating the potentiality of novel discoveries by enhancing the success, feasibility, and efficiency of discovery validation. In other words, its ultimate goal lies in identifying in the process of clinical testing of human disease (through direct observation) what the obstacles are (Mankoff et al. 2004; Hörig et al. 2005) and allowing basic scientists and physicians to share their expertise to identify and compare the challenges at the interface between basic and clinical investigation, proposing integrated and integrating solutions to increase the efficiency of the process (Wang and Marincola 2012). T-Med's success encompasses not only scientific and operational but also financial, ethical, social, regulatory, and legislative contingencies (Littman et al. 2007).

Recently there has been a growing awareness of the role played by T-Med culture in the improvement of patient outcomes (Olson and Claiborne 2012; Wang 2012). Reducing human disease and mortality is, in fact, the end purpose toward which T-Med is generally and commonly recognized to be oriented (Anastasio et al. 2013; Hutton 2007). In this sense the followed figure shows how in a Translational Medicine process clinical and molecular data are collected, aggregated, and integrated in easily retrievable standardized formats in order to provide opportunities for the biomedical discoveries that can impact cancer research and treatment (see Fig. 18.5).

Despite the importance of T-Med for the patient, it has a wider view because it produces different values for different actors examining various aspects of this medical approach (Littman et al. 2007). For academics, it represents a chance to confirm and validate novel concepts or to discover new ones with the hope that they could become effective clinical applications and be relevant to human disease (Marincola 2003). For patients and for clinicians, it refers to the need to accelerate the capture of biomedical research benefits, wishing the gap between "what we know and what we practice" to be bridged (Davis et al. 2003; Littman et al. 2007). For those who invested in it, T-Med provides financial returns (Wang and Marincola 2012). Therefore, we distinguish a number of steps for this integration process, considering the single nodes first and thereafter specific relationships with

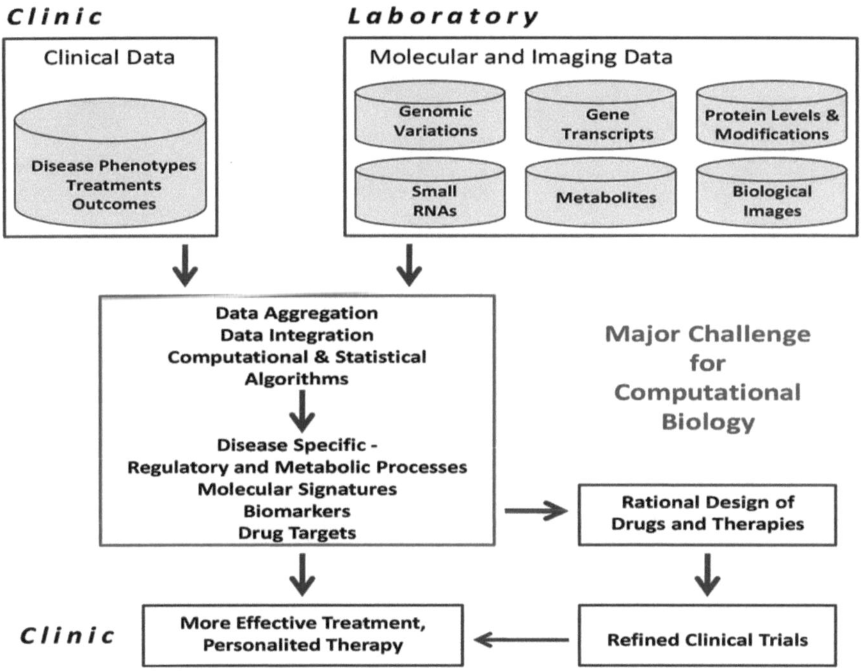

Fig. 18.5 Integration between clinical and laboratory data in Translational Medicine (*Source*: Mathew et al. 2007)

one another, moving toward their visible usefulness for reciprocal interests (starting from studies and going toward testing and matching phases, as in Fig. 18.6).

Coherently with suggestions proposed by the above-described service and systems theories, T-Med is also characterized by a variegated list of benefits and stakeholders. Nevertheless, it seems possible to identify a unifying purpose capable of complying with the expectations and needs of all involved actors (Marincola 2003) once we elevate the level of observation and analyze its beneficial effects on society.

This last aspect is very important when we consider that the provision of healthcare service requires the active participation of various actors, including institutional actors (local health authorities, hospitals, districts, nursing homes, municipalities, and volunteer associations) that are responsible for the care and the provision of services; other national and local agencies of planning and control (region, state, and local entities), which collaborate in the support and delivery of services; actors who are currently in charge of medical and scientific training (public administrations, professional associations, scientific societies, trade unions of appropriate categories, and universities); citizens; providers of goods and services of health organizations; and others (Polese and Capunzo 2013; Polese and Carrubbo 2014).

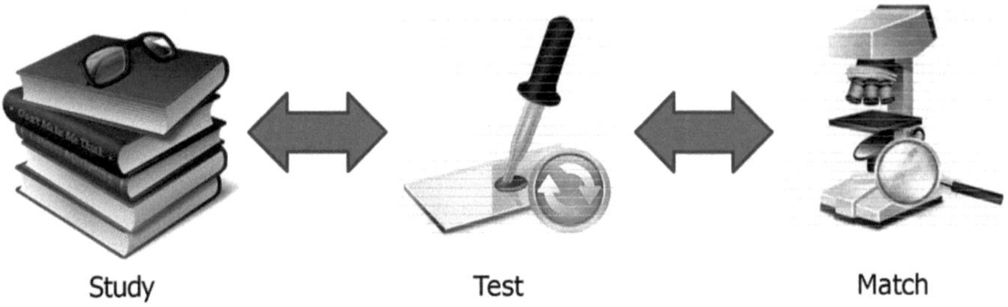

Fig. 18.6 T-Med steps

Each of the actors involved in this area pursues a different interest. Each mentioned actor participates in health creation and dissemination by exchanging resources and information.

The work all actors do, aimed at affirming a collective orientation as the recognition of health as a public value, necessarily requires the involvement and awareness of such heterogeneity of actors as health system stakeholders. Involved and engaged actors contribute, by sharing their resources, to the creation of public health through the sharing of goals and pathways, transforming the paradigm of clinicians and doctors from passive recipients of patients' needs to proactive actors engaging with patients for their benefit (Nambisan and Nambisan 2009).

These numerous actors involved in T-Med success appear to be interconnected in value cocreation networks, in which value and service for the patient (and the other actors) is the outcome of joint activities within the same system. From this perspective, patients, clinicians, private and public hospitals, pharmaceutical industries, and institutions are sources of and contributors to the system's performance.

T-Med contexts are demanding contexts in which organizations ought to pursue continuous improvement and change. This situation, in systems terms, implies that health systems are open and strongly dynamic (Barile et al. 2014). Service research system theories offer interesting insights and contribute to the understanding of value cocreation exchanges in healthcare. According to systems theories, service logic may in fact be the enabler of harmonic interactions and satisfactory exchanges among involved actors.

The rapid progress of science, powered by new technologies and tools for basic and clinical medical research, has allowed a continuous update of diagnostic and therapeutic procedures that contain and integrate elements from different scientific fields. On the one hand, the large amount of scientific information acquired from different clinical, experimental, and technology sectors has enabled the creation of a very developed and constantly and rapidly growing network. On the other hand, it has raised the question of what is the best way to integrate all this information (Carrubbo et al. 2015). In view of this growing success, however, is important to consider the high cost of research and, in particular, the real practical effect of research on the improvement of health, disease, and not least, the patient's quality of life. The integration of scientific and clinical research is thus becoming a crucial topic in the scientific community. T-Med success is directly correlated to innovation because innovation and technology transfer affect health services, linking the quality of care to continuous improvement and to T-Med research contexts. Innovation in health contexts results from both scientific and technological progress and is often strictly dependent on their reciprocal inferences.

18.5 Managerial Implications

Service Innovation, service approaches, and T-Med have a relevant role in emerging innovation in healthcare (and in society) because they collectively provide new frameworks for different levels of exchanging resources. These frameworks are able to provide new implications for management and strategy at different levels of healthcare organizations.

The first step in healthcare management, able to improve the relationships and manage the complexity, is to develop a smart organization that bases its management, strategies, and the whole system itself on advanced technological solutions that encourage resource exchanging inside and outside a service system.

Organizations in healthcare necessarily consider the existence of many actors, including the patients themselves, involved in various ways in the process of creating health value and, in particular, how their actions/interactions affect company results; the strengthening of relationships requires an increasingly significant involvement of stakeholders (before defined "empowerment"), mutual recognition, and continuity (Capunzo et al. 2013). In terms of management, (1) economic, (2) interpersonal, and (3) cultural smart "efforts" to realize and promote effective integration means investing time and capital (including human) in the reeducation in a logic of mutual benefit and sustainability. However, this logic is, to date, the fundamental basis for the system enhancing the healthcare business with efficient and profitable experiences and feelings held by the beneficiaries of the offered service (IBM 2009).

As mentioned, healthcare should be organized in a more functional way to maximize its usefulness for the different needs of users and the availability of resources of the actors involved in the process of service delivery.

At a macrolevel, all of the above implies and requires the coordination of government, managerial, and social systems and therefore the introduction of strong mechanisms of integration to ultimately allow operation in a new logical, contemporary, interactive, and collaborative but not more sequential exchange.

Smarter healthcare (intended as complex service ecosystem) brings technological, interconnected, and intelligent solutions helpful to emerging innovation. Smarter healthcare begins with better, more detailed, and faster communication; the integration of data so that each actor is involved in the system has access to these data and has access to a network of collaborative care, reducing errors, and increasing efficiency. The T-Med framework is able to help emerging innovation in this direction by contributing to a new approach to healthcare, in order to support the development of a new management of healthcare based on dynamic innovations, proposing solutions to embrace all of the actors involved in the supply chain, including the end user (the patient).

Thus, we talk about a service system that operates with integrated technologies, a new approach in the healthcare sector that uses information to create real knowledge for the assisted, and a more efficient operation of available equipment and tools (De Santo et al. 2011).

18.6 Conclusions and Findings

Service Innovation is increasingly related to innovative and technological solutions intertwined with organizational, relational, and managerial new protocols and methods, coherent with ever-demanding correlations between evolving external changes and internal performance/results. This led to a continuous learning process that helps to understand the role of innovation in service and its progress in managing constraints, emerging opportunities, new ideas, and solutions.

Service Innovation in healthcare contexts results from scientific and technological progress and emerges from the common perceptions of all actors involved; this kind of approach emerges from the VSA and SS perspectives and could be a new way to approach the relationships among actors in T-Med.

The healthcare (as complex service ecosystem) can be intended as a context in which cocreation occurs and systems shape themselves into networks offering a shared and diffuse value for all involved actors.

In terms of Service Innovation in healthcare, many propositions are underway. Under the lens of SS and VSA, T-Med appears to be more scalable and useful for many applications and contributes to an increase in performance:

- It affects the strategic role of competences as the basis of a good service encouraging the exchange of knowledge and the resources integration.
- It fosters relationships between actors operating in healthcare linking them using the technology.
- Its aim is focused on final service-user needs in healthcare, approaching to the patient empowerment logic.
- It highlights the final goal of a structured service ecosystem, of which healthcare is an example.
- It concerns the efforts made by many actors individually as leverage for the improvement of the service ecosystem as a whole.

The side effects of network cooperation, the win-win logic, and value cocreation represent some of the evidence for these sentences, supporting the comprehension of events and the complexity reduction. The mutual exchange between service providers and service users forces the logic of value cocreation for the investigated processes, intending health as the most important valuable element for a collectivity.

Today, the real involvement of patients inside healthcare service provision represents effective patient empowerment in practice; the solutions (e.g., organizational, technological, managerial, operational, and educational) must be smart enough to support that empowerment in an adequate manner, i.e., to adapt the functions of any service systems in which they operate to the evolving modern environment.

Ultimately, the modern process of using the healthcare service could be called a "cooperative game," aimed at the promotion, implementation, and coordination of multi-actors.

In a value cocreation logic, technology becomes not only an essential tool in researching new ways to interconnect different players in the innovation processes in service but also a key element in the emergence of new potential innovation. Service Innovation is not the result of a "value proposer" but the result of a continuous exchange of information and resources, in particular within T-Med.

References

Abraham, E., Marincola, F. M., Chen, Z., & Wang, X. (2012). Clinical and translational medicine: Integrative and practical science. *Clinical and Translational Medicine, 1*(1), 1.

Anastasio, A., Armenante, A., Gimigliano, A., Moscarino, A., Panzera, S., Romano, R., et al. (2013). To be on not to be translational. *Translational Medicine, 5*(3), 5–6.

Anderson, R. (2012). Clinical and translational medicine in Europe—Horizon 2020 and beyond. *Journal of Translational Medicine, 10*(Suppl. 2), A5.

Barile, S. (Ed.). (2008). *L'impresa come sistema. Contributi sull'approccio sistemico vitale* (IIth ed.). Torino: Giappichelli.

Barile, S., Polese, F., & Carrubbo, L. (2012a). Il Cambiamento quale Fattore Strategico per la Sopravvivenza delle Organizzazioni Imprenditoriali. In S. Barile, F. Polese, & M. Saviano (Eds.), *Immaginare l'innovazione* (pp. 2–32). Torino: Giappichelli Editore. e-book.

Barile, S., Polese, F., Saviano, M., Carrubbo, L., & Clarizia, F. (2012b). Service research contribution to healthcare networks' understanding. In J. Mickelsson & A. Helkulla (Eds.), *Innovative service perspectives* (p. 71). Hanken: Hanken School of Economics.

Barile, S., Pels, J., Polese, F., & Saviano, M. (2012c). An introduction to the viable systems approach and its contribution to marketing. *Journal of Business Market Management, 5*(2), 54–78.

Barile, S., Polese, F., Antonucci, E., & Carrubbo, L. (2014). Al confine dell'innovazione tecnologica in sanità: la medicina traslazionale. In S. Aleo, R. De Matteis, & G. Vecchio (Eds.), *Le responsabilità in ambito sanitario* (pp. 1233–1267). Padova: Cedam.

Bjerrum, O. J., & Linde, H. H. (2011). European new safe and innovative medicine initiatives: History and processes (Through December 2009). In B. H. Littman & R. Krishns (Eds.), *Translational medicine and drug discovery* (pp. 265–288). New York: The Cambridge University Press.

Bryson, J. R., & Monnoyer, C. (2004). Understanding the relationship between services and innovation: The RESER review of the European service literature on innovation. *The Services Industries Journal, 24*, 205–222.

Capra, F. (1997). *The web of life*. London: Flamingo.

Capunzo, M., Polese, F., Boccia, G., Carrubbo, L., Clarizia, F., & De Caro, F. (2013). Advances in service research for the understanding and the management of service in healthcare networks.

In E. Gummesson, C. Mele, & F. Polese (Eds.), *Service-dominant logic, network and systems theory and service science*. Napoli: Giannini Editore.
Carrubbo, L., Clarizia, F., Hysa, X., & Bilotta, A. (2013). New 'smarter' solutions for the healthcare complex service system. In E. Gummesson, C. Mele, & F. Polese (Eds.), *System theory and service science: Integrating three perspectives in a new service agenda*. Napoli: Giannini Editore.
Carrubbo, L., Bruni, R., Cavacece, Y., & Moretta Tartaglione, A. (2015). Service system platforms to improve value co-creation: Insights for translational medicine. In E. Gummesson, C. Mele, & F. Polese (Eds.), *System theory and service science: Integrating three perspectives in a new service agenda*. Napoli: Giannini Editore.
Castiaux, A. (2012). Developing dynamic capabilities to meet sustainable development challenges. *International Journal of Innovation Management, 16*(6), 1–16.
Chakraborty, S., & Dobrzykowski, D. (2013). Supply chain practices for complexity in healthcare: A service-dominant logic view. *The IUP Journal of Supply Chain Management, 10*(3), 53–75.
Chakraborty, S., & Dobrzykowski, D. (2014). Examining value co-creation in healthcare purchasing: A supply chain view. *Verslas: Teorija ir praktika/Business: Theory and practice, 5*(2), 179–190.
Coombs, R., & Miles, I. (2000). Innovation, measurement and services: The new problematique. In S. J. Metcalfe & M. Ian (Eds.), *Innovation systems in the service economy* (pp. 85–103). New York: Springer.
Davis, D., Evans, M., & Jadad, A. (2003). The case of knowledge translation: Shortening the journey from evidence to effect. *British Medicine Journal, 327*, 33–35.
De Santo, M., Pietrosanto, A., Napoletano, P., & Carrubbo, L. (2011). Knowledge based service systems. In E. Gummesson, C. Mele, & F. Polese (Eds.), *System theory and service science: Integrating three perspectives in a new service agenda*. Napoli: Giannini Editore.
Den Hertog, P. (2000). Knowledge-intensive business services as co-producers of innovation. *International Journal of Innovation Management, 4*, 491–528.
Gadrey, J., & Gallouj, F. (2002). *Productivity, innovation and knowledge in services, new economic and socio-economic approaches*. Cheltenham: Edwrad Elgar.
Golinelli, G. M. (2000). *L'approccio sistemico al governo dell'impresa. L'impresa sistema vitale* (Vol. I). Padova: Cedam.
Golinelli, G. M. (2005). *L'approccio sistemico al governo dell'impresa. L'impresa sistema vitale*. Padova: Cedam.
Golinelli, G. M. (2010). *Viable Systems Approach (VSA). Governing business dynamic*. Padova: Kluwer Cedam.
Golinelli, G. M., Pastore, A., Gatti, M., Massaroni, E., & Vagnani, G. (2002). The firm as a viable system: Managing inter-organisational relationships. *Sinergie, 58*, 65–98.
Gronhaug, K., & Kaufmann, G. (1988). *Innovation: A cross-disciplinary perspective*. Oslo: Norwegian University Press.
Gummesson, E. (2008). Extending the new dominant logic: From customer centricity to balanced centricity. *The Journal of the Academy of Marketing Science, 36*(1), 15–17.
Gummesson, E., & Polese, F. (2009). B2B is not an island! *Journal of Business and Industrial Marketing, 24*(5/6), 337–350.
Hörig, H., Marincola, E., & Marincola, F. M. (2005). Obstacles and opportunities in translational research. *Nature Medicine, 11*, 705–708.
Hutton, J. (2007). Transforming translation—Harnessing discovery for patient and public benefit. US. *Report of the Translational Research Working Group of the National Cancer Advisory Board*.
Iandolo, F., Calabrese, M., Antonucci, E., & Caputo, F. (2013). Towards a value co-creation based healthcare system. In E. Gummesson, C. Mele, & F. Polese (Eds.), *Service dominant logic, network and systems theory and service science: Integrating three perspectives for a new service agenda*. Napoli: Giannini.

IBM. (2009). *Patient centered medical home*. Cambridge: University of Cambridge Institute for Manufacturing.

IBM. (2012). *The value of building sustainable health system*. Cambridge: University of Cambridge Institute for Manufacturing.

Littman, B. H., Di Mario, L., Pleban, M., & Marincola, F. M. (2007). What's next in translational medicine? *Clinical Science, 112*, 217–227.

Maglio, P. P., Srinivasan, S., Kreulen, J. T., & Spohrer, J. (2006). Service systems, service scientists, SSME, and innovation. *Communications of the ACM, 49*(7), 81–85.

Mankoff, S. P., Brander, C., Ferrone, S., & Marincola, F. M. (2004). Lost in translation: Obstacles in translational medicine. *Journal of Translational Medicine, 2*(1), 14.

Marincola, F. M. (2003). Translational medicine: A two-way road. *Journal of Translational Medicine, 1*(1), 1.

Mathew, J. P., Taylor, B. S., Bader, G. D., Pyarajan, S., Antoniotti, M., & Chinnaiyan, A. M. (2007). From bytes to bedside: Data integration and computational biology for translational cancer research. *PLoS Computational Biology, 3*(2), 12.

McColl-Kennedy, J. R., Vargo, S. L., Dagger, T. S., & Sweeney, J. C. (2009). Customers as resource integrators: Styles of customer co-creation. In E. Gummesson, C. Mele, & F. Polese (Eds.), *Service science, S-D logic and network theory*. Napoli: Giannini Editore.

McColl-Kennedy, J. R., Vargo, S. L., Dagger, T. S., Sweeney, J. C., & van Kasteren, Y. (2012). Healthcare customer value co-creation practice styles. *Journal of Service Research, 15*(4), 370–389.

Mele, C., & Polese, F. (2011). Key dimensions of service systems: Interaction in social and technological networks to foster value co-creation. In H. Demirkan, J. Spohrer, & V. Krishna (Eds.), *The science of service systems* (pp. 37–59). London: Springer.

Meyer, M., & Müller, I. (2006). Networked healthcare: A practical guide to understanding influence networks in the healthcare industry. *Journal of Medical Marketing, 6*(4), 250–259.

Nahapiet, J., & Ghoshal, S. (1997). Social capital, intellectual capital and the creation of value in firms. *Academy of Management Review, 23*(2), 242–266.

Nambisan, P., & Nambisan, S. (2009). Models of consumer value co-creation in health care. *Health Care Management Review, 34*(4), 344–354.

Nordgren, L. (2009). Value creation in health care services—Developing service productivity: Experiences from Sweden. *International Journal of Public Sector Management, 22*(2), 114–127.

Olson, S., & Claiborne, A. B. (2012). *Strengthening a workforce for innovative regulatory science in therapeutics development*. Washington, DC: The National Academies Press.

Payne, A., Storbacka, K., & Frow, P. (2008). Managing the co-creation of value. *Journal of the Academy of Marketing Science, 36*(1), 83–96.

Polese, F. (2013). *Management sanitario in ottica sistemico vitale*. Torino: Giappichelli Editore.

Polese, F., & Capunzo, M. (2013). The determinants of translational medicine success. *A managerial contribution. Traslational Medicine @ UniSa, 6*, 29–34.

Polese, F., & Carrubbo, L. (2014). A managerial view of translational medicine. *Annals of Translational Medicine and Epidemiology, 1*(1), 1–3.

Polese, F., Carrubbo, L., & Russo, G. (2009a). Managing business relationships: Between service cultural and viable systems approach. *Esperienze d'Impresa, 2*, 121–144.

Polese, F., Sarno, S., & Carrubbo, L. (2009b). Service science and innovation management: Sustainable service and quality performance in the value co-creating age. In *Proceedings of 12th QMOD and Toulon-Verona Conference*, Italy, Verona.

Polese, F., Pels, J., & Brodie, R. (2011). Theoretical underpinning to successful value co-creation. In E. Gummesson, C. Mele, & F. Polese (Eds.), *Service-dominant logic, network and systems theory and service science*. Napoli: Giannini Editore.

Qiu, R. G., Fang, Z., Shen, H., & Yu, M. (2007). Towards service science, engineering and practice. *International Journal of Services Operations and Informatics, 2*(2), 103–113.

Rehman, M., Dean, A. M., & Pires, G. D. (2012). A research framework for examining customer participation in value co-creation: Applying the service dominant logic to the provision of living support services to oncology day-care patients. *International Journal of Behavioural and Healthcare Research, 3*(3–4), 226–243.

Rullani, E. (2005). Intelligenza terziaria e sviluppo economico: Dalla prima alla seconda modernità. In E. Rullani, P. Barbieri, M. Paiola, & R. Sebastiani (Eds.), *Intelligenza terziaria. Motore dell'economia* (pp. 13–60). Milano: Franco Angeli.

Sarkar, I. N. (2010). Biomedical informatics and translational medicine. *Journal of Translational Medicine, 3*, 8–22.

Saviano, M., Bassano, C., & Calabrese, M. (2010). A VSA-SS approach to healthcare service systems the triple target of efficiency, effectiveness and sustainability. *Service Science, 2*(1/2), 41–61.

Saviano, M., Parida, R., Caputo, F., & Datta, S. K. (2013). Insights on Italian and Indian health care systems from VSA perspective. In D. Vrontis, Y. Weber, & E. Tsoukatos (Eds.), *6th Annual EuroMed Conference of the EuroMed Academy "Confronting Contemporary Business Challenges through Management Innovation"*. EuroMed Press, 1586–1603.

Saviano, M., Parida, R., Caputo, F., & Datta, S. K. (2014). Health care as a worldwide concern. Insights on the Italian and Indian health care systems and PPPs from a VSA perspective. *EuroMed Journal of Business, 9*(2), 198–220.

Schilling, M. A. (2008). *Gestione dell' Innovazione*. Milano: McGraw-Hill.

Spohrer, J., Maglio, P. P., Bailey, J., & Gruhl, D. (2007). Steps toward a science of service systems. *Computer, 1*, 71–77.

Sundbo, J., & Gallouj, F. (2000). Innovation as a loosely coupled system in services. *International Journal of Services Technology and Management, 1*, 15–36.

Tether, B. S. (2003). The sources and aims of innovation in services: Variety between and within sectors. *Economics of Innovation and New Technology, 12*, 481–505.

Turner, M. (2011). Embracing change: A pharmaceutical industry guide to the 21st century. In B. H. Littman & R. Krishns (Eds.), *Translational medicine and drug discovery* (pp. 328–343). New York: The Cambridge University Press.

Vargo, S. L., & Lusch, R. F. (2004). Evolving to a new dominant logic for marketing. *Journal of Marketing, 68*, 1–17.

Vargo, S. L., & Lusch, R. F. (2006). Service-dominant logic: What it is, what it is not, what it might be. In R. F. Lusch & S. L. Vargo (Eds.), *The service-dominant logic of marketing: Dialog, debate, and directions* (pp. 43–56). Armonk: ME Sharpe.

Vargo, S. L., Maglio, P. P., & Akaka, M. A. (2008). On value and value co-creation: A service systems and service logic perspective. *European Management Journal, 26*(3), 145–152.

Von Glasersfeld, E. (1984). An introduction to radical constructivism. In P. Watzlawick (Ed.), *The invented reality* (pp. 17–40). New York: Norton.

Wang, X. (2012). A new vision of definition, commentary, and understanding in clinical and translational medicine. *Clinical and Translational Medicine, 1*, 1–5.

Wang, X., & Marincola, F. M. (2012). A decade plus of translation: What do we understand? *Clinical and Translational Medicine, 3*, 1–3.

Wieland, H., Polese, F., Vargo, S., & Lusch, R. (2012). Toward a service (eco) systems perspective on value creation. *International Journal of Service Science, Management, Engineering and Technology, 3*(3), 12–25.

Woolf, S. H. (2008). The meaning of translational research and why it matters. *Journal of the American Medical Association, 299*, 211–213.

Zerhouni, E. A. (2005). Translational and clinical science-time for a new vision. *New England Journal of Medicine, 353*(15), 1621–1623.

Chapter 19
Stakeholder Engagement in Sustainability Innovation: Experiences in the Jewellery Business

Alessandra De Chiara

Abstract The chapter aims to study engagement as the basis for innovative projects that respond to an objective of social and environmental sustainability and describes the multi-stakeholder committee as the ideal container for realizing such engagement. The chapter explores the experience of the Ethics Committee of Coloured Gemstones, by Assogemme, in Italy. The goal of the Committee is to define "a procedural code concerning the production and marketing of the coloured gemstones in accordance with the principles of ethical conduct". The stakeholder engagement approach and the founding of a multi-stakeholder committee for the creation of ethical guidelines are widespread in the jewellery business, which presents many ethical problems: child labour, environmental havoc and social inequality. But the case analysed is the first experience in the coloured gemstone jewellery industry, which is characterized by a plurality of ethical aspects related to the variety of these gemstones. The study of this Committee, which is engaged in the challenge to create a sustainable innovation process, a managerial system for the traceability of jewellery products made with coloured gemstones, underlines opportunities and problems of the stakeholder engagement approach.

19.1 Introduction

Why are there in the jewellery sector examples of multi-stakeholder committees that are working on the conception of new procedures aimed at guaranteeing a responsible production of jewels?

The answer to this question lies in the features of the jewellery production chain, which is very fragmented and characterized by businesses located in different areas of the world; especially regarding the first manufacturing processes, such

A. De Chiara (✉)
University of Naples "L'Orientale", Naples, Italy
e-mail: adechiara@unior.it

T. Russo-Spena et al. (eds.), *Innovating in Practice*,
DOI 10.1007/978-3-319-43380-6_19

businesses are based in areas with a poor economy. The production chain is characterized by the presence of multinational corporations that control it and small production operators, who in the mining phase can be craftsmen or single individuals that work with a very simple instrumentation.

Differing cultural systems, differing economies, rich and poor countries, differing operators, multinational corporations and craftsmen are peculiarities in the sector that have fostered serious environmental crimes, human rights violations and social inequality. In fact wealth produced through the business has hardly affected those countries and communities where raw materials, diamonds, gold and gemstones can be found.

The representativeness of the several needs of the industries, countries, local institutions, NGOs, communities and scientific world appears as a necessary condition to try defining some guidelines that can suggest an ethical behaviour in the entire jewellery supply chain. The different guidelines help the firms implement innovation processes in accordance with the principles of sustainability in order to avoid fostering even indirectly serious violations, and at the same time they define new behavioural models that can affect the welfare of everybody.

These are the reasons why for their definition they require a wide participation of the parties involved.

This chapter investigates the case of the Ethics Committee of Coloured Gemstones founded in Italy, which represents an original case because it is the only one that is working internationally on the implementation of sustainability in the industrial sector of jewels made using gemstones that, due to the diversity of the stones, presents high levels of complexity. Through the study of the composition and operation of the Committee, the chapter aims to study the engagement as the basis for sustainable innovation process.

The chapter is so articulated: the first part highlights the importance of engagement in the studies about innovation and sustainability and describes the multi-stakeholder committees as enlarged participation models for sustainable projects.

The topics of this section are the importance of these models to study sustainable innovation for the common good and the necessity of a wide representativeness and sharing to make the different parties work together; the second part of the chapter presents the empirical analysis, which clarifies the objectives and the methodology, describes some cases of multi-stakeholder committees in the jewellery sector and investigates the case of the Ethics Committee of Coloured Gemstones, by Assogemme, in Italy, to examine in depth the themes of composition and sharing within the Committee. The second part will be followed by the conclusions.

19.2 The Stakeholder Engagement for Innovation and Sustainability

The stakeholder engagement approach asserts a core value: the position of responsibility required of firms and stakeholders in order to engage in the sharing of information, the construction of a dialogue and an ongoing commitment to solve problems together (Svendsen and Laberge 2005). The relational aspect is therefore a key element; the creation and subsequent consolidation of the "social capital" of the company is strategic, because it represents the "glue" that holds together the social relations (Burt 1997). Engagement is necessary both for innovation and for sustainability.

Innovation literature has deepened the strategic importance of collaboration with regard to three main aspects: effects, actors involved and characteristics of innovation strategies.

The positive effects of accessing complementary technological resources (Becker and Dietz 2004) are related to a faster development of innovations, an improved market access, economies of scale and scope, cost sharing and risk spreading (Ahuja 2000; Cassiman and Veugelers 2002; Hagedoorn 2002; López 2008). A higher overall performance level (Abramovsky et al. 2009) and a higher R&D intensity (Becker and Dietz 2004; Sampson 2007) can be obtained too, since companies are able to share investment and costs and may take advantage of partners' resources and capabilities, increasing the profitability of R&D (Belderbos et al. 2003).

Literature underlines that all types of companies and different stakeholders can be involved (Hagedoorn et al. 2000). Large firms can manage spillovers and be involved in R&D cooperation, and they can have a high share of R&D expenditures (Fritsch and Lukas 2001). Some studies have underlined that vertical spillovers, associated with suppliers and customers, have a more significant effect on the R&D performance and welfare than horizontal spillovers, which are linked to universities, research institutes and competitors (Atallah 2002). In addition, cooperation with customers and public sector institutions is positively related to the success of product innovations, and the cooperation with suppliers and universities has a more significant influence on the success of process innovations (Freel and Harrison 2006).

Regarding the development of innovation strategies, some researches distinguish those companies that follow an incremental path built on the firm's existing internal knowledge from those that provide knowledge to aid defining trajectories that are new to the firm (Bercovitz and Feldman 2007). In the latter case, collaboration with universities and government labs is widespread, and cooperative arrangements for innovation are more common among those firms that introduced innovations that were new not only to the firm but also to the market. Furthermore, the intensity of R&D activities tends to increase the likelihood of a firm having cooperative arrangements for innovation with external partners (Tether 2002). The search for external partners is normally associated with more complex innovation processes, e.g. those that combine both process and product innovation (Piga and Vivarelli 2004). Regarding the complexity of innovation process, Bayona et al. (2001) and Miotti and Sachwald (2003) provide evidence that firms operating in more technology-intensive sectors have a greater propensity to establish cooperative R&D and innovation agreements.

At the same time, the issue of sustainability is no doubt an approach that requires an action of engagement of stakeholders (Molteni et al. 2006). Acting responsibly means supporting investments and costs, as well as acquiring appropriate skills, which are often unaffordable actions for a single actor, especially for a small and medium size firm. But also regarding the impacts, the wide participation of more actors will enable the development of larger projects.

Literature has underlined the benefits of a stakeholder approach: both inside, for the companies belonging to sustainability projects, and outside for the beneficiaries of projects (Cesaretti and Azzurra 2011; Piciocchi et al. 2009); the intensity of the "mechanism of contamination" between different actors belonging to the network and the "modalities of communication" affect the ability of the individual components to implement structured systems of sustainability and influence the degree of social responsibility of other actors in the network (Caroli and Tantaldo 2011).

The expanded involvement stimulates the exchange of skills, but also of instances and needs; it legitimates projects and provides them with the necessary conditions to be developed and effectively achieve the estimated results (Caroli and Tantaldo 2011). They are the conditions that are defined by some authors as a relational state (Albareda et al. 2004; Mendoza 1996; Midttun 2005), or by others, from a different perspective, collaborative governance or even corporate responsibility and responsible competitiveness clusters (Zadek et al. 2006).

A particularly study approach regards SMEs and supply chains. Regarding SMEs, literature suggests acting through an approach based on the local network, including the institutions that can play a key role as intermediaries for the promotion of joint strategies inspired by social responsibility (Antoldi et al. 2008; De Chiara 2012). The creation of a network also seems to be an appropriate way to develop the necessary skills to help the interpenetration of social responsibility in defining business strategies, thereby making sustainability an integral part of the strategy also pursued by individual companies (Molteni et al. 2006).

Regarding the supply chain, logistics social responsibility, purchasing social responsibility (Carter and Jennings 2002) as well as sustainable supply chain management (Teuscher et al. 2006) are some approaches that emphasize the effectiveness of the adoption of CSR practices in a collaborative approach to management, aimed at involvement of stakeholders at different stages of the implementation process (Maignan et al. 2002; Vurro et al. 2009). The collaboration appears to be functional to the creation of a climate of confidence, reduction of costs associated with trade and improvement in the effectiveness of the actions (Cox 2004; Russo-Spena and De Chiara 2012; van Tulder et al. 2009), as well as it is essential to successfully implement the strategy of corporate social responsibility

(Andersen and Skjoett-Larsen 2009; Carter and Jennings 2004; Carter and Rogers 2005; De Chiara and Russo-Spena 2011; Murphy and Poist 2002).

At the same time, the international ethic codes and the guidelines (OECD, ILO, EC, UN, etc.) stress the necessity to act with a wider participation and involvement of different stakeholders' perspective.

19.2.1 The Multi-stakeholders Committees for Sustainable Projects

Literature has enhanced different aspects of engagement for sustainable projects: from the right perspective of the engagement, which should be based on exchange and involvement, free from strategic motivations and individual interests (Greenwood 2007; Noland and Philliphs 2010), to the characteristic of partners and to their individual capabilities, for what concerns the composition of the network (Zucchella 2007; Stam et al. 2010; Wang and Fesenmaie 2007).

An interesting perspective is the study of enlarged participation models. They have been defined as *multi-stakeholder processes*, "aimed to bring together all major stakeholders in a new form of communication, decision-finding (and possibly decision-making) structure on a particular issue; based on recognition of the importance of achieving equity and accountability in communication between stakeholders; involving equitable representation of three or more stakeholder groups and their views; based on democratic principles of transparency and participation and aimed to develop partnerships and strengthened networks between and among stakeholders" (Hemmati 2002: 19), or defined as *global action networks* (GAN), a "civil society initiated multi-stakeholders arrangements that aim to fulfill a leadership role in the protection of a global commons or the production of global public goods" (Waddell 2007), or *institutional innovators* (Kuhnert 2001).

The distinctive aspects of these networks can be tracked back in the composition, in the values that must inspire the governance and in the dialogue and functional processes. "Global and multi-level" as "multi-stakeholder and cross-sectoral" have been related to a composition, which must guarantee a complete representativeness of more levels (global, national, local) and actors, whose different business memberships (Glasbergen 2010)—"democracy, equity and justice transparency, accountability, corporate social responsibility, solidarity, good governance, economic justice, gender equity, and so on" (Hemmati 2002: 40)—are the values that should inspire the governance. Some studies underline that the governance of a committee should develop a model in line with the OECD principles on corporate governance, that is to say, "distribution of duties and responsibilities among different supervisors; regulatory and enforcement authorities; ensuring equitable treatment of all shareholders, including minority and foreign shareholders; recognition of the rights of stakeholders established by law or through mutual agreements; encouraging active co-operation between corporations and stakeholders in creating wealth, jobs, and the sustainability of financially sound enterprises" (OECD 2004: 17).

Many of these principles refer to a concept of "social responsibility" that should animate the behaviour of the members of the multi-stakeholder committee. Literature has defined CSR as one of the pillars on which to build corporate governance (Ahmmed and Mohammad 2005), or a model of extended corporate governance (Sacconi 2007), and vice versa; corporate governance has been considered a theory concerned with the alignment of management and shareholder interests. Therefore, if the governance model covers a range of issues for the protection of the interests of shareholders and stakeholders, like OECD shows, it embraces the principles of CSR; ergo the committee must and can operate not only in the interests of the business community but also for those of the whole community.

19.3 Realizing a "Learning Alliance" for a Common Good

Collaborative models, aimed at finding new solutions inspired by the principles of sustainability—that is to say, those cardinal principles of environmental protection and human rights, declined in the international codes of conduct (Green Paper of EC, Guidelines of OECD, etc.)—must not only be understood as a determinant of the decision-making process, guarantor of the different demands of the stakeholders, but are also essential elements to implement the strategies, to get sustainability initiatives off and to produce an effective repercussion in terms of sustainable development.

The decision-making process that concerns a sustainable innovation, addressing to create new "ethical" products/services and/or processes that modify the way to operate, requires a contribution of heterogeneous skills, experiences and requests obtained also by expanding the range to include the market, the institutions and the entire community. The aim must be to create a "learning alliance", in which a range of stakeholders, typically located at different levels and within different domains but connected by a common interest, come together as a group to optimize relations and break down barriers to learn. It is an opportunity to create new relationships and networks in which a group takes collective action to address an issue, to solve a problem or to assess new opportunities in the field of sustainability. Participating in decision-making processes appears already as a first important result, since it represents a learning experience for those that can then affect the action planning. The process of participating in a multi-stakeholder committee—sharing experiences and taking part in decision-making processes and debating on the selection of good practices and the direction of the project—will be an empowering process and a learning experience for all members. As a consequence, the group will increase its ability to work together through the process of working together and identifying change and solutions.

Besides, the joint action can allow to conceive the best solutions, but it also allows to obtain the development of legitimated and major projects, because of the existence of necessary conditions in order to be efficaciously developed (Caroli and Tantaldo 2011). Especially with regard to the actors of the supply chain, liabilities, costs and benefits can be well balanced throughout the production chain.

The multi-stakeholder perspective is essential to implement sustainable innovations and to produce an effective repercussion in terms of sustainable development. In fact the credibility of a sustainability project depends on how it affects the whole community. Therefore, a wide participation—in the first instance by the actors of the supply chain and in a broader perspective by the market, institutions and the community—will prevent the occurrence of any dystonias in the behaviour of participants that could undermine the effectiveness of the project and the achievement of objectives. In other words, a wide participation reduces the risks that sustainable actions by a company can be minimized or even cancelled by other actors' unethical actions in the same chain or chains connected. So collaboration in sustainability projects must go through a co-sustainability approach as much as possible, a co-creation of sustainable solutions (new ethical products/services, new way to operate) in order to reach a common good.

Participation, sharing, organization and coordination are essential to accomplish an act of collaborative sustainability.

The sharing is the perspective chosen in this chapter to study the operation of these multi-stakeholder committees. Undoubtedly, this aspect has elements of complexity, given the heterogeneity of the participants in these committees, but it is an important aspect for achieving their purposes.

A preliminary aspect concerns the process of setting up a multi-stakeholder Committee as following: the identification of the purpose and of the objectives on the short and medium term; the process of discovery, but also of imagination, of the type of knowledge and expertise that the committees need; and the definition of the role of the group that has formed, distinguishing a task support and supervisor process role (Advisory Committee) from a more complex task of defining the decision and approval of proposals brought to the discussion (Steering Committee). Then the governance structure can be defined by identifying the roles and responsibilities of the individual participants, considering how each member can contribute, the rules and processes by which decisions are made, but above all the principles that inspired the components, which ensure an effective collaboration.

The sharing among members is a crucial aspect to the operation of a multi-stakeholder committee. The Glasbergen's model (2010) highlights specific management capabilities for the reaching of the sharing, related to emotional and transactional aspects and the creation of an enabling environment (Table 19.1), and for facilitating the operation of these committees. The emotional component concerns the creation of a psychological space, a space where all the participants can feel co-ownership and responsibility for the network as a whole. Networks must be sources of inspiration, unity and moral support. "Participants must be motivated to participate actively" (Glasbergen 2010: 3). While the creation of a transaction space concerns a space in which the participants can productively combine knowledge exchange and action-oriented functions. This function requires some formalization of the relationships between the participants while keeping the flexibility of

Table 19.1 Management capabilities for a collective action of global action networks

	Fixation	Flexibility
Psychological space	Emotional and cognitive coherence ⟷	Emotional and cognitive diversity
Transaction space	Central rule system and decision-making ⟷	Self-organization of participants
Enabling environment	Internally closed ⟷	Externally open

bottom-up processes. Finally, it is necessary to facilitate the realization of an enabling environment, where the participants are both sources of change and enablers of change through their activities directed to participants outside the network and they aim to act as change agents in their issue field (ibidem).

All three capabilities are crucial for ensuring a collective action in a networking process. Besides, it is necessary to find an equilibrium in the orientation between choices of *fixation* and of *flexibility*, that is, to say, between the necessity "to stabilize participants' orientations, expectations and rules of conduct, while maintaining a dynamic innovative potential"; between the necessity "to reduce variety by central rules, while facilitating self-organization of participants"; and between the necessity "to induce joint activities in a formalized internal structure, while keeping the interfaces with the outer world open and constructive" (ibidem).

19.4 Empirical Research: Objectives and Methodology

Empirical research has focused on the study of the Ethics Committee of Coloured Gemstones of Assogemme (Confindustria) in Italy, with the aim to analyse the operation of a multi-stakeholder committee which is developing a sustainable innovation: the creation of new standards for the traceability of coloured gemstones used in jewellery. The choice of this business depends on the ethical problems presented in the jewellery business, while the choice of this Committee has an explanation in the fact that it is a new experience for the supply chain of precious stones that appeared in Italy.

The research focuses on the function of this committee and on solutions and tools put in place to facilitate the collaboration among participants, to manage the diversity of its members and to create the sharing, in order to answer the following research questions: is it effective to build a multi-stakeholder committee to create new processes related to sustainability? Does the multi-stakeholder approach allow an effective collaboration among different individuals? Is the sharing a strategic element to reach within the multi-stakeholder committee in order to motivate participants to collaborate actively?.

The empirical investigation is organized into two phases. The first highlights the ethical issues in the field of jewellery and describes the main multi-stakeholder experiences for the creation of a sustainable supply chain and the definition codes of conduct for the companies. This research phase is realized through a documentary study. The second phase concerns the study of an Italian multi-stakeholder experience, the Ethics Committee of Coloured Gemstones of Assogemme (Confindustria), and it is realized through a direct interview with the president of this committee.

19.4.1 Ethical Issues in the Jewellery Business: The Multi-stakeholder Approach for Achieving a Responsible Chain

The jewellery business is exposed to strong ethical risks, in the dual role of environmental pollution and human rights violations.[1] The high ethical issues in part go back to the structural characteristics of the jewellery industry:

(a) A strong dispersion of the activities of the value chain—from the goldmines, diamond and platinum mines situated in Africa, Canada, Australia and Russia to cleaning and processing activities in India, China and Turkey and to distribution in the United States, the European Union, Japan and in the emerging markets of China and India.
(b) A strong separation between the decision-making activities that, especially in the diamonds industry, are concentrated in the hands of a few large firms of developed countries, and the operations/production activities, in which prevails a strong fragmentation with a localization of production sites especially in emerging countries and in the developing ones.

In addition to fragmentation of activities and dislocation of the same in strongly contrasting areas, there are two more kinds of problems: the presence of operators belonging to different cultural value systems, which, therefore, affects the interpretation of the significance of the principle of social and environmental sustainability, and the heterogeneity in the composition of the operators, which generates a diversity of patterns of behaviour, as well as the risk of establishing a breeding ground for the onset of action of power by large corporations against small operators of the supply chain (De Chiara 2013).

To achieve a responsible chain of gold, of diamond jewellery and other precious stones, we find examples of multi-stakeholder committees, which are designed and developed envisaging the enlarged participation of stakeholders and the sharing of objectives and actions for the development of standards/certification systems of gold and diamonds.

[1]Gold, cassiterite, wolframite and coltan are conflict minerals, minerals mined in conditions of armed conflict and human rights abuses, and are regulated by the US Conflict Minerals law.

So is the experience of the "Kimberley Process", the agreement is finalized and approved by the joint efforts of the governments of many countries involved in the extraction of diamonds, multinational manufacturers of diamonds and civil society. Institutionalized in November 2002, following the approval of the United Nations (13 March 2002), the Kimberley Process Certification Scheme (KPCS)—which takes its name from the town of Kimberley in South Africa where, in May 2000, the representative members of the different categories of stakeholder met—is a certification system to trace diamonds and to ensure that they come from legitimate mining equipment. Currently, this certification system is the most effective tool to protect the diamond industry and above all to ensure that the profits generated by this activity can also benefit local communities and the individuals who work there, assuming that the diamond is a natural resource and should bring benefits to all stakeholders. The system also allows to avoid that the extraction can finance the purchase of weapons and fuel conflicts and civil wars and cause serious human rights violations, which was a rather common situation prior to the agreement, so much so to coin the definition of blood diamond or "conflict diamonds" (UN 2006).

Even the experience of the Responsible Jewellery Council—the international non-profit organization founded in 2005 that contains a growing number of companies operating in the sector of jewellery (diamonds, gold, platinum group metals)—has the aim to build a responsible supply chain for diamonds, gold and platinum group metals, "from the mine to the consumer". The organization has developed a reference framework for the sector by creating a series of standard "Code of Practices" and "Chain of Custody" to promote responsible business practices and standards. These results have been achieved through the work of the RJC Standards Committee: a committee composed of 14 members "elected RJC Member representatives, two from each Member Forum, and up to 12 external participants from NGOs, related standards organizations and other experts".

The committee designs, implements and continuously improves the RJC Certification and at the same time "provides policy advice on the consultation and engagement of stakeholders during standards development activities". The Multi-stakeholder Standards Committee, composed of each part of the supply chain, civil society, other standards organizations and independent experts, works in collaboration with Alliance for Responsible Mining, Diamond Development Initiative, Solidaridad and Swiss Better Gold Initiative on responsible gold sourcing from mine to jewellers, London Bullion Market Association (LBMA) and Conflict-Free Smelter program—audit harmonization for gold refiners.

Another important experience is that of the Organization for Economic Cooperation and Development (OECD), which drafted the Due Diligence Guidance for Responsible Supply Chains of Minerals from Conflict-Affected and High-Risk Areas: "a guide that offers suggestions for the responsible management of global supply chains of minerals, in order to help companies to respect human rights and do not contribute to conflict through their activities". The OECD has developed a specific supplement to the mining activity in the gold sector, and in this case as well, it works in the form of a working group, formed to "OECD, ICGLR and other partner countries, international organisations, industry at several levels of the gold supply chain, international and local civil society organisations, expert consultancy groups and other independent experts" (OECD 2011).

As declared in the official documents: "this work is the first example of a collaborative government-backed multi-stakeholder initiative on responsible supply chain management of minerals from conflict-affected areas". It was endorsed by 11 Heads of State and Governments of the International Conference on the Great Lakes Region in the Lusaka Declaration adopted on 15 December 2010. Furthermore, the UN Security Council on the DRC supported taking forward the due diligence recommendations contained in the report of the UN Group of Experts on the DRC, which rely on the OECD Due Diligence Guidance (OECD 2011: 1).

The multi-stakeholder committee had the task to discuss the first draft of the "Supplement on Gold" and has received great appreciation from the participants for its mode of constitution: "Participants noted that the first meeting of the OECD-hosted working group on gold as one of the first gatherings of such a diverse range of stakeholders convened specifically to explore due diligence on gold from conflict-affected and high-risk areas" (OECD 2011: 3).

The importance of a broad participation for the success of the project has been repeatedly reaffirmed by the same committee that requested: "Participants felt that some stakeholders that would be vital to the success of due diligence efforts for conflict-free gold supply chains should be invited to participate in the OECD-hosted working group on gold. Participants considered it a priority to invite key non-member countries and companies in the gold supply chain based in those countries. It was clarified that relevant non-OECD countries had been invited. Participants also noted that all the relevant agencies within the participating countries, such as customs authorities, finance ministries and central banks, should participate to share their relevant expertise. Participants agreed to set up a Drafting Committee composed of willing members of the OECD-hosted working group on gold, which would take responsibility for integrating comments from members of the working group and [make]any subsequent revisions to the Supplement" (OECD 2011: 5).

And yet, with reference to the composition of the Committee that had to collect all the additions to the first document, it is stated that "The Drafting Committee will have a balanced tripartite structure reflecting the composition of the stakeholders in the working group, would be small and nimble enough to communicate regularly and easily, and be responsive to issues and discussions as they arise" (OECD 2011: 9).

All these experiences clearly show that not only the multi-stakeholder approach is very used to develop innovation processes related to sustainability in the jewellery business, but it is useful because all the experiences cited have reached the formalization of a standard.

19.4.2 The Experience of the Ethics Committee of the Coloured Gemstones, Assogemme

The Ethics Committee of the Coloured Gemstones has been established on 17 July 2013 and promoted by Assogemme (at Confindustria, Rimini) in Italy (see Fig. 19.1). Its main objective is to prevent that in the supply chain, exploitation of minors and workers occurs in general, while ensuring fair wages and healthy workplaces. These objectives are only apparently simple, because their achievement is made very complex by the immense extension of the supply chain and production areas. In fact, the supply chain of coloured gemstones, from the mine to the market, is highly fragmented with a high degree of opacity. The high value of the gems increases the likelihood of illegal activities, which are often characterized by the involvement of criminal groups (UNICRI 2013).

The composition of this Committee, in its building phase, is an expression almost exclusively of the jewellery industry. Among its participants, there were testimonials from the leading companies such as Bulgari, Gucci, Pomellato, Bruni, some members of the small and medium enterprises for the production of jewellery, members of the scientific world of gemmology and finally the representative

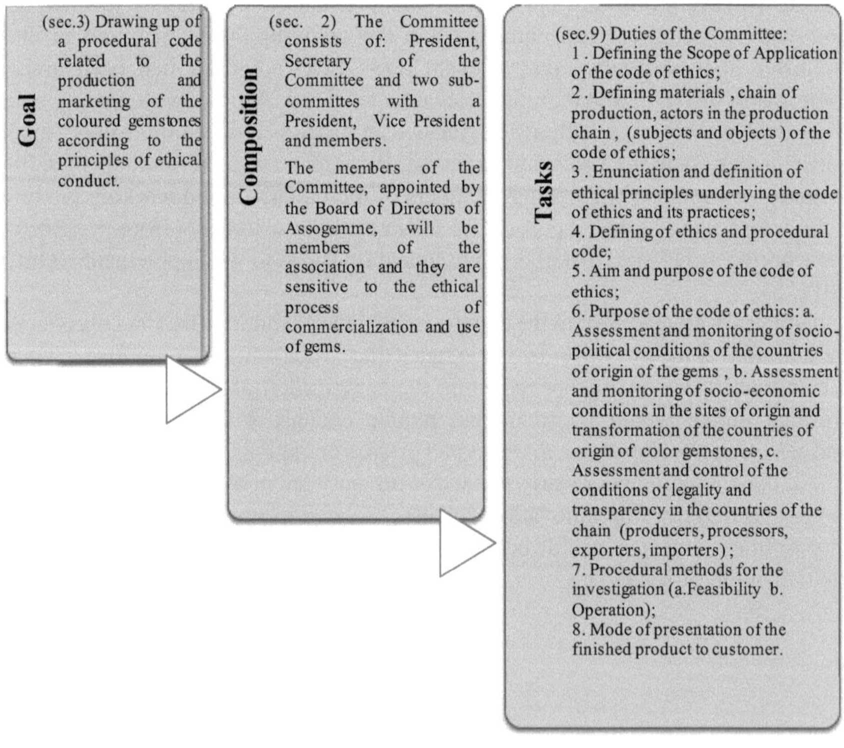

Fig. 19.1 Features of the Ethics Committee of the Coloured Gemstones of Assogemme

exponents of the major industry associations, Assogemme, Responsible Jewellery Council, Federorafi.

This expertise can be classified into three types: the specific business, gemmological scientific research and business institutions with characterizing features of the jewellery industry. Although it was a mono business in a first phase of constitution—unlike the other committees that are working to achieve an ethically correct jewellery production chain—the Committee's works are already leading to the redefinition of composition, of procedures and of working goals.

In regard to the composition, the research has observed a tendency to an enrichment of the team of participants for a possible enlargement of the representativeness of the Committee, as the inclusion of researchers in geopolitical and economic fields, while further contacts have been taken with the institution UNICRI (United Nations Research Institute on Crime and Justice), which is currently working on a protocol to trace the paths of gems, suspected to incite crime, and with Save the Children and the Responsible Jewellery Council.

The Committee consisted of the Chairman of the Ethics Committee, the secretariat and two subcommittees: one called "scientific committee", with a president, a vice-chairman and members, and another, unnamed, which included the major brands of jewellery, also organized with a chairman and a vice-president. Nowadays, there are some changes in the structure of subcommittees: the one represented by jewellery companies has added new members following the line of specialization, to ensure better operational organization, while the scientific subcommittee, which already assigned more competences, has further enhanced the diversity of skills, whose contribution in scientific research is considered a prerequisite for the implementation of the code of ethics.

Also from an organizational point of view, we can observe this attitude. In fact, these changes underline that also this Committee is moving towards a greater representation of the various instances and the creation of a multi-stakeholder group, like the other experiences working to the ethical problems in this business.

This approach, while presents difficulties for the interaction between different subjects, as underlined by its President, was deemed preferable in consideration of the same aims of the Committee, set in sec. 3 (Purpose) of the statute, "Drafting of a procedural code related to the production and marketing of the coloured gemstones according to the principles of ethical conduct". This purpose gives to Committee a role of steering committee, namely, the definition of the decisions and approval of proposals brought to discussion by the working groups, which actually requires a wider and wider participation to perform its functions.

In fact, the themes that the Committee must deliberate are broad and diversified (set in sec. 9—Duties of the Committee). They include the formulation of a code of ethics and the statement of the purpose and objectives; the materials and actors (subjects and objects of the code of ethics) have been preliminarily defined with the guiding principles, together with the mode of assessment and verification of conditions: (a) sociopolitical ones in the countries of origin of coloured gemstones, (b) socio-economic ones in the sites/countries of origin and transformation of coloured gemstones, (c) legality and transparency in the countries of the chain (producers, processors, exporters importers). The Committee shall define the detailed procedural rules for investigations of the conditions mentioned above, considering the feasibility and the operability, and the manner of presentation of the finished product to the customer.

Besides, among the tasks of the Committee, a number of new initiatives have been recently added that could be implemented within the code "(...) action of information given to the audience of operators, program implementation of ethical criteria for correct naming of gemmological materials, dissemination of 'fundamentals of gemmology', supervision of the sources and agencies engaged in the correct nomenclature, a brief guide to best practice in order to comply with the principles of 'fair disclosure', a FAQ and a phone line for answers to questions".

Taking into account the role of the steering committee and the characteristics of the production chain, the presence of different competences and needs within the committee becomes a preliminary condition to work effectively.

The investigation has shown that diversity in the participants is a critical element for the effectiveness of the operations and that it has to be wisely managed in order to build a fertile environment for the exchange and the interaction.

The management of this element has led to the definition of some decisions and operative tools. Firstly, the commitment of the Committee has been addressed to strengthen the *transaction space* (see Fig. 19.1). In fact, procedures and sub-objectives of the work have been refined, and the changes to the structure of the subcommittees have been carried out to ensure the specific needs for knowledge and research: the scientific committee works on the technical aspects related to the traceability of the stones, with respect to the scientific characteristics of the same; the subcommittee of management has its objective in the identification of a traceability system and takes also into account how dangerous the mining regions can be, ensuring that the end customer purchases an ethical gemstones. The attention of the latest subcommittee focuses on this stakeholder, drawing inspiration from the Standard "Code of Practices" of the Responsible Jewellery Council.

Although the field of action of the two subcommittees takes its origin from different stakeholders—the first, extraction/mining firms, for studying the characteristics of the stones and then back to the origins of the same, and the other, the customer, to satisfy the consumers' need for purchasing an ethical product—the integration between the two groups is strongly encouraged by the Bureau of the Committee. The presidents of the two subcommittees should define business "ad hoc" meetings, they must support the exchange of ideas, materials and results, and they must engage in the construction of a shared dialogue. Dialogue and participation are sought with a series of organizational and information tools. With respect to subcommittees, "the president acts as a speaker on matters falling within the area assigned to him, organizes the preparatory work, refers to the Committee and suggests resolutive proposals" (sec. 9). Otherwise regarding the inner workings, the subcommittees are free to organize their work among participants in order to ensure the necessary flexibility in the working groups.

Besides, to facilitate the exchange among all participants, a "digital platform" has been created, and the construction of a digital archive for the collection of the work of the Committee and of all the necessary documentation has also been planned. The creation of a community among participants has also been planned.

To allow the functioning of the Committee, the statute "Regulation for the operation of the Ethics Committee of the Coloured Gemstones" regulates other issues: the Committee operates for 3 years, and its components continue to perform their functions until the appointment of the new organization (sec. 4); the Committee members meet in an ordinary session, usually at least once a year (sec. 6); and the President convenes and chairs the meetings, establishing the agenda on the basis of the indications of the components, and coordinates the work (sec. 5).

The emotional aspects linked to the creation of a sense of belonging and unity between the participants are elements deemed as crucial too by the Bureau of the Committee to boost motivation and assure an active collaboration. About these aspects, the research has shown that the ability of the Committee to create a *psychological space* is not solid.

In fact, besides the "Regulation" for coordinating the professional relationship between all participating members, the presidency is determined to build on existing basic principles of the OECD on corporate governance, principles which have already informally animated the constitution of the same committee in the invitation to members to participate. Actually the management of the *psychological space* is realized mostly informally: convivial meetings are organized among the participants with the opportunity to visit their locations (companies, associations, etc.) and exchange experiences.

The solutions analysed above concern the exchange and interaction management among the members of the Committee, but some instruments to allow exchange outside of the Committee have also been created (Table 19.2).

For creating an *enabling environment*, some actions are governed by the statute, as the task of the Committee to draw up, for each year, a report on the situation of the projects carried out (sec. 10), which will be presented to the Board of Directors of Assogemme at the final statement of the association (sec. 10). The Committee deals with the Association in a consistent and effective cooperation. It can request data, documents and information related to matters of competence, and it provides information and/or projects that are useful to the bodies of the Association, who have the task of formulating proposals and implementing interventions related to topics and subjects including those pertaining to the same Committee (sec. 11). These actions, albeit in the statute, do not have a specific discipline, as the participation in conferences and seminars for the disclosure of the Committee's work in contribution to sustainable development.

Table 19.2 Management capabilities of the Ethic Committee of Coloured Gemstones

	Fixed	Flexible
Psychological space	Reference (informal) to the OECD principles on corporate governance	Meeting convivial and itinerant at the locations of the participants Informal contacts of the President of the committee with the chairmen of subcommittees
Transaction space	Organizational solutions: two sub-committees with the distribution of tasks Technical solutions: ICT (digital platform, digital archive, community)	Auto-internal organization of the subcommittees
Enabling environment	Relations with Assogemme: drafting, for each year, a report on the situation of the projects submitted to the Executive Council of Assogemme	Participation in conferences and round tables

19.5 Discussion

The implementation of a participatory process to achieve a goal of sustainability requires first of all an enlarged representation of the different components/activities of the production chain. This representation makes possible to meet the demands and concerns from each stage of production and distribution and, therefore, to collect the specific skills related to each phase of operation and harvest the specific contribution of the different actors to the "solution" of the problem. But, above all, it allows feasible solutions to be proposed and gives credibility to the project. In fact, the participation of the actors in the supply chain, on the one hand, should minimize the risk of seeing thwarted the efforts made by an operator because of the unethical behaviour of other actors in the supply chain, and on the other, it should maximize the impact of a sustainable project in terms of an actual effect on the whole community.

The enlarged participation and the cultural diversity of the participants present a problem of management of diversity. This statement, which seems to be a contradiction, really means to emphasize that it doesn't represent an obstacle but it is a problem to manage the complexity of the interaction between the members of the multi-stakeholder committees.

In the analysed Committee, it appears as evident the awareness of the importance of a broader representation in the composition. The major presence of members of the jewellery industry, despite belonging to different professional fields, is meant as a temporary condition, which characterizes the start-up of the Committee. But the address followed by the presidency is towards a heterogeneous identity of the Committee. Besides, the Ethics Committee featured both a decision-making and a guiding role, as it works on sustainable innovation, so it would be an initiative which must be addressed towards a major representativeness of parties. If the sharing of ideas, skills and experiences occurs within a multi-stakeholder committee—and if this committee is credited a decision-making and guiding role—then it is even more necessary to ensure a broader representation in it. At the same time, the committee must be a representative of all stakeholders if its works are sustainability oriented, without having to recall again the concepts of richness of diversity in innovation processes, of value of the exchange and of importance of stakeholder engagement.

The research shows that the Presidency attributes to diversity an undeniable value, but it is also seen as a factor of complexity of engagement, but not an obstacle to reach the finality of the project.

To facilitate the exchange and contamination of experiences among participants, from a functional point of view, the Committee analysed responds above all with a series of organizational and technological tools: the creation of sub-working groups responsible for specific "work packages", as well as investments in ICT, is just an example.

The outlined path for the constitution and the functioning of this committee doesn't have in itself a deterministic nature, and therefore its outcome is not obvious. The extent and fragmentation of the supply chain, from mining areas located almost always in poor areas of the world, makes difficult and complex the predefined objective. As in the words of the President, the Committee is developing a path of experimentation, which has no other examples. Therefore, the result of this project is not obvious, but rather it must be built in the attempt to find a satisfactory solution for all participants.

So the outlined characteristics can be considered as preconditions in a multi-stakeholder committee that can facilitate the investigation of a solution in relation to a predefined goal. Complementary expertise in different professional fields; behaviours inspired by the principles of mutual respect, honesty, open-mindedness and commitment to the project; and the definition of initiatives and instruments that can foster the *psychological and transaction space* and also facilitate the exchange with the external area should have ensured multi-stakeholder committee to reach an objective of common interest (see Fig. 19.2).

19.6 Conclusions

The experience of the Ethics Committee of the Coloured Gemstones is crucial because it is the only experience in Italy—and also on an international level among recognized associations—which was set up and is working to develop innovative solutions, which are based on ethical principles in the process of procurement and supply chain management of coloured gemstones. It manifests the willingness of firms, even before the intervention of the legislature, to regulate the behaviour of the economic actors to give rise to a sustainable supply chain for coloured gemstones.

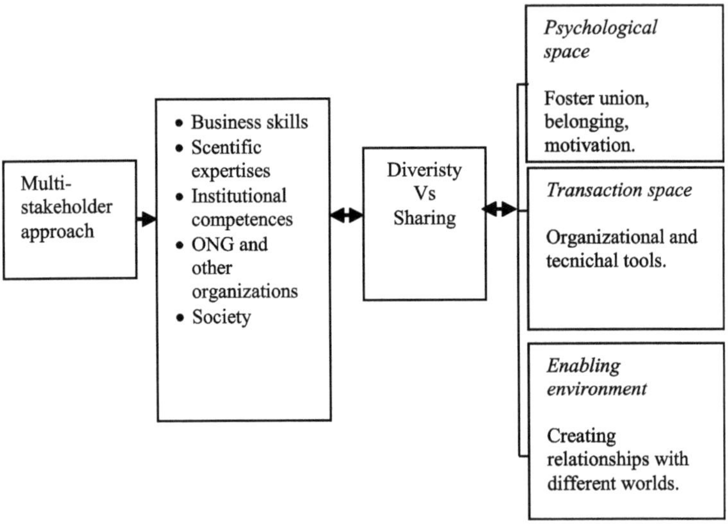

Fig. 19.2 Pre-conditions for sustainable process innovation in a multi-stakeholder committee

Devising systems for traceability of precious stones are a real innovative process, which must be designed within multi-stakeholder committees in which it is ensured a broad and representative participation of the different instances of the actors.

Innovation and sustainability appear to be two paths that are increasingly obliged to build the competitiveness of enterprises in developed countries. Strategic pathways, as suggested by the literature but also as evidenced by managerial practice, require collaborative models for the participation of complementary skills. The multi-stakeholder committee can be a container in which to realize the exchange and integration of different types of knowledge, but at the same time the representation of the different skills among the participants; it is necessary to ensure a sharing of the principles that govern individual behaviour and guarantee initiatives and instruments to make collaboration and work between different stakeholders possible. The first should be guided by the concept of social responsibility and values that it supports: mutual respect, openness, honesty, trust and commitment to society. Then it is important that the governance of the committee should develop a governance model in line with some of OECD principles on corporate governance. The latter should be created to facilitate the internal exchange between the participants and the external dialogue with the public.

Principles, behaviours and actions could guarantee an operation that is not only in the interests of the business world but also of the whole community.

Acknowledgments A heartfelt thanks to Dr. Luisa La Via, chairman of the Committee.

References

Abramovsky, L., Kremp, E., López, A., Schmidt, T., & Simpson, H. (2009). Understanding co-operative R&D activity: Evidence from four European countries. *Economics of Innovation and New Technology, 18*, 243–265.

Ahmmed, M. H., & Mohammad, A. (2005). The cost and management. *Corporate Governance: Bangladesh Perspective, 33*(6), 18–26.

Ahuja, G. (2000). The duality of collaboration: Inducements and opportunities in the formation of interfirm linkages. *Strategic Management Journal, 21*(3), 317–343.

Albareda, L., Ysa, T., & Lozano, J. M. (2004, October 22–23). The role of public policies in promoting CSR: A comparison among the EU-15. In *proceedings of Interdisciplinary CSR Research Conference, ICCSR*. University of Nottingham.

Andersen, M., & Skjoett-Larsen, T. (2009). Corporate social responsibility in global supply chain. *Supply Chain Management: An International Journal, 14*(2), 75–86.

Antoldi, F., Cerrato, D., & Todisco, A. (2008). La responsabilità sociale delle PMI distrettuali'. In M. Molteni & A. Todisco (Eds.), *Responsabilità sociale d'impresa* (pp. 115–142). Milano: Guida del Sole 24 Ore.

Atallah, G. (2002). Vertical R&D spillovers, cooperation, market structure, and innovation. *Economics of Innovation and New Technology, 11*(3), 179–209.

Bayona, C., Garcia-Marco, T., & Huerta, E. (2001). Firms' motivations for cooperative R&D: An empirical analysis of Spanish firms. *Research Policy, 30*(8), 1289–1307.

Becker, W., & Dietz, J. (2004). R&D cooperation and innovation activities of firms—Evidence for the German manufacturing industry. *Research Policy, 33*(2), 209–223.

Belderbos, R., Carree, M., Diederen, B., Lokskin, B., & Veugelers, R. (2003, March 14–15). The determinants of R&D cooperation: Evidence from Dutch CIS Data 1996–1998. In *Proceedings of ZEW Workshop on the Empirical Economics of Innovation and Patenting*, Mannheim, Germany.

Bercovitz, J., & Feldman, M. (2007). Fishing upstream: Firm innovation strategy and university research alliances. *Research Policy, 36*(7), 930–948.

Burt, R. (1997). The contingent value of social capital. *Administrative Science Quarterly, 42*(2), 339–365.

Caroli, M., & Tantaldo, C. (2011). (a cura di). *Le Reti d'Imprese come strumento privilegiato per la diffusione di una Sostenibilità di Sistema*, Luiss Business School, 2011. Accessed February 2015, from http://www.pcnitalia.it

Carter, C. R., & Jennings, M. M. (2002). Logistics social responsibility: An integrative framework. *Journal of Business Logistics, 23*(1), 145–180.

Carter, C. R., & Jennings, M. M. (2004). The role of purchasing in corporate social responsibility: A structural equation analysis. *Journal of Business Logistics, 25*(1), 145–186.

Carter, C. R., & Rogers, D. (2005). A framework of sustainable supply chain management: Moving toward new theory. *International Journal of Physical Distribution and Logistics Management, 38*(5), 360–387.

Cassiman, B., & Veugelers, R. (2002). R&D cooperation and spillovers: Some empirical evidence from Belgium. *American Economic Review, 92*(4), 1169–1184.

Cesaretti, G., & Azzurra, A. (Eds.). (2011). *Strategie e strumenti per la valorizzazione sostenibile delle produzioni agroalimentari di qualità*. Milano: Franco Angeli.

Cox, A. (2004). The art of the possible: Relationship management in power regimes and supply chains. *Supply Chain Management, 9*(5), 346–356.

De Chiara, A. (2012). La competitività responsabile nelle PMI: Strategia vincente nei territori "nemici". *Rivista Piccola Impresa/Small Business, 1*, 29–52.

De Chiara, A. (2013). Commercio etico dei diamanti: Iniziative normative o autoregolamentazione in un'ottica di network? *Micro and Macro Marketing, 3*, 521–546.

De Chiara, A., & Russo-Spena, T. (2011). CSR strategy in multinational firms: Focus on human resources, suppliers and community. *Journal of Global Responsibility, 2*, 60–74.

Freel, M., & Harrison, R. (2006). Innovation and cooperation in the small firm sector: Evidence from 'Northern Britain'. *Regional Studies, 40*(4), 289–305.
Fritsch, M., & Lukas, R. (2001). Who cooperates on R&D? *Research Policy, 30*(2), 297–312.
Glasbergen, P. (2010). Global action networks: Agents for collective action. *Global Environmental Change, 20*(1), 130–141.
Greenwood, M. (2007). Stakeholder engagement: Beyond the myth of corporate responsibility. *Journal of Business Ethics, 34*, 315–327.
Hagedoorn, J. (2002). Inter-firm R&D partnership: An overview of major trends and patterns since 1960. *Research Policy, 31*(4), 477–492.
Hagedoorn, J., Albert, N. L., & Vonortas, N. S. (2000). Research partnerships. *Research Policy, 29* (4–5), 567–586.
Hemmati, M. (Ed.). (2002). *Multi-stakeholder processes for governance and sustainability*. London: Earthscan Publications.
Kuhnert, S. (2001). An evolutionary theory of collective action: Schumpeterian entrepreneurship for the common good. *Constitutional Political Economy, 12*, 13–29.
López, A. (2008). Determinants of R&D cooperation: Evidence from Spanish manufacturing firms. *International Journal of Industrial Organization, 26*(1), 113–136.
Maignan, I., Hillebrand, B., & McAlister, D. (2002). Managing socially responsible buying: How to integrate non-economic criteria into the purchasing process. *European Management Journal, 20*(6), 641–648.
Mendoza, X. (1996). *Las transformaciones del sector público en las sociedades avanzadas. Del estado del bienestar al estado relacional'* (Papers de Formació No. 23). Diputació de Barcelona.
Midttun, A. (2005). Realigning business, government and civil society: Emerging embedded relational governance beyond the (Neo)Liberal and welfare state models. *Corporate Governance: International Journal of Business in Society, 5*(3), 159–174.
Miotti, L., & Sachwald, F. (2003). Cooperative R&D: Why and with whom? An integrated framework of analysis. *Research Policy, 32*(8), 1481–1499.
Molteni, M., Pedrini, M., & Bertolini, S. (Eds.). (2006). *La responsabilità sociale d'impresa nelle aziende familiari italiane*. Milano: Franco Angeli.
Murphy, P. E., & Poist, R. F. (2002). Social responsible logistics: An exploratory study. *Transportation Journal, 41*(4), 23–35.
Noland, J., & Philliphs, R. (2010). Stakeholder engagement, discourse ethics and strategic management. *International Journal of Management Reviews, 12*(1), 39–49.
OECD. (2004). Principles of corporate governance. Accessed May 5, 2007, from www.oecd.org/dataoecd/32/18/31557724
OECD. (2011, May 4). First meeting of the OECD-hosted working group on gold. Summary report. http://www.oecd.org/daf/inv/investmentfordevelopment/48582867.pdf
Piciocchi, P., Vollero, A., & Palazzo, M. (2009). Towards the integration between corporate community relations and CSR communication strategies. *The Enel Case Study. Esperienze d'impresa, 2*, 61–75.
Piga, C., & Vivarelli, M. (2004). Internal and external R&D: A sample selection approach. *Oxford Bulletin of Economics and Statistics, 66*(4), 457–482.
Russo-Spena, T., & De Chiara, A. (2012). CSR, Innovation strategy and supply chain management: Toward an integrated perspective. *International Journal of Technology Management, 58*, 83–108.
Sacconi, L. (2007). A social contract account for CSR as an Extended Model of corporate governance (II): Compliance, reputation, and reciprocity. *Journal of Business Ethics, 75*, 77–96.
Sampson, R. (2007). R&D Alliances and firm performance: The impact of technological diversity and alliance organization on innovation. *Academy of Management Journal, 50*(2), 364–386.
Stam, D. A., van Knippenberg, D., & Wisse, B. (2010). The role of regulatory fit in visionary leadership. *Journal of Organizational Behavior, 31*(4), 499–518.

Svendsen, A. C., & Laberge, M. (2005). Convening stakeholder networks. *Journal of Corporate Citizenship, 19*(Autumn), 91–104.

Tether, B. (2002). Who co-operates for innovation, and why: An empirical analysis. *Research Policy, 31*(6), 947–967.

Teuscher, P., Grüninger, B., & Ferdinand, N. (2006). Risk management in Sustainable Supply Chain (SSCM): Lessons learnt from the case of GMO-free soybeans. *Corporate Social Responsibility and Environmental Management, 13*(1), 1–10.

UN. (2006). *United Nations Department of Public Information. Conflict Diamonds*. Accessed online December 2006.

Unicri. (2013). Coloured gemstones. Accessed March 2014, from http://www.viji.eu/wp-content/uploads/2013/04/Coloured-gemstones-presentation-UNICRI.pdf

van Tulder, R., van Wijk, J., & Kolk, A. (2009). From chain liability to chain responsibility. *Journal of Business Ethics, 85*(Suppl. 2), 399–412.

Vurro, C., Russo, A., & Perrini, F. (2009). Shaping sustainable value chains: Network determinants of supply chain governance models. *Journal of Business Ethics, 90*(Suppl. 4), 607–621.

Waddell, S. (2007). Realising global change: Developing the tools: Building the infrastructure. *The Journal of Corporate Citizenship, 26*, 69–84.

Wang, Y., & Fesenmaie, D. R. (2007). Collaborative destination marketing: A case study of Elkhart country. *Indiana Tourism Management, 28*, 863–875.

Zadek, S. (2006). *The logic of collaborative governance: Corporate responsibility, accountability, and social contract* (Corporate Social Responsibility Initiative, Working Paper No. 17). Cambridge: School of Government, Harvard University.

Zucchella, A. (2007). La responsabilità sociale nelle reti d'impresa. Symphonya. *Emerging Issues in Management, 2*, 69–77.

Part VI
Experiences: Tools

Chapter 20
Co-development and Retooling as New Roles for Applied Research

Eveliina Saari, Mervi Hasu, Laura Honkaniemi, Katri Kallio, Tiina Tuominen, and Mikko Lehtonen

Abstract Knowledge creation dynamics is in transition in modern information-loaded society. We need to see a researcher's role in a new way: not only as a constructor of scientific facts but as a co-developer of practical tools. We analyse how co-development occurs between researchers and practitioners as a learning process in the context of a multi-actor application-orientated project. The co-development of a practical tool for supporting co-innovation in the public sector was organised as a learning journey. The chapter explicates the researchers' various methods for enhancing learning between actors and between practice contexts. The co-development unfolds as a temporary, fragile path through different artefacts in a knowledge creation process guided by the researchers. It may also be characterised as an interplay between local practice-specific and generalised knowledge, which enables retooling practitioners for their future.

20.1 Introduction

In the era of the World Wide Web, social media and mobile phones, we can say that we live in the middle of constant growth of new information, which is just waiting to be googled for different purposes. However, for example, management scientists have noticed that practitioners do not easily learn from scientific evidence, if they have not been part of the knowledge creation process themselves

E. Saari (✉) • M. Hasu • L. Honkaniemi
Finnish Institute of Occupational Health, Helsinki, Finland
e-mail: eveliina.saari@ttl.fi; mervi.hasu@ttl.fi; laura.honkaniemi@gmail.com

K. Kallio
VTT Technical Research Centre of Finland Ltd, Espoo, Finland
e-mail: katri.kallio@vtt.fi

T. Tuominen • M. Lehtonen
Aalto University, Espoo, Finland
e-mail: tiina.tuominen@aalto.fi; mikko.h.lehtonen@aalto.fi

T. Russo-Spena et al. (eds.), *Innovating in Practice*,
DOI 10.1007/978-3-319-43380-6_20

(Rynes et al. 2007). Although we are surrounded with excessive masses of information, and probably just because of that, reaching the level of understanding and for applying generalised knowledge for solving practical and societal problems remains a challenge. We should reconsider the role of scientists and experts in this situation, in which new information is instantly reachable for every layman and practitioner. We propose shared knowledge and tool creation between researchers and practitioners as a way of improving the use of scientific knowledge in the development of practices. We introduce a new position for researchers in applied research projects and open up co-development tools as artefacts in shared knowledge creation.

Symmetrical communication between scientists and practitioners is not something self-evident. Management scientists have identified several tensions in academic-practitioner relationships, such as different, deviating logics in operating, time horizons, communication conventions, interests and incentives. These may either hinder or become resources in shared knowledge creation. Due to the scarce empirical research on the topic, the relationship between researchers and practitioners has been defined by the management scholars mainly by normative opinion statements, either compatible or incompatible (Bartunek and Rynes 2014).

The role of innovation research is currently in transition for many reasons. The concept of innovation has become wider, as its focus today covers not only technological inventions and novel products but also services and organisational processes. Everyday organisational practices, the involvement of layman users and service workers as well as leaders' and managers' actions come under the scrutiny of innovation researchers, as incremental innovation processes seem to be becoming more typical in service innovations (Sundbo and Toivonen 2011). Innovations in service provision are often only recognised as innovations a posteriori (Toivonen and Tuominen 2009). The analysis of these emergent and intangible innovation processes becomes a useful starting point for reflection and learning among the participators themselves. Innovation researchers are not expected to merely produce ethnographic analyses on how scientific facts are constructed, as in laboratory studies (Latour and Woolgar 1979; Lynch 1985). Instead they become co-developers, influential mediators, facilitators and network builders, while studying the processes occurring among users, employees, managers and policymakers in inclusive, practice-based innovations.

Organising applied research projects is becoming increasingly complex. National and international research projects are conducted in consortia, with several research organisations as partners. Typically, several organisations or firms are expected to engage themselves as research objects and co-developing partners. This raises continuous demand to organise the way of communicating and collaborating between researchers and practitioners. Fine-tuned, microlevel and sensitive data can only be collected in an atmosphere of mutual trust and understanding (Hasu 2005). Researchers have to organise their projects as co-development processes between several stakeholders as opposed to traditional academic procedures.

Funding agencies expect societal impact, and client organisations expect practical solutions from innovation projects. At the same time, the researchers try to create and accumulate scientific knowledge and unfold the research object as if it was something epistemic (Knorr Cetina 2001). However, the timelines and priorities of producing these different forms of knowledge differ considerably, and researchers have to find a way in which to combine all the efforts.

This chapter analyses how the co-development between researchers and practitioners actually took place during an ambitious research project aimed at enhancing innovation management capabilities by creating a co-innovation tool for the Finnish public sector organisations. The analysis does not only see a success story but also takes a critical view on omnipotent expectations towards modern applied research. The chapter illustrates how the researchers designed a learning arena for the organisations, while trying to understand their subtle innovation management practices. We describe the complex organising of a multi-actor project and analyse how the co-development of both scientific and practical artefacts occurs between researchers and practitioners.

20.2 Relationship Between Research and Practice

Classic constructivist studies on research work have usually been rich ethnographic studies in laboratory environments (Latour and Woolgar 1979; Lynch 1985). Their focus has been on how the scientific facts are created, and the analysis of the orientation to societal applications has often been neglected. The symbolic interactionists (Fujimura 1987) saw scientists' work as problem driven and as moving between different levels of work organisation. They described the researchers' task as making the problems doable by aligning the different levels: the experiment, the laboratory and the social world. This dynamic of research work at the local level is still partly valid; however, it does not yet capture the nuances of action researchers, who try to enhance learning in several organisations.

The dialogue between researchers and their "research objects" has been a neglected topic in science and technology studies. Jay Labinger (1995) and Ullica Segestråhle (1993) have criticised that the sociologists of scientific knowledge have not been interested in what the (natural) scientists actually think of the analysis of their work. However, when studying innovation management phenomena, in which learning and co-development in the innovation network are in focus, it would be unfair to exclude the role of the researcher who makes the phenomenon visible. Practical elaborations and scientific papers are potential artefacts between researchers and practitioners for dialogue and intervention, and their contents and meaning should also be the focus of critical analysis (Hasu and Miettinen 2006).

Management scientists have raised the topic of a knowledge gap between management academia and practice in scientific articles, particularly since 2000 (Bartunek and Rynes 2014). The plausible explanation for this is that economic and political conditions have been changing in ways that have made both academics and practitioners more receptive to allying with each other. On the practitioner side, intensified competition in global markets has increased efficiency pressures and made practitioners more amenable to any ideas—academic or practical. The movement towards "evidence-based management" has emerged as an ideal way to shift the professional decisions of managers away from unsystematic experience towards the best available scientific evidence (Rousseau 2006). On the academic side, declining public funding for universities and research organisations has made these institutions dependent on increasing private funding and thus more eager to connect with practitioners (Rynes et al. 2001).

Interestingly, the discussion on the benefits of research-practitioner collaboration has been polarised. On the positive side, practitioners are argued to be excellent sources of stimulating problems, and collaboration has been seen as producing unique insights, leading to important new scientific discoveries (Amabile et al. 2001). On the other hand, it has been claimed that collaboration with practitioners makes projects narrow, short term and commercially profitable. Difficulties in overcoming corporate restrictions on data collection, interpretation or dissemination have also been pointed out (Rynes, et al. 2001).

Several tensions between practitioners and academics have been identified in collaborative research and shared knowledge creation. However, although different logics in operating and communicating present severe obstacles to integrating knowledge between researchers and practitioners, they may also be sources of generative and valuable research outcomes. The time horizons of academics and practitioners differ and may cause problems for conducting high-quality research, but if collaboration is seen more broadly than actions in one single project, the relationship between research and practice may unfold into several time dimensions (Bartunek and Rynes 2014).

Management scientists have discussed whether research is guided by rigour or relevance. Some academics fear that emphasising (practical) relevance interferes with the conduct of high-quality basic research. Prevailing understanding in science is that the researcher should be an objective observer, without interfering to practical problem solving. It is also harder to find high-quality scientific publication channels for practice-oriented research. The evidence-based management movement explains that rigorous research can and should be very relevant. However, some scholars refer to Stokes' (1997) examples from the natural sciences; that in his research, Pasteur focused on both "understanding and use". As in Pasteur's quadrant, Tushman and O'Reilly (2007) argued that academic research in business schools should contain both fundamental understanding and considerations of use (Bartunek and Rynes 2014).

Being aware of the identified challenges, and being influenced by these abovementioned pressures on both researchers and practitioners to conduct generative research for, we have observed a lack of empirical analysis of these two "in action". Even the co-development between the actors is usually treated as a black box, not opened up in traditional scientific reporting. We claim that in order to understand and expand co-development between partners, collaboration methods should be documented and reflected on from the perspective of learning and tool creation. Previous research in the field of management sciences has revealed that scientific findings hardly ever convert easily into use, at least not by practitioners reading scientific articles (Rynes et al. 2007). This kind of learning from scientific evidence, although not being a part of knowledge production itself, somehow treats the practitioners as mere passive consumers of scientific knowledge. Therefore, researcher roles as activating practitioners to become co-innovators of their own forthcoming tools should be favoured as a more generative path leading to the co-creation of novel knowledge.

Design researchers have developed inspiring participative methods and tools through which various stakeholders participate in concrete co-creation and innovation activities (Binder and Brandt 2008; Mattelmäki and Lehtonen 2006). In designing the workshops aiming at learning between researchers and practitioners, we complement our learning framework with these tools often used in developing services, such as blueprinting the service, prototyping the future service and storytelling to communicate and enrich ideas in a playful, uncritical atmosphere.

20.3 Organisational and Practice-Based Learning in Research Context

Innovations in service organisations can be recognised as dispersed, informal learning processes and knowledge creation rather than a process of technological progress or actions of a specific research and development unit (Nilsen and Ellström 2012). From the point of view of historical forms of work (Victor and Boynton 1998), we can identify that public services are developing through the phases of industrial production and craft work, mass production, process enhancement and mass customisation towards co-configuration. Co-configuration setting requires continuous learning from interactions between the user and the producers in service processes. A new type of expansive learning, which is horizontal and dialogical between activity systems, is needed in this kind of work (Engeström 2004). We believe that co-configuration will be increasingly a common type of interaction, both between public sector organisations and their users and between these organisations and researchers. In the research project that we will discuss in this chapter, we created tools that could be used in both of these interaction arenas.

Interaction between service providers and the users can also be understood in terms of the cyclical process of learning in practice-based innovation. Learning and innovation processes are seen as driven by an interplay between the explicit and implicit dimensions of work (Ellström 2010). Thus, local and tacit knowledge from service workers and users flows towards the basis of officially prescribed and formally organised, explicit knowledge and vice versa. Ellström (2010) calls the move from an abstract idea to practical action a process of adaptive or reproductive learning. Adaptive learning is about learning to handle certain tasks or to master the norms, practices and routines in an organisation. Developmental learning may in turn transform established practices in the organisation into new solutions or ways of working (Ellström 2010; Engeström 1999).

Practice-based innovation and the co-configuration phase of work need novel artefacts as mediators. In this chapter, we use Wartofsky's (1979) theory of artefacts in order to analyse their role in co-development processes. He defined artefacts as historically developed tools that have evolved in practice. The creation of artefacts can be seen as particularly typical of human beings when producing and reproducing the conditions of species' existence or survival (1973: 200). In a sense, artefacts transmit the mode of action across generations. Primary (*what*) artefacts are directly used in production. Secondary (*how, why*) artefacts are representations that transmit or preserve the skills or praxis in the production as well as the use of primary artefacts. They are externally embodied representations in socially shared practices containing how and why something should function. Tertiary (*where-to*) artefacts can be defined as imaginative, future-orientated representations of possibilities, going beyond present actualities. They represent visions and anticipated changes that may be used to change the world (Wartofsky 1973). The artefact is thus something that is produced jointly between the researcher and the practitioner in the research-action process. However, it also becomes a resource for learning and identifying future problems (Cerf 2011). As Cerf (2011) points out, this process is proceeding in constant uncertainty of its outcome. However, trust and intersubjectivity between partners, belief that the process will produce something new, is needed to keep it going. The theory of primary, secondary and tertiary artefacts is used to expand understanding of how the co-innovation tool was co-developed as an interplay between researchers and practitioners. The project as a whole can be seen as an iterative learning and co-creative process, gradually constructing the co-innovation tool for the public sector, as well as scientific artefacts for the researchers.

Science's typical learning process is conceptualisation and movement from concrete (observations) to abstract (concepts) and vice versa. According to Roth and Hwang (2006), who has carried out ethnographic research among mathematicians, rather than abstracting from something (e.g. constituting a class that connects separate empirical objects together), abstraction and concretisation occur at the same instance. "They are two moments of the same process engendered in the repeated self-movement of the conscious mind from itself to the object and back" (Roth and Hwang 2006: 324). In our study, the process of making abstraction and concretisation is even more complex, because it takes place in a complex social context: among researchers from different disciplines and between researchers and practitioners from different practice contexts.

In the following, we introduce the Inno-wellbeing project and its analysis as a co-development process, which created practical benefits, a generalised co-innovation model and scientific publications. It is unfolded as a learning process through interactive learning arenas and as a co-construction process towards a materialised form of a co-innovation tool.

20.4 Case of Inno-Wellbeing Project as an Example of Co-development and Retooling

20.4.1 Aims and Partners of the Project

The project background is built on the widely shared observation that the inefficient implementation and slow distribution of innovations present problems for the public sector, which at the same time must improve productivity and maintain personnel wellbeing and motivation. New innovation management practices are essential for new solutions when producing public services. To date, service innovation and personnel wellbeing have been treated separately in leadership discourses, but focus on employee engagement is crucial for service innovation. The aim of the Inno-wellbeing project was to find ways to combine the processes of innovation management, personnel management and personnel wellbeing.

Through the research project, public sector leaders and employees were expected to gain new insights into employee- and user-driven innovation management. We also aimed to provide recommendations on how to develop innovation activities, processes and management while simultaneously improving productivity and wellbeing. Participants were expected to gain new theoretical and practical knowledge regarding innovation management in the public sector.

The research focused on seven public organisations in Finland, from various fields of public services, for example, municipal organisations, state-owned companies and civil service departments. The study cases were chosen due to their advancement in employee-driven innovation practices. The project applied various qualitative research methods and aimed at creating a network for mutual learning between case organisations and a wider network of public sector organisations.

From seven public service organisations, we picked 16 cases for analysis. We used multiple data collection methods to form rich understanding of the cases. The data was collected during 2010–2012/2013 and comprised of 113 semi-structured interviews, ethnographic observations and documents. The interviewees represented all hierarchical groups and nearly all organisational roles: from top management to front-line service employees. A few service users were also interviewed. The division of cases and their service contexts are presented in Table 20.1.

The 3-year project was conducted by a consortium consisting of the Finnish Institute of Occupational Health (FIOH), the Aalto University School of Science and the Technical Research Centre of Finland (VTT).

The 16 cases represented different phases of an innovation process. Inspired by expansive learning theory and practice-based learning concepts, we identified four phases in these processes: (1) cultivating, (2) resource integrating, (3) coordinating and (4) generalising. The analysis of the 16 innovation process cases in terms of their phase in trajectories of learning has been reported in another paper (Hasu et al. 2015a, b).

Table 20.1 Practitioner organisations and their various service contexts

Public organisation	Service area
Finnish Tax Administration	Corporate taxation
City of Vantaa	Children and youth
City of Hämeenlinna	Children Elderly
City of Oulu	Elderly
Hospital District of Helsinki and Uusimaa, Service Centre HUS-Servis	Support services
City of Helsinki, Service Centre Palmia	Catering/cleaning
Agrifood Research Finland	Applied research

20.4.2 Data and Methods for Studying Co-development Process

We analysed how the phases of co-development between the researchers and practitioner organisations took place during the Inno-wellbeing project. We organised four practical phases of co-development and interaction, which expanded the scope of learning from a local level to a wider context: (1) *feedback workshops* for each organisation about the preliminary analyses; (2) *learning workshops*, during which the researchers set up a learning situation in which the organisations would learn from each other; (3) *a prototype testing event* for the co-innovation tool for all the organisations together; and (4) *a developmental evaluation* of all partners, organised a few months after the end of the project, in which the impact of the practical and epistemic artefacts as an outcome of the project were pondered and evaluated. All but Phase 2 were video recorded. Notes on the researchers' reflections in each phase were collected. The authors of this paper are the researchers of the studied project; thus, part of the analysis consists of self-reflection on promises and pitfalls after the project.

The points of interaction between researchers and organisations were modelled with the aid of practice-based and organisational learning theories. In this chapter, we first describe how the co-development between researchers and organisations emerged as a reflective narrative of the entire research process. Then we analyse at two illustrative vignettes of what researchers and practitioners encountered in the shared learning arena and how they co-developed elements of new knowledge and artefacts. We use the service-blueprint method to model these processes. The interdependency of the epistemic artefacts and the creation of practical tools are analysed with the aid of Wartofsky's theory on artefacts.

Fig. 20.1 Co-development between researchers and practitioners producing why, what, how and where-to artefacts

20.5 Findings: Analysis of Co-development Process and Tool Creation

Figure 20.1 depicts the phases of the Inno-wellbeing project as two intersecting learning trajectories of the practitioners and researchers. For the sake of clarity and simplicity, the practitioner organisations are described as following only one trajectory. However, all seven organisations evolved at their own pace in terms of their own learning and tool creation. The Fig. 20.1 focuses on outlining learning events in which shared knowledge creation and tool-(artefact) making took place facilitated by the researchers.

Interestingly enough, both practitioners and researchers aim for the dual object of activity. Practitioner organisations tackle how to produce efficient services and simultaneously renew them. Application-orientated researchers try to simultaneously create useful tools for the practitioners and gain scientific understanding of the studied phenomena. Both activities thus live in continuous tension-laden circumstances, trying to synchronise these dual motives.

20.5.1 Research Plan as an Embryo of All the Knowledge Creation Artefacts

Before the collaboration began, every practitioner organisation and their researchers had to find compatible motives. A negotiation was organised in each practitioner organisation on why innovation management, innovations and wellbeing should be studied in their organisation. In these negotiations, preliminary ideas as regards the focus of study in each organisation were discussed. Usually it was fairly easy to identify several employee- or user-driven innovation activities in each organisation—the difficulty was to choose between them. On the basis of the preliminary ideas, the organisations had to write a short, one-page commitment letter to the research funding agency, as well as invest a small amount of money in the project. This required arguing the benefits of the forthcoming project to their top directors, a process that remained hidden from the researchers. The researchers wrote the research plan, which was mainly based on the theoretical need to combine the processes of innovation management, personnel management and personnel wellbeing. In a sense, the research plan is a future-oriented artefact itself that anticipates why the research should be carried out, what is to be studied, how new (practical and theoretical) knowledge is created and where it all leads us. It contains all the artefacts Wartofsky refers to as being in their embryonic form. (And, by discovering this we understand why constructing a research plan is currently such an enormous, challenging job!)

20.5.2 What and How Artefacts in Gaining Case-Specific Data and Understanding of Innovation Process

The researchers started collecting ethnographic data in seven organisations, dividing themselves into pairs or triads so that each case was studied by researchers from at least two organisations. The data from each promising employee- or user-driven innovation or innovation practice was brought to a database shared by the researchers. For the researchers, the multiple data functioned as a primary (*what*) artefact in their research activity.

When the researchers started to analyse and combine different perspectives of the interviewed actors into a coherent narrative or description, the data turned into a secondary (*how*) artefact. The first material form of the *how* artefact was a power point slideshow, which the researchers prepared for the first collaborative learning event between practitioners and researchers.

The first actual co-development method between researchers and practitioners was *case feedback*, which was organised in each organisation as a half-day workshop. The researchers presented the analysis of each identified employee- or user-driven innovation process or innovation practice. If there were two cases in one organisation, the group work enhanced knowledge transfer and learning between them. For the practitioners, the preliminary case descriptions, constructed from various actors' interviews, from employees to top managers, unfolded the multi-faceted process as a whole and turned it from dispersed implicit knowledge into explicit knowledge. The case descriptions worked as *how* artefacts, integrating knowledge of the practitioners' partial views on the case. A detailed example of how case feedback from one of the organisations worked as a reflective arena between researchers and practitioners will follow later in the text (see Vignette 1).

20.5.3 Learning Across Service Contexts and Between Researchers

The second set of co-development methods aimed for *learning across organisations*. In this phase of the project, tentative descriptions of each of the 16 cases were written. The project group organised a public seminar in which a method of a scripted panel was used for making the role of each stakeholder in the innovation process visible. The researchers prepared the panel by choosing *a model case* and made a script that would reveal the chronological evolution of the innovation studied. The script contained only questions prepared for different stakeholders of the process. The stakeholders were invited to the front to answer the questions from their own perspective, and this mixture of a play and a panel was then presented in front of the audience. The richness of the viewpoints of the different stakeholders regarding the same innovation, as well as the chronological phases, formed a collectively remembered narrative. The event was video recorded and later used as material for a video, presenting a *how artefact* for an even wider audience. Later on in the project, videos were also made to describe the construction process of the co-innovation tool and finally to illustrate employees and middle managers as innovators.

Another learning event across organisations was *pair learning* between practitioner organisations. The researchers organised workshops for the practitioners, in which the cases were set in complementary pairs in such a way that they could learn innovative practices or service ideas from each other. The key innovators in each case were asked to present highlights of their innovation case to the other organisation and vice versa. Then they returned to their own original groups and tried to apply the lessons learnt. The researchers worked as memo writers and observed what was learnt.

In addition, in these learning workshops, the researchers constructed a table of theses or working principles derived from successful cases for the practitioners. This method, and how the theses were used later on as material in a workbook for employee innovators, is explained in more detail later in the text (see Vignette 2).

The internal seminars of the interdisciplinary research group were also an arena for co-development between researchers from different disciplines: psychology, adult education, engineering and economics. During the data collection, analysis and co-development, the researchers worked without the practitioners, figuring out the elements, structure and functioning of the co-innovation model. The process was managed by the project leader, and the arena for this was called model seminars. Researchers organised nine 1-day seminars from 2010 to 2011, which started by digging into the theoretical literature on public sector leadership, work-related wellbeing and service innovation. After the collection of the empirical data and the practitioners' learning events, the seminars became more practice-driven reflections on the cases and aiming at generalisation. Researchers organised ten of these model building seminars from 2012 to 2013.

20.5.4 Creating Where-to Artefacts for the Practitioners

In the early phases, there was tangible anxiety among researchers whenever they discussed the future idea and form of the co-innovation tool. But suddenly the pieces of the puzzle seemed to fit each other. The researchers realised that all they had co-developed with the practitioners—case descriptions, theses on innovation management, and the tools they had either found in the practitioner organisations or developed during the research project—could be and should be the content of the co-innovation model. The slogan for the first annual seminar: "Everyone is needed to complete an innovation!", became a vision for the model, and although it sounded simple and even naïve, it contained something that did not yet exist in the majority of public sector organisations. The future vision was that leaders should foster innovations from below, from employees and citizens, and not only implement strategies from above. Using Wartofsky's definitions of artefacts, the visionary slogan worked as a *where-to* artefact in the project.

The researchers decided that the media for the co-innovation tool should be a website, but this had to be tested as a tool in the public sector context before its dissemination. The project manager encouraged the researchers to write the case descriptions and tools in a reader-friendly way, and herself created an extra element that contained the roles of the different actors who should be involved: managers, supervisors, employees and HR developers.

The co-development method in this third phase of the process was a prototype piloting seminar for the co-innovation tool. The researchers invited all the participant organisations to co-innovate a future phase of Finnish public sector. The participants were given a playful task to co-innovate the future of the Finnish public sector. They were encouraged to use a prototype of the tool when carrying out this task. The co-innovation tool was in written format (a 70-page booklet), available for the group work. Different groups designed future ideas for different services: health care services, tax services, children's services and support services. The researchers were observers in the groups and were not allowed to guide the discussions. The model prototype, despite its clumsy written form, seemed to inspire some groups but was totally neglected by others. The most interesting moment was when the manager representatives from the groups gathered for a board meeting. The chair, a manager practitioner, concluded all the group talk by drafting and drawing a future model for innovation management referring to the important actors who should be the active actors of innovations: shop-floor workers and clients. Although she was in some way repeating the message that the researchers had earlier observed in the cases, it appeared as if she was doing it in her own way, based on her personal insight at that moment.

After the prototype piloting, the researchers edited the contents of the model into web pages. When wondering how to persuade shop-floor workers to use the web-based model, the researchers had an idea to design a printed workbook that would work like an easy interface and introduction to the web model. In the final seminar of the project, the outcome was published: the web-based model, the workbook for employee innovators and a video of the prototype seminar.

20.5.5 Creating Where-to Artefacts for the Researchers and Practitioners: Ending the Co-development?

Six months after the Inno-wellbeing project ended, the researchers organised the last learning event for the practitioners. A *developmental evaluation* took place from four perspectives: what kind of societal impact the project had created, what had been learnt in the practitioner organisations, what kind of research knowledge had been created and what kinds of project ideas had been derived from the Inno-wellbeing project. The researchers presented a chart of alternative future directions for research. The workbook for employee innovators had already become a tool and a focus in the following intervention project concerning the developmental agency of employees in elderly care services. The reflections in the evaluation seminar tried to construct a *where-to* artefact. However, in writing their scientific articles, the researchers were confined by several scientific discussions, thus constructing deeper theoretically grounded *why* artefacts. New research plans also pushed researchers to think of new research questions. The trustful relationships that were created with the practitioners during the project formed a solid basis for new projects.

20.6 Summary

The artefacts that were created during the Inno-wellbeing project are summarised in Table 20.2. Each phase leveraged the use value of the co-innovation model under construction.

The Inno-wellbeing project was a multi-phased co-development process between researchers and practitioners aiming at retooling the public service employees into their co-configurative future work in which innovating with their

Table 20.2 Co-development methods and artefacts created in the Inno-wellbeing project

Co-development	Artefact for praxis	Artefact for research	Use value
Negotiations of cases	WHY: A commitment paper for the project	WHY: Research plan	Creating shared commitment and trust
Data collecting	WHAT: Consent to be interviewed	WHAT: Data	Practical involvement in gaining knowledge embedded in practice
Case feedback	HOW: Analysis in slide show	HOW: Case descriptions	Practice-specific knowledge creation, validation, learning from own practice
Learning across organisations – Model cases – Pair learning – Theses – Data and model seminars	HOW: Video of co-created narrative of an innovation as an example Theses as principles of innovation management Group work notes on what to learn from each other's innovations	WHERE-TO: "Everyone is needed to complete an innovation"—slogan HOW: Cases as sources of understanding the phenomena of innovation management HOW: Theses as building bricks in co-innovation model	Conceptualising towards generalised knowledge, learning between practice contexts, preparing model prototype for retooling public sector innovations
Prototype pilot seminar	WHERE-TO: A vision of the Finnish public sector	HOW: Feedback on how the co-innovation tool could be used	Imaginary and playfully testing the model in action
Reader feedback	HOW: A web-based model containing all the artefacts HOW: A workbook for employee innovators		Improving the model for future use, spreading it to new users
Developmental evaluation	WHERE-TO: Evaluation of the use of the co-innovation model—four perspectives in group work	WHERE-TO: A chart of future projects WHY: Scientific articles and new research plans	Back to the purpose of the model, contextual use-value reflections

customers could be part of their mundane work. The material outcome from each co-development phase formed the co-innovation model (www.inno-vointi.fi) for the public sector actors. It included case descriptions, participatory tools, new role descriptions, videos, blogs and instructions how to use innovators' work book to be used as tools in future learning events. We now investigate two vignettes from this process as examples on how practical and theoretical insights were intertwined during the project.

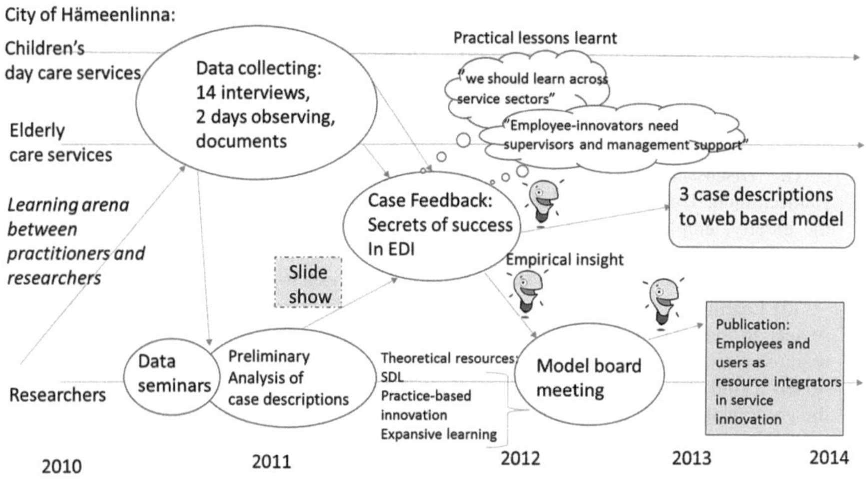

Fig. 20.2 An example of researchers' and practitioners' learning path

20.6.1 Vignette 1: From Data to Theoretical Constructions

We now unfold how interaction between researchers and one of the practitioner organisations, namely, the City of Hämeenlinna, produced practical insights into and ideas as to how the empirical cases would contribute to theoretical discussions and finally a scientific publication. (For the case of clarity, only one path leading to a single co-authored scientific publication was chosen here as the focus, although simultaneous parallel processes existed in reality. The service innovation cases in the City of Hämeenlinna have so far produced at least three other scientific manuscripts.) The following rough service-blueprint model provides a tool for making this path visible (Fig. 20.2).

In the early project negotiations in October 2010, the middle managers suggested that innovation processes in elderly care services and in children's day care would be interesting targets for the Inno-wellbeing project. It was agreed that the researchers could decide on the specific processes and local sites after first interviewing the managers and policy-level actors in the City. Although the top level interviews tended to be loaded with strategy talk, the managers could identify which employee innovators should be interviewed. Thus the choice of what to study was specified (and extended) during the course of interviews.

From 2010 to the end of 2011, the researchers conducted 15 interviews altogether, observed for 2 days, and collected about 50 documents related to service innovating in elderly care and children's day care.

Researchers held several data seminars among themselves to make sense of the data. The feedback workshop for the practitioners pushed the researchers to formulate their preliminary findings into a form that could be easily presented and discussed. Researchers designed the feedback workshops according to a common structure. In Hämeenlinna, the structure consisted of the following:

(1) Overview of the entire project and presentation of findings concerning issues that were seen as hindering or enhancing (employee- and user-driven) development work at the seven organisations.

(2) Discussing four different employee-driven innovation cases identified at Hämeenlinna. Three of these derived from elderly care: segmented open-day activity for the elderly, employee-driven innovation in project form and involving employees in strategy-driven development. The last case derived from early childhood education and showed how the Forest preschool evolved from a novel idea to a city-wide service concept.

(3) Learning and generalising from the cases in group work sessions: what were the secrets to success in the studied employee-driven innovation models? This question required seeing local observations on a more general level, thus demanding of the practitioners a kind of pre-generalisation capability. The researchers worked only as observers in the group work. Comparison and learning across service sectors was enhanced.

These differences in employee's coping, in management and in the societal circumstances of the services provoked new questions in the workshop. In the concluding discussion, the practitioners and researchers tried to determine together how employee- and user-driven innovation could be made easier in organisations. The following excerpt demonstrates a discussion on the role of the power and trust of individual innovators and managerial actions.

Innovator 1: "If I think it (an endeavour to create something new) will succeed, it helps me carry on, compared to if I think it won't succeed anyway".

Innovator 2: "I'd like to add, that if I face obstacles, I think I can overcome them. And when I do, I'm empowered".

Top manager: "May I ask, where do you get this enormous power? You two (laughing)".

Innovator 2: "I believe that I've achieved something good. I have to believe in that".

Innovator 1: "I start something like a brave, mad person, then when I succeed, it carries me forward..."

Researcher: "Although I wasn't previously familiar with the elderly care cases, it appears to me that these innovations have received very different social support. I understand that the Forest preschool has been supported much more by the management, while open elderly care services are only gradually approaching the question of how to spread the innovation. The Forest preschool was probably in this fragile situation earlier too. If management had decided to give up all our specialised services and provide families with only one standard service, then its story would've ended. So even one action from the manager, deciding that—yes this is something we will start to spread!—is a very important part of innovation management. So innovations don't only depend on individuals, interplay between hierarchical levels is needed as well".

Innovator 1: "Yes (Nodding all the time during the speech turn. Begins to give an example of continuous piloting, which did not lead to implementation)". (video from case feedback in Hämeenlinna 24.11.2011)

The differences in the case suddenly became visible to both practitioners and researchers, as one of the researcher's notes reveal:

"Group work that dealt with two similar kinds of employee-driven innovations across children's day care services and open day services for the elderly, was very fruitful. It appeared to me that the employee-innovators from elderly care and early childhood services were very similar as people; energetic, positive, socially capable, and knowledgeable. However, their experiences of how the surrounding organisation, work community and leadership had begun to support their innovative efforts were totally different. The elderly care worker had experienced tough times; when she focused on planning "fun" activities for the elderly, she had faced prejudices from the nurses' side; her supervisors had changed continuously, 17 times in 11 years; and her job was constantly on a temporary basis".

"On the other hand, the kindergarten teacher received support and care from her superiors. The Forest preschool was developed in peace over 4–5 years, and now even the top management has taken powerful actions to spread the novel service concept into other areas of the town. Although there have been many obstacles, she has coped with them as challenges that she could overcome" (Field report, group work 1, the reflective part 11/11/24).

These reflections were e-mailed to the research group after the workshop. The researcher had inspiring discussions with the project leader after the case feedback, as reported shortly in an email:

> "The workshops in the organisations were successful, and we researchers acquired a new kind of realisation, and a taste of being in dialogue with our practitioner organisations. I just spoke to M (the project leader) on the phone. We planned the next phase of comparing the cases and how our conference papers could promote this. When we analyse the innovation trajectories in parallel, we gain an opportunity for both generalising and learning from each other. We can see the significance of the innovation context, social support and management when we analyse the cases in parallel" (E-mail 25.11.2011 among researchers).

The case feedback event provided insights and "kind of brought the cases to life". When the project leader ideated a new topic for a publication in a model management meeting in the spring of 2012, the idea of using two cases for demonstrating the theories travelled from the previous discussions. It became obvious that the researchers should continue with these two innovation trajectories, but now experiment theoretical modelling.

The senior researchers realised that there seemed to be a need for bringing expansive learning and practice-based innovation discussion into the service-dominant logic discussion. The writing of a co-authored book chapter entitled Employees and users as resource integrators in service innovation (Hasu et al. 2015b) began. A more practical outcome was that the case descriptions were developed as parts of the web-based co-innovation tool. This meant a different writing process, which aimed for a readable, short description on a website.

20.6.2 Vignette 2: Practice-Driven Theses as Tools for Learning and Generalising

Figure 20.3 describes how the empirical case descriptions provided a rich source for generalising principles on good innovation management. Working with the generalisations provided learning tools for the researchers and resulted in both scientific and practical outcomes, as the following figure and narrative indicate.

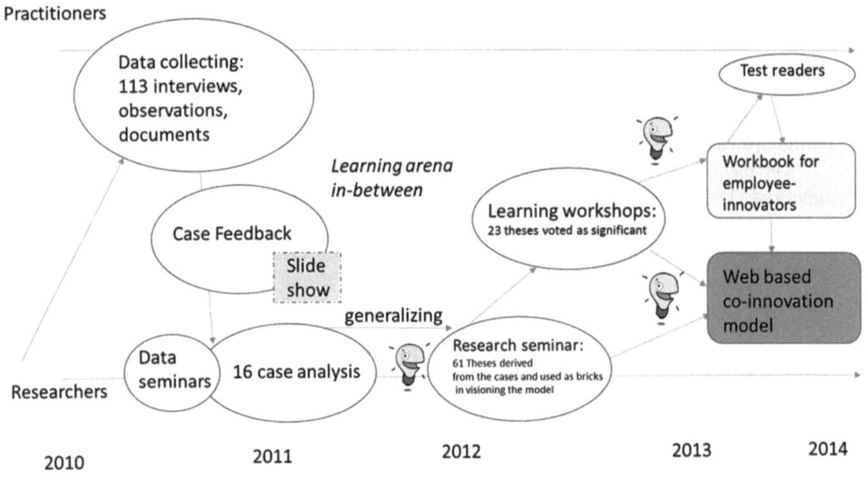

Fig. 20.3 Theses as tools of learning and generalising

After collecting the data from seven different organisations, the researchers started to analyse the case studies. They had to find a way to discuss the findings, interpretations and insights from them together. One way that was agreed upon was to search for key actors, actions and activities that promoted the innovations from each case. In addition, in order to present the cases to each other, the project manager asked them to form 1–3 essential theses from each 16 case descriptions. This phase can be defined as generalising from rich empirical data. The theses included the main findings of the entire study in a very concise form. One- or two-sentence long normative arguments or rules of thumb contained lessons learnt from each case, for example, slogans on how to succeed in an innovation process, such as *Let your staff experiment and don't worry too much about obstacles or failures in advance*, or on how to enhance innovation or the prerequisites of innovative behaviour, such as *Increase your expertise by making visits to other workplaces. Then apply lessons learnt to your own work.*

The researchers gathered together for a 3-day workshop in Lapland, which provided an opportunity to devote all their attention to the data analysis. There, the researchers presented the case descriptions and action analyses to each other and then formulated the theses from the cases together. The theses were clarified, defended and sometimes even reformulated as a collective effort. In the research workshop, each thesis was written on a separate paper. At this stage, altogether 61 theses were developed.

The researchers were divided into two teams (3 and 4 persons) and used the theses as material to envisage the vision of the co-innovation tool. The teams prepared two competitive visualisations with a marketing speech. This was a playful, cheerful session, but also demanded hard joint thinking about the main plot, purpose and form of the tool. The theses provided an opportunity to discuss the challenges of innovation management at the general level, but represented abstractions of the empirical findings. Both teams used the theses as tools or "building bricks" to reach a meta-level abstraction. The theses provided a chance to perceive the concrete empirical findings and generalisations at the same time (see Roth and Hwang 2006). The aim was to visualise a model for the practitioners, to guide them to the future phase of innovation management in the public sector. This collective reorganising work with the theses nourished the idea that the co-innovation tool should contain abstractions, rich contextual descriptions and co-development methods. An ambitious idea was that the different practitioners could use all these as material, inspiration and a kind of learning platform and source to start co-innovating in their own networks, in the same way as the researchers had experimented among themselves.

After the seminar, the researchers decided to create a common Excel table including all the theses and to continue refining them by (1) pondering whose activities this may concern (employee, supervisor, manager or user), (2) determining the case sources and (3) finding what research literature repeats the finding or related theoretical background. It was decided that the theses were to be material for the practitioners' learning workshops. Thus, afterwards, the practitioners' comments and votes were added to the same table.

In the practitioners' learning workshops, group work was carried out for all the 61 theses. The theses were printed on separate papers and arranged in random order on tables in a large workshop room. Each practitioner chose one favourite thesis: the one most meaningful for him/her at that moment. In all, 23 theses were picked as favourites. Then each practitioner presented his/her favourite, explained why this was so meaningful and also gave examples of how the thesis manifests in their everyday work. The other practitioners commented and gave more examples of the same theme. All in all, this procedure was once again something between the abstract (thesis) and the concrete (my own experience of such actions in my own organisation), activating a sort of innovation capability among the practitioners.

At this point of the project, the researchers had to decide on how to best utilise the theses. One of the main themes of the whole research project was employee-driven innovation. Therefore, they decided to use the theses in an innovation-enhancing workbook for employees. Negotiations with the publisher began, and the size of the workbook (approx. 40 pages = approx. 20 theses) was decided.

The theses were arranged by themes, e.g. activating the employee, user participation, the value of target setting, etc. Four researchers formed two pairs and the themes were divided between pairs. The pairs selected the most central theses from each theme (according to all the material collected on the theses: examples, votes, theories, applicability).

The influence of the practitioners was shown in two ways. For example, thesis number 17: *Even a minor improvement is better than no improvement at all*, was chosen for the workbook: although the researchers had not written any scientific reference to prove it, it had gained three votes in the learning workshop. However, some theses that did not receive any votes in the learning workshops were also chosen, e.g. Thesis number 1. *The societal need and purpose of work should be explicitly expressed.* This finding was derived from two case studies and was recognised as being explicitly communicated and measured by the middle and top managers. For example, the City of Vantaa set a target to reduce the risk of exclusion from secondary education or work life by stating the aim: (we should have) zero outcast pupils after primary school. The voice of the praxis was thus embedded in the theses in two ways.

The researcher pairs designed the first draft of the workbook. Each two-page spread included a thesis, a case example or theoretical information and a very practical exercise(s) on how to apply this theme to one's own job. Three test users from the target group of public service workers (a school social worker, a public services advisor and a practical nurse student) commented on the first draft. The feedback was mainly encouraging, but there were also some critical comments. The draft was carefully edited on the basis of the feedback. The rest of the publishing process (linguistic editing, graphics, printing, marketing) was then completed.

A few months after the Inno-wellbeing project ended, a developmental evaluation between the researchers and practitioners was organised. Mostly middle and senior managers participated in the seminar. The participants were asked to comment on the impact of the whole research project and its practical outcome, including the workbook. Some of the participants made rather critical comments about the workbook: it was considered too "light". Others had not distributed the workbooks in their organisation at all. The researchers were disappointed with the feedback.

However, the workbook was launched in new research fields: to four elderly care service organisations, to further study if it would encourage shop-floor workers to develop their own work. Already in the early negotiations with the research sites, the attitudes regarding the workbook were enthusiastic and curious (Fig. 20.4).

Finally, the researchers prioritised the refined theses. The five core principles, which were modified for the web-based model, were the following: (1) make the relevancy of the renewal visible, (2) involve the employees, (3) involve the user, (4) enhance encounters and (5) experiment.

20.7 Conclusions and Discussion

Although we live in a knowledge society in which both increasing amount of evidence-based and experience-based new knowledge is easily reached via the Internet, it does not solve the problem of how this knowledge becomes a shared understanding and how it could be converted into use, e.g. mobilise a new kind of employee-driven innovation management. As our chapter indicated, scientific knowledge can be observed as co-development of knowledge tools, artefacts as we called, between researchers and practitioners. We analysed how the co-development process between researchers and multiple practitioner organisations was planned and implemented. In doing this, we opened the position of researchers in a new way, as tool-makers (not just scientific fact-makers) for leveraging the practitioners' understanding of their present activity and retooling

Fig. 20.4 Published innovators' workbook

their actions for the future. As argued by the management scientists (Bartunek and Rynes 2014), these dialogical processes are rarely revealed and analysed as empirical cases; they are often later only speculatively reflected on. This is understandable, because collecting data on how the research is actually conducted as a collective sensemaking and knowledge creation process by researchers and practitioners is hardly ever designed as part of the study. This paper gives an example of how analysing artefacts produced for either practitioners or for the scientific community provides important points of departure to reflect on. The valuable point is not in showing how well we accomplished the co-development itself but in revealing the research process as a collaborative tool-making process aiming for a meta-tool for the practitioners.

In the Inno-wellbeing project, the co-development process with the practitioners was planned very roughly in the research plan. Although two-stage feedback seminars and learning opportunities were promised to the practitioners, the researchers did not know in advance how these seminars should be organised and what their outcome would be. The co-development was a truly creative process, which inspired the designing of the forthcoming phase. The researchers, who came from different disciplines, often had very different ideas at the beginning on how the workshops should be organised and what would be learnt from the outcome. As the analysis of the deviating learning trajectories of the researchers and practitioners indicated, we agree with Cerf (2011) in that the co-development of artefacts in an action-research process is always a temporary phase, because the practitioners and researchers stick to their own professional worlds. This makes intersubjective skills and trust between partners particularly important throughout the process, although our analysis did not yet empirically reach these nuances.

The first vignette illustrated how a learning event between researchers and practitioners produced both practical local insights among the practitioners and inspired the researchers to generalise the phenomena. The second vignette showed how practice-driven theses, and playing with classifying and prioritising them, led to the idea of using them as the core of a workbook for employee innovators. Without the co-development between researchers and practitioners, neither the practical tool nor the scientific article idea would have emerged. This indicates that in an action-research process, local practical tool development is interconnected with theoretical concept formation, as well as the development of a generalised practical tool (see also Cerf 2011).

This paper is a representation of the new mode of research activity and should carry a message about its complexity and fragility for funding agencies as well. There is a tendency that in application-orientated research, researchers are expected to foresee and market bold visions of future innovations, which solve societal problems and involve all the actors needed in the value creation. However, organising an even partially successful learning process between actors is a demanding effort that consists of constant risk of misunderstandings, communication gaps, repetition and frustration, but also fortunately insights and outstanding results. In other words, co-development does not usually go as planned; it is always a surprising, creative and unpredicted process. Even co-created artefacts, such as the web-based co-innovation model, often remain underused if they do not become part of the new activity either in practice or in research. Retooling practical activities is a continuous open-ended process enhancing renewal of the activities. In such a way, the new co-developed tools benefit partners from the organisations more than pieces of new understanding in the form of scientific facts published in journals.

The funding agencies of the applied research should realise that learning among partners is not something that just happens; organising the communication and collaboration takes time and involvement. However, even then, there is no guarantee of a sustainable outcome and of how the practical and epistemic artefacts travel into wider use. Co-developed artefacts, such as the project innovators' workbook, may become a tool and solution for new audiences (lacking employee empowerment) rather than for the practitioners that were involved in developing it. Developmental impact evaluation (Saari and Kallio 2011) after the project has officially ended and is a potential method for becoming aware of problems, but also for finding new paths for re-innovating and implementing the co-developed tool with new potential partners.

References

Amabile, T. M., Patterson, C., Mueller, J., Wojcik, T., Kramer, S. J., Odomirok, P. W., & Marsh, M. (2001). Academic-practitioner collaboration in management research: A case of cross-profession collaboration. *Academy of Management Journal, 44*(2), 418–431.

Bartunek, J. M., & Rynes, S. L. (2014). Academics and practitioners are alike and unlike: The paradoxes of academic-practitioner relationships. *Journal of Management, 40*(5), 1181–1201.
Binder, T., & Brandt, E. (2008). The design: Lab as platform in participatory design research. *CoDesign. International Journal of CoDesign in Design and the Arts, 4*(2), 115–129.
Cerf, M. (2011). Is participatory research a scientific practice? *Journal of Rural Studies, 27*(4), 414–418.
Ellström, P.-E. (2010). Practice-based innovation: A learning perspective. *Journal of Workplace Learning, 22*(1/2), 27–40.
Engeström, Y. (1999). Innovative learning in work teams: Analyzing cycles of knowledge creation in practice. In Y. Engeström, R. Miettinen, & R.-L. Punamäki (Eds.), *Perspectives on activity theory* (pp. 377–404). Cambridge: Cambridge University Press.
Engeström, Y. (2004). New forms of learning in co-configuration work. *Journal of Workplace Learning, 16*(½), 11–21.
Fujimura, J. H. (1987). Constructing 'do-able' problems in cancer research: Articulating alignment. *Social Studies of Science, 17*(2), 257–293.
Hasu, M. (2005). In search of sensitive ethnography of change: Tracing the invisible handoffs from technology developers to users, mind, culture, and activity. *An International Journal, 12*(2), 90–112.
Hasu, M., & Miettinen, R. (2006). *Dialogue and intervention in science and technology studies: Whose point of view?* (Working Papers 35/2006). University of Helsinki, Center for Activity Theory and Developmental Work Research.
Hasu, M., Saari, E., Honkaniemi, L., Tuominen, T., Lehtonen, M. H., Kallio, K., & Toivonen, M. (2015a). Trajectories of learning in practice-based innovation—Organizational roles at play in sustainable innovation management. In M. Elg, P.-E. Ellström, M. Klofsten, & M. Tillmar (Eds.), *Sustainable development in organizations* (pp. 127–152). Cheltenham: Edward Elgar.
Hasu, M., Toivonen, M., Tuominen, T., & Saari, E. (2015b). Employees and users as resource integrators in service innovation—A learning framework. In R. Agarwal, W. Selen, R. Green, & G. Roos (Eds.), *Handbook of service innovation* (pp. 169–192). London: Springer.
Knorr Cetina, K. D. (2001). Objectual practice. In T. R. Schatzki, K. Knorr Cetina, & E. von Savigny (Eds.), *The practice turn in contemporary theory* (pp. 175–188). London: Routledge.
Labinger, J. A. (1995). Science as Culture: A view from the petri dish. *Social Studies of Science, 25*(2), 285–306.
Latour, B., & Woolgar, S. (1979). *Laboratory life. The social construction of scientific facts.* Beverly Hills, CA: Sage.
Lynch, M. (1985). *Art and artifact in laboratory science. A study of shop work and shop talk in a research laboratory*. London: Routledge& Kegan Paul.
Mattelmäki, T., & Lehtonen, T. (2006). Designing alternative arrangement for ageing workers. In *Proceedings of the Participatory Design Conference* (pp. 101–104). Palo Alto, CA: CPSR.
Nilsen, P., & Ellström, P. E. (2012). Fostering practice-based innovation through reflection at work. In H. Melkas & V. Harmaakorpi (Eds.), *Practice-based innovation: Insights, applications and policy implications* (pp. 155–172). Berlin: Springer.
Roth, W.-M., & Hwang, S. W. (2006). On the relation of abstract and concrete in scientists' graph interpretations: A case study. *Journal of Mathematical Behavior, 25*(4), 318–333.
Rousseau, D. M. (2006). Is there such a thing as "evidence-based management"? *Academy of Management Review, 31*, 256–269.
Rynes, S. L., Bartunek, J. M., & Daft, R. L. (2001). Across the great divide: Knowledge creation and transfer between practitioners and academics. *Academy of Management Journal, 44*(2), 340–355.
Rynes, S. L., Giluk, T. L., & Brown, K. G. (2007). The very separate worlds of academic and practitioner periodicals in human resource management: Implications for evidence-based management. *Academy of Management Journal, 50*(5), 987–1008.
Saari, E., & Kallio, K. (2011). Developmental impact evaluation for facilitating learning in innovation network. *American Journal of Evaluation, 32*(2), 227–245.

Segeståhle, U. (1993). Bringing the scientists back in: The need for an alternative sociology of scientific knowledge. In T. Brante, S. Fuller, & W. Lynch (Eds.), *Controversial science. From content to contention*. New York: State University of New York Press.
Stokes, D. E. (1997). *Pasteur's quadrant: Basic science and technological innovation*. Washington, DC: Brookings Institution Press.
Sundbo, J., & Toivonen, M. (2011). *User-based innovation in services*. Cheltenham: Edward Eldgar Publishing.
Toivonen, M., & Tuominen, T. (2009). Emergence of innovations in services. *The Service Industries Journal, 29*(7), 887–902.
Tushman, M. L., & O'Reilly, C. A., III. (2007). Research and relevance; Implications on Pasteur's quadrant for doctoral programs and faculty development. *Academy of Management Journal, 50* (4), 769–774.
Victor, B., & Boynton, A. (1998). *Invented here: Maximizing your organization's internal growth and profitability*. Boston, MA: Harvard Business School Press.
Wartofsky, M. W. (1973). Perception, representation, and the forms of action: Towards an historical epistemology. In M. Wartofsky (Ed.), *Models: Representation and the scientific understanding*. Dordrecht: Reidel.
Wartofsky, M. W. (1979). *Models. Representation and the scientific understanding*. London: D. Reidel Publishing Company.

Chapter 21
Inter-professional Collaboration: An Evaluation Study

Monica Pedrazza, Riccardo Sartori, and Sabrina Berlanda

Abstract Both conventional wisdom and some of the literature in this area tell us that public organizations and services find it hard to innovate. We present a case study on the impact of a third sector intra-service innovation on the interservices' network of a regional child welfare agency. The innovation was implemented in 2010 and the follow-up presented in this chapter was completed in 2014. The innovation consisted of the introduction of a new tool for the direct, systematic observation of children's behaviour and attitudes. Residential youth workers, an important part of the child welfare agency's workforce, introduced the new tool in order to improve the quality of their contribution to the information flow on young people in residential care. The follow-up research allowed us to recognize that the introduction of a new tool, while improving team work performance and reducing interpersonal and team conflicts at the intra-service's educative service level, highlighted an important asymmetry between service providers in the child welfare agency network with regard to the provision of information.

21.1 Introduction

As Edvardsson et al. (2011) argue, value co-creation should be understood as value-in-social context, and it is shaped by social forces, is reproduced in social structures and can be asymmetric for the actors involved.

Mutual gain in bilateral exchanges is only one possible result. Service exchange between customers or users and service providers can be asymmetric (Edvardsson et al. 2011). We argue that asymmetries between different stakeholders can also arise in the service delivery process, where information and data have to be transferred and shared in order to achieve the common goal of innovation implementation and dissemination. The decision-making process in the public child welfare service is a complex interdisciplinary process, requiring purposeful

M. Pedrazza (✉) • R. Sartori • S. Berlanda
University of Verona, Verona, Italy
e-mail: monica.pedrazza@univr.it; riccardo.sartori@univr.it; sabrina.berlanda@univr.it

T. Russo-Spena et al. (eds.), *Innovating in Practice*,
DOI 10.1007/978-3-319-43380-6_21

interdependence between social workers, who guide and monitor the process, and all other professionals: health professionals, educational professionals and third sector care service providers. One group of residential youth workers introduced in 2010 a new tool (for development and implementation of the tool, see Chap. 10) for the systematic and direct observation of the behaviour of children or young people in residential care.

In Italian state child welfare, there is no formal and shared framework enabling these practitioners to gather, organize and share their enormous store of information on children's behaviour and attitudes. There is a structural lack of formalized tools and shared protocols.

The tool in question was introduced to improve the flow of information, and mutual understanding, within a third sector organization working for the care service (Leung 2007, 2009). The tool was implemented by the use of software because of the great amount of information it permits team members to gather. At any moment one can turn to the summary diagram which shows the trends in a child's behaviour. It came about in a planned way (Djellal et al. 2013) during a psychologist-led training course for residential youth workers and is concerned with important knowledge (Wallin and von Krogh 2010) that needs to be shared between all the actors.

The innovative development and implementation of a tool for the systematic observation of children allowed a smooth transition to improved collaboration and enhanced trust between colleagues and to better practice in the sharing of information with social workers. Our aim is multi-faceted: on the one hand we want to explore how employee-driven tool sustains prevailing professional practices, and on the other hand we want to explore which type of impact its implementation has on relationship between professionals. We therefore present the results of follow-up activities which identified the impact of innovation on the whole regional child welfare agency that is involving also health professionals and school staff.

21.2 Theoretical Background

21.2.1 Innovation in Public Services

Both conventional wisdom and some of the literature in this area tell us that public organizations and services find it hard to innovate. Furthermore innovation in the public sector has often been neglected in the mainstream of innovation studies. Public organizations are considered less autonomous than private sector organizations; they are subject to continual scrutiny (Dierkes et al. 2001; La Palombara 2001) by outsiders ready to intervene such as mass media, political parties and civil society and tend to be thereby conservative rather than innovative. Sørensen and Torfing (2012) edited a special issue of *The Innovation Journal* on collaborative innovation in public sector.

Even though AA remind us of Weber's (1922) definition of bureaucracy in terms of routines, repetitiveness and order, they clearly state that there are some important drivers of innovation in the public sector: citizen's and public opinion's pressure on politicians and well-educated employees, who want to improve social services responding to new social problems and citizen's needs. In addition new laws often require regional, local or national adjustments, which in turn often activate innovation processes. Public sector organizations tend to be reactive rather than proactive and conservative rather than innovative so that efforts to reform them fail more often than not. Changes are thus seldom embraced because of the high magnitude of inertial forces (Kaufman 1981; La Palombara 2001; Miller 1978). Furthermore, public sector employees and bodies show the propensity to selectively interact with and learn from individuals seen as similar to themselves (Rivera and Rogers 2004).

21.2.2 Public Sector Innovation: Barriers and Drivers

One of the most frequently cited barriers to innovation in public organization is concerned with the fact that they typically have to protect and handle confidential data. According to Lee et al. (2012) in fact, public services often give priority to the outside-in type of open innovation, rather than to the inside-out type (Chesbrough 2003, 2006). Also the lack of resources is identified as a barrier to the activation of innovation processes, but, along with Sørensen and Torfing's (2012) interpretation, this can be easily compensated by the huge size and complexity of public services which enables them to absorb costs and even innovation failures.

On the one hand, innovation is seen as positive despite the type of organization and resources; in fact, it often seems to legitimate and increase the perceived reputation and value of public organizations itself (De Vries et al. 2014); on the other hand, professional groupings which are typically well established in public sector are often inherently resistant against innovation (Cunningham et al. 2005).

According to the principle of subsidiarity, the extensive use of contracting out (Sørensen and Torfing 2012) increases public organizations' responsiveness to users and a more effective relationship with external private partners, stimulating thereby innovation. Scholars (Corwin et al. 2012; Cunningham et al. 2005; Dossou-Yovo and Tremblay 2012; Sørensen and Torfing 2012) address the fact that collaboration between actors, managers' and employees' communicative skills and reflexivity are crucial competences which can foster both development and adoption of innovations. The legitimation of forms of collaboration (Aagaard 2012) through formal procedures (Metze and Levelt 2012) as also the support of cross institutional collaboration between professionals from different public services may foster a virtuous circle of collaborative inter- and intra-organizational innovation.

Cunningham et al. (2005) listed an interesting number of barriers and drivers to innovation. The latter can be hindered by a shortage in variable of individual differences such as employees' relevant skills or by contextual lack of resources.

Public services are usually organized around highly specialized technocrats and technicians (i.e. physicians, social workers), who adhere to different and compartmentalized established roles and are reticent to engage in innovative practices or processes. That's why the horizontal and vertical dialogue between different professional groups remains a difficult and sometimes unachievable task for public institutions. According to Cunningham et al. (2005) and despite Sørensen and Torfing's (2012) considerations, risk aversion still hinders a large number of innovation opportunities. According to De Vries et al. (2014), innovation in public sector occurs mainly at local and regional level; in fact public institutions' dimensions, their complexity and different hierarchical layers prevent public sector from engaging in development, adoption and implementation of innovations. However, the high sensitivity to negative media exposure and the tendency towards a blame culture induce public service managers and politicians to provide resources and incentives for service improvements and innovation (Cunningham et al. 2005). On the other hand, they exert a relevant administrative and organizational pressure to enhance public service accountability (Heinrich 2002). Almost all public service innovation is also aimed at economic efficiency; it is therefore quite difficult to enact virtuous innovation processes which improve service delivery only as a function of, for example, increased practitioners' wellbeing, that is, as second-order consequences of innovation along Rogers's taxonomy (1998).

21.2.3 Innovation as Knowledge-Based Outcome

According to De Vries et al. (2014), outcomes of innovation are seldom reported because of the self-serving, pro-innovation bias affecting researchers (Rogers 2003), institutions and organization which typically rest on the increased public positive consideration accorded to them.

Rogers' s (2003) taxonomy of innovation consequences comprises desirable vs. undesirable, direct vs. indirect (i.e. first- and second-order changes) and anticipated vs. unanticipated consequences. Scholars suggest studying the innovation outcome over time, that is, after a congruent period from its adoption and implementation. Quintane et al. (2011) suggest one could benefit an individual model of knowledge, focused on tacit and explicit types (Nonaka 1994; Tsoukas and Vladimirou 2001), being innovation an artefact which contains the knowledge needed to sustain all its phases. We should therefore pay special attention to knowledge availability and to knowledge sharing practices rather than to the innovation outcome itself. An interesting aim of this book is to identify under which conditions innovation can emerge and which elements speed it; usually up innovation is thus seen as an ongoing activity which arises from a complex interaction among several individual and social context variables.

In order to ensure knowledge availability, participation (Mergel and Desouza 2013) should be granted at each innovation stage. In fact knowledge availability is also a matter of power: more attention should be therefore paid to systematic monitoring of inequalities, perceived status differences and perceived self-efficacy in the exercise of personal agency in creation, adoption, implementation and cross-functional dissemination of innovation.

21.2.4 *Innovation as New Interpretation of Work Practices*

Innovation is often perceived as new in local contexts (Ellström 2010: 28) in which it is develop, even though it's novelty may be limited in a broader context. This type of innovation is often focused on reinterpretation or new interpretation of practices. This is aligned with Pettigrew and Fenton's (2000) and Schumpeter's (1980) view that innovation can also be a new combination of known elements.

According to Ellström (2010), as also shown in Fig. 21.1, implicit work processes are autonomously interpreted relative to prescribed processes. This autonomy is often expressed in the form of subjective interpretations of work practices and subsequent potential improvisations (Point 1 in Fig. 21.1), for example, deviations from the prescribed procedures or from their meanings. Improvisations are often related to new subjective interpretations of routine or work practices.

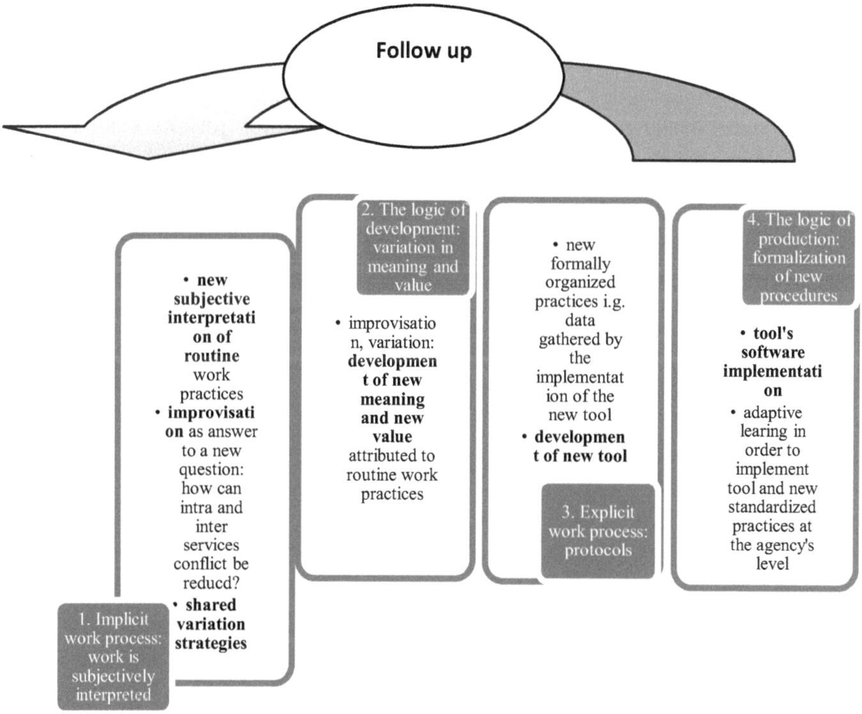

Fig. 21.1 Flow chart of the practice-based innovation and follow-up

Improvisation is crucial for any type of innovation process. In our case study practitioners offered a subjectively improvised new interpretation of their recursive evaluation practices: they assigned to it a new meaning and a subsequent new value (Point 2 in Fig. 21.1). Protocols and officially prescribed procedures are pertinent to the explicit work process. The logic of production implies that professionals need to adapt and learn in order to implement developed tools and standardized performances (Points 3 and 4 in Fig. 21.1).

As described in the previous chapter, during training sessions run by a psychologist, residential youth workers' new interpretation of their practices was identified. Practitioners focused on the origins of interpersonal and team conflicts relating to the content of narrative reports on children behaviour. The meanings and interpretations of children's behaviour differed: convergent evaluations were usually viewed positively, divergent evaluations negatively. These practitioners developed a new interpretation of the prescribed task (Point 1 in Fig. 21.1), and a new meaning was assigned to divergences: instead of considering divergent evaluations as opposite views, they chose to regard them as complementary views of the complex shared task of enhancing children's social competences (see details and discussion in Chap. 10).

In order to be able implement this new perspective emerging from their reflective activities on implicit knowledge in their recursive work practices (Points 3 and 4 in Fig. 21.1), residential youth workers developed a tool for the measurement and formal recording of children's behaviour. The innovation entailed:

- New form and structure for the recording of evaluative procedures
- New meaning assigned to the procedure for the evaluation of outcomes
- New value assigned to the content of the procedures for the evaluation of outcomes

The tool was first implemented in documentation procedures at the third sector level, involving residential youth workers and social workers, and later at the level of the agency, thereby including health professionals and school staff too.

According to the flow chart presented in Fig. 21.1, reflective training section allowed residential youth workers to ascribe a new meaning to routine work practices. This ascription is part of a subjectively interpreted procedure (Point 1 of Fig. 21.1). Subsequently, improvisation ('Hazard' as meant by Harré 1993) introduces a new meaning to prescribed practices (Point 2 in Fig. 21.1). Explicit work processes, organized in protocols and procedures, are the context in which the 'new meaning' is assigned and new tools can be implemented (Points 3 and 4 in Fig. 21.1).

According to practice-based innovation practices, the new meaning arises from practice: professionals interpret in new ways their daily work activities.

21.3 Knowledge Production and Knowledge Sharing in the Child Welfare Agency

The epistemological premises guiding knowledge production practices through the use of assessment tools vary under the following headings:

- Epistemological premises and theoretical frameworks
- Setting
- Method
- Focus of outcome evaluation

The different perspectives we summarize and report in Table 21.1 are needed by the service as a whole in order that it can gather and organize data and information on all aspects of the life and situation of the young people in its residential care. While within each service the assessments and outcome evaluations conducted are congruent with its epistemological premises and methods, we often notice that the final interservices outcome evaluation is not agreed by all those contributing: health, social, educational and care.

Tacit knowledge is linked to the implicit premises that each professional adopts, and in each case it orients interpretation practices by reference to 'what others know'. As Tsoukas and Vladimirou (2001) argue, it is difficult to verbalize such knowledge.

According to Leung (2009), practitioners apply, modify and share tools according to their needs. As shown in Table 21.1, we argue that data and

Table 21.1 Interservices knowledge production: tools and data

Child welfare services	Recorded expert knowledge	Setting of data collection	Method	Focus	Protocols
Medical professionals	Evidence-based practice (EBP)	Formal (objective clinical descriptive)	Quantitative	Physical and psychological health	Formal
Teachers	School records and written reports (related to educational standards)	Formal (objective comparative evaluation)	Quantitative	Learning cognitive and affective development	↓
Social workers	Explanatory statements and memo evidence-informed practice (EIP)	Both formal and informal (relational)	Quantitative Qualitative	Safety and social Integration	↓
Residential youth workers	Narrative reports (user-centred practice)	Both formal and informal (relational)	Qualitative	Autonomy and wellbeing	Informal

Table 21.2 Types of social work knowledge (Leung 2009: 702)

Knowledge as object	Knowledge in midway	Knowledge as process
Information/data Procedural knowledge Theoretical knowledge	Tools Stories Mental models	Skill enactment Practice reasoning process Work process knowledge Value sharing process

information on young people in residential care, the production of which is filtered through the work experience of each professional, can be located on a continuum flowing from 'vivid' evidence (Munro 1999) resulting from rigid and officially prescribed protocols to more 'abstract and dull' evidence (Munro 1999). The latter derives from relational practices and subjectively interpreted evaluations where improvisation is often required and was usually overlooked (Ellström 2010). It is this latter type, the integrative part of the innovation process, whose impact is to be assessed.

Leung (2009) offers a description of different types and processes in the production and sharing of knowledge in social work agencies and organizations. The author describes knowledge as lying on a continuum from knowledge as object to knowledge as process (Table 21.2).

Moreover according to Leung's (2009) analysis of organizational and practitioner's knowledge in social work, the innovation process we are reporting on occurs at 'knowledge in midway' (Table 21.2) in the third sector, and its potential positive effects should include an impact on the child welfare agency as a whole. This could be expected to increase work process knowledge through the development of new shared values.

The crucial elements of child welfare (Boutanquoi et al. 2013) are the initial assessment, the evaluation of users' social situations before intervention, the diagnostic evaluation *in itinere* and then the evaluation of outcomes, which typically allows team members to make decisions on the ending of intervention by the social service. In order to perform these tasks, professionals develop tools intended to be a guide for collecting data, organizing information and creating knowledge. Data and information from a variety of services—health, care, educational and social—need to be shared in a meaning-making dialogue where service knowledge is produced. This dialogue is not a simple exchange of information at a cognitive level; it implies interpersonal perceptions of emotions and attitudes (Leung 2009).

Tools facilitate the organization and collection of data and equip the observation activities. Boutanquoi et al. (2013), however, argue that the use of tools raises several issues as in the following:

- The items used are related to cultural and social norms not always shared by both families and professionals.
- Tools often reinforce individual and specific professional competences rather than facilitating the co-construction of meaning and significance in data on children or young people in residential care.

- Data and information on young people in residential care are the starting point, but one should contextualize them and thereby ascribe meaning to them.

Social workers and practitioners are committed to the care of the user (Golia and Pedrazza 2014; Pedrazza et al. 2013), avoiding judgement: as Leung (2007: 192) argues, they 'are committed to the core values of care and social justice and the belief that they should not participate in any oppressive practice'.

Social workers' knowledge resides in their routines and patterns of behaviour (Wallin and von Krogh 2010) triggered in response to the need to produce evaluations and to the problems connected with data collection and transfer.

The agency's staff, as with any child protection organization, is burdened by procedural documentation (Gambrill and Shlonsky 2001). The problem of large amounts of documentation is exacerbated by a high caseload and high staff turnover. Last but not least, a high degree of sensitivity to negative media exposure leads to training being valued more highly than continuity of expertise (Munro 2009).

However, in being responsible for the whole decision-making process on children, social workers are often left alone, and thus the role of other professionals shifts from one of teammate to that of spectator (Feng et al. 2010).

An extensive, multidisciplinary review of team decision-making in child welfare allows Nouwen et al. (2012) to propose a framework as a guideline to the evaluation and analysis of the whole process. These authors emphasize that very few studies have analysed the team decision-making process per se. They propose an effective decision-making team framework which encompasses both team architecture—which entails committed professionals, trust and alignment—and team learning, which incorporates collective information processing and shared mental models.

21.4 Objectives and Methods

21.4.1 Research Context

We look at an Italian regional child welfare agency as a community of practice involving actors at three different levels of complexity (Bronfenbrenner 1979):

- Macro: policymakers
- Meso: managers of public child welfare agencies (health and social services) and the boards of third sector organizations (residential and daytime homes for young people in care)
- Micro: public sector professionals and practitioners (health professionals and social workers, educators and school officers) and third sector practitioners (residential youth workers)

Health professionals, social workers, psychologists, teachers, educational practitioners and residential youth workers are service providers within, or working with, the agency. They should work interdependently, share responsibility for final outcomes and be able to manage their relationships across organizational boundaries (Cohen and Bailey 1997: 241). They have to work collaboratively across disciplinary and methodological boundaries in order to address the complexity of most of the problems they deal with.

In the present study researchers were asked to mediate between the agency's management and their professionals and practitioners. In order to achieve this goal, we started an exploratory qualitative study to identify the features of the information flow involving the various professionals and practitioners involved in the agency.

The mandate requesting follow-up research, from the institution and from the political leadership (Mergel and Desouza 2013), specified the involvement of academic researchers and staff of the regional child welfare agency.

The development and implementation of the new tool, introduced in 2010, allowed residential youth workers to achieve two important goals as first-order consequences of innovation (Tan 2004):

- They could assign new meaning to their daily working practices.
- They could improve intra-services and interservices communication.

They also became confident that they could contribute in a more systematic and shared way to interservices data transfer on young people in residential care.

The residential youth workers hoped that the process would result in both the families of origin and other professionals trusting them more and relying on them more.

Organization, data gathering and evaluation all involve tacit knowledge on the part of social workers. This knowledge becomes explicit and concrete, only when it is put to its purpose, that is, in narrative reports on the positive and negative behaviour of children or in data collection procedures using systematic observation tools such as grids or inventories of behaviours and attitudes.

The provision of information and data by residential youth workers to different professionals of the agency can be regarded as the application of the residential youth workers' knowledge to the whole process of regulation, administration and shared elaboration of data and information on the young people in their care.

The innovation had two consequences which enriched the complexity of the institutional-social background of our follow-up:

1. The introduction of the new tool (see also Chap. 18) activated a subtle competition between the various groups of professionals contributing to the work of the agency's organized residential and daytime care. The agency's survival depends on their retaining the trust of the local department of the public social service in their reliability and effectiveness. The economic crisis puts pressure on state budgets and induces changes and innovations which seek to cut costs while increasing the quality of the service provided (Gallouj and Zanfei 2013).
2. The local department of child welfare expressed interest in the agency's innovation and asked researchers and the team of social workers to start

dissemination activities and follow up actions in order to optimize the flow of information on young people in residential care at a provincial level, involving different professionals and all third sector organizations. A sort of domino effect was begun, and many third sector organizations shifted from a 'pure' narrative and qualitative approach to a more quantitative, systematic way of evaluating their interventions on young people in residential care, so that similar new tools were introduced and utilized.

21.4.2 Research Design

Follow-up actions were initiated following the request by the regional government that all third sector organization implement the prescribed accreditation process: each organization had to account for its practices, objectives, procedures and evaluation outcomes.

Follow-up actions were intended to help it and its professionals to optimize inter-professional data handling and reporting procedures, which are both functional to the decision-making process.

After the implementation (2010) of the new tool in residential home, researchers started a qualitative study in order to assess the impact (2014) of the implementation of the new tool in the service and of associated dissemination activities.

The service as a whole can be viewed from two different levels of complexity:

1. The intra-service perspective: the residential care service provides relevant and significant information for the family of origin and provides for the recovery, autonomy and wellbeing for young people in residential care (their 'moral career' to use the term coined by Harré (1993).
2. The interservices perspective: the residential care service provides relevant and significant information on young people in residential care to the health service, the social service and the educational service, in order to allow each of them to cooperate in the decision-making process. This cooperation is effective when all team members recognize that they need each other if their common task is to be successfully accomplished.

The main objective of the local social service department is to integrate the form and content of the newly explicit knowledge from the residential youth workers with the overall flow of information on the young people in residential care. The integration of practitioners' knowledge has to be checked and supported in order to avoid the dispersion and loss of important information.

In the present study, authors involved professionals in a procedure of reflective, co-constructive follow-up in order to analyse how their tacit knowledge affected their perception of the tool's implementation.

Nine focus groups were carried out, involving 97 professionals (36 % socio-educative service, 24 % health service, 21 % educational service, 19 % social work). Each focus group involved from 10 to 12 participants at a time. Participants were selected according to the following criteria: (1) high intrinsic personal motivation to take part in the research, (2) 10-year experience in the local child welfare agency, (3) their distribution between the services belonging to the agency (health service, social service, care service and educational service) and (4) being familiar with the implementation and use of the new tool.[1]

The main objective of this exploratory, co-constructive procedure was to take implicit, tacit knowledge and render it explicit and reflective.

Loosely structured focus groups were carried out in order to explore the following aspects:

- The systematic observation of young people's state, attitudes and behaviours according to the residential youth workers (indexes of the wellbeing of the young people in residential care) as a measure of alignment between all professionals and practitioners (residential youth workers)
- Characteristics of the information flow: formal communications (planned meetings) vs. informal (e.g. phone calls, unrecorded meetings) and technologically mediated vs. face-to-face encounters
- Critical issues emerging in the information flow (Flanagan 1954)

21.4.3 Data Analyses

We followed Smith's (1995) procedural and methodological recommendations for analysing qualitative data drawn from focus group transcripts.

According to Dey (1993), qualitative analysis involves breaking down the data into smaller units and then reassembling them in new ways. During the classification process (NVivo 8 software), we assigned the data to categories on the basis of relevant characteristics and meanings. Each category gathers responses based on the principle of content equivalence/similarity. We generated categories ad hoc for each theme via an inductive process. Each minimum unit of meaning is classified by attributing one or more categories to it.

Analysis was structured around three conceptually progressive coding operations (Strauss and Corbin 2008): (1) open or substantive coding, (2) axial coding and (3) selective coding.

According to the qualitative analysis through NVivo software, we identified the most important categories and the hierarchical relationship between categories, including the analyses of the subordinate and superordinate categories.

[1]The costs of the investigation were partly covered by a local private foundation (CARITRO, Trento). Data were collected in February–May 2014.

21.5 Results

21.5.1 Residential Youth Workers' Focus on Behaviour and Wellbeing of Young People in Residential Care

According to the Nouwen et al. (2012) model of effective decision-making in child welfare teams, alignment is a central feature of team architecture. We looked for alignment opportunities by distinguishing the focus of each professional's observation practices. The highest potential level of alignment corresponds to the highest degree of overlap between these variables and the issues relevant to young people in residential care. Redundancy offers a wide range of alignment opportunities.

The informal domestic context of the residential care service allows practitioners to focus on the autonomy and wellbeing of the young people in daily interactions. Social competence in domestic and informal contexts, trust in adults and peers and domestic routines were codified as important focuses of practitioners' daily observation (Table 21.3). Table 21.4 shows the issues observed as relevant to young people in residential care accorded highest percentage ratings by residential youth workers.

The core category of residential youth workers' information (gathered by means of the new tool) is referred to as what we called 'the material and social aspects of the young people's everyday life'.

The core category is linked to the following five domains:

- Time management
- Private and common space management
- School attendance
- Trust in peers
- Trust in adults

The identification of the core category and of five related domains relates to the ability of practitioners to draw distinctions in the process of carrying out their work in residential homes and day centres for young people in residential care (Tsoukas and Vladimirou 2001).

Table 21.3 shows that residential youth workers focus mainly on the following areas: adjustment and maladjustment strategies and children's social competence. Table 21.4 refers to the specific indicators of children's comfort and discomfort

Table 21.3 Coding categories (superordinate categories)

Coding categories (parent nodes)	Number of coding references	Percentage of coding references
Adjustment	377	29.16 %
Maladjustment	592	45.79 %
Self-confidence, self-care and trust (towards peers and adults)	324	25.05 %
Total	1293	100 %

Table 21.4 Coding categories (subordinate categories) with the highest number of coding references

Coding categories (parent nodes)	Number of coding references	Percentage of coding references
Emotional distress, discomfort and unease	447	34.6 %
School attendance	321	24.8 %
Social competence	304	23.5 %
Other categories	221	17.1 %
Total	1293	100 %

analysed in the most depth. Residential youth workers are mainly concerned with relational issues; they have access to the children's affective reactions and to their responses to, and sensitivity to, both internal and external experiences and events. Their contribution to the agency's knowledge about the children is often about emotional and socio-relational matters which are often perceived as too subjective and raw.

21.5.2 *Formal vs. Informal Information Flow*

We also analysed the communication flow and its characteristics: its formality or informality and redundancy, where high redundancy corresponds to high levels of need for informal communication.

The formal settings of communication are typically regulated by the agency's protocols. These protocols stipulate that the settings and opportunities for interservices communication should offer practitioners and professionals sufficient time and space for exchanging data and for integrating their own knowledge with other colleagues' knowledge and perspectives on the young people in residential care.

In accordance with Nouwen et al.'s model (2012), we distinguished formal from informal communication opportunities and settings constituting the topos where mental models are shared and discussed. In the communications by residential youth workers within the agency, there are no significant proportional differences between formal and informal (formal, 54.54 %, vs. informal, 45.46 %). It is interesting to note that the formal protocols shared by all professionals and practitioners are clearly not able to reduce information redundancy in the inter-professional information flow. In fact the professionals arrange a large number of informal interpersonal direct contacts. The high level of redundancy noted in the communication flow, evidenced by the large number of informal meetings, seems simply to multiply the number of meetings without increasing mutual understanding and without satisfying all the professionals' needs for information and cooperation.

21.5.3 Critical Incidents in the Information Flow

To identify the difficulties that professionals have to cope with every day in interservices communications, we adopted the Critical Incident Technique (Flanagan 1954). An incident is any event or situation that is recent, sufficiently complete and representative of one's work. An incident is critical when it makes a significant contribution to work and challenges one's abilities. This technique has been successfully applied to different domains (Butterfield et al. 2005; Flanagan 1954; Gremler 2004).

The 'perceived and reported critical incidents in interservices communications' (Fig. 21.2) relate to three main issues[2]: lack of trust and recognition on the part of other professionals in the agency (38.8 %), problematic sharing practices (37.3 %) and the heterogeneity of the structure of the service (15 %) and lack of intrinsic motivation (8.9 %).

In our case the largest group of critical issues perceived by residential youth workers were related to trust (Fig. 21.2). They perceive a large degree of overlap between their own contributions to information on the young people in residential care and those of other professionals. They often sense a lack of recognition of their role as competent and reliable teammates, especially vis-à-vis health practitioners and teachers. The high level of confidence vested in documents communicating diagnoses, certifications and clinical assessments leads to health professionals' reports being regarded as more important than reports from residential youth workers about the emotional, interpersonal and social life events of the children.

A second critical issue relates to sharing practices (Fig. 21.2). The residential youth workers feel unable to persuade other professional to share their commitment to creating a support network for users and their families. They also perceive a certain lack of interest in developing a shared language for discussing the young people in residential care. As a consequence they also perceive a lack of engagement on the part of the other professionals in the joint development of an individual programme for each child. The professionals often fail to identify a common goal, and they then retreat into their own scientific perspective, thereby losing an opportunity to support and to enhance a sense of common identity within the agency. Only 15 % of critical issues are related to different formalized protocols and procedures (86.18 %) and to structural ergonomic differences in the service (13.82 %). Interpersonal relations related to professional roles and trust issues are perceived as being crucial in the information flow on young people in residential care. Structural features are regarded as less important (Vulliamy and Sullivan 2000).

[2]Data referred to open, axial and selective coding of focus group transcripts.

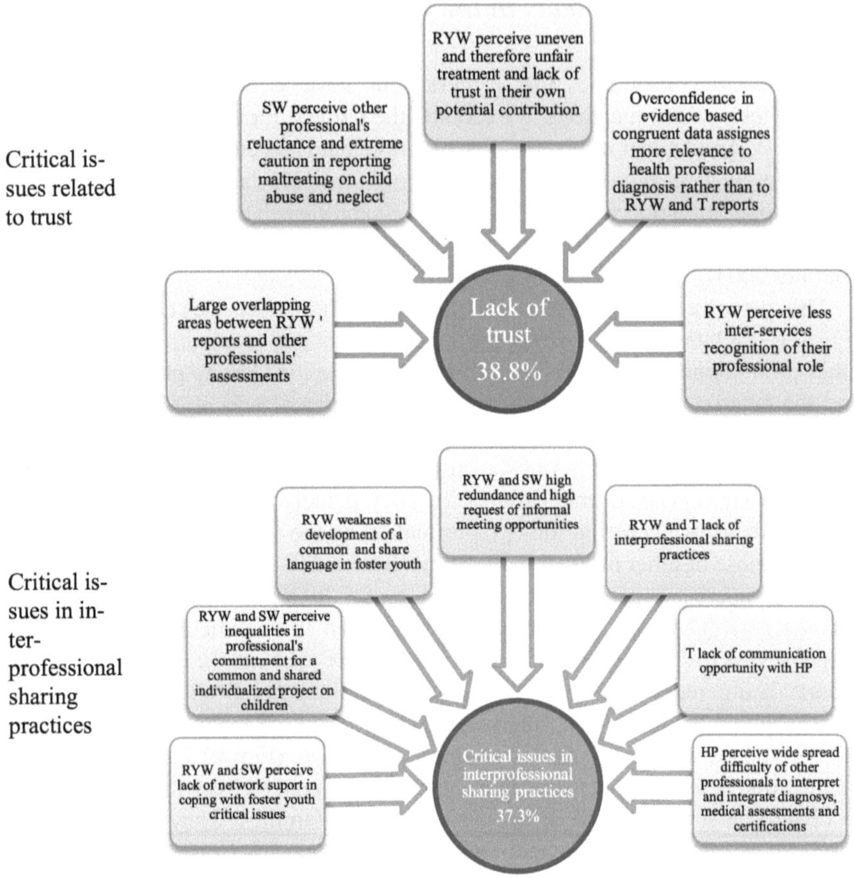

Legend: RYW = Residential youth workers; T = Teachers; SW = Social Workers; HP = Health Professionals

Fig. 21.2 Perceived and reported critical incidents in interservices communications

21.6 Conclusion and Implications

Public service's innovations are often concerned with value creation and thereby with the creation of intangible goods such as perceived quality on the part of users, trust and better interservices communication. Requirements of outcome-based performance management and lack of financial resources are increasing performance evaluation activities at all government levels (Yen et al. 2012).

One of the most frequent consequences of this widespread situation is the need to ensure effective innovation. In our study the regional government required follow-up action on the innovation concerning the development and implementation of a new tool for systematic observation of children's behaviours and attitudes in a third sector organization providing residential care services for the local child welfare agency.

The innovation we refer to had two first-order, anticipated consequences (Rogers 2003; Tan 2004):

1. Decrease in the number of interpersonal and intra-team contrapositions. The complex innovation process for the development of the tool increased thereby practitioners' wellbeing. In fact positive affect seems to increase the scope both of attention and cognition (Fredrickson 1998); it also makes cognitive additional material available for processing and for increasing flexibility (Isen 1999). In Csikszentmihalyi's (1996) and Hennessey and Amabile's view, innovation can be the result of intense intrinsic motivation. In this regard economic recognitions or promotions are of limited interest (Mergel and Desouza 2013).
2. Increased appropriability (Sole and Edmondson 2002) on the part of social workers, physicians, mental health professionals, school officers, court members of the knowledge, created and disseminated through the implementation of the new tool within the third sector organization.

As the outcome of effective reflective practices, the new tool contributes to a reduction in the number of potential conflict issues at team level. It supports the reinterpretation of differences of perception between co-workers of threats that is of negative children's behaviours, viewing them as complementary ways of applying their youth social work skills. It supports practitioners in regarding their individual differences as differing ways of perceiving adolescents' behaviour, deserving equal dignity rather than uncontrolled, biased rejection. Thus, the interservice information flow and knowledge production on the young people were both improved: data gathering methods, data exploitation and interservices data dissemination were all cleared and thereby simplified.

As the evaluation of the outcomes of educational treatment could be shared by means of the new tool, the number of sources of conflict for teammates decreased, but residential youth workers developed enhanced expectations of recognition for the centrality of their role in knowledge production on the young people, including at the agency's organizational level. In this study, in fact, an unanticipated and even negative outcome of innovation (De Vries et al. 2014; Piening 2011) manifested itself in the paradoxical effect that relying only on the effects of adoption/implementation of the tool at intra-service level, the residential youth workers' needs were unmet: their reactive response consisted in an augmented expectation of recognition and respect from other professionals, e.g. physicians, most of whom typically enjoy a higher status, above all remuneration. According to Cunningham et al. (2005), the 'not invented here attitude' can in fact enhance the unwillingness to accept new ideas from outside their own professional group.

The emergence of this paradoxical effect as second-order consequence of innovation (Rogers 2003; Tan 2004) activated rapid and concerned actions on the part of the regional agency management focused on the re-equilibration of the unequally distributed power among professionals and practitioners within the agency. So even 'negative' unexpected and paradoxical effects of innovation implementation can trigger virtuous interpersonal and organizational circles of communication, supporting cross-functionally the dissemination phase of innovation over time.

We argue the implementation of the tool was affected by the effects of inertia (Miller 1965) in the persisting asymmetry in the power relations between professionals and practitioners and thus in the mutual recognition of the importance of each others' subjective contribution to the decision-making process.

Sole and Edmondson (2002: 31) found that dispersed team knowledge is a high valuable resource, 'but varied in its ease depending on differences in teams' member awareness of relevant knowledge and differences in its appropriability'. Sole and Edmondson (2002) suggest that manager of dispersed teams, in our case social workers, should be able to enhance both awareness and appropriability of knowledge. Awareness should be heightened through both technological and social means. New tools and development of network in support of inter-institutional communication could trigger recognition and easy recollection of relevant knowledge offering new knowledge sharing opportunities.

Follow-up actions revealed that the implementation of the new tool is not the only means on which one can rely on in order to ensure appropriability of knowledge at interservice level: social competences, individual responsibility, trust and professionals' awareness also have to be enhanced. Thus innovation can be the result of an individual's successful cognitive process, but it has to be supported at group and organizational level (Quintane et al. 2011). The positive effect of the implementation of the new tool is also concerned with that fact that it lightens up areas where practitioners have to increase their commitment and engagement at cross-functional level.

Quintane et al. (2011: 939) argue that '"conceptualizing the innovation process"... "identifying the process leading to the creation of innovation that allows an understanding of how the innovation has been created"... "should permit the replication of the results of the innovation process without having to repeat the knowledge generation process itself"'. Moreover AA state that a fundamental feature of the innovation process lies in knowledge 'availability' rather than in its creation. Thus, we argue that comparative studies should help in understanding the relationship between practitioners' wellbeing, as a result of practitioners' justice evaluations (De Cremer and Sedikides 2005) and innovation outcome.

Several studies (Dolan et al. 2008) devote special attention to wellbeing as an independent variable of innovation creation. In order to ensure a deeper understanding of what successful innovation really means, we argue that this variable should be measured and controlled over time at any stage of the innovation process: at its adoption, implementation and dissemination stage.

Acknowledgements This study was supported by the CARITRO Foundation, Trento, Italy, in 2013: Trento University (principal investigator Prof. Olga Bombardelli), Verona University (investigator Prof. Monica Pedrazza), CeRP Milan and Trento (Dr. Simona Taccani and Dr. Cristina Zorzato).

References

Aagaard, P. (2012). Drivers and barriers of public innovation in crime prevention. *The Innovation Journal, 17*(1), article 6.

Boutanquoi, M., Bournel-Bosson, M., & Minary, J.-P. (2013). Evaluating situations in child welfare: From tools to workgroups. *Children and Youth Services Review, 35*(7), 1152–1157.

Bronfenbrenner, U. (1979). *The ecology of human development: Experiments by nature and design*. Cambridge: Harvard University Press.

Butterfield, L. D., Borgen, W. A., Amundson, N. E., & Maglio, A. T. (2005). Fifty years of the critical incident technique: 1954–2004 and beyond. *Qualitative Research, 5*(4), 475–497.

Chesbrough, H. (2003). The era of open innovation. *MIT Sloan Management Review, 44*(3), 35–41.

Chesbrough, H. (2006). *Open innovation: The new imperative for creating and profiting from technology*. Boston, MA: Harvard Business School Press.

Cohen, S. G., & Bailey, D. E. (1997). What makes teams work: Group effectiveness research from the shop floor to the executive suite. *Journal of Management, 23*(3), 239–290.

Corwin, L., Corbin, J. H., & Mittelmark, M. B. (2012). Producing synergy in collaborations: A successful hospital innovation. *The Innovation Journal, 17*(1), article 5.

Csikszentmihalyi, M. (1996). *Creativity: Flow and the psychology of discovery and invention* (pp. 107–126). New York: HarperCollins.

Cunningham, P., Grant-Pearce, C., Green, L., Miles, I., Rigby, J., & Uyarra, E. (2005). In sickness, in health and in innovation: NHS DIRECT—A health sector innovation study. *Administration, 53*(3), 42–65.

De Cremer, D., & Sedikides, C. (2005). Self-uncertainty and responsiveness to procedural justice. *Journal of Experimental Social Psychology, 41*(2), 157–173.

De Vries, H. A., Bekkers, V. J. J. M., & Tummers, L. G. (2014). *Innovation in the public sector: A systematic review and future research agenda*. Speyer: EGPA Conference.

Dey, I. (1993). *Qualitative data analysis. A user-friendly guide for social scientists*. London: Routledge.

Dierkes, M., Berthoin Antal, A., Child, J., & Nonaka, I. (Eds.). (2001). *Handbook of organizational learning and knowledge*. Oxford: Oxford University Press.

Djellal, F., Gallouj, F., & Miles, I. (2013). Two decades of research on innovation in services: Which place for public services? *Structural Change and Economic Dynamics, 12*(27), 98–117.

Dolan, P., Metcalfe, R., Powdthavee, N., Beale, A., & Pritchard, D. (2008, September 12). *Innovation and well-being*. Paper prepared for NESTA.

Dossou-Yovo, A., & Tremblay, D.-G. (2012). Public policy, intermediaries and innovation system performance: A comparative analysis of Quebec and Ontario. *The Innovation Journal, 17*(1), article 3.

Edvardsson, B., Tronvoll, B., & Gruber, T. (2011). Expanding understanding of service exchange and value co-creation: A social construction approach. *Journal of the Academy of Marketing Science, 39*, 327–339.

Ellström, P.-E. (2010). Practice-based innovation: A learning perspective. *Journal of Workplace Learning, 22*(1/2), 27–40.

Feng, J.-Y., Fetzer, S., Chen, Y.-W., Yeh, L., & Huang, M.-C. (2010). Multidisciplinary collaboration reporting child abuse: A grounded theory study. *International Journal of Nursing Studies, 47*(12), 1483–1490.

Flanagan, J. C. (1954). The critical incident technique. *Psychological Bulletin, 51*(4), 327–358.

Fredrickson, B. L. (1998). What good are positive emotions? *Review of General Psychology, 2*(3), 300–319.

Gallouj, F., & Zanfei, A. (2013). Innovation in public services: Filling a gap in the literature. *Structural Change and Economic Dynamics, 12*(27), 89–97.

Gambrill, E., & Shlonsky, A. (2001). The need for comprehensive risk management programs in child protective services. *Children and Youth Services Review, 23*(1), 79–107.

Golia, S., & Pedrazza, M. (2014). Assessing the relational competence as the core dimension of social worker's perceived self efficacy through the Rasch model. *Electronic Journal of Applied Statistical Analysis, 7*(1), 58–80.
Gremler, D. D. (2004). The critical incident technique in service research. *Journal of Service Research, 7*(1), 65–89.
Harré, R. (1993). *Social being* (2nd ed.). Oxford: Basil Blackwell.
Heinrich, C. J. (2002). Outcomes-based performance management in the public sector: Implications for government accountability and effectiveness. *Public Administration Review, 62*(6), 712–725.
Isen, A. M. (1999). Positive affect. In T. Dalgleish & M. J. Power (Eds.), *Handbook of cognition and emotion* (pp. 521–539). New York: Wiley.
Kaufman, H. (1981). *The administrative behavior of Federal Bureau Chiefs*. Washington: The Brookings Institution.
La Palombara, J. (2001). Power and politics in organizations: Public and private sector comparisons. In M. Dierkes et al. (Eds.), *Handbook of organisational learning and knowledge* (pp. 557–581). Oxford: Oxford University Press.
Lee, S. M., Hwang, T., & Choi, D. (2012). Open innovation in the public sector of leading countries. *Management Decision, 50*(1), 147–162.
Leung, L. C. (2007). Reflective practices: Challenges to social work education in Hong Kong. *Social Work Education, 26*(6), 632–644.
Leung, Z. C. S. (2009). Knowledge management in social work: Types and processes of knowledge sharing in social service organizations. *The British Journal of Social Work, 39*(4), 693–709.
Mergel, I., & Desouza, C. K. (2013). Implementing open innovation in the public sector: The case of Challenge.gov. *Public Administration Review, 73*(6), 882–890.
Metze, T., & Levelt, M. (2012). Barriers to credible innovations: Collaborative regional governance in the Netherlands. *The Innovation Journal, 17*(1), article 8.
Miller, J. G. (1965). Living systems: Basic concepts. *Behavioral Science, 10*(3), 193–237.
Miller, J. G. (1978). *Living systems*. New York: McGraw Hill.
Munro, E. (1999). Common errors of reasoning in child protection work. *Child Abuse and Neglect, 23*(8), 745–758.
Munro, E. (2009). Managing societal and institutional risk in child protection. *Risk Analysis, 29*(7), 1015–1023.
Nonaka, I. (1994). A dynamic theory of organizational knowledge creation. *Organization Science, 5*(1), 14–37.
Nouwen, E., Decuyper, S., & Put, J. (2012). Team decision making in child welfare. *Children and Youth Services Review, 34*, 2101–2116.
Pedrazza, M., Trifiletti, E., Berlanda, S., & di Bernardo, G. A. (2013). Self-efficacy in social work: Development and initial validation of the self-efficacy scale for social workers. *Social Sciences, 2*, 191–207.
Pettigrew, A. M., & Fenton, E. M. (Eds.). (2000). *The innovating organization*. London: Sage.
Piening, E. P. (2011). Insights into the process dynamics of innovation implementation. *Public Management Review, 13*(1), 127–157.
Quintane, E., Casselman, R. M., Reiche, B. S., & Nylund, P. A. (2011). Innovation as a knowledge-based outcome. *Journal of Knowledge Management, 15*(6), 928–947.
Rivera, M. A., & Rogers, E. M. (2004). Evaluating public sector innovation in networks: Extending the reach of the National Cancer Institute's web-based health communication intervention research initiative. *The Innovation Journal, 9*(3), 1–5.
Rogers, M. (1998). The definition and measurement of innovation. *Melbourne Institute Working Paper, 10*, 5–27.
Rogers, E. M. (2003). *Diffusion of innovations* (5th ed.). New York: Free Press.
Schumpeter, J. A. (1980). *Theory of economic development. An inquiry into profits, capital, credit, interest and the business cycle*. Somerset, NJ: Transaction Publishers.

Smith, J. A. (1995). Semi-structured interviewing and qualitative analysis. In J. A. Smith, R. Harrè, & L. van Langenhove (Eds.), *Rethinking methods in psychology* (pp. 9–26). London: Sage.

Sole, D., & Edmondson, A. (2002). Situated knowledge and learning in dispersed teams. *British Journal of Management, 13*(S2), S17–S34.

Sørensen, E., & Torfing, J. (2012). Introduction: Collaborative innovation in the public sector. *The Innovation Journal, 17*(1), 1–14.

Strauss, A., & Corbin, J. (2008). *Basics of qualitative research: Grounded theory procedures and techniques* (3rd ed.). Newbury Park, CA: Sage.

Tan, B. S. (2004). The consequences of innovation. *The Innovation Journal, 9*(3), 1–42.

Tsoukas, H., & Vladimirou, E. (2001). What is organizational knowledge. *Journal of Management Studies, 38*(7), 973–993.

Vulliamy, A. P., & Sullivan, R. (2000). Reporting child abuse: Pediatricians' experiences with the child protection system. *Child Abuse and Neglect, 24*(11), 1461–1470.

Wallin, M. W., & von Krogh, G. (2010). Organizing for open innovation: Focus on the integration of knowledge. *Organizational Dynamics, 39*, 145–154.

Weber, M. (1922). *Economy and society: An outline of interpretive sociology*. Berkley, CA: U. California Press.

Yen, R. H., Wang, W. K., Wei, C.-P., Hsu, S. H.-Y., & Chiu, H.-C. (2012). Service innovation readiness: Dimensions and performance outcome. *Decision Support Systems, 53*(4), 813–824.

Chapter 22
Investigating Elderly Care Contexts According to Stakeholders' Needs

Elena Luppi

Abstract This chapter aims to present an innovative and participatory model for assessment, improvement of quality and innovation in residential facilities for fragile elderly people. Starting from a theoretical and practical reflection on the value of the care relationship in these organisations, this chapter focuses on innovating care practices in these organisations. In this model care is understood not only as a practice but as an attitude of the organisation: an approach to assistance which goes beyond the mere satisfaction of physical needs, looking at the person as a whole and working within a framework of constant promotion of their well-being. The chapter presents a method for innovating elderly care services through evaluation, self-evaluation and participatory planning. This method was developed through two case studies, the research design and tools of which are presented. The research is guided by an approach of Educational Evaluation [Scriven (The methodology of evaluation. Rand McNally, 1967; Evaluation thesaurus. Edge Press, 1981; Evaluation thesaurus. Sage, 1999)] and Action Research: the professionals are actively involved both during the definition of the quality standards and their relative evaluation. Some specific evaluation tools are developed, and the data gathered is constantly fed back to the professionals in order to stimulate their reflexive practice and foster innovation by improving the quality of care.

22.1 Introduction

Today, the ageing population represents one of the most significant changes Western countries have seen in the last few centuries. The increase in average life expectancy and reduction in causes of mortality have progressively generated a sharp increase in the elderly population, accompanied by a significant drop in birth rates (Luppi 2009; 2010). These phenomena are changing the demographic face of

E. Luppi, PhD (✉)
Department of Education Studies, University of Bologna, Bologna, Italy
e-mail: elena.luppi@unibo.it

T. Russo-Spena et al. (eds.), *Innovating in Practice*,
DOI 10.1007/978-3-319-43380-6_22

many countries in Europe and around the world. The World Health Organisation (WHO 2013) reports that between 1980 and 2010 there was a 30 % increase in the population aged over 65 in Europe and estimates that this group will count for 25 % of the total European population by 2050.

The increase in the elderly population, and particularly that of the "oldest old", brings with it a parallel increase in the need to manage this segment of the population, increasingly having to tackle conditions of fragility. The first to answer care needs are the family, mainly spouses/partners, and adult children. Consequently, difficult situations arise which place caregivers at risk in terms of both health and the reconciliation of family life and work. This situation is greatly affected by the social policies through which welfare systems recognise and answer the needs of the elderly population. Various welfare models exist, leading to as many different service systems, which vary according to the level of care provided outside the family, with the needs of the elderly served by public, private or—more often—mixed forms of care (Esping-Andersen 1990; Esping-Andersen et al. 2002; Lewis and Ostner 1994).

Unfortunately the question of elderly care still takes second place, in the policies of family-work reconciliation and social (rather than family) welfare, to the debate on other services (e.g. early childhood) across Europe, even in countries with more inclusive welfare systems (Lechner and Neal 1999). This project springs from the intention to develop a model for innovation in elderly care that moves from evaluation and educational evaluation in particular. Attempting to combine the theories of innovation in services and the approach of evaluation and formative assessment, this model aims to increase the quality of elderly care services, particularly focusing on care relations.

The model for innovation presented within this contribution moves, on one side, from the theories on *educational evaluation research* (Becchi and Bondioli 1994; Husén and Postlethwaite 1994; Kellaghan and Stufflebeam 2003; Scriven 1967, 1981, 1999; Walberg and Haertel 1990). In particular, we refer to *evaluation from an educational viewpoint* as described by Bondioli and Ferrari (2004) as a process in which scientifically rigorous tools are used to gather information on a given event and to implement a dynamic process of data interpretation, which impacts the studied situation in terms of change, improvement and innovation. The research was also oriented towards training in order to involve the whole team as directly as possible, stimulating towards reflection and the implementation of innovation and positive changes.

In this approach, evaluation is conceived as an opportunity to reflect on practices and innovative services starting from an evidence-based assessment of quality in care that involves both staff and users.

On the other hand, this approach is inspired by the theories on service innovation (Sundbo and Toivonen 2011). In particular we start from Ellström's definition of innovation as relating to "some form of specific change that is new (at least locally), and that leads to what is in some sense a better accomplishment of goals at the system level (the local unit or the larger organisation/system of which it is a part). The goals do not necessarily have to be financial or production-oriented in nature but may relate to other values that we want to achieve by means of certain operations, for example meeting the needs for healthcare, care or education" (Ellström 2010: 28).

In the proposed model of evaluation, a cycle is designed that, according to a set of indicators for quality in elderly care, allows us to assess care practices, leads staff to self-assessment and accompanies them in designing some potential innovations that will then be assessed again. This evaluation cycle is supposed to transform potential innovations to effective innovations.

A potential innovation is considered as a proposed change that is in some sense considered as positive but has not yet been demonstrated to be adequate or effective in reaching the organisation's goals. An actual innovation, on the contrary, is an innovation whose effectiveness has been demonstrated and can therefore be applied (Ellström 2010).

This educational evaluation research for innovation was conducted in parallel with two large residential elderly care facilities in the city of Bologna (Italy): Villa Ranuzzi and Villa Serena. Villa Ranuzzi, founded in 1938, is a sheltered residential care home for the elderly, accredited with the Regional Health Service, with a total of 240 beds, 150 with NHS funding and another 70 private places.

Villa Serena, founded in 1973, offers the following services: residential home for the elderly (CRA) with 82 beds with NHS funding and temporary care facilities with direct access from the hospital wards, with 18 NHS-funded beds, and a housing community with 5 housing units.

These services were used as a case study and to experiment and validate the innovative evaluation practices developed. Data was gathered using different methodologies, aiming to gather elements for analysing the context, the organisational dynamics and the care actions.

The results of this research were used to define precise (observable and measurable) care quality indicators and to validate evaluation tools designed to measure these indicators (questionnaires, observation grids and self-assessment tools).

The two care facilities (RSA—"residenza sanitaria assistenziale") where the research was conducted acted as contexts for processing and assessing the evaluation tools and methods, which can be extended to any facility providing care services to fragile elderly people or those who are no longer self-sufficient.

This chapter presents the method for evaluation and innovation that has been implemented and some of the assessment tools developed during the project.

22.2 Theoretical Background on Elderly Care

The implemented model for evaluation and innovation applies an inductive-hypothetico-deductive method (Cattel 1966) to build a process of intentional value cocreation (Vargo and Lusch 2004; Vargo et al. 2008) as first step. The inductive hypothetico-deductive method (Cattel 1966) is used to cross-reference theoretical and empirical data: literature on elderly care and evidence-based information on care and services through research and investigation tools. Value cocreation is considered as a constructivist-interpretive (Berger and Luckmann 1967; Blumer 1969; Goffman 1963; Levin and Levin 1988) or dialogical (Tronvoll et al. 2011) process which defines the framework value of an organisation or a service.

In this research, value represents the premise for assessment and evaluation and the horizon for innovation. As Edvardsson et al. (2011) state, value is cocreated in a complex process which involves not only customers and services but is also influenced by several social variables. This is particularly true in elderly care field: an environment that is dealing with an ageing population, the biggest demographic revolution occurring today. The needs of the elderly population are evolving very quickly, and intensive care for this target is asked to become more and more flexible and innovating, according to the brand new needs of their customers.

In educational evaluation, the approach to evaluation adopted for this study, value cocreation, is the premise for assessment, self-assessment and innovation. The starting point for defining value in elderly care in this project is an analysis of the most recent and innovative evidence-based approaches on residential care for the elderly. From the deductive point of view, this research refers to some paradigms and models that are empirically validated and focused on theoretical bases and values. These models overcome adult-centric stereotypes of the elderly to focus on the person, their actual needs and the promotion of quality of life in the third age. Care in the third age, particularly that provided in institutionalised care contexts, is historically characterised by a prevailing medical-health approach, often with no consideration for psychosocial needs. Today many people are voicing the need to reintegrate the various aspects of formal and informal care, answering both medical and health needs and psychosocial and human needs.

Particular inspiration came from the models of the *Culture Change Movement* (Brune 1995, 2011; Calkins 2002; Rahman and Schnelle 2008; Shura et al. 2011; Thomas 2006) and the principles of quality of care described by Coons and Mace (1996). All these authors underline the need to see elderly well-being from a holistic perspective, consequently designing care services that respond to individual needs in physical, psychological, social, intellectual, emotional and spiritual spheres.

The *Culture Change Movement* was born in the United States, with the transition of elderly care facilities from a medical and managerial model to a social-humanistic model, and spread throughout the world, bearing witness to a real change in the elderly care culture. This movement, which adopts a range of approaches, is in contrast to the management models of residential facilities established in the 60s (LTC, *long-term care management models*) based on organisational optimisation, using a bureaucratic managerial system focusing on efficiency, standardisation of services and cost optimisation, moulding services for the elderly in a very similar manner to those of hospitals. According to these models, the facilities were designed as places shut off from the outside world, governed by strict routines, marked by spaces and times in which the role of the staff focused far more on control than on support.

The *Culture Change Movement*, based on person-focused values and practices, places the care of the elderly and their families back in the centre of the organisation in order to design organisations fostering authentic well-being. In the United States, the facilities which adopted organisational changes by changing the paradigms proposed by these models rapidly recorded significant drops in depression and mortality rates, an increase in social interaction and a reduction in staff absenteeism and turnover (Calkins 2002).

The ultimate objective is to achieve the best possible quality of life, both for the guests and the members of staff. To do so, the importance of continuous training for staff and coordinators is emphasised in a perspective of what is defined as *wise leadership* (Brune 1995; Thomas 2006), a care culture focusing on people and an individualised approach to organisational and project choices. This individualisation does not only concern the guests but also considers the importance of the care relationship between guests and workers, including the latter in the needs analysis and the consequent design of interventions.

These person-focused approaches, which recognise and safeguard freedom, autonomy and dignity, privilege all possible strategies to reduce the mediatisation or measures of containment of the disorders the guests suffer from. At the same time, the organisational style promoted revolves around dialogue among the staff, to reduce distances, vertical relations and excessive bureaucracy, to foster the most direct possible dialogue between the staff and the elderly and their loved ones. It moves clearly away from the adult-centric perspective in which the needs of the elderly are established a priori or independently of the actual conditions and wishes of the elderly themselves. It is interesting to underline how all these models highlight the close interdependence between the quality of life of the guests and the quality of care, on one hand, and between the quality of professional relations and the interpersonal dynamics among staff members, on the other.

The idea is therefore of total quality, resulting from the interdependence between quality of life of the guests, quality of the care relationships between guests and staff and the quality of the working life of each member of staff. The innovation (s) leading to increased quality of care obligatorily require an organisational restructuring to bring about quality in working dynamics, harmony among team members and, consequently, the prevention of the risk of professional *burn out* (Maslach 1982) or *rest out* (Norman and Redfern 1997) that the professionals working in these contexts can suffer from.

This perspective is close to that of the *organisational social capital* model, which underlines how relationships and the quality of communication in a working context, particularly in the personal services sector, affect both the social climate of an organisation and professional performance (Andrew 2010; Kiss et al. 2014).

All these approaches highlight elements underlying quality of care in a residential facility and have offered many inputs for defining the quality evaluation path described below.

22.3 Methodology

The key concept of this research path is that of the quality of care, in particular, from an educational viewpoint, the quality of relationships with the elderly in the various moments of care and in the educational and recreational activities.

Evaluation is understood here as a moment for reflection and self-reflection, leading to a decisional stance, starting from the information collected systematically and intersubjectively. Evaluation becomes a permanent action, connoted by monitoring, i.e. the methodical and systematic collection of qualitative and quantitative data, which is analysed to formulate opinions on the value and pertinence of the actions analysed. The outcome of this evaluation leads to thought on practices, activating a process of innovation, through the participatory design of actions, changes and interventions. These innovations, in turn, become the subject of evaluation, based on the established quality indicators, thus monitoring their implementation, with a view to the constant improvement of the quality of service. In this way, a virtuous circle is created in the organisation, and its operators implement a permanent process of self-diagnosis, reflection, designing solutions, improvements and creative innovations.

We refer to an evaluation and monitoring process which is far removed from the meaning generally attributed to these actions in the managerial field: as a means to check the conformity of what was done with what was planned or designed. In this case, evaluation is designed as a learning path, an opportunity to describe a set of common indicators, based on which it is possible to monitor the implemented processes and actions, using objective data gathered systematically.

The concept of *quality* understood here is a complex, intersubjective concept, a value which must necessarily be cocreated in order to include the viewpoints of all stakeholders in the care facility: various grades of workers, management, the users, their families, etc.

The definition of the concept of quality in care and its description in terms of operational indicators is the result of an inductive-hypothetico-deductive (Cattel 1966) empirical investigation. This process moves from the above-cited literature to investigate the care environments through the analysis of documentation, observation of the practices adopted, interviews, focus groups and the administration of specific questionnaires.

In the perspective of *educational evaluation,* particular importance is given to the intersubjective reading of the gathered data, as it is considered that only an ecological vision and a reflective, interpretative reading of the data by the stakeholders of the studied context can comprehend the complexity in defining the idea of underlying quality.

The evaluation processes developed require the analysis of a reality defined on the basis of scientifically valid indicators and instruments but which cannot be considered "neutral" and which therefore require an intersubjective, interpretative reading in order to be able to formulate judgements and take decisions. Such indicators and instruments need to be negotiated by users, according to the notion of value cocreation (Sørensen et al. 2013), where value is fundamentally derived and determined in use. For this reason, the people being evaluated become active participants in the process of analysis and reflection on their own professional practices, provided that the evaluation indicators are agreed on by the whole group.

The first stage of the project aimed to precisely outline the concept of *quality* within the service, from the viewpoint of the workers, users and their families. The work was based on the conviction that quality of care in a residential facility for the elderly cannot be defined a priori or definitively but must consider the points of view and needs of the different stakeholders in these delicate care contexts. To achieve this objective, we analysed:

- The documentation available in the two facilities concerning the organisation of work
- The activity guidelines in the two facilities
- The satisfaction questionnaires completed by the guests' families

A series of observations were organised in order to outline the key aspects of the routines in the two facilities. Parallel to the observations, some semi-structured interviews were organised with reference figures who could be considered significant witnesses for each of the two facilities.

The perceptions of *quality* of the stakeholders gathered during the observations and interviews, together with the analysis of the documentation, were cross-referenced with the referred literature in order to define specific indicators to evaluate quality.

Reference was made to the approaches to elderly care that focus on the person, recognising and safeguarding their freedom, autonomy and dignity. Particularly, we were inspired by models that were traceable to the "culture change movement" (Brune 1995, 2011; Calkins 2002; Rahman and Schnelle 2008; Thomas 2006) and the principles of care quality identified by Coons and Mace (1996) which outline a series of purposes which a long-term care service must develop:

- Guaranteed freedom of choice (as much control as possible over their own lives and involvement in decision-making)
- Individualisation: recognition of personal differences and preferences and design of bespoke responses (personal differences as a source of enrichment)
- Privacy and dignity
- Continuity with the past and continuation in normal social roles (activities, hobbies, relations, roles in social relations)
- Sensory and visual stimulation through activities and organisation of the environment
- Creation of domestic, neighbourhood and community environments
- Opportunities and moments of entertainment, humour and creativity (Fig. 22.1)

On the basis of the observations, various areas were defined which lead to or have an impact on the quality of the care relationship: (verbal and nonverbal) communication with the guest, communication with relatives, communication among staff (formal and nonformal moments), relations between staff (among

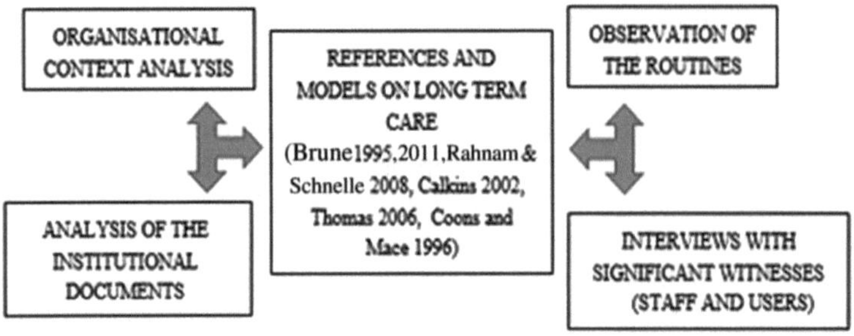

Fig. 22.1 Process for defining standards for quality in care

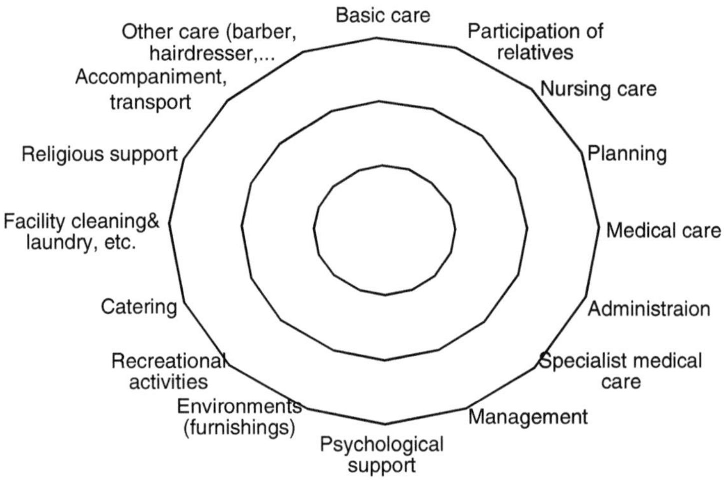

Fig. 22.2 Care service activity map

uniform or nonuniform professional groups), collaboration, discussion and sharing of opinions among staff, adoption of problem-solving strategies by the team and reflection on their work.

Each of these dimensions was present to different extents depending on the care sector and service. A care service activity map was therefore drafted in order to define appropriate quality indicators for each area according to the aforementioned standards. The map is presented in Fig. 22.2.

In the second phase of the project (Fig. 22.3), the guidelines drafted were described in a series of indicators, subsequently validated in the various focus groups involving all staff.

The defined indicators were expressed as observable descriptors, actions and behaviour, starting from which a series of tools for evaluation, self-evaluation and reflection were produced with a view to increasing care quality, in particular:

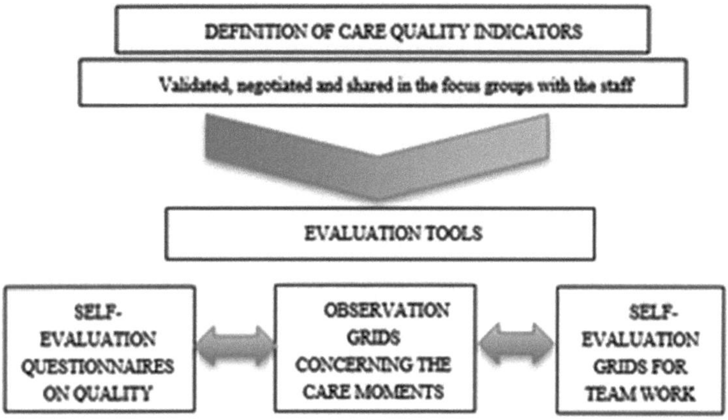

Fig. 22.3 From the definition of indicators to the construction of evaluation tools

- A questionnaire used by the staff to self-evaluate quality
- Grids to observe quality during care activities
- Grids to analyse and reflect on the data collected

These tools are presented in the following paragraph.

The evaluation questionnaire was administered to all members of staff. The observation grids were used by the researcher and by some suitably trained members of staff. This aimed to help the members of the organisation to develop observation and self-evaluation skills useful for developing methods of reflecting on their own professional practice.

Each tool was then validated and applied in order to obtain data which, case by case, was fed back to the staff in order to foster the awareness of their strengths and weaknesses and help produce the innovations needed to overcome any criticalities. Specific meetings were organised to discuss and reflect on the results, acting as training sessions for the workers. The researcher presented the data to uniform professional groups of workers, fostering joint reflections within the group and offering the stimuli required to facilitate decision-making on significant topics and where possible design targeted interventions.

Figure 22.4 offers a schematic overview of the project phase focusing on evaluation, discussion, training, reflection and designing innovation.

In the discussion sessions, tools were created for analysing, reflecting on and taking decisions, organised to offer a guide for analysis and some elements for promoting dialogue, discussion and, above all, the creative identification of possible innovations (De Bono 1969).

Generally, the method of designing educational evaluation involves a continuous process of evaluation for innovation, in which the phases of evaluating in order to self-evaluate, reflect, decide and design innovations follow on from each other, and are then subjected once more to the evaluation process, with a view to constantly increasing quality. The evaluation is considered in its educational transformational potential.

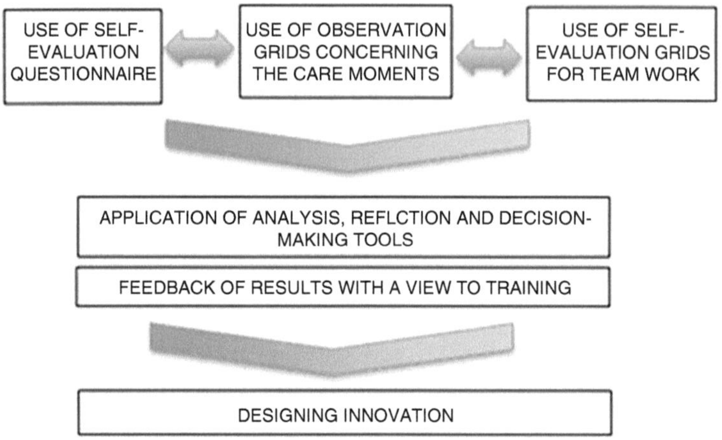

Fig. 22.4 From the evaluation to the innovation

22.4 Results

22.4.1 Common Understanding of Quality in Elderly Care

The first outcome of this evaluation research project was the definition and validation of a series of quality indicators for elderly care services. To produce a valid and reliable framework of indicators and descriptors, it was first of all necessary to contextualise the indications and strategic areas identified through the analysis of literature, cross-referencing them with the legislative references and the documentation on the RSAs. This application of the guidelines to the context led to the definition of a series of general purposes and orientations, which were practically applied to observable actions, situations and behaviour. This in turn led to the definition of a preliminary descriptive model of the elements defining quality in a residential care service for the elderly. The guidelines drafted were described in a series of indicators which were then validated in the various focus groups involving all staff.

In a subsequent phase, the indicators were articulated further to ensure a precise definition in the various care contexts, based on the results of a series of systematic observations conducted by persons inside and outside of the care team.

The service quality indicators were discussed and defined in order to investigate the organisation of care as a multidimensional context, each closely correlated to the others, to determine quality of service as a whole.

Table 22.1 in the followings presents the list of indicators, with some key concepts which offer general guidance.

The indicators validated with the group of staff involved in the focus groups led to the development of a range of evaluation, self-evaluation and reflection tools, aiming to increase the quality of care. Each tool was then validated and applied in

Table 22.1 Box indicators for quality in elderly care

General guidance
Presence of a culture of respect, mutual support and care Nonmedical but social and domestic organisational model which encourages warm relations between staff and guests and includes care of the quality of life
Service quality indicators
Organisational aspects The organisation focuses on individuality and the recognition of personal differences The staff roles are broadened to include their involvement in activity planning The organisation helps the guests to continue their life styles as long as possible (e.g. when to go to bed or get up, participation or nonparticipation in activities) Care and communication (verbal/nonverbal) Staff place importance on the psychosocial needs of the guests Staff are trained in verbal communication with the patients Staff are trained in nonverbal communication with the patients Staff use a relaxed, nonauthoritarian communication style Communication with relative
Staff roles Staff learn to accept and interpret difficult behaviour as a sign of distress Staff are aware of the individual differences, potential and needs of the guests Staff help the guests to continue their life styles as long as possible (e.g. when to go to bed or get up, participation or nonparticipation in activities) Staff make an effort to encourage the guests to maintain their independence and personal care as long as possible Staff encourage guests to take decisions Communication between staff Staff share knowledge and information Staff support each other in tackling difficult situations Staff work to solve problems together Staff recognise the value of the work of others and offer supporting and assistance Staff are respected, considered, listened to and encouraged to use their talents to enrich the care activities The resources and creativity of the staff are held in esteem and the staff are stimulated to use them
Training The staff are trained in the most appropriate strategies for reducing difficult behaviour (avoiding as far as possible the use of physical restraints or medication) The staff are trained to understand the psychosocial needs of the guests The staff are trained in strategies to satisfy the psychosocial needs of the guests Training is organised in order to: –Allow the staff to improve their own skills –Allow the staff to identify problem-solving strategies –Allow the staff to reduce the risk of burn out
Setting Attention is paid to the tranquillity of the context, silence and respecting moments of quiet The setting is structured to resemble a domestic environment, to reinforce the sense of belonging Sensory stimulation is fostered, activating interest through objects, colours and sounds Safety measures are designed to offer as much freedom as possible for the guests Bedrooms contain the guests' own furniture or objects Bathrooms are designed to guarantee privacy The physical context includes familiar, comfortable objects helping every resident to maintain continuity with the past

order to obtain data which, case by case, was fed back to the staff in order to foster the awareness of their strengths and weaknesses and help produce the innovations needed to overcome any criticalities.

22.4.2 Elderly Care Quality Evaluation Tools

Here we present a summary of the tools developed on the basis of the previously presented indicators. Each tool is an opportunity to foster evaluation and self-evaluation processes, not to formulate an opinion but rather to develop awareness of the distance from or vicinity to the identified quality standards and to take decisions and design innovations to improve the quality of service with a view to care.

Each tool proposed is an object of joint reflection on the quality of the care service related to the shared quality indicators or criteria. These indicators were defined, as was seen, through an inductive-hypothetico-deductive method (Cattel 1966) in which empirical and theoretical research data were cross-referenced: on one hand, the data concerning the literature on care quality in residential facilities for the elderly and on the other hand, the analysis of the organisation and documentation and also the results of the interviews with all the staff in defining the concept of the quality of care. The indicators were recognised by all the staff as shared quality criteria for the care services.

Staff Self-Evaluation Questionnaires

The questionnaire presented below responds to a logic of team self-evaluation. Through this tool, each member of staff is asked to offer an evaluation that does not refer to the individual but rather to the whole team: it asks them to express their evaluation of the group and not themselves (see Table 22.2).

The questionnaire presents a series of topics indicating elements of quality in elderly care. Each topic includes statements which present "quality" situations concerning the staff's way of working. To respond, the staff allocate a score of 1–4 to each of these elements as per the instructions in the introduction.

The results of the questionnaire can be processed and presented, indicating for each item the average and standard deviation, and are then discussed in order to compare the opinions and perceptions of the team members, reflecting together on the strengths and criticalities.

Observation Tools

During the project a number of tools were developed to gather the data from the observation of significant moments or routine activities. These semi-structured observation grids were used by the staff or by external observers to focus their attention on the quality indicators pertaining to given moments of professional practice. Here below are two of the observation grids produced: the first (Table 22.3) concerns the observation of collective moments (such as team meetings) and the second (Table 22.4) analyses the management of problematic situations.

Table 22.2 Self-evaluation questionnaire on quality of care

Self-evaluation questionnaire on quality of care				
The questionnaire includes statements presenting situations of "quality" in the team's way of working: to respond, allocate a score to each of those elements to indicate how far the way of working of the team I belong to is close to or far from the indicated quality criterion In particular if you think that the statement indicates a situation: – That the team is still very far from → indicate a score of 1 – That the team is still fairly far from → indicate a score of 2 – That the team is getting closer → indicate a score of 3 – That it coincides with the actual situation of the team → indicate a score of 4				
Organisational aspects	*1*	*2*	*3*	*4*
Clinical records are considered an important working tool, they are consulted regularly and in any case whenever necessary				
Each patient's clinical records are easily accessible to all professional figures				
The available documentation contains information that helps to understand the personal characteristics of the guest (information on their life, tastes, preferences)				
Communication with the guest	*1*	*2*	*3*	*4*
The members of the team are attentive to their verbal communication with the guest (calling the guest by name, using informal or formal forms of speech according to preference, using an appropriate tone of voice, using an understandable language)				
The members of the team are attentive to their nonverbal communication with the guest (the approach to care is relaxed and not authoritarian, gestures are delicate and not brusque)				
The members of the team dedicate time and attention when explaining every care activity to the guest				
The members of the team always try, as far as possible, to put the guest in a position where they can choose				
The members of the team organise their work in order to dedicate time and attention to communication with the guests				
Communication with relatives is managed in order to offer information and reassurance, reducing anxiety and conflict as far as possible				
Care relationships	*1*	*2*	*3*	*4*
The members of the team are able to accept and interpret difficult behaviour as a sign of distress				
The members of the team are familiar with the individual differences of each guest, their residual capacities and needs (personal care plan), reading the information sheet at the end of the bed, listening to the guest				
The members of the team help the guests to continue their lifestyles as long as possible (e.g. when to go to bed or get up, when to go to the bathroom, participation or nonparticipation in activities)				
The members of the team make an effort to encourage the guests to maintain their independence and personal care as long as possible				
The members of the team identify, manage and encourage positive relations among the guests (relations between roommates, other guests during lunch, during moments of socialisation, etc.)				

(continued)

Table 22.2 (continued)

Self-evaluation questionnaire on quality of care				
The members of the team are attentive to respecting the guests' dignity and privacy during care and communication (the staff communicate with each guest in a manner that does not disclose personal information, the staff pay attention to respecting privacy, during personal hygiene, e.g. by closing the door)				
The members of the team pay attention to the tranquillity of the context, silence, respecting moments of quiet (the light is kept on at night only when strictly necessary, noise is avoided during rest hours)				
Relations between team members	*1*	*2*	*3*	*4*
The members of the team are respected, considered, listened to and encouraged to use their talents and resources to enrich the care activities				
The members of the team support each other in tackling difficult situations and work to solve problems together				
Group activities are organised and managed to allow all figures involved to participate				
The moments and methods of handing over on shift changes are organised in order to inform all staff clearly, offering a full vision of all the activities concerning the guest				
Every member of the team is appropriately involved in the design and decision-making of all care activities concerning each guest				
The members of the team recognise the value of work and professionalism of all colleagues; they collaborate and offer support and assistance				
Training	*1*	*2*	*3*	*4*
The training sessions are organised in order to allow the team members to improve their skills				
The training sessions are organised in order to allow the team members to identify problem-solving strategies				
The training sessions are organised in order to allow the team members to reduce the risk of those forms of work stress that affect people working in care and assistance professions				
The members of the team are supported and guided in the adoption of the most appropriate strategies for reducing difficult behaviour (avoiding as far as possible the use of physical restraints or medication)				
The members of the team are encouraged to pay attention to their verbal and nonverbal communication with the guests				

Both tools aim to foster reflection both when the observation takes place, requiring the grid compiler to use an analytical, objective and targeted standpoint, and when the results are discussed in the group. Each event is observed and recorded by several observers in order to obtain information that is as reliable as possible, but also to allow the team, during the group analysis of the results, to reflect on the triangulation of the viewpoints and opinions as a necessary strategy for self-analysis and for the design of innovations.

Instruments for Analysis, Reflection and Designing Innovation

The evaluation instruments designed and validated for the project were used to evaluate the elderly care services recognising the strengths and criticalities, with

Table 22.3 Quality observation of collective moments

Observation Of Collective Moments

Date________ Professionals involved________________(Nurse______Nursing Assistant______Nursing Assistant Coordinator_____Charge Nurse_____Recreation____ Physiotherapist)

Communication subject:___________

Are the personal care plan and/or documentation consulted? □ yes □ no

Horizontal communication (between professional figures of the same level)

From the point of view of the contents, the communication is:

Complete	*1*	*2*	*3*	*4*	*5*	*6*	*Incomplete*
Clear	*1*	*2*	*3*	*4*	*5*	*6*	*Unclear*

From the point of view of the form, the communication is characterised by:

Calm	*1*	*2*	*3*	*4*	*5*	*6*	*Aggressiveness*
Cooperation	*1*	*2*	*3*	*4*	*5*	*6*	*Opposition*

Vertical communication (between managers and staff or nurses)

From the point of view of the contents, the communication is:

Complete	*1*	*2*	*3*	*4*	*5*	*6*	*Incomplete*
Clear	*1*	*2*	*3*	*4*	*5*	*6*	*Unclear*

From the point of view of the form, the communication is characterised by:

Calm	*1*	*2*	*3*	*4*	*5*	*6*	*Aggressiveness*
Cooperation	*1*	*2*	*3*	*4*	*5*	*6*	*Opposition*

Transversal communication (among different professional figures):

Calm	*1*	*2*	*3*	*4*	*5*	*6*	*Aggressiveness*
Cooperation	*1*	*2*	*3*	*4*	*5*	*6*	*Opposition*

In communication:

Turns are respected	1	2	3	4	5	6	Are not respected
All participants are consulted	1	2	3	4	5	6	Nobody is consulted

Communication with relatives (where present):

The personal care plan is explained	1	2	3	4	5	6	Is not explained
The tone is calm	1	2	3	4	5	6	Agitated
The expression is smiling	1	2	3	4	5	6	Elusive

Do the staff act as if the relative was not present? □ yes □ no

Problems or criticalities met in drafting the personal care plan? □ yes □ no

Time dedicated to discussing the problem________________________________

Did everyone involved express their point of view? □ yes □ no

Was a solution to the problem found? □ yes □ no

Other comments____

Table 22.4 Observation grid of critical events

Observation Of Critical Events							
date:______ time:_______ figures involved______________________________							
Description of the situation							
Description of the episode___________________							
Solution identified?							□ yes □ no
Individually	1	2	3	4	5	6	Collectively
Time dedicated to discussing the problem_____________________________							
Time dedicated to identifying responsibilities ____________________							
Did everyone involved express his or her point of view?							□ yes □ no
Other comments____							

relation to fixed standards defined in the form of indicators. Even more than definitive or certifying evaluation tools, these devices were designed, as explained, with a view to educational evaluation, as an element of a procedure for the progressive and constant increase in quality of care. The perspective of *educational evaluation* (Becchi and Bondioli 1994; Husén and Postlethwaite 1994; Kellaghan and Stufflebeam 2003; Scriven 1967, 1981, 1999; Walberg and Haertel 1990), in which this project lies, implies the implementation of organisational practices of evaluation and self-evaluation, aiming to raise awareness of procedures and dynamics, reflect on them and design actions for improvement and innovation, increasingly approaching the quality standards defined as a group.

Responding to the need to make the questionnaires, the observation grids presented and the related practices drivers of organisational change and improvement, aiming to increase quality, a further instrument was developed to accompany the team in their analysis of the gathered data, reflection on it and in the decision-making process leading to innovation. This need responds to the desire to create an evaluation mechanism which goes beyond the mere expression of judgement, creating a virtuous circle in which the evaluation is followed by a moment of discussion, reflection on the data, generating the production of innovation, which in turn is subjected to evaluation and reflection, and so on (Bondioli and Ferrari 2004). In this way the evaluation becomes part of the organisational procedures, actively involving all the stakeholders in the care service, in a dynamic of awareness, the assumption of responsibility and openness towards innovation.

The instrument proposed in Table 22.5 was used as a guide for the team in the analysis, reflection and decision-making process, based on the results emerging from the questionnaires administered and the observations carried out. This instrument aims to help the group to adopt metacognitive strategies for analysis, with a view to self-evaluation, promoting reflection on their practices (Dewey 1933; Schön 1983) and designing new solutions to problems or strategies to increase the quality of service. The instrument accompanies the group in an analytical

Table 22.5 Instrument for analysis, reflection and designing innovation

Instrument for analysis, reflection and designing innovation based on the data gathered through the RSA care quality evaluation tools

The results gathered with the instrument must be analysed collectively with a view to decision making.

The group will refer to the principles constituting the background on which the idea of quality is placed and which guides the evaluation tools.

-Instructions for analysis

-Activities to be performed in a group, in a series of stages to analyse each area

-Appoint a group leader/moderator to guide the stages of the proposed methodology.

-Follow the stages described below

On the basis of the results of the grid, briefly identify strengths, weaknesses and aspects for further study. List them briefly in the table.

Strengths	Weaknesses	Aspects for further study

Then analyse each column:

-Extrapolate the main elements to be enhanced from the strengths column

-Starting from the weaknesses column, identify the possible causes of the problems

-From the points for further study column, identify some points for further analysis or data collection

Elements to be enhanced	Possible causes of criticalities	Points for further study

According to the overview formulated, considering particularly the criticalities recorded, try to envisage possible strategies for improvement.

The group must identify the strategies freely and creatively, accepting all the proposals offered spontaneously during the discussion, without analysing them in detail in this initial phase.

Having listed all the strategies for improvement the group analyses them one after the other, identifying the relative advantages and disadvantages, in order to analyse the ideas that emerged and select the most effective and feasible among them.

Improvement innovation strategies	Advantages	Disadvantages

reading of the results emerging, through a process based on the concept of de Bono's mind mapping (1969) in which thought is guided in analysing the various constituent elements of a situation.

All the instruments presented were used, not only to evaluate quality of care but also, and above all, to raise awareness among the elderly care organisations of the elements determining total quality of service, offering procedures and practices for self-evaluation, reflection, redesign and consequently an increase in quality.

This approach aims to stimulate the team members to work with an attitude of empirical and experimental research in analysing the problems observed, trying to identify the correlations between variables and, where possible, the links of cause and effect. The data analysis sheet uses some teaching and learning methods aiming to foster discussion, reflection and description of problems, as well as identify innovative and creative solutions.

The path of construction and validation of the instruments used in both facilities involved in the study allowed them to pinpoint and critically analyse a series of problems which drastically affect the quality of life in an elderly care home, with direct repercussions on care and work in the organisation. Using this approach, it was possible to design and implement improvement actions which had not been thought of before. The involvement of all the working team members in the educational evaluation helped the group to develop *empowerment*, awareness of potentials and internal resources, mutual trust and greater confidence in tackling and solving problematic situations together.

This path promotes an approach that enhances the organisation's potential for designing and implementing innovation: we can define this attitude as an organisational mindset for innovation that is promoted through the proposed educational evaluation model. This model can be also seen as related to the notion of learning organisations (Senge 1990) where organisations are conceived as creative and adaptable systems, promoting commitment and the ability to learn at all levels. Indeed, active staff participation is a fundamental element of the educational evaluation model presented to foster reflection on professional practices, learning and analysing contexts in a critical and objective manner, formulating hypotheses for intervention, implementing them and checking their effectiveness.

22.5 Conclusion and Implications

The educational evaluation research path presented aims to offer a theoretical and practical contribution to foster practices which help to increase the quality of care in facilities for fragile elderly people who are no longer self-sufficient, promoting innovation in this direction. An in-depth, constant reflection on the value of the care relationship in these care organisations is now necessary and even more so is the involvement of all the stakeholders of these care contexts in this reflection.

The paradigms of well-being in the third age, quality of life and elderly care relations must be redefined and negotiated in a constant dialogue between users, families and professionals, pursued by care organisations in a virtuous circle which progressively increases quality. In this perspective, we think of the concept of quality as a model to move towards, constantly striving for improvement. The metaphor of the horizon is particularly fitting: quality is a dot on the horizon we must reach, which continues to move further away, because we progressively increase the standards we set, continuously redefining and renegotiating the very idea of quality of care. Today's elderly people have different characteristics and needs to those of yesterday, and those of tomorrow will have others again. Care organisations must become places of innovation, to promote well-being for the elderly and their families. If such striving for innovation is to be accompanied by an orientation towards authentic, measurable and evaluable quality, it is fundamental to use controlled innovation models, based on the paradigms of empirical research. In this perspective, care organisations are asked to become learning organisations (Senge 1990): contexts which reflect on their own practices, which define and validate quality standards, formulate causal hypotheses to understand the difficulties, envisaging solutions or innovations, designing, implementing and monitoring these innovations and continuing the cycle of educational evaluation.

The processes of innovation that has been developed can also understood with a view to practice-based change process (Sørensen et al. 2013) According to Fuglsang and Sørensen (2011), practice-based change processes can be defined as processes related to daily practices occurring in smaller, unintentional daily changes which can have an impact in terms of learning, creating knowledge and changing services. This is also defined as small steps in an innovation process (Fuglsang and Sørensen 2011; Sundbo, and Toivonen 2011).

The innovation process developed can be seen as a process that helps care organisations to promote, implement and monitor practice-based change processes.

It is precisely in this perspective that this research project was designed and implemented, to provide a concrete basis for reflections on the care relationship with fragile elderly people who are no longer self-sufficient, to define the concept of quality of care and be able to observe and evaluate it. The project offers a methodologically and empirically substantiated model of educational evaluation and controlled innovation.

The evaluation method and instruments developed during the project aim to offer a methodological contribution which can be applied to other elderly care contexts in order to trigger mechanisms of self-evaluation, reflection and design of innovation, progressively increasing the quality of service.

This type of path aims to innovate the processes, as well as the results of the care action, in a vision in which quality comes from a synergy of intentions and actions among the care professionals, the users and their families, with a concrete impact on the places and times of care.

The evaluation tools developed constantly refer to the founding values of an idea of care which goes beyond the adult-centric or assistential-healthcare perspective, consolidating a model which sees the third age as a phase of life with its own characteristics and with its own plans and projects. Restoring dignity to long-term care is one of the key objectives of this work. Here the organisation and its stakeholders are accompanied and educated to take an analytical stance, constantly evaluating their own work and innovating their practices to foster a culture of elderly care which responds to all the needs of ageing and in which all elderly people tackling the existential challenge of fragility are accompanied and supported.

References

Andrew, R. (2010). Organizational social capital, structure and performance. *Human Relations, 63* (5), 583–608.

Becchi, E., & Bondioli, A. (1994). La valutazione: Una pratica in via di definizione. In M. Ferrari (Ed.), *La valutazione di contesti prescolari*. Bergamo: Junior.

Berger, P. L., & Luckmann, T. (1967). *The social construction of reality: A treatise in the sociology of knowledge*. London: Penguin.

Blumer, H. (1969). *Symbolic interactionism: Perspective and method.* Englewood Cliffs, NJ: Prentice Hall.

Bondioli, A., & Ferrari, M. (a cura di) (2004). *Verso un modello di valutazione formativa.* Bergamo: Junior.

Brune, K. B. (1995). *Eden Alternative certification, pilot group*. New Berlin, NY: Lindengrove.

Brune, K. B. (2011). Culture change in long term care services: Eden-greenhouse-aging in the community. *Educational Gerontology, 37*(6), 506–525.

Calkins, M. P. (2002). *The nursing home of the future: Are you ready*?. Nursing Homes Long-Term Care Management.

Cattel, R. B. (1966). *Handbook of multivariate experimental psychology*. Chicago: Rand McNall & Company.

Coons, D., & Mace, N. (1996). *Quality of life in long term care*. New York: Haworth Press.

De Bono, E. (1969). *The mechanism of mind*. London: Penguin.

Dewey, J. (1933). *How we think*. Boston, MA: Heat.

Edvardsson, B., Tronvoll, B., & Gruber, T. (2011). Expanding understanding of service exchange and value co-creation: A social construction approach. *Journal of the Academy of Marketing Science, 39*, 327–339.

Ellström, P. E. (2010). Practice-based innovation: A learning perspective. *Journal of Workplace Learning, 22*(1/2), 27–40.

Esping-Andersen, G. (1990). *The three worlds of welfare capitalism*. Princeton, NJ: Princeton University Press.

Esping-Andersen, G., Gallie, D., Hemerijck, A., & Myles, J. (2002). *Why we need a new welfare state*. Oxford: Oxford University Press.

Fuglsang, L., & Sørensen, F. (2011). The balance between bricolage and innovation: Management dilemmas in sustainable public innovation. *The Service Industries Journal, 31*(4), 581–595.

Goffman, E. (1963). *Behavior in public places: Notes on the social organization of gatherings*. New York: Free Press.

Husén, T., & Postlethwaite, T. N. (1994). *International encyclopedia of education* (2nd ed.). New York: Pergamon Press.

Kellaghan, T., & Stufflebeam, D. L. (2003). *International handbook of educational evaluation*. Dordrecht: Kluwer.

Kiss, P., De Meester, M., Kristensen, T. S., & Braeckman, L. (2014). Relationships of organizational social capital with the presence of "gossip and slander", "quarrels and conflicts", sick leave, and poor work ability in nursing homes. *International Journal of Occupational and Environmental Health, 87*(8), 929–936.

Lechner, V. M., & Neal, M. B. (1999). *Working and caring for the elderly. International perspectives*. Philadelphia, PA: Brunner/Mazel, Taylor & Francis.
Levin, J., & Levin, W. C. (1988). *The human puzzle: An introduction to social psychology*. Belmont, CA: Wadsworth.
Lewis, J., & Ostner, I. (1994). *Gender and evolution of European social policies* (ZeS-Arbeitpaper. No. 4). Centre for Social Policy Research, University of Bremen.
Luppi, E. (2009). Education in old age: An exploratory study. *International Journal of Lifelong Education, 28*, 241–276.
Luppi, E. (2010). Autonomy and citizenship in the late third age. *Educational Gerontology, 36*, 872–892.
Maslach, C. (1982). *Burnout: The cost of caring*. New York: Prentice Hall.
Norman, I., & Redfern, S. (1997). *Mental health care for elderly people*. Edinburgh: Churchill Livingstone.
Rahman, A. N., & Schnelle, J. F. (2008). The nursing home culture change movement: Recent past, present, and future directions for research. *The Gerontologist, 48*, 142–148.
Schön, D. A. (1983). *The reflective practitioner. How professionals think in action*. London: Temple Smith.
Scriven, M. (1967). *The methodology of evaluation*. Chicago: Rand McNally.
Scriven, M. (1981). *Evaluation thesaurus*. Inverness, CA: Edge Press.
Scriven, M. (1999). *Evaluation thesaurus*. Newbury Park, CA: Sage.
Senge, P. M. (1990). *The fifth discipline: The art and practice of the learning organization*. New York: Doubleday.
Shura, R., Siders, R. A., & Dannefer, D. (2011). Culture change in long-term care: Participatory action research and the role of the resident. *Gerontologist, 51*(2), 212–225.
Sørensen, F., Sundbo, J., & Mattsson, J. (2013). Organisational conditions for service encounter-based innovation. *Research Policy, 42*, 1446–1456.
Sundbo, J., & Toivonen, M. (2011). *User-based innovation in services*. Cheltenham: Edward Elgar Publishing.
Thomas, W. H. (2006). *In the arms of elders: A parable of wise leadership and community building*. Acton, MA: VanderWyk & Burnham.
Tronvoll, B., Brown, S. W., Gremler, D. D., & Edvardsson, B. (2011). Paradigms in service research. *Journal of Service Management, 22*(5), 560–585.
Vargo, S. L., & Lusch, R. F. (2004). Evolving to a new dominant logic for marketing. *Journal of Marketing, 68*(1), 1–17.
Vargo, S. L., Maglio, P. P., & Akaka, M. (2008). On value and value co-creation: A service systems and service logic perspective. *European Management Journal, 26*, 145–152.
Walberg, H. J., & Haertel, G. D. (Eds.). (1990). *International enciclopedia of educational evaluation*. Oxford: Pergamon Press.
World Health Organization. (2013). *The European health report 2012: Charting the way to well-being*. Copenhagen. http://www.euro.who.int/__data/assets/pdf_file/0003/184161/The-European-Health-Report-2012,-FULL-REPORT-w-cover.pdf

MIX
Papier aus verantwortungsvollen Quellen
Paper from responsible sources
FSC® C105338

If you have any concerns about our products,
you can contact us on
ProductSafety@springernature.com

In case Publisher is established outside the EU,
the EU authorized representative is:
Springer Nature Customer Service Center GmbH
Europaplatz 3, 69115 Heidelberg, Germany

Printed by Libri Plureos GmbH
in Hamburg, Germany